THE CARNITINE SYSTEM

# Developments in Cardiovascular Medicine

VOLUME 162

*The titles published in this series are listed at the end of this volume.*

# The Carnitine System

## *A New Therapeutical Approach to Cardiovascular Diseases*

*Edited by*

J.W. DE JONG
*Thoraxcenter, Erasmus University, Rotterdam, The Netherlands*

and

R. FERRARI
*Cattedra di Cardiologia, Università degli Studi di Brescia, Fondazione Clinica del Lavoro di Pavia, Gussago/Brescia, Italy*

Springer Science+Business Media, B.V.

**Library of Congress Cataloging-in-Publication Data**

The carnitine system : a new therapeutical approach to cardiovascular
diseases / edited by J.W. de Jong and R. Ferrari.
p. cm. -- (Developments in cardiovascular medicine ; v. 162)
Includes bibliographical references and index.
ISBN 978-94-010-4122-5 ISBN 978-94-011-0275-9 (eBook)
DOI 10.1007/978-94-011-0275-9
1. Carnitine--Therapeutic use. 2. Cardiovascular system--Diseases--Chemotherapy. 3. Acetylcarnitine--Therapeutic use.
4. Carnitine--Derivatives--Therapeutic use. I. Jong, Jan Willem de, 1942- . II. Ferrari, R. (Roberto) III. Series.
[DNLM: 1. Carnitine--therapeutic use. 2. Heart--drug effects.
3. Cardiovascular--drug therapy. W1 DE997VME v. 162 1995 / QU 187
C2895 1995]
RC684.C37C37 1995
616.1'061--dc20
DNLM/DLC
for Library of Congress 94-46324

ISBN 978-94-010-4122-5

*Printed on acid-free paper*


Originally published by Kluwer Academic Publishers in 1995
Softcover reprint of the hardcover 1st edition 1995

# Table of contents

# List of contributors

ARDUINO ARDUINI
Istituto di Scienze Biochimiche, Facoltà di Medicina, Università degli Studi Gabriele D'Annunzio, Via dei Vestini, I-66100 Chieti, Italy
*Chapter 12 co-authors*: Secondo Dottori, Francesco Molajoni, Ruth Kirk and Edoardo Arrigoni-Martelli

JON BREMER
Institute of Medical Biochemistry, University of Oslo, P.O. Box 1112, Blindern, Oslo 0317, Norway
*Chapter 2*

SALVATORE CAPONNETTO
Department of Internal Medicine, Division of Cardiology, University of Genoa, Viale Benedetto XV, 6, I-16132 Genoa, Italy
*Chapter 24 co-author*: Claudio Brunelli

NERINA CORSICO
Pharmacology Department, Sigma-Tau S.p.A., Via Pontina Km. 30.400, I-00040 Pomezia (Rome), Italy
*Chapter 26 co-author*: Edoardo Arrigoni-Martelli

JAN WILLEM DE JONG
Cardiochemical Laboratory, Thoraxcenter, Ee 2371, Erasmus University Rotterdam, P.O. Box 1738, 3000 DR Rotterdam, The Netherlands
*Chapter 19 co-authors*: Alessandro Mugelli, Fabio Di Lisa and
Roberto Ferrari
*Chapter 27 co-author*: Roberto Ferrari

FABIO DI LISA
Università degli Studi di Padova, Dipartimento di Chimica Biologica, Via Trieste, 75, I-35121 Padova, Italy

*Chapter 3 co-authors*: Roberta Barbato, Roberta Menabò and Noris Siliprandi

ROBERTO FERRARI
Cattedra di Cardiologia, Università degli Studi di Brescia, c/o Spedali Civili, P.le Spedali Civili, 1, I-25123 Brescia, Italy
*Chapter 1 co-author*: Jan Willem de Jong
*Chapter 15 co-author*: Odoardo Visioli
*Chapter 23 co-author*: Inder Anand

DANIELLE FEUVRAY
URA-CNRS 1121, Laboratoire de Physiologie Cellulaire, Université Paris XI, Bât. 443, F-91405 Orsay Cédex, France
*Chapter 14*

WILLIAM R. HIATT
Section of Vascular Medicine, University of Colorado School of Medicine, 4200 E. Ninth Ave, Box B-180, Denver, CO 80262, USA
*Chapter 25 co-author*: Eric P. Brass

JOHAN FOKKE KOSTER
Department of Biochemistry I, Ee 642, Faculty of Medicine and Health Science, Erasmus University Rotterdam, P.O. Box 1738, 3000 DR Rotterdam, The Netherlands
*Chapter 9*

JOS M.J. LAMERS
Department of Biochemistry I, Cardiovascular Research Institute COEUR, Erasmus University Rotterdam, P.O. Box 1738, 3000 DR Rotterdam, The Netherlands
*Chapter 7*

A. JAMES LIEDTKE
Section of Cardiology, Hospital and Clinics, Highland Avenue, H6/339, University of Wisconsin, Madison, WI 53792–3248, USA
*Chapter 20*

BRETT O. SCHÖNEKESS
Cardiovascular Disease Research Group, 423 Heritage Medical Research Centre, University of Alberta, Edmonton, Alberta, Canada T6G 2S2
*Chapter 4 co-author*: Gary D. Lopaschuk

JEANIE B. McMILLIN
Department of Pathology & Laboratory Medicine, University of Texas Medical School, Health Science Center at Houston, 6431 Fannin, Houston, TX 77030, USA
*Chapter 6*

ROSELLA MICHELETTI
Prassis Istituto di Ricerche Sigma Tau, Via Forlanini 3, I-20019 Settimo Milanese (Milano), Italy
*Chapter 22 co-authors*: Antonio Schiavone and Giuseppe Bianchi

JOSEF MORAVEC
Lab. d'Energétique et de Cardiologie Cellulaire, INSERM, Faculté de Pharmacie, Université de Bourgogne, 7, Bvd Jeanne d'Arc, F-21033 Dijon, France
*Chapter 10 co-authors*: Zainab El Alaoui-Tablibi and Christian Brunold

ALESSANDRO MUGELLI
Dipartimento di Farmacologia, Università degli Studi di Firenze, Viale G.B. Morgagni, 65, I-50134 Firenze, Italy
*Chapter 18 co-authors*: Elisabetta Cerbai and Mario Barbieri

DENNIS J. PAULSON
Department of Physiology, Midwestern University, 555 31st Street, Downers Grove, IL 60515, USA
*Chapter 13 co-author*: Austin L. Shug

CARL J. PEPINE
Division of Cardiology and Center for Exercise Science, University of Florida, Box 100277, Gainesville, FL 32610, USA
*Chapter 16 co-author*: Michael A. Welsch

HANS MICHAEL PIPER
Physiologisches Institut, Justus-Liebig-Universität, Aulweg 129, D-35392 Giessen, Germany
*Chapter 8 co-authors*: Thomas Noll and Berthold Siegmund

VERA REGITZ-ZAGROSEK
Department of Cardiology, German Heart Institute Berlin, P.O. Box 650505, D-13305 Berlin, Germany
*Chapter 11 co-author*: Eckart Fleck

PAOLO RIZZON
Institute of Cardiology, Policlinico, University of Bari, Piazza G. Cesare, I-70124 Bari, Italy
*Chapter 17 co-authors*: Matteo Di Biase, Giuseppina Biasco and Maria Vittoria Pitzalis

GER J. VAN DER VUSSE
Department of Physiology, Faculty of Medicine, University of Limburg, P.O. Box 616, 6200 MD Maastricht, The Netherlands
*Chapter 5*

PIETER D. VERDOUW
Experimental Cardiology, Thoraxcenter, Ee 2351, Erasmus University Rotterdam, P.O. Box 1738, 3000 DR Rotterdam, The Netherlands
*Chapter 21 co-authors*: Loes M.A. Sassen, Dirk J. Duncker and Jos M.J. Lamers

# 1. Introduction

ROBERTO FERRARI and JAN WILLEM DE JONG

## Introduction

Carnitine was discovered in the first part of this century. It turned out to be essential for fatty acid metabolism. The compound is also known as vitamin $B_T$ because the mealworm *Tenebrio molitor* cannot synthesize it. Although carnitine is vital for man [1], it is not actually a vitamin. Carnitine can, however, be considered a vitamin for heart muscle, which relies on its synthesis taking place in other organs [2]. The compound has been used successfully in the treatment of a variety of diseases. In the last few years, derivatives of L-carnitine such as acetyl-L-carnitine and propionyl-L-carnitine have been made available to doctors for treatment of specific pathologies. The effects of this family of related carnitine compounds on cardiovascular systems and diseases constitute the major issue addressed in this book.

## Aim of this book

In the last two decades, several books on carnitine have been published. Some deal with carnitine in general [3–7], others describe specific aspects, for instance, its effect on morphology [8] and carnitine deficiency [9]. Recently, also a video on L-carnitine and cardiac metabolism was released [10]. A treatise focussing on experimental and clinical aspects of the carnitine family and cardiovascular diseases was lacking. We believe that the present book provides the reader with a concise update in this field. Hopefully, clinicians and basic scientists will appreciate the information collected from experts on various aspects of the fascinating compound, carnitine.

## Nomenclature: What's in a word?

Carnitine is a relatively simple molecule with a carboxylic acid group and an alcohol group (Figure 1). Esters can thus be formed between the acid group

*J.W. de Jong and R. Ferrari (eds): The carnitine system*, 1–3.

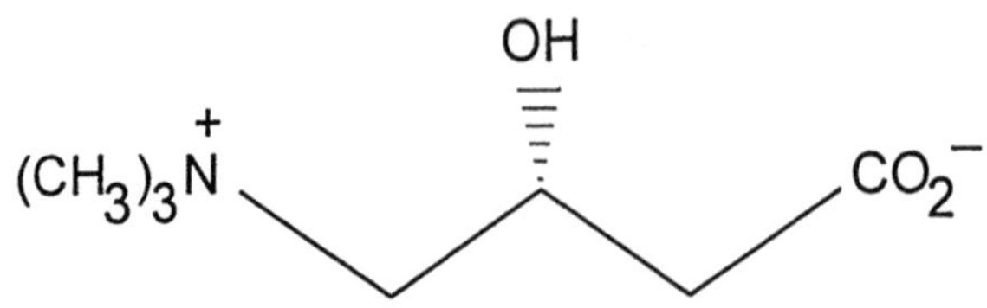


*Figure 1.* Structure of carnitine (L-3–hydroxy-4–*N*-trimethylaminobutyric acid), showing a carboxylic acid group and an alcohol group. Naturally occurring esters form between the alcohol group of carnitine and a (fatty) acid.

of carnitine and an alcohol, or between the alcohol group of carnitine and an acid. The naturally occurring esters, acetyl-L-carnitine, propionyl-L-carnitine and the acylcarnitines with a longer fatty acid chain, all belong to the second category (the first should be called ethyl-L-carnitine, propyl-L-carnitine and medium/long-chain alkylcarnitines). In this book, reference is made to L-palmitylcarnitine, palmityl-L-carnitine, L-palmitoylcarnitine, palmitoyl-L-carnitine. All refer to the ester between palmitic acid and L-carnitine, although palmitylcarnitine is, strictly speaking, less correct. However, this nomenclature is so widespread in the literature that we have asked the Chapter authors of this book to be consistent in the terminology they use in their own chapter.

Carnitine and its derivatives exist in the form of two steroisomeres (enantiomeres). Only the L-form is active with certain enzymes (see Chapter 26). The D- and the L-form show similar non-enzymatic effects on free-radical scavenging, for example, as discussed in Chapter 12 (see also [11, 12]). We refer to Roden's recent editorial for the advantage of enantiospecific therapy [13].

## Editorial comment

Each chapter was reviewed by the editors and adaptations were made in agreement with the authors. We take this opportunity to thank the authors for their constant intellectual input, patience and collaboration. In addition, we as editors have taken the liberty of highlighting some important aspects of each chapter.

**Acknowledgements**

We gratefully acknowledge the secretarial contribution of Carina D.M. Poleon-Weghorst and the editorial assistance of Dr Bill Dotson Smith.

We would also like to thank Dr A. Koverech (Pomezia/Rome) for his encouragement and support towards the publication of this volume.

**References**

1. Pande SV, Murthy MSR. Carnitine: vitamin for an insect, vital for man. Biochem Cell Biol 1989; 67: 671–3.
2. Siliprandi N, Sartorelli L, Ciman M, Di Lisa F. Carnitine: metabolism and clinical chemistry. Clin Chim Acta 1989; 183: 3–11.
3. Frenkel RA, McGarry JD, editors. Carnitine biosynthesis, metabolism, and functions. New York: Academic Press, 1980.
4. Gitzelman R, Baerlocker K, Steinmann B, editors. Carnitine in der Medizine. Stuttgart: Schattauer, 1987.
5. Kaiser J, editor. Carnitine – its role in lung and heart disorders. Munich: Karger, 1987.
6. Ferrari R, DiMauro S, Sherwood G, editors. L-Carnitine and its role in medicine: From function to therapy. London: Academic Press, 1992.
7. Carter AL, editor. Current concepts in carnitine research. Boca Raton: CRC Press, 1992.
8. Laschi R. L-Carnitine and ischaemia – A morphological atlas of the heart and muscle. Pomezia: Biblioteca Scientifica/Fondazione Sigma-Tau, 1987.
9. Borum PR, editor. Clinical aspects of human carnitine deficiency. New York: Pergamon Press, 1986.
10. Bayés de Luna A, Rizzon P, Hugenholtz PG, editors. L-Carnitine and cardiac metabolism: Modern concepts. Eur Video J Cardiol 1994; 2(2).
11. Di Giacomo C, Latteri F, Fichera C et al. Effect of acetyl-L-carnitine on lipid peroxidation and xanthine oxidase activity in rat skeletal muscle. Neurochem Res 1993; 18: 1157–62.
12. Hülsmann WC, Peschechera A, Arrigoni-Martelli E. Carnitine and cardiac interstitium. Cardioscience 1994; 5: 67–72.
13. Roden DM. Mirror, mirror on the wall . . . Stereochemistry in therapeutics. Circulation 1994; 89: 2451–3.

*Corresponding Author*: Dr Jan Willem de Jong, Erasmus University Rotterdam, Cardiochemical Laboratory, Thorax Centre EE2371, P.O. Box 1738, 3000 DR Rotterdam, The Netherlands

PART ONE

# The carnitine system in the heart: molecular aspects

# 2. Carnitine-dependent pathways in heart muscle

*Dedicated to the memory of E. Jack Davis* (1930–1993)

JON BREMER

---

"Malonyl-CoA in the heart probably participates in the regulation of fatty acid oxidation since the heart carnitine palmitoyl-CoA transferase I is extremely sensitive to malonyl-CoA."

---

## Introduction

Heart and skeletal muscle normally cover most of their energy needs by oxidizing fatty acids. When rat heart is offered both glucose and high concentrations of fatty acids, fatty acids are the preferred substrate even in the presence of insulin. Under such conditions 90% of the $CO_2$ produced will derive from fatty acids [1, 2].

In animal tissues the oxidation of long-chain fatty acids in mitochondria depends on the presence of carnitine. The heart therefore depends on carnitine for most of its energy production. This follows from the localization of the long-chain acyl-CoA synthetase in the outer membrane of the mitochondria [3] while the $\beta$-oxidation enzymes are found in the matrix. Since the inner mitochondrial membrane is impermeable to CoA esters, the carnitine palmitoyltransferases I and II (CPT I and CPT II), and the carnitine translocase are required to transport the activated long chain fatty acids as carnitine esters into the mitochondria for oxidation (Figure 1).

In the heart short-chain fatty acid oxidation does not depend on carnitine because these fatty acids are activated in the matrix of the mitochondria by a butyryl-CoA synthetase able to activate fatty acids up to hexanoate [4]. In this tissue therefore octanoate and longer fatty acids depend on carnitine for their oxidation. In the liver the distribution of fatty acid activation depending on chain length is somewhat different. In the matrix of liver mitochondria there is an octanoate activating enzyme active with acids of chain lengths up to $C_{10}$ or $C_{12}$ [5]. In this tissue therefore also medium chain length fatty acids can be oxidized without participation of carnitine.

With the key role of carnitine in fatty acid metabolism in mind it is not

*J.W. de Jong and R. Ferrari (eds): The carnitine system*, 7–20.

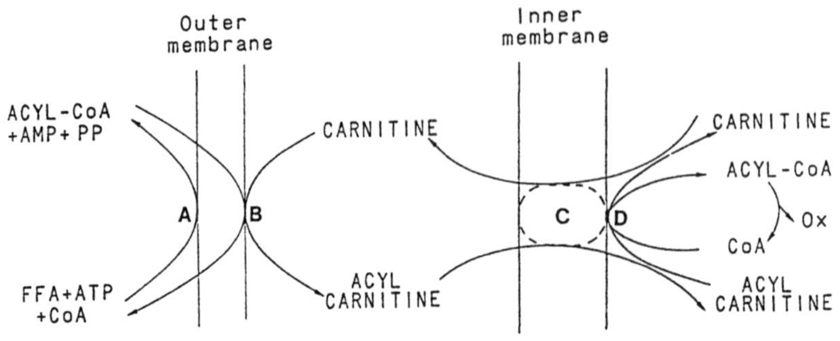


*Figure 1.* The organization of activated fatty acid transfer in mitochondria. A, acyl-CoA synthetase; B, carnitine palmitoyltransferase I (CPT I); C, carnitine translocase; D, carnitine palmitoyltransferase II (CPT II); FFA, free fatty acids; Ox, oxidation; PP, inorganic pyrophosphate.

surprising that loss of carnitine leads to accumulation of triacylglycerol in the heart. Bressler and Wittels [6] found that the heart lipidosis in diphtheria may be explained by a loss of carnitine, and Mølstad and Bøhmer [7] found that diphtheria toxin probably promotes carnitine loss by inhibiting the synthesis of the carnitine carrier in the cell membrane. Later a series of inborn errors with loss of heart carnitine and lipidosis have been reported, the first case in 1973 [8].

## Carnitine acyltransferases in the heart

In animal tissues at least four different carnitine acyltransferases have been identified, carnitine acetyltransferase in the inner membrane of the mitochondria [9] and in the peroxisomes [10], a carnitine medium-chain acyltransferase in the peroxisomes [11], carnitine palmitoyltransferase I in the outer membrane of the mitochondria [12], and carnitine palmitoyltransferase II in the inner membrane of the mitochondria [13]. All these enzymes except CPT I have been purified from liver. The CPT I and CPT II are membrane bound and require detergents for their solubilization. However, CPT I is labile to detergents, and this lability has prevented its purification and characterization in active form.

In liver carnitine acyltransferases are found also in the endoplasmic reticulum, but these enzymes are still insufficiently characterized and their possible functions are unknown.

Carnitine acetyltransferase has a high activity in heart and muscle [14]. In rat liver this enzyme is found in both mitochondria and peroxisomes with low activity, but it is induced by clofibrate [15]. The same carnitine acetyltransferase seems to be present in both organelles [9].

In heart the carnitine acyltransferases so far have been found only in the mitochondria. However, it is possible that also heart peroxisomes contain carnitine acyltransferases, although this has not been directly demonstrated.

Careful enzyme purification and chain length specificity studies have led to the isolation of two purified carnitine acyltransferases from heart mitochondria, one carnitine short-chain acyltransferase (the acetyltransferase) and one carnitine medium-chain acyltransferase [14]. The latter enzyme most likely is identical with the CPT II, and it is the same in all tissues [16]. It is interesting that this enzyme shows different chain length specificities depending on reaction direction. In the direction of acylcarnitine formation the optimum chain length of the acyl-CoA is decanoyl-CoA while in the direction of acyl-CoA formation the optimum chain length of the acylcarnitine substrate is myristoyl- or palmitoyl-carnitine [14]. This change in chain length specificity depending on reaction direction may be connected with a kinetic peculiarity of the enzyme. Long-chain acyl-CoA is, beside being substrate for the enzyme, also a competitive inhibitor to the second substrate carnitine. Therefore the apparent Km for carnitine increases with increasing concentrations of long-chain acyl-CoA [17]. When the freely reversible reaction is run in the opposite direction the long-chain acyl-CoA becomes a product in the reaction, and now a strong product inhibition is observed [17]. It is not established whether this product inhibition has any physiological significance. However, it should be kept in mind that the normal reaction direction for CPT II is acyl-CoA formation in the mitochondria (Figure 1). The product inhibition by long-chain acyl-CoA therefore may limit the acyl-CoA level in the matrix of the mitochondria when the acyl-CoA/CoA ratio is high in the cytosol, thus saving free CoA for the thiolase- and other CoA dependent reactions in the mitochondria.

The carnitine medium-chain acyltransferase has been localized to the peroxisomes in liver [11]. This enzyme as well as carnitine acetyltransferase may have functions in fatty acid shortening in the heart and other tissues (see later). Recent studies have shown that this peroxisome enzyme of liver, like CPT I, is sensitive to malonyl-CoA [18] and to tetradecylglycidoyl-CoA (TDG-CoA) [19]. However, the properties of the putative heart peroxisomal enzymes have not been studied.

## Carnitine palmitoyltransferase I and its regulation

Beside the two purified carnitine acyltransferases of the inner membrane, heart mitochondria contain the malonyl-CoA sensitive CPT I in the outer membrane. This enzyme is rapidly inactivated when extracted with detergents [20]. Chain length specificity studies on purified malonyl-CoA sensitive CPT I therefore have not been performed, but early chain length studies on whole mitochondria suggest that the CPT I has a broad chain length specificity including both medium-chain- and long-chain fatty acids [15]. Its lability suggests that CPT I and CPT II are different enzymes. In the fetal heart CPT I has a low activity – or is latent [21]. The switch from carbohydrate to fatty acid oxidation in the newborn coincides with an increase in CPT I activity [22].

There is still no general agreement about the nature of CPT I and its regulation. We originally suggested that CPT I is regulated by a separate malonyl-CoA binding peptide in the mitochondrial membrane, and we found that malonyl-CoA binding protein(s) could be separated from the enzyme [23]. In liver a malonyl-CoA binding fraction can confer malonyl-CoA sensitivity on CPT II [24]. Similar results have been obtained with heart extracts [25]. From heart mitochondria a CPT/$\beta$-oxidation enzyme complex has also been extracted which is sensitive to malonyl-CoA [26]. This complex loses its sensitivity to malonyl-CoA when it is treated with salts. The isolation of such a complex does not agree well with the established localization of the CPT I in the outer membrane of mitochondria [12]. In other studies Woeltje and coworkers [16] have identified a protein in liver, heart and skeletal muscle mitochondria which binds tetradecylglycidic-CoA (TDG-CoA), a CPT I inhibitor. They assume that this protein is CPT I, while Chung and coworkers [25] assume that it is the malonyl-CoA binding regulatory unit. From binding studies Woeltje and coworkers [16] concluded that substrate (palmitoyl-CoA), malonyl-CoA, and tetradecylglycidyl-CoA (TDG-CoA) all bind to the same site on this putative CPT I. However, Zierz and Engel [27] concluded from kinetic studies on normal and mutant muscle CPT I that malonyl-CoA inhibits the enzyme by binding to a site different from the acyl-CoA binding substrate site. The heart TDG-CoA binding protein has a lower molecular weight than the corresponding liver protein (approximately 86 kDa against 90–94 kDa), but they are both bigger than CPT II (approximately 68 kDa). The amino acid sequence of the liver protein is related to that of CPT II with 30% overlap in identity. In some parts (substrate binding sites?) of liver CPT I and CPT II, the amino acid sequence is virtually identical [28] a finding which supports the hypothesis that this protein represents CPT I. This identity is also supported by Kolodziej and coworkers [29] who have obtained an antibody against the liver protein which inactivates liver CPT I and which also interacts weakly with the smaller heart enzyme.

Further studies are needed to elucidate the exact nature of the malonyl-CoA regulation of CPT I.

The heart CPT I is more sensitive to malonyl-CoA than is liver CPT I [30, 31]. Also, the liver enzyme is less sensitive to malonyl-CoA in fasted animals while no such change in sensitivity is observed in the heart [32, 33].

**Distribution of carnitine and coenzyme A**

More than 90% of the total carnitine in the body is found in skeletal muscle and heart in mM concentrations (the concentration varies in different species). Almost all the carnitine in the heart is found in the cytosol while the mitochondria contain only a small pool corresponding to the small matrix space [34]. This small pool is in equilibrium with the great cytosolic pool via the carnitine translocase in the inner membrane of the mitochondria. This translocase permits a rapid 1:1 exchange of carnitine and acylcarnitines across the membrane, but only a slow one-way net transport of carnitine [35]. Part of the carnitine is found as long-chain acylcarnitines and acetylcarnitine in both pools. Small amounts of other carnitine esters (e.g. propionylcarnitine and branched chain acylcarnitines formed from branched chain amino acids) may also be present [36]. All the carnitine esters are more or less in equilibrium with the corresponding CoA esters via CPT I and II and carnitine acetyltransferase.

CoA has a distribution very different from that of carnitine in the heart. Most of the CoA (90%) is localized in the mitochondrial matrix while only about 10% is found in the far more voluminous cytosol [34]. These distributions of carnitine and CoA in the tissue probably facilitate the creation of an inward gradient of activated fatty acids with a rapid flux into the mitochondria for oxidation. More details on the distribution of carnitine and CoA are present in Chapter 5.

**Interdependence of $\beta$-oxidation and the citric acid cycle**

The complete oxidation of acylcarnitines in the mitochondria can be seen as the result of two oxidation cycles, the $\beta$-oxidation cycle in which a fatty acid is shortened by two carbons with the formation of one acetyl-CoA per turn, and the citric acid cycle in which one acetyl-CoA per turn is oxidized to $CO_2$ and water (Figure 2). In liver the $\beta$-oxidation cycle can operate independent of the citric acid cycle because the acetyl-CoA formed can be converted to ketone bodies, but in heart mitochondria the two cycles have to operate almost like two cogwheels because almost all the acetyl-CoA formed has to be disposed of in the citric acid cycle. A little acetyl-CoA can be hydrolysed to free acetate [38], and another small amount of acetyl groups can be transferred to carnitine in the intact cell by carnitine acetyltransferase (the physiological significance of this acetylcarnitine formation will be discussed below). This domination of the citric acid cycle in the disposal of acyl-CoA in heart mitochondria is easily demonstrated. If the citric acid cycle is blocked by inhibition of the aconitase with fluorocitrate (formed from fluoroacetylcarnitine), the oxidation of palmitoylcarnitine is almost completely blocked except for a slow oxidation due to formation of some free

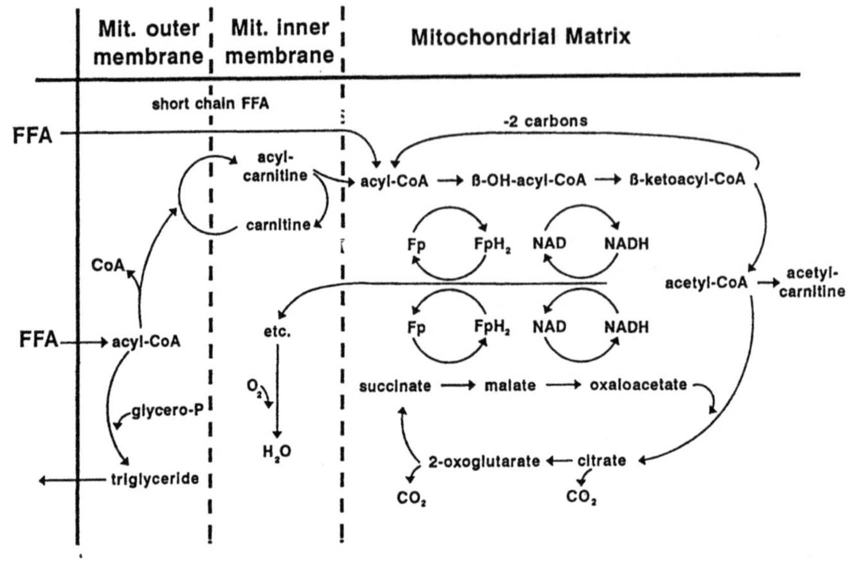


*Figure 2.* The interaction of fatty acid β-oxidation with the citric acid cycle in the mitochondria. FFA, free fatty acids; glycero-P, glycerophosphate; etc., electron transport chain; Fp, flavoprotein; β-OH-acyl-CoA, β-hydroxyacyl-CoA; Mit., Mitochondrial.

acetate by acetyl-CoA hydrolase [37]. In isolated heart mitochondria a further loosening of the two cycles can be obtained by addition of free carnitine to the reaction medium. This permits disposal of acetyl-CoA by acetylcarnitine formation [38]. However, it must be kept in mind that in the intact tissue the amount of acetyl groups which can be disposed of as acetylcarnitine formation is limited by the size of the carnitine pool. This pool is about 10-fold bigger than the CoA pool of the mitochondria, but it is still small in relation to the total flux of acetyl groups in heart metabolism. There is no quantitatively important extramitochondrial use of acetylcarnitine in heart or skeletal muscle, and uptake and release of carnitine and its esters are relatively slow processes in these tissues [39]. High intensity exercise may lead to a 5-fold increase of the acetylcarnitine level in skeletal muscle with only an insignificant increase in the blood plasma level in spite of the great muscle mass in relation to plasma volume [40]. Acetylcarnitine in the cytosol therefore has to be transferred back to the matrix of the mitochondria for oxidation in the citric acid cycle.

Since the citric acid cycle is so dominating in the disposal of acetyl-CoA and acetylcarnitine in the heart, it may be physiologically important that fatty acids and other efficient precursors of acetyl-CoA and acetylcarnitine seem to "secure" their own total oxidation in heart and skeletal muscle by increasing the level of citric acid cycle intermediates [41]. The acetyl-CoA-stimulated pyruvate carboxylase may represent an important mechanism by which the level of cycle intermediates increases [42, 43]. However, by extreme accumulation of long chain acylcarnitines they can be excreted from perfused rat hearts [44].

## Respiratory control of fatty acid oxidation

The utilization of fatty acids (acylcarnitines) and other substrates (glucose, pyruvate, lactate) in mitochondria is controlled by the work load on the heart, i.e. when the utilization of ATP is slow, a more reduced state in the mitochondria will develop and slow down the rate of substrate oxidation (respiratory control). Pyruvate oxidation and the citric acid cycle reactions (except succinate oxidation) are rapidly inhibited when the NADH/NAD ratio increases in the mitochondria [45, 46]. This means that these reactions are under efficient respiratory control. This is not the case for fatty acid $\beta$-oxidation, either in the liver or in the heart. In liver mitochondria fatty acid $\beta$-oxidation coupled to ketogenesis continues almost unabated at unphysiologically high NADH/NAD ratios [47]. In this tissue the rate of fatty acid oxidation therefore is determined mainly by the level of acylcarnitine in the tissue, i.e. by the extramitochondrial acyl-CoA/CoA ratio and the activity of CPT I, not by the redox state. Neither in heart is the fatty acid $\beta$-oxidation under direct respiratory control. This is demonstrated in Figure 3 which is a graphical presentation of experiments previously published [48]. When [1-$^{14}$C] palmitoyl-CoA is oxidized in heart mitochondria in the presence of free carnitine, ADP and phosphate (respiratory state 3), radioactive $CO_2$ is rapidly formed (Figure 3). The isolated heart mitochondria contain sufficient citric acid cycle intermediates to permit this $CO_2$ formation. When ADP is left out (respiratory state 4), the $CO_2$ formation disappears, showing that the citric acid cycle is completely inhibited. However, the formation of acid soluble products, mainly acetylcarnitine, is now increased (Figure 3), showing that fatty acid $\beta$-oxidation is poorly suppressed [48]. When radioactive palmitoylcarnitine is the substrate (with no free carnitine present) its oxidation is suppressed in state 4, even in the presence of malate. Concomitantly almost all the free CoA in the mitochondria is converted to acetyl-CoA [38], again showing that the oxidation of acetyl-CoA in the citric acid cycle is suppressed because of respiratory control. Addition of free carnitine now accelerates $\beta$-oxidation showing that the slow oxidation of palmitoylcarnitine was due to a lack of free CoA for CPT II and/or the thiolase, not because of a direct respiratory redox control of $\beta$-oxidation. These and other studies show that once fatty acids are available for oxidation in the mitochondria, the rate of oxidation is determined by the availability of free CoA more than by the redox state (NADH/NAD ratio) in the mitochondria [48].

Observations by Oram and coworkers [49] suggest a mechanism through

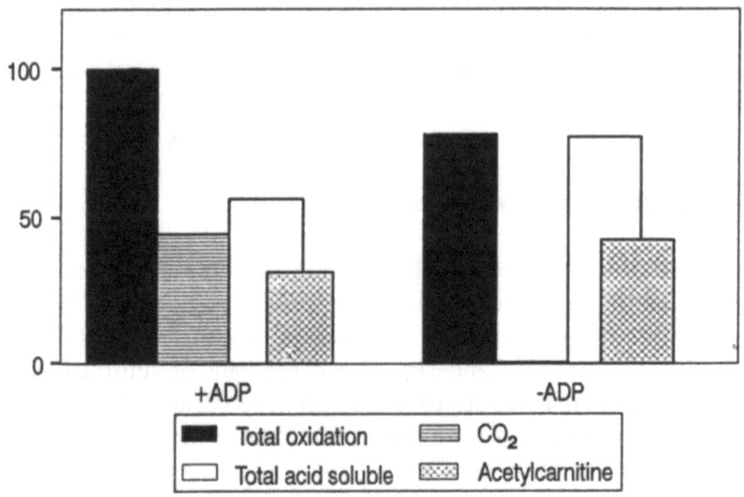


*Figure 3.* The effect of respiratory state on the oxidation of palmitoyl-CoA in the presence of carnitine in rat heart mitochondria. Incubation conditions: [1–$^{14}$C]-palmitoyl-CoA, 37 μM; (−)carnitine, 1 mM; and rat heart mitochondria, 0.5 mg of protein; were incubated at 25° with Hepes, pH 7.4, 25 mM; potassium phosphate, 5 mM; mannitol, 60 mM; Kcl, 50–60 mM; and where noted, ADP, 2.5 mM (respiratory state 3), or without ADP (respiratory state 4). The maximum rate of β-oxidation in the presence of ADP ($CO_2$ + acid soluble products) equals 100. The fraction of "acid soluble products" recovered as acetylcarnitine was measured in a second experiment as radioactivity precipitated as Reinecke salt. This figure is a graphical presentation of previously published experiments [48].

which fatty acid oxidation in the heart is under indirect respiratory control when high concentrations of fatty acids are present. The formation of acetylcarnitine mentioned above is probably of importance in this indirect control. In perfused rat hearts they found that the levels of acetyl-CoA and acetylcarnitine are higher with a small workload than with a high workload, while the level of long-chain acyl-CoA and long-chain acylcarnitines paradoxically increased with a high workload, in spite of an increased rate of oxidation. These results show that the workload primarily regulates the rate of acetyl-CoA oxidation in the citric acid cycle. When the citric acid cycle slows down because of a more reduced state in the mitochondria, acetyl-CoA accumulates. This leads to a low level of free CoA and to increased acetylation of carnitine. The carnitine translocase will equilibrate this high acetylcarnitine/carnitine ratio in the mitochondria with the greater carnitine pool in the cytosol, and the low level of free carnitine will lead to the observed drop in long-chain acylcarnitine [49]. The combination of a low acylcarnitine and a low free CoA in the mitochondria will slow down the transfer of activated fatty acids into the mitochondria.

Wang and coworkers [38] have suggested an additional mechanism to control the fatty acid oxidation rate. The accumulated acetyl-CoA in the mitochondrial matrix may inhibit 3-ketoacyl-CoA thiolase, and the accumulated 3-ketoacyl-CoA will then inhibit the acyl-CoA dehydrogenase(s). This mechanism may contribute to the slowdown of $\beta$-oxidation, although Latipää [50] found no accumulation of 3-hydroxy-acyl-CoA and enoyl-CoA in arrested oleate perfused hearts (no work load). He suggested that the transport of fatty acids into the mitochondria was inhibited, more in agreement with the mechanisms suggested by Oram and coworkers [49].

### Malonyl-CoA in heart

Heart is not an organ with significant synthesis of fatty acids. It is surprising therefore that rat hearts contain about as much malonyl-CoA as rat liver [51]. The rate of turnover of heart malonyl-CoA is unknown, but it may have a function in fatty acid elongation, and evidently it may be important in regulation of fatty acid oxidation since heart CTP I is extremely sensitive to malonyl-CoA. Like in the liver, the level of malonyl-CoA decreases in fasting animals [51]. Variations in malonyl-CoA may thus explain that glucose plus insulin is more inhibitory on fatty acid oxidation in normal heart than in diabetic heart [52]. However, the role of malonyl-CoA as a regulator of fatty acid oxidation in the heart is not established. When isolated heart mitochondria oxidized palmitoyl-CoA in the presence of carnitine we found less inhibition by malonyl-CoA than with liver mitochondria [48]. However, exposure of CPT II in damaged mitochondria may explain this result since the CPT II is insensitive to malonyl-CoA.

### The function of peroxisomal β-oxidation in heart

When rape seed oil with high levels of erucic acid is fed to rats, the animals get a temporary lipidosis of the heart [53]. Hydrogenated fish oil with a high content of brassidic acid and other monoene 22 carbon fatty acids cause a similar although weaker lipidosis [54]. In heart cells in culture erucic acid is slowly oxidized after a primary chain shortening which is rate limiting [55]. The basic cause of the lipidosis probably is a poor ability of the mitochondria to handle these long chain fatty acids. The erucoyl-CoA and erucoylcarnitine are poor substrates for CPT and for oxidation in mitochondria [56], and for the acyl-CoA dehydrogenase [57].

The heart lipidosis due to rape seed oil is temporary. In continued feeding an adaptation takes place and the lipidosis disappears. This adaptation is mainly caused by increased peroxisomal $\beta$-oxidation, mainly in the liver, but also in the heart [58]. The peroxisomal oxidation of erucic acid and other fatty acids is incomplete. In the intact cell fatty acids usually undergo only 1 to 3 $\beta$-oxidation cycles, thus erucic acid is converted to oleic acid. Most studies suggest that peroxisomal oxidation is carnitine independent [59]. However, the peroxisomes, at least in liver, contain both a carnitine acetyltransferase and a carnitine medium chain acyltransferase. The function of these acyltransferases have not been established. A suggestion is that carnitine is required for transfer of the fatty acids into the peroxisomes [60], but it is an equally reasonable assumption that they are active in transfer of peroxisomal products (acetyl groups and shortened fatty acids) to the mitochondria for complete oxidation. Thus, in the liver the peroxisomal shortening of erucic acid seems to be independent of carnitine while its complete oxidation does depend on carnitine [59]. For shorter fatty acids it has turned out to be difficult to establish such a mechanism because after induction of peroxisomal activity there is also increased activity of acyl-CoA hydrolases [61]. Thus peroxisomal $\beta$-oxidation in liver leads to formation of free acetate [62], possibly also to free shortened fatty acids. The putative function of carnitine in transport of acyl groups to and from the peroxisomes in the heart so far remains unestablished.

## Conclusions

The formation of carnitine esters of fatty acids has an established function in the mitochondrial oxidation of medium and long-chain fatty acids in the heart and other tissues.

The formation of acetylcarnitine in heart is probably involved in the regulation of the rate of fatty acid oxidation which depends on workload [49]. Carnitine as an acceptor of acetyl and other acyl groups may also "buffer" the acetyl-CoA/free CoA ratio of the mitochondria.

Malonyl-CoA in the heart [51] probably participates in the regulation of fatty acid oxidation since the heart CPT I is extremely sensitive to malonyl-CoA [30, 31].

Propionylcarnitine and branched chain acylcarnitines may be formed in the heart [36]. The physiological significance is uncertain. Release of carnitine or acylcarnitines is slow from muscle tissues [39]. The increased formation of these carnitine esters in inborn errors of metabolism probably leads to loss of carnitine primarily from kidneys, blood plasma and liver [63, 64]. The resulting low serum carnitine level then leads to a slow depletion of carnitine from skeletal muscle and heart.

This chapter has been dedicated to the memory of E. Jack Davis (1930–93), Department of Biochemistry and Molecular Biology, Indiana University. Many of the considerations in this chapter come from our discussions and common publications on heart and muscle metabolism through more than 20 years.

## References


1. Evans JR, Opie LH, Shipp JC. Metabolism of palmitic acid in perfused rat heart. Am J Physiol 1963; 205: 706–7.
2. Saddik M, Lopaschuk GD. Myocardial triglyceride turnover and contribution to energy substrate utilization in isolated working rat hearts. J Biol Chem 1991; 266: 8162–70.
3. De Jong JW, Hülsmann WC. A comparative study of palmitoyl-CoA synthetase activity in rat-liver, heart and gut mitochondrial and microsomal preparations. Biochim Biophys Acta 1969; 197: 127–35.
4. Webster Jr LT, Gerowin LD, Rakita L. Purification and characteristics of a butyryl-CoA synthetase from bovine heart mitochondria. J Biol Chem 1964; 240: 29–33.
5. Mahler HR, Wakil SJ, Bock RM. Studies on fatty acid oxidation. I. Enzymatic activation of fatty acids. J Biol Chem 1953; 204: 453–67.
6. Bressler R, Wittels B. The effect of diphtheria toxin on carnitine metabolism in the heart. Biochim Biophys Acta 1965; 104: 39–45.
7. Mølstad P, Bøhmer T. The effect of diphtheria toxin on the cellular uptake and efflux of L-carnitine. Evidence for a protective effect of prednisolon. Biochim Biophys Acta 1981; 641: 71–8.
8. Engel AG, Angelini C. Carnitine deficiency of human skeletal muscle with associated lipid storage myopathy: a new syndrome. Science 1973; 179: 899–902.
9. Miyazawa S, Ozasa H, Furuta S, Osumi T, Hashimoto T. Purification and properties of carnitine acetyltransferase from rat liver. J Biochem 1983; 93: 439–51.
10. Farrell SO, Bieber LL. Carnitine octanoyltransferase of mouse liver peroxisomes: properties and effect of hypolipidemic drugs. Arch Biochem Biophys 1983; 222: 123–32.
11. Maxwell MAK, Tolbert NE, Bieber LL. Comparison of the carnitine acyltransferase activities from rat liver peroxisomes and microsomes. Arch Biochem Biophys 1976; 176: 479–88.
12. Murthy MSR, Pande SV. Malonyl-CoA binding site and the overt carnitine palmitoyltransferase activity reside on the opposite sides of the outer mitochondrial membrane. Proc Natl Acad Sci USA 1987; 84: 378–82.
13. Norum KR, Bremer J. The localization of acyl coenzyme A: carnitine acyltransferase in rat liver cells. J Biol Chem 1967; 242: 407–11.
14. Clarke RH, Bieber LL. Effect of micelles on the kinetics of purified beef heart mitochondrial carnitine palmitoyltransferase. J Biol Chem 1981; 256: 9861–8.
15. Solberg HE. Acyl group specificity of mitochondrial pools of carnitine acyltransferases. Biochim Biophys Acta 1974; 360: 101–12.
16. Woeltje KF, Esser V, Weis BC et al. Inter-tissue and inter-species characteristics of the mitochondrial carnitine palmitoyltransferase enzyme system. J Biol Chem 1990; 265: 10714–9.
17. Bremer J, Norum KR. The mechanism of substrate inhibition of palmityl coenzyme A: carnitine acyltransferase in rat liver cells. J Biol Chem 1967; 242: 1744–8.
18. Derrick JP, Ramsay RR. L-Carnitine acyltransferase in intact peroxisomes is inhibited by malonyl-CoA. Biochem J 1989; 262: 801–6.
19. Skorin C, Nechochea, C, Johow V, Soto U, Grau AM, Bremer J, Leighton F. Peroxisomal fatty acid oxidation and inhibitors on the mitochondrial carnitine palmitoyltransferase I in rat isolated hepatocytes. Biochem J 1992; 281: 561–7.
20. Lund H. Carnitine palmitoyltransferase: characterization of a labile detergent-extracted malonyl-CoA-sensitive enzyme from rat liver mitochondria. Biochim Biophys Acta 1987; 918: 67–75.
21. Brosnan JT, Fritz IB. The oxidation of fatty-acyl derivatives by mitochondria from bovine fetal and calf hearts. Can J Biochem 1971; 49: 1296–300.
22. Warshaw JB, Terry ML. Cellular energy metabolism during fetal development. II. Fatty acid oxidation by the developing heart. J Cell Biol 1970; 44: 354–60.

23. Bergseth S, Lund H, Bremer J. Is carnitine palmitoyltransferase inhibited by a malonyl-CoA binding unit in the mitochondria? Biochem Soc Trans 1986; 14: 671–2.
24. Ghaddiminejad I, Saggerson ED. Carnitine palmitoyltransferase ($CPT_2$) from liver mitochondrial inner membrane becomes inhibitable by malonyl-CoA if reconstituted with outer membrane malonyl-CoA binding protein. FEBS Lett 1990; 269: 406–8.
25. Chung CH, Woldegiorgis G, Dai G, Shrago E, Bieber LL. Conferral of malonyl coenzyme A sensitivity to purified rat heart mitochondrial carnitine palmitoyltransferase. Biochemistry 1992; 31: 9777–83.
26. Kerner J, Bieber, L. Isolation of a malonyl-CoA sensitive CPT/$\beta$-oxidation enzyme complex from heart mitochondria. Biochemistry 1990; 29: 4326–34.
27. Zierz S, Engel AG. Different sites of inhibition of carnitine palmitoyltransferase by malonyl-CoA, and by acetyl-CoA and CoA, in human skeletal muscle. Biochem J 1987; 245: 205–9.
28. Esser V, Britton CH, Weis BC, Foster DW, McGarry JD. Cloning, sequencing, and expression of a cDNA encoding rat liver carnitine palmitoyltransferase I. Direct evidence that single polypeptide is involved in inhibitor interaction and catalytic function. J Biol Chem 1993; 268: 5817–22.
29. Kolodziej MP, Crilly PJ, Corstophine CG, Zammit VA. Development and characterization of a polyclonal antibody against rat liver mitochondrial overt carnitine palmitoyltranferase (CPT I). Distinction of CPT I from CPT II and of isoformes of CPT I in different tissues. Biochem J 1992; 282: 415–21.
30. Saggerson ED, Carpenter CA. Carnitine palmitoyltransferase and carnitine octanoyltransferase activities in liver, kidney cortex, adipocyte, lactating mammary gland, skeletal muscle and heart. Relative activities, latency and effect of malonyl-CoA. FEBS Lett 1981; 129: 229–32.
31. Mills SE, Foster DW, McGarry JD. Interaction of malonyl-CoA and related compounds with mitochondria from different rat tissues. Relationship between ligand binding and inhibition of carnitine palmitoyltransferase I. Biochem J 1983; 214: 83–91.
32. Cook GA, Otto DA, Cornell NW. Differential inhibition of ketogenesis by malonyl-CoA in mitochondria from fed and starved rats. Biochem J 1980; 192: 955–8.
33. Bremer J. The effect of fasting on the activity of liver carnitine palmityltransferase and its inhibition by malonyl-CoA. Biochim Biophys Acta 1981; 665: 628–31.
34. Idell-Wenger JA, Grotyohann LW, Neely JR. Coenzyme A and carnitine distribution in normal and ischemic hearts. J Biol Chem 1978; 253: 4310–8.
35. Pande SV, Parvin R. Characterization of carnitine acylcarnitine translocase system of heart mitochondria. J Biol Chem 1976; 251: 6683–91.
36. Davis EJ, Bremer J. Studies with isolated surviving rat hearts. Interdependence of free amino acids and citric acid cycle intermediates in muscle. Eur J Biochem 1973; 38: 86–97.
37. Bremer J, Davis EJ. Fluoroacetylcarnitine: metabolism and metabolic effects in mitochondria. Biochim Biophys Acta 1973; 326: 262–71.
38. Wang H-Y, Baxter CF, Schulz H. Regulation of fatty acid $\beta$-oxidation in rat heart mitochondria. Arch Biochem Biophys 1991; 289: 274–80.
39. Brooks DE, McIntosh JEA. Turnover of carnitine by rat tissues. Biochem J 1975; 49: 1296–300.
40. Hiatt WR, Regensteiner JG, Wolfel EE, Ruff L, Brass EP. Carnitine and acylcarnitine metabolism during exercise in humans. Dependence on skeletal muscle metabolic state. J Clin Invest 1989; 84: 1167–73.
41. Bowman RH. Effect of diabetes, fatty acids, and ketone bodies on tricarboxylic acid cycle metabolism in the perfused rat heart. J Biol Chem 1965; 240: 2308–21.
42. Lee S-H, Davis EJ. Carboxylation and decarboxylation reactions. Anaplerotic flux and removal of citrate cycle intermediates in skeletal muscle. J Biol Chem 1979; 254: 420–30.

43. Davis EJ, Spydevold Ø, Bremer J. Pyruvate carboxylase and propionyl-CoA carboxylase as anaplerotic enzymes in skeletal muscle mitochondria. Eur J Biochem 1980; 110: 255–62.
44. Hülsmann WC, Schneijdenberg CTWM, Verkleij AJ. Accumulation and excretion of long-chain acylcarnitine by rat hearts; studies with aminocarnitine. Biochim Biophys Acta 1991; 1097: 263–9.
45. Bremer J. Pyruvate dehydrogenase, substrate specificity and product inhibition. Eur J Biochem 1969; 8: 535–40.
46. Lumeng L, Bremer J, Davis EJ. Suppression of the mitochondrial oxidation of (−)-palmitylcarnitine by the malate-aspartate and $\alpha$-glycerophosphate shuttles. J Biol Chem 1976; 251: 277–84.
47. Bremer J, Wojtczak AB. Factors controlling the rate of fatty acid ß-oxidation in rat liver mitochondria. Biochim Biophys Acta 1972; 280: 515–30.
48. Bremer J, Davis EJ, Borrebæk B. Factors influencing the carnitine-dependent oxidation of fatty acids in the heart. In: Ferrari R, Katz AM, Shug A, Visioli O, editors. Myocardial ischemia and lipid metabolism. New York: Plenum Press, 1984: 15–26.
49. Oram JF, Bennetch SL, Neely JR. Regulation of fatty acid utilization in isolated perfused rat hearts. J Biol Chem 1973; 248: 5299–309.
50. Latipää PM. Energy-linked regulation of mitochondrial fatty acid oxidation in the isolated perfused rat heart. J Mol Cell Cardiol 1989; 21: 765–71.
51. Singh B, Stakkestad JA, Bremer J, Borrebæk B. Determination of malonyl-CoA in rat heart, kidney and liver: A comparison between acetyl-CoA and butyryl-CoA as fatty acid synthase primers in the assay procedure. Anal Biochem 1984; 138: 107–11.
52. Randle PJ, Garland PB, Hales CN, Newsholm EA, Denton RM, Pogson CI. Interaction of metabolism and the physiological role of insulin. Recent Progr Hormone Res 1966; 22: 1–44.
53. Abdellatif AM, Vles RO. Pathological effects of dietary rapeseed oil in rats. Nutr Metabol 1970; 12: 285–95.
54. Beare-Rogers JL, Nera EA, Heggtveit HA. Cardiac lipid changes in rats fed oils containing long-chain fatty acids. Can Inst Food Technol J 1971; 4: 120–24.
55. Pinson A, Padieu P. Erucic acid oxidation by beating heart cells in culture. FEBS Lett 1974; 39: 88–90.
56. Christophersen BO, Bremer J. Erucic acid – an inhibitor of fatty acid oxidation in rat liver mitochondria. Biochim Biophys Acta 1972; 280: 506–14.
57. Heijkenskjöld L, Ernster L. Studies on the mode of action of erucic acid on heart metabolism. Acta Medica Scand 1975; Suppl 585: 75–86.
58. Norseth J. The effect of feeding rats with partially hydrogenated marine oil or rapeseed oil on the chain shortening of erucic acid in perfused heart. Biochim Biophys Acta 1979; 575: 1–9.
59. Christiansen RZ, Christiansen EN, Bremer J. The stimulation of erucate metabolism in isolated rat hepatocytes by rape seed oil and hydrogenated marine oil-containing diets. Biochim Biophys Acta 1979; 573: 417–29.
60. Buechler KF, Lowenstein JM. The involvement of carnitine intermediates in peroxisomal fatty acid oxidation: a study with 2–bromofatty acids. Arch Biochem Biophys 1990; 281: 233–238.
61. Katoh H, Kawakima Y, Watanuki H, Kozyka H, Isono, H. Effects of clofibric acid and tiadenol on cytosolic long-chain acyl-CoA hydrolase and peroxisomal $\beta$-oxidation in liver and extrahepatic tissues of rats. Biochim Biophys Acta 1987; 920: 171–9.
62. Leighton F, Bergseth S, Rørtveit T, Christiansen EN, Bremer J. Free acetate produced by rat hepatocytes during peroxisomal fatty acid and dicarboxylic acid oxidation. J Biol Chem 1989; 264: 10347–50.

63. Hokland BM, Bremer J. Formation and excretion of branched-chain acylcarnitines and branched-chain hydroxy acids in the perfused rat kidney. Biochim Biophys Acta 1988; 961: 30–7.
64. Hokland BM. Uptake, metabolism and release of carnitine and acylcarnitines in the perfused rat liver. Biochim Biophys Acta 1988; 961: 234–41.

*Corresponding Author*: Dr Jon Bremer, Institute of Medical Biochemistry, University of Oslo P.O. Box 1112, Blindern, 0317 Oslo, Norway

# 3. Carnitine and carnitine esters in mitochondrial metabolism and function

FABIO DI LISA, ROBERTA BARBATO, ROBERTA MENABÒ and NORIS SILIPRANDI

"Under physiological conditions, long-chain acylcarnitines are formed mostly in the mitochondrial intermembrane space and *imported* into the matrix, whereas short-chain acylcarnitines are formed within the matrix space and *exported* into the cytosol."

**Abbreviations:** ANT – adenine nucleotide translocase, CAT – carnitine acetyl transferases, CoASH – free (unesterified) coenzyme A, COT – carnitine octanoyl transferase, CPT – carnitine palmitoyl transferase, CT – carnitine translocase, FFA – free fatty acid, IVC – isovalerylcarnitine, LCACar – long-chain acylcarnitines, LCACoA – long-chain acyl-CoAs, MTP – membrane transition pore, PDH – pyruvate dehydrogenase, PLC – propionylcarnitine, SCACar – short-chain acylcarnitines, SCACoA – short-chain acyl-CoAs, TCA – tricarboxylic acid(s)

## Introduction

Under physiological conditions, myocardial energy production is almost entirely dependent on mitochondrial oxidative phosphorylation [1]. At rest, in the post-absorptive state, the major fuel for mitochondrial oxidation is represented by long-chain fatty acids [2]. The degradation of these substrates occurs in mitochondria and carnitine is required for the transport of activated acyls, namely acyl-CoAs, across the inner mitochondrial membrane [3, 4]. The alteration of this process may result in cytosolic accumulation of triglycerides leading eventually to heart failure [5, 6]. Thus, it is hardly surprising that cardiologists focused their attention on carnitine, linking this factor exclusively to lipid metabolism. However, by its interaction with coenzyme A (CoA), carnitine exerts a role in any CoA-dependent process [7–9]. The modulation of the ratio between free CoA (CoASH) and esterified CoA can be seen as the main task accomplished by carnitine. An increase in CoASH availability or a decrease in acyl-CoA levels expands the roles of carnitine to substrate choice [10], removal of inhibitory metabolites [11] or modulation of key enzymatic steps [7, 9, 12, 13]. In addition, some carnitine esters appear to specifically modify cell metabolism and function [7, 14].

*J.W. de Jong and R. Ferrari (eds): The carnitine system*, 21–38.

## Carnitine-dependent enzymes

The reversible exchange of acyl moieties between CoA and carnitine is catalyzed by several carnitine acyl transferases. The differences among these enzymes can be described in terms of cellular localization, substrate specificity, structure and reactivity with inhibitors. Generally, these transferases are classified on the basis of their affinity for acyl-CoA [8, 15]. *Carnitine acetyl transferases* (CAT) catalyze those reactions involving short-chain acyl esters with a chain length ranging from 2 to 10 carbon atoms. Regarding the transferases for the long-chain acyl esters (>10 C), the term COT (*carnitine octanoyl transferase*) is used for the extramitochondrial proteins, whereas CPT (*carnitine palmitoyl transferase*) is generally adopted to indicate the mitochondrial enzymes. Despite the different names, these two subgroups of enzymes show a similar broad range of specificity towards medium-chain and long-chain acyl-CoA with the highest affinity for decanoyl-CoA [8].

### *Carnitine palmitoyl transferase*

The topology, the structure and even the number of possible isoforms of CPT are still under debate (Figure 1). Presently one of the two major hypotheses proposed (reviewed in [16]) suggests that the overall CPT activity results from the integrated activity of two different proteins: an oligomer made up of 68 kDa subunits [17–19] which is inserted in the inner mitochondrial membrane (CPT II). Another protein (CPT I) which is responsible for malonyl-CoA inhibition is present on either the outer side of the inner mitochondrial membrane or on the outer mitochondrial membrane [20]. In the alternative hypothesis only one protein possessing CPT activity is regulated by interacting with one or more malonyl-CoA binding proteins [21]. Nevertheless, these different schools of thought share the following well-established points: i) in the inner membrane of mitochondria there is at least one form of CPT (classified as CPT II) which has been isolated [17], cloned and sequenced [18, 19], and ii) the inhibition of CPT by malonyl-CoA is a crucial step in the overall control of intracellular lipid metabolism [22]. More recently the relevance of this regulation, crucial in liver metabolism, has been emphasized for the heart [23]. In addition, CPT has been reported to be stimulated by a cyclic AMP-dependent phosphorylation [24] and, accordingly, by phosphatase inhibition [25].

### *Carnitine acetyl transferase*

CAT is a monomer associated with the inner mitochondrial membrane. Up to this point no regulatory mechanism has been described, so CAT appears to function as a simple Michaelis-Menten enzyme. It was the first carnitine-dependent enzyme to be isolated [26]. Its commercial availability made the enzymatic assay of carnitine possible, thus disclosing the clinical interest for

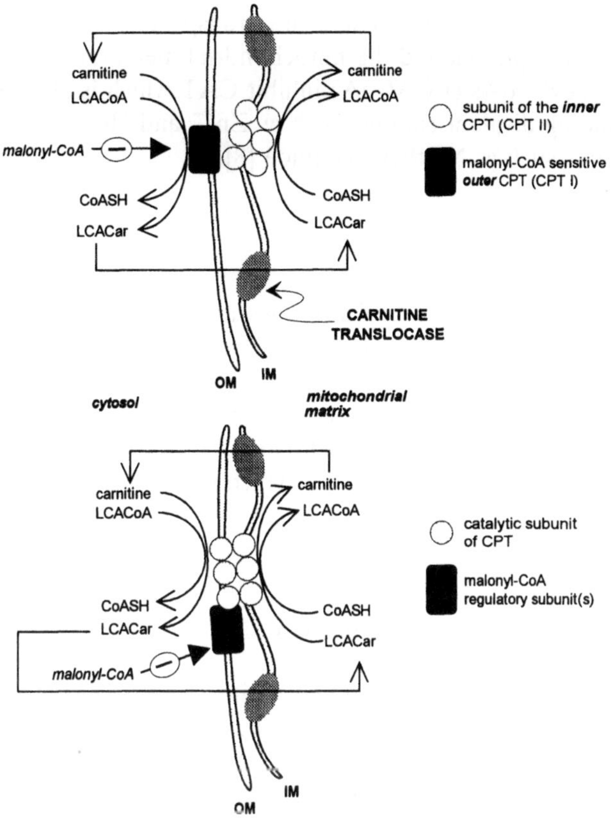


*Figure 1.* Long-chain acyl esters traffic across the mitochondrial membranes. The two major hypotheses concerning the structure and the function of carnitine palmitoyl transferase(s) are summarized in the two panels (adapted from [8, 16]). The upper panel shows the point of view supported mainly by McGarry for which the inner and the outer CPT are completely different proteins with different catalytic activities. The outer CPT should be responsible for malonyl-CoA sensitivity. Conversely, as shown in the lower panel, according to Bieber et al., there is only one CPT which is regulated by interacting with malonyl-CoA binding protein(s). CPT, carnitine palmitoyl transferase; IM, inner membrane; LCACar, long-chain acylcarnitine; LCA-CoA, long-chain acyl-CoA; OM, outer membrane. Note inhibition by malonyl-CoA.

this compound and allowing the discovery of carnitine related defects [27, 28]. Among CoA esters, CAT affinity for propionyl CoA is slightly higher than that for acetyl-CoA and butyryl-CoA. A progressive decrease in its activity is observed when chain length is increased to about $C_8$ or $C_{10}$. The following $K_{0.5}$ values for the four substrates (2 in each direction) have been reported: carnitine 0.1–0.3 mM; acetyl-CoA 32–70 μM; acetylcarnitine 0.3–0.7 mM; CoASH 10–30 μM [8]. Although the enzyme binds both carnitine isomers, only R-carnitine, the natural (−) isomer (commonly denominated

L-carnitine), is active in enzyme-catalyzed acetyl transfer. The same applies for CPT, even though both isomers are transported with similar kinetic patterns across the plasma and the mitochondrial membranes.

Long-chain acyl-CoAs (LCACoA) inhibit CAT which is also inhibited by sulfhydryl-binding divalent cations, such as $Zn^{2+}$ and $Hg^{2+}$, and sulfhydryl-specific reagents such as N-ethylmaleimide [8].

### *Carnitine translocase*

The transport of carnitine or acylcarnitines across the inner mitochondrial membrane is catalyzed by an antiport exchange system, namely *carnitine translocase* (CT), which operates with a rigorous 1:1 stoichiometry [29, 30] and is inhibited by the reactants of −SH groups [31, 32]. The transport capacity of CT, which in heart mitochondria is >100 nmol/min/mg protein [33, 34] rules out a rate-limiting role for which was previously proposed [35].

## **Acylcarnitines and mitochondrial function**

Both CPT and CAT are abundant in heart mitochondria. The former enzyme promotes fatty acid oxidation by translocating activated long-chain fatty acids into the mitochondrial matrix space. The latter allows the utilization of a large variety of substrates accounting for the efflux of various short-chain acylcarnitines from isolated mitochondria incubated in the presence of carnitine (Table 1).

Under physiological conditions, long-chain acylcarnitines (LCACar) are formed mostly in the intermembrane space and *imported* into the matrix, whereas short-chain acylcarnitines (SCACar) are formed within the matrix space and *exported* into the cytosol (Figure 2). All these carnitine-dependent processes implicate modulation of the acyl-CoA/CoASH ratio and the buffering of CoASH availability [7–9]. This role of carnitine is especially evident within the matrix space, where the scarce cellular amounts of CoA (120 nmol/g wet weight, 2.5 nmol/mg mitochondrial protein) are almost entirely compartmentalized (95% in cardiac myocytes) [36]. In contrast, the total carnitine content of the cell is >7-fold higher, but due to the equilibrium imposed by CT, the concentrations of CoA and carnitine in the matrix are similar, whereas in the cytosol the carnitine/CoASH ratio is always very high [36]. This may result in a different extent of carnitine and CoA esterification inside and outside the mitochondrion [9]. Thus, changes in the cell content of carnitine are accompanied by an immediate adjustment of the acyl-CoA/CoASH ratio in the matrix space. Accordingly, it has been widely demonstrated that the addition of carnitine to isolated mitochondria induces a profound decrease of the acyl-CoA/CoASH ratio [37, 38].

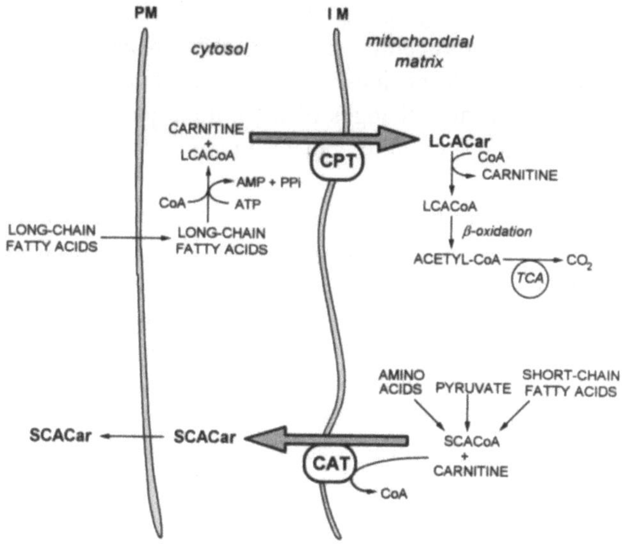


*Figure 2.* Movement of substrates and acylcarnitines across cellular membranes. Under physiological conditions long-chain acylcarnitines (LCACar) and short-chain acylcarnitines (SCACar) are likely to move in opposite directions. LCACoA, long-chain acyl-CoA; SCACoA, short-chain acyl-CoA; CPT, carnitine palmitoyl transferase; CAT, carnitine acetyl transferase; TCA, tricarboxylic acid cycle; PM, plasma membrane; IM, inner mitochondrial membrane; PPi, inorganic pyrophosphate.

## *Long-chain acylcarnitines*

Profound alterations of mitochondrial physiology are produced by LCACoA [39] which accumulate during ischemia [6, 36, 40]. These CoA esters are amphipatic molecules which can insert themselves in the phospholipid bilayer altering both membrane architecture and permeability [39, 41]. These changes are more likely to occur at LCACoA concentrations above the critical micellar concentration which in the case of palmitoyl-CoA is about 30 μM. At lower concentrations, LCACoA are able to specifically affect the activity of various transport systems of the inner mitochondrial membrane without perturbing its permeability [42]. The most classical example of these modifications is the inhibition of adenine nucleotide translocase (ANT) [43]. The inhibitory effects of LCACoA are exacerbated by $Ca^{2+}$ and blunted by several cations such as $Mg^{2+}$ and polyamines [42, 44] which are present in millimolar concentrations within the matrix space. The protective effect exerted by these cations, which are usually present in mitochondria, may explain the lack of ANT inhibition which was obtained by increasing LCACoA matrix content [45].

Carnitine addition is able to restore ANT function and hence, oxidative phosphorylation, by changing LCACoA to LCACar which are devoid of inhibitory effects. More recently, LCACoA have been added to the long list of promoters of the cyclosporine-sensitive membrane transition pore (MTP) [46, 47]. Also, the abrupt changes of membrane permeability and function brought about by MTP opening can be prevented or partially restored by carnitine [47].

LCACoA accumulation might play a central role in the well-documented free fatty acid (FFA) myocardial toxicity [6, 41, 48, 49]. However, besides the $Mg^{2+}$ effect, it is worthwhile considering that the $K_i$s for transport inhibitions (>5 μM) are well above the mitochondrial content of total CoA (free + esterified). Furthermore, the concentration of "free" LCACoA is likely to be reduced by the more or less specific binding with various proteins [50], although it is possible that these esters do not accumulate homogeneously, reaching very high local concentrations within the hydrophobic core of the membranes.

*Short-chain acylcarnitines*

The CoA-carnitine relationship is pivotal for energy metabolism. CoA is required for β-oxidation, for the catabolism of several amino acids, for the detoxification of organic acids and xenobiotics, for pyruvate dehydrogenase (PDH) [51], for α-ketoglutarate dehydrogenase [12, 52] and thus for the TCA (tricarboxylic acid) cycle [7]. A reduced availability of carnitine induces a decrease of matrix CoASH and a parallel increase of the acyl-CoA/CoASH ratio, both of which are inhibitory in the aforementioned mitochondrial dehydrogenases.

Unlike the corresponding acyl-CoAs, acylcarnitines, especially SCACar, are capable of diffusing across cellular membranes and may be eliminated in the urine. The urinary excretion of specific acylcarnitines is relevant for the diagnosis of several inborn errors of metabolism [53].

Large amounts of acetylcarnitine are produced by isolated rat heart mitochondria utilizing pyruvate as a substrate (Table 1). When mitochondria are incubated in the absence of a TCA cycle sparker, such as malate, and in the presence of a large carnitine availability, the rate of acetylcarnitine formation is almost superimposable to that of pyruvate decarboxylation and oxygen consumption [54]. Under these conditions, carnitine induces a >10-fold decrease in the acyl-CoA/CoASH ratio [38] and stimulates pyruvate oxidation even more than dichloroacetate, the well-known inhibitor of PDH kinase [7].

The effect of carnitine on pyruvate utilization is less evident in liver mitochondria which have intrinsically low amounts of CAT. It is conceivable that in the liver the utilization of acetyl-CoA for ketogenesis renders the formation of acetylcarnitine less compelling.

In the isolated mitochondria, the process of acetylcarnitine production

*Table 1.* Short-chain acylcarnitine production from different substrates in rat heart mitochondria.

| Substrate | Acylcarnitine produced | Production rate |
|---|---|---|
| Pyruvate (2.5 mM) | acetyl | 93.0 ± 8.2 |
| Pyruvate + malate (1 mM) | acetyl | 48.0 ± 4.2 |
| Decanoyl-CoA (40 μM) | acetyl | 25.4 ± 3.7 |
| Palmitoyl-CoA (40 μM) | acetyl | 18.3 ± 2.9 |
| α-Ketoisovalerate (1 mM) | isobutyryl + propionyl | 11.9 ± 1.5 |
| α-Ketoisocaproate (1 mM) | isovaleryl | 2.6 ± 0.5 |
| α-Methyl-β-ketovalerate (1 mM) | isovaleryl + propionyl | 4.5 ± 0.9 |

Mitochondria were incubated in the presence of 5 mM [$^3$H]-L-carnitine. Values are nmol/min/mg mitochondrial protein, mean ± S.D. of at least 4 different experiments.

from pyruvate shows an apparent $K_m$ for carnitine of about 1 mM which is higher than that reported for the isolated enzyme [54]. Although so far no mechanism has been postulated to explain this difference, it may be suggested that in vivo CAT activity is not saturated by carnitine.

Carnitine and CoA availability may also influence the choice of the substrate utilized for energy production. Although the heart can oxidize either carbohydrate, amino acids or fat, it is generally accepted that free fatty acids (FFA) are the preferred metabolic fuels [2]. Fatty acids can predominate over carbohydrates by means of a coordinated inhibition at three control sites: i) glucose entry, ii) hexokinase-phosphofructokinase, iii) PDH [1]. Neverthcless, after carbohydrate feeding, the heart switches its fuel preference from fatty acids to glucose. Brosnan and Reid [55] demonstrated that in the isolated rat heart oleate oxidation is inhibited by either pyruvate or lactate addition to the perfusate. Lactate was less efficient than pyruvate, but the difference disappeared in the presence of dichloroacetate. In fact, unlike lactate, pyruvate activates the complex per se and makes dichloroacetate activation superfluous. The question then arises as to how acetyl-CoA produced by pyruvate decarboxylation may reduce FFA oxidation. The most plausible mechanism could be envisaged as an intramitochondrial competition for CoASH at the level of both the CPT and the thiolase reaction (Figure 3). Indeed the latter enzyme is inhibited by a high acetyl-CoA/CoASH ratio [56]. As shown by Table 2, in rat heart mitochondria pyruvate decreases both decanoyl-CoA utilization (reduction of the radioactivity incorporated in metabolic products) and acetylcarnitine formation from the decanoyl moiety. The former can be interpreted as a consequence of a reduced CPT activity, whereas a reduced rate of β-oxidation might explain the decrease in acetylcarnitine. Hence CoASH and carnitine availability could be considered a major regulating factor in the decision making concerning the substrate to be used. These mitochondrial processes could reinforce the inhibition of FFA oxidation operated by malonyl-CoA in the cytosol [23].

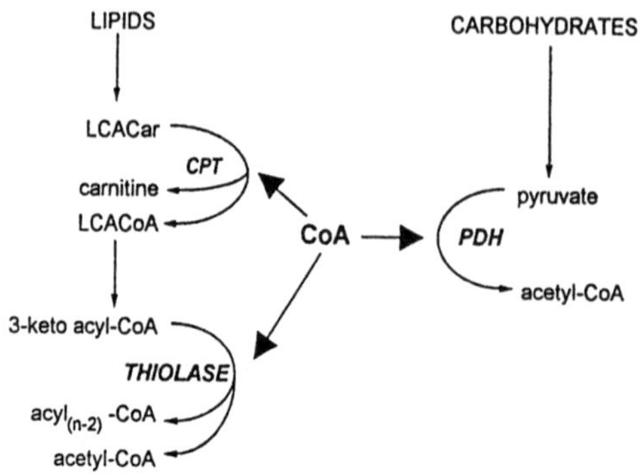


*Figure 3.* Competition between mitochondrial CoA-requiring reactions related to lipid and carbohydrate oxidation. LCACar, long-chain acylcarnitine; LCACoA, long-chain acyl-CoA; CPT, carnitine palmitoyl transferase; PDH, pyruvate dehydrogenase.

*Table 2.* Effect of pyruvate on the metabolism of decanoyl-CoA in rat heart mitochondria.

| Product | −Pyruvate | +Pyruvate |
|---|---|---|
| Acetylcarnitine | 25.34 | 12.27 |
| Decanoylcarnitine | 2.99 | 13.20 |
| ASM | 3.18 | 1.84 |
| $CO_2$ | 2.20 | 0.97 |
| Total | 33.71 | 28.28 |

Values are nmol/min/mg mitochondrial protein. Mitochondria were incubated in an isoosmotic medium containing 0.1 mM [1−$^{14}$C]decanoyl-CoA and 5 mM L-carnitine in the absence or in the presence of 0.5 mM pyruvate. ASM, acid-soluble metabolites.

The inhibition of several mitochondrial dehydrogenases by an elevated acyl-CoA/CoASH ratio prevents substrate utilization leading to a sort of energy starvation. For instance, the oxygen uptake for $\alpha$-ketoglutarate oxidation is blunted by the addition of acetoacetate which traps the available CoASH in the form of acetoacetyl- and acetyl-CoA [12]. On the other hand, pyruvate oxidation is inhibited by acetyl-CoA which accumulates when the TCA cycle is blocked by malonate [57, 58]. In both these conditions, carnitine addition restores mitochondrial function by converting the inhibitory metabolites into their corresponding acylcarnitines [12, 57, 58].

## Acylcarnitines and myocardial function

In isolated mitochondria as well as in the intact heart, LCACoA toxicity could be ascribed, at least in part, to a decreased CoASH availability. The

*Table 3.* Effect of propionate perfusion on the metabolic profile of isolated rat hearts.

| Metabolite | −Propionate | +Propionate | +Propionate + Carnitine |
|---|---|---|---|
| CoASH | 46.7 ± 3.9 | *6.9 ± 1.3 | *29.0 ± 2.2 |
| SCACoA | 19.0 ± 2.6 | *51.2 ± 5.7 | *32.4 ± 3.8 |
| LCACoA | 12.7 ± 2.7 | *20.5 ± 4.0 | 11.9 ± 2.4 |
| Carnitine | 661.3 ± 29.9 | *276.0 ± 39.4 | |
| SCACar | 85.1 ± 7.0 | *471.1 ± 31.2 | |
| LCACar | 38.5 ± 3.8 | 40.4 ± 4.1 | |
| ATP | 2.6 ± 0.1 | *1.8 ± 0.2 | 2.6 ± 0.1 |
| PCr | 3.5 ± 0.2 | *1.9 ± 0.2 | 3.6 ± 0.2 |
| Lactate release | 23.1 ± 5.6 | *82.0 ± 13.1 | 49.7 ± 14.4 |

Hearts were perfused for 90 min with a saline buffer containing 11 mM glucose under normoxic conditions. Propionate (2 mM) was added in the absence or in the presence of 5 mM carnitine to the perfusion medium. Values are nmol/g wet weight for CoA and carnitine fractions, μmol/g for ATP and phosphocreatine (PCr) and μmol/g/15 min for lactate release (mean ± S.E., n = 6). *$p < 0.05$ vs control (without propionate). Carnitine contents were omitted in the group receiving carnitine, since in the presence of massive amounts of exogenous carnitines the evaluation of endogenous ones appeared unreliable. For abbreviations of esters, see legend to Figure 2.

role of CoASH/esterified-CoA and carnitine/esterified carnitine ratios in the evolution of ischemic damage has been investigated in isolated rat hearts perfused with a saline solution containing glucose in both the presence and absence of propionate [14]. During post-ischemic reperfusion contractile recovery was strongly impaired by propionate. The damaging effect of propionate was prevented by carnitine, indicating an impairment of mitochondrial processes dependent on CoA and carnitine. Even in normoxic hearts, CoASH was severely depleted by propionate perfusion. Propionyl-CoA and methylmalonyl CoA, barely present in control hearts, were produced in amounts accounting for CoASH disappearance. Also, free carnitine was severely reduced (about 30% of control values) and concomitantly a large increase in propionylcarnitine was observed (Table 3). The consequent imbalance of energy metabolism was reflected by a significant decrease of tissue content of ATP and phosphocreatine. As expected, the increase in acyl-CoA/CoA ratio inhibited PDH as demonstrated by the increase in myocardial lactate efflux to values comparable to those induced by the inhibition of the respiratory chain. Despite these large changes of CoA and carnitine status, both inotropism and chronotropism were unaffected, suggesting that in the Langendorff model glycolysis alone can sustain the contractile activity. This hypothesis was tested directly by measuring left ventricular pressure using pyruvate as substrate and inhibiting glycogenolysis with iodoacetate. Under these conditions, propionate resulted in a gradual decrease in the developed

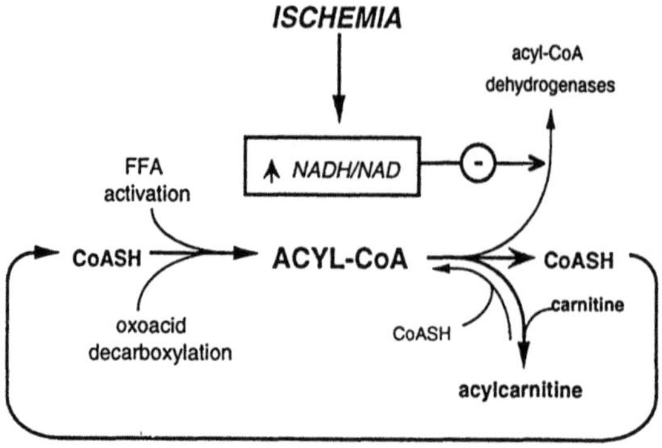


*Figure 4.* Effect of ischemia on the principal reactions producing and utilizing acyl-CoA. The width of the arrows indicates the estimated quantitative importance of the various reactions.

pressure which was followed by cardiac arrest [59]. Both the high rates of lactate release and the contractile failure were prevented by carnitine.

Thus, in propionate perfused hearts, both the lowered tissue ATP content plus free CoA create unfavorable conditions which worsen ischemic injury. These impairments in energy metabolism may in turn be responsible for the decline in cardiac contractility observed in the reperfusion phase.

The relevance of CoASH and carnitine availability for a regular myocardial function is also supported by other reports. In working hearts, a model which implies a higher energy consumption, propionate affects cardiac performance even under aerobic conditions and in the presence of glucose [60]. In addition, in working hearts perfused with acetoacetate, a fall in TCA cycle intermediates induced by inhibition of $\alpha$-ketoglutarate dehydrogenase was associated with contractile failure [52].

Anoxia or ischemia also result in an increase of esterified/free carnitine ratio [36]. This metabolic shift, which reproduces the analogous modification of CoA status, is caused by the inhibition of mitochondrial dehydrogenases consequent to the excess of reduced flavin and pyridine coenzymes. The reduced rates of $\beta$-oxidation and the TCA cycle freeze CoA in the form of LCACoA or SCACoA, and available carnitine acts as a scavenger of acyl moieties in order to liberate CoA (Figure 4). This action of carnitine appears to be pertinent for pyruvate utilization. In fact, in hypoxic tissues pyruvate is mostly converted to lactate due to PDH inhibition. By decreasing the acetyl-CoA/CoA ratio, carnitine might stimulate PDH, thus diverting pyruvate from its reduction to lactate and causing its oxidation to acetyl-CoA and then into acetylcarnitine (Figure 5). Experimental and clinical evidence support these concepts. As reported elsewhere in this book, Ferrari et al. [61] were the first group to demonstrate that carnitine administration can

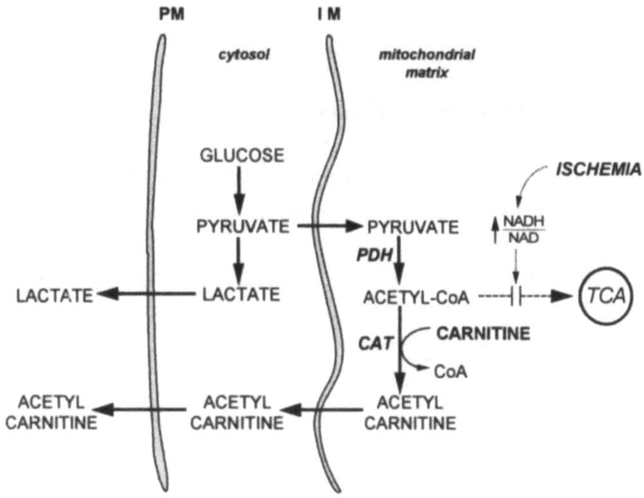


*Figure 5.* Alternative pathways of pyruvate utilization depending on carnitine availability in ischemic tissues (alanine is omitted). PDH, pyruvate dehydrogenase; CAT, carnitine acetyl transferase; TCA, tricarboxylic acid cycle; PM, plasma membrane; IM, inner mitochondrial membrane.

reduce lactate formation in subjects suffering from coronary artery disease. Similar results were obtained also in patients suffering from intermittent claudication [62]. In normal persons subjected to maximal cycle ergometer exercise, we demonstrated that plasma lactate reduction is linearly correlated with an increase in acetylcarnitine [63].

The possibility that carnitine might be able to influence muscle metabolism has been argued [64] by considering steady-state carnitine contents rather than metabolic fluxes. Due to the low rates of carnitine absorption into plasma and eventually into muscles, during the time of our exercise protocol, muscle carnitine content would have been increased by less than 2% [64]. Obviously such an increase is unlikely to modify CAT activity. However, carnitine transport across the sarcolemma seems to occur mainly by a 1:1 exchange with intracellular carnitine or carnitine esters [65]. Thus when acetylcarnitine accumulates in the cytosol, a large availability of extracellular carnitine could promote the exchange and consequently the washout of carnitine esters without major changes in carnitine tissue content [66]. This also appears to be the case in inborn errors of metabolism treated with carnitine. Indeed carnitine, given orally to patients with isovaleric acidemia, rapidly induced a large increase in plasma and urine isovalerylcarnitine [67]. This increase is most likely the result of an exchange of exogenous free carnitine with tissue isovalerylcarnitine formed from the excess isovaleryl-CoA.

## Acylcarnitines as pharmacological agents

The ability of SCACar to cross cellular membranes suggests the possibility that administered SCACar could reach the mitochondrial matrix space. Here, their transformation into the corresponding acyl-CoA may contribute to a useful integration of the acyls in mitochondrial metabolic pathways. Furthermore, SCACar administration could add some degree of specificity to carnitine therapy. In fact, at least theoretically, the more a tissue is endowed with CAT (myocardium being one of the richest) the more it should benefit from SCACar administration.

### *Acetylcarnitine*

The oxidation of either endogenous fatty acids or added acetate, which is impaired in aged mitochondria, was promptly restored by the addition of catalytic amounts (1 μmole) of acetylcarnitine [68]. Conceivably, this minute quantity supplied the initial amount of ATP needed for the activation and subsequent oxidation of fatty acids. Once started, the process proceeds autocatalytically. This capability of acetylcarnitine to spark mitochondrial energy-linked processes, together with its property to be promptly transported into cardiac cells, might be relevant to the restoration of dampened functions in deenergized tissue.

### *Propionylcarnitine*

Detailed descriptions of various effects of this carnitine ester on the cardiovascular system are reported in the last section of this book. The biochemical rationale for propionylcarnitine (PLC) administration concerns the possibility of feeding the TCA cycle (anaplerosis) with the carbon skeleton of propionate without altering the energy metabolism. These effects were initially documented in rat liver mitochondria [69]. When oleate was the oxidizable substrate, propionate induced a decrease of both $CO_2$ production and ATP content with a concomitant AMP increase, as is expected from propionate activation into propionyl-CoA. Unlike propionate, PLC, which is converted into propionyl-CoA without energy expenditure, did not alter the adenine nucleotide pool and increased the oxidation rate of oleate. This latter effect might depend upon both the accelerated flux in the TCA cycle, promoted by the anaplerotic effect of the newly formed propionyl-CoA and on the larger availability of intramitochondrial carnitine derived from PLC in the CAT-catalyzed reaction.

More recently the anaplerotic effect was demonstrated in rat heart mitochondria utilizing [2–$^{14}$C]pyruvate in the presence of either PLC or carnitine [70]. $CO_2$ production, which in these conditions is a tracer of acetyl-CoA utilization in the TCA cycle, was increased more by PLC than by equimolar amounts of carnitine. The utilization of the propionyl moiety emerges as a conceivable mechanism underlying PLC action. Propionyl-CoA may be rapidly transformed into oxaloacetate which in turn would stimulate pyruvate oxidation by promoting the entry of acetyl units in the TCA cycle. The inhibition of this PLC effect by malonate further supports this mechanism. On the other hand, acetylcarnitine formation was much lower in PLC treated mitochondria, ruling out the possibility that the decrease in acetyl-CoA/CoA ratio could be involved.

Opposing effects of propionate and PLC were also obtained in isolated and perfused rat hearts [14, 71]. In agreement with analogous results obtained by other authors on different experimental models [72–75], we demonstrated a protective effect of PLC on the ischemic heart, which is in sharp contrast with the above mentioned exacerbation of the ischemic damage induced by propionate. Concurrently, when propionate was replaced by equimolar amounts (2 mM) of PLC in the perfusing medium, adenine nucleotide, CoA and carnitine contents as well as lactate release were not different from control hearts. Thus, the protective effect of PLC may be a consequence of the anaplerotic utilization of propionate in the presence of optimal amounts of ATP, CoASH and carnitine. Obviously, the involvement of mitochondrial function in PLC effects does not exclude other mechanisms. It is likely that the protective action exerted by PLC might result from a positive interaction between i) improved mitochondrial function; ii) iron chelation [75]; iii) preservation of vascular patency [76].

### *Isovalerylcarnitine*

Both $\alpha$-ketoisocaproate and its parent amino acid, leucine, are known to inhibit lysosomal proteolysis in rat liver perfused in the absence of amino acids [77]. This raises the question as to whether leucine is responsible for the inhibition per se, or via some of its catabolites. We focused our attention on isovalerylcarnitine (IVC) which proved to reproduce the leucine effect with a similar dose-dependency [78]. Since only a negligible amount of added IVC is detectable within liver cells, either in perfused liver or in isolated hepatocytes, the inhibition of proteolysis could be mediated by a receptor located on the plasma membrane [79].

Successively we considered possible effects of IVC on some of the cytosolic proteolytic systems, particularly that of calpains which are ubiquitous $Ca^{2+}$-dependent proteases present in mammalian cells and subject to regulation by a variety of mechanisms [80].

Surprisingly IVC proved to be a potent activator rather than an inhibitor of calpain in different tissues including muscles [81]. The D isomer of IVC and both L-isobutyryl- and L-methylbutyryl-carnitine, derivatives of valine and isoleucine respectively, were almost ineffective [82]. The activating action of IVC on calpain has been envisaged in a remarkable increase in the affinity of calpain for $Ca^{2+}$. Furthermore it was shown that IVC acts synergistically with the cytoskeleton activator and that the two activators bind to different sites of the calpain molecule [82].

We also verified this stimulation in vivo by studying erythrocyte calpain in a patient suffering from isovaleric acidemia who is currently under carnitine treatment [83]. As already mentioned, in this metabolic error carnitine administration induces a marked rise of IVC in both plasma and urine. The continuous exposure of erythrocytes to high IVC concentrations was associated with a stimulation of calpain activity which returned to control levels when carnitine was replaced with glycine.

The circumstance that IVC inhibits lysosomal proteolysis and stimulates the limited $Ca^{2+}$-dependent proteolysis, mediated by calpains, appears as a paradox. Hypothetically, IVC might initiate a process of limited proteolysis in order to promote a remodelling rather than a degradation of cytosolic proteins. For this purpose the activation of calpains and the inhibition of lysosomal proteases are equally well-suited. Finally, since the proteins most susceptible to the calpain attack are those of the cytoskeleton fibres, which make extensive connections to mitochondria and other organelles [84], it is also conceivable that cytoskeleton disconnection promoted by calpain activation could render lysosomal proteases less accessible to the target proteins. Considering the mechanisms herein discussed, IVC might represent a specific signal to initiate these structural and functional modifications.

## Acknowledgements

This work was supported by CNR target project "BTBS" (Grant nr. 9201257; PF 70) and Sigma Tau (Pomezia, Italy). The authors thank Dr A. Cessario for critically reading the manuscript.

## References

1. Randle PJ, Tubbs PK. Carbohydrate and fatty acid metabolism. In: Berne RM, editor. Handbook of physiology. Circulation. Washington DC: Am Physiol Soc, 1978: 804–44.
2. Vary TC, Reibel DK, Neely JR. Control of energy metabolism of heart muscle. Annu Rev Physiol 1981; 43: 419–30.
3. Fritz IB. Carnitine and its role in fatty acid metabolism. Adv Lipid Res 1963; 1: 285–334.
4. Bremer J. Carnitine in intermediary metabolism. The metabolism of fatty acid esters of carnitine by mitochondria. J Biol Chem 1962; 237: 3628–32.
5. Liedtke AJ. Alterations of carbohydrate and lipid metabolism in the acutely ischemic heart. Progr Cardiovasc Dis 1981; 23: 321–36.
6. Van der Vusse GJ, Glatz JFC, Stam HCG, Reneman RS. Fatty acid homeostasis in the normoxic and ischemic heart. Physiol Rev 1992; 72: 881–940.

7. Siliprandi N, Sartorelli L, Ciman M, Di Lisa F. Carnitine: metabolism and clinical chemistry. Clin Chim Acta 1989; 183: 3–11.
8. Bieber LL. Carnitine. Annu Rev Biochem 1988; 57: 261–83.
9. Ramsay RR, Arduini A. The carnitine acyltransferases and their role in modulating acyl-CoA pools. Arch Biochem Biophys 1993; 302: 307–14.
10. Broderick TL, Quinney HA, Lopaschuk GD. Carnitine stimulation of glucose oxidation in the fatty acid perfused isolated working rat heart. J Biol Chem 1992; 267: 3758–63.
11. Chalmers RA, Roe CR, Tracey BM, Stacey RE, Hoppel CL, Millington DS. Secondary carnitine insufficiency in disorders of organic acid metabolism: modulation of acyl-CoA/CoA ratios by L-carnitine in vivo. Biochem Soc Trans 1983; 11: 724–5.
12. Hülsmann WC, Siliprandi D, Ciman M, Siliprandi N. Effect of carnitine on the oxidation of $\alpha$-oxoglutarate to succinate in the presence of acetoacetate or pyruvate. Biochim Biophys Acta 1964; 93: 166–8.
13. Brass EP, Beyerinck RA. Interactions of propionate and carnitine metabolism in isolated rat hepatocytes. Metabolism 1987; 36: 781–7.
14. Di Lisa F, Menabò R, Barbato R, Miotto G, Venerando R, Siliprandi N. Biochemical properties of propionyl- and isovaleryl-carnitine. In: Carter A, editor. Current concepts in carnitine research. Boca Raton FL: CRC Press, 1992: 27–36.
15. Bremer J. Carnitine: metabolism and functions. Physiol Rev 1983; 63: 1420–80.
16. Brady PS, Ramsay RR, Brady LJ. Regulation of the long-chain carnitine acyltransferases. FASEB J 1993; 7: 1039–44.
17. Clarke PR, Bieber LL. Isolation and purification of mitochondrial carnitine octanoyltransferase activities from beef heart. J Biol Chem 1981; 256: 9861–8.
18. Woeltje KF, Esser V, Weis B et al. Cloning, sequencing, and expression of a cDNA encoding rat liver mitochondrial carnitine palmitoyltransferase II. J Biol Chem 1990; 265: 10720–5.
19. Finocchiaro G, Taroni F, Rocchi M et al. cDNA cloning, sequence analysis, and chromosomal localization of human carnitine palmitoyltransferase. Proc Natl Acad Sci USA 1991; 88: 661–5.
20. Esser V, Britton CH, Weis BC, Foster DW, McGarry JD. Cloning, sequencing, and expression of a cDNA encoding rat liver carnitine palmitoyltransferase I. Direct evidence that a single polypeptide is involved in inhibitor interaction and catalytic function. J Biol Chem 1993; 268: 5817–22.
21. Chung CH, Woldegiorgis G, Dai G, Shrago E, Bieber LL. Conferral of malonyl coenzyme A sensitivity to purified rat heart mitochondrial carnitine palmitoyltransferase. Biochemistry 1992; 31: 9777–83.
22. McGarry JD, Foster DW. Regulation of hepatic fatty acid oxidation and ketone body production. Annu Rev Biochem 1980; 49: 395–420.
23. Saddik M, Gamble J, Witters LA, Lopaschuk GD. Acetyl-CoA carboxylase regulation of fatty acid oxidation in the heart. J Biol Chem 1993; 268: 25836–45.
24. Harano Y, Kashiwagi A, Kojima H, Suzuki M, Hashimoto T, Shigeta Y. Phosphorylation of carnitine palmitoyltransferase and activation by glucagon in isolated rat hepatocytes. FEBS Lett 1985; 188: 267–72.
25. Guzman M, Geelen MJ. Activity of carnitine palmitoyltransferase in mitochondrial outer membranes and peroxisomes in digitonin-permeabilized hepatocytes. Selective modulation of mitochondrial enzyme activity by okadaic acid. Biochem J 1992; 287: 487–92.
26. Fritz IB, Schultz SK, Srere PA. Properties of partially purified carnitine acetyl transferase. J Biol Chem 1963; 236: 2509–17.
27. Engel AG, Angelini C. Carnitine deficiency of human skeletal muscle with associated lipid storage myopathy: a new syndrome. Science 1973; 179: 899–902.
28. DiMauro S, DiMauro PM. Muscle carnitine palmitoyltransferase deficiency and myoglobinuria. Science 1973; 182: 929–31.
29. Pande SV. A mitochondrial carnitine acylcarnitine translocase system. Proc Natl Acad Sci USA 1975; 72: 883–7.

30. Ramsay RR, Tubbs PK. The mechanism of fatty acid uptake by heart mitochondria: an acylcarnitine-carnitine exchange. FEBS Lett 1975; 54: 21–5.
31. Noel H, Goswami T, Pande SV. Solubilization and reconstitution of rat liver mitochondrial carnitine acylcarnitine translocase. Biochemistry 1985; 24: 4504–9.
32. Pauly D, Yoon SB, McMillin JB. Carnitine-acylcarnitine translocase in ischemia: evidence for sulfhydryl modification. Am J Physiol 1987; 253: H1557–65.
33. LaNoue KF, Bryla J, Williamson JR. Feedback interactions in the control of citric acid cycle activity in rat heart mitochondria. J Biol Chem 1972; 247: 667–79.
34. Lysiak W, Toth PP, Suelter CH, Bieber LL. Quantitation of the efflux of acylcarnitines from rat heart, brain, and liver mitochondria. J Biol Chem 1986; 261: 13698–703.
35. Ramsay RR, Tubbs PK. The effects of temperature and some inhibitors on the carnitine exchange system of heart mitochondria. Eur J Biochem 1976; 69: 299–303.
36. Idell-Wenger JA, Grotyohann LW, Neely JR. Coenzyme A and carnitine distribution in normal and ischemic hearts. J Biol Chem 1978; 253: 4310–8.
37. Hansford RG, Cohen L. Relative importance of pyruvate dehydrogenase interconversion and feed-back inhibition in the effect of fatty acids on pyruvate oxidation by rat heart mitochondria. Arch Biochem Biophys 1978; 191: 65–81.
38. Lysiak W, Lilly K, Di Lisa F, Toth PP, Bieber LL. Quantitation of the effect of L-carnitine on the levels of acid-soluble short-chain acyl-CoA and CoASH in rat heart and liver mitochondria. J Biol Chem 1988; 263: 1151–6.
39. Brecher P. The interaction of long-chain acyl-CoA with membranes. Mol Cell Biochem 1983; 57: 3–15.
40. Shug AL, Thomsen JH, Folts JD et al. Changes in tissue levels of carnitine and other metabolites during myocardial ischemia and anoxia. Arch Biochem Biophys 1978; 187: 25–33.
41. Katz AM, Messineo FC. Lipid-membrane interactions and the pathogenesis of ischemic damage in the myocardium. Circ Res 1981; 48: 1–16.
42. Siliprandi N, Di Lisa F, Sartorelli L. Transport and function of carnitine in cardiac muscle. In: Berman MC, Gevers W, Opie LH, editors. Membranes and muscle. Oxford: IRL Press, 1985: 105–19.
43. Pande SV, Blanchaer MC. Reversible inhibition of mitochondrial adenosine diphosphate phosphorylation by long chain acyl coenzyme A esters. J Biol Chem 1971; 246: 402–11.
44. Toninello A, Dalla Via L, Testa S, Siliprandi D, Siliprandi N. Transport and action of spermine in rat heart mitochondria. Cardioscience 1990; 1: 287–94.
45. LaNoue KF, Watts JA, Koch CD. Adenine nucleotide transport during cardiac ischemia. Am J Physiol 1981; 241: H663–71.
46. Gunter TE, Pfeiffer DR. Mechanisms by which mitochondria transport calcium. Am J Physiol 1990; 258: C755–86.
47. Siliprandi D, Biban C, Testa S, Toninello A, Siliprandi N. Effects of palmitoyl CoA and palmitoyl carnitine on the membrane potential and $Mg^{2+}$ content of rat heart mitochondria. Mol Cell Biochem 1992; 116: 117–23.
48. Oliver MF, Kurien VA, Greenwood TW. Relation between serum-free-fatty acids and arrhythmias and death after acute myocardial infarction. Lancet 1968; 1: 710–4.
49. Opie LH. Effect of fatty acids on contractility and rhythm of the heart. Nature 1970; 227: 1055–6.
50. Glatz JFC, Van der Vusse GJ. Cellular fatty acid-binding proteins: current concepts and future directions. Mol Cell Biochem 1990; 98: 237–51.
51. Kerbey AL, Randle PJ, Cooper RH, Whitehouse S, Pask HT, Denton RM. Regulation of pyruvate dehydrogenase in rat heart. Mechanism of regulation of proportions of dephosphorylated and phosphorylated enzyme by oxidation of fatty acids and ketone bodies and of effects of diabetes: role of coenzyme A, acetyl-coenzyme A and reduced and oxidized nicotinamide-adenine dinucleotide. Biochem J 1976; 154: 327–48.
52. Russell III RR, Taegtmeyer H. Coenzyme A sequestration in rat hearts oxidizing ketone bodies. J Clin Invest 1992; 89: 968–73.

53. Chalmers R, Roe C, Stacey T, Hoppel C. Urinary excretion of L-carnitine and acylcarnitines by patients with disorders of organic acid metabolism: evidence for secondary insufficiency of L-carnitine. Pediatr Res 1984; 18: 1325–8.
54. Di Lisa F, Menabò R, Siliprandi N. Propionate as a CoA and carnitine trap in perfused rat hearts. J Mol Cell Cardiol 1989; 21 (Suppl 2): S30 (Abstr).
55. Brosnan JT, Reid K. Inhibition of palmitoylcarnitine oxidation by pyruvate in rat heart mitochondria. Metabolism 1985; 34: 588–93.
56. Wang HY, Baxter Jr CF, Schulz H. Regulation of fatty acid ß-oxidation in rat heart mitochondria. Arch Biochem Biophys 1991; 289: 274–80.
57. Pande SV, Parvin R. Characterization of carnitine acylcarnitine translocase system of heart mitochondria. J Biol Chem 1976; 251: 6683–91.
58. Ferri L, Valente M, Ursini F, Gregolin C, Siliprandi N. Acetyl-carnitine formation and pyruvate oxidation in mitochondria from different rat tissues. Bull Mol Biol Med 1981; 6: 16–23.
59. Di Lisa F, Menabò R, Barbato R, Siliprandi N. Contrasting effects of propionate and propionyl-L-carnitine on energy-linked processes in ischemic hearts. Am J Physiol 1994; 267: 455–61.
60. Bolukoglu H, Nellis SH, Liedtke AJ. Effects of propionate on mechanical and metabolic performance in aerobic rat hearts. Cardiovasc Drugs Ther 1991; 5 (Suppl 1): 37–43.
61. Ferrari R, Cucchini F, Visioli O. The metabolical effects of L-carnitine in angina pectoris. Int J Cardiol 1984; 5: 213–6.
62. Brevetti G, Chiariello M, Ferulano G et al. Increases in walking distance in patients with peripheral vascular disease treated with L-carnitine: a double-blind, cross-over study. Circulation 1988; 77: 767–73.
63. Siliprandi N, Di Lisa F, Pieralisi G et al. Metabolic changes induced by maximal exercise in human subjects following L-carnitine administration. Biochim Biophys Acta 1990; 1034: 17–21.
64. Hultman E, Cederblad G, Harper P. Carnitine administration as a tool of modify energy metabolism during exercise. Eur J Appl Physiol 1991; 62: 450 (Lett to the Ed).
65. Sartorelli L, Ciman M, Siliprandi N. Carnitine transport in rat heart slices. I The action of thiol reagents on the acetylcarnitine/carnitine exchange. Ital J Biochem 1985; 34: 275–81.
66. Siliprandi N, Di Lisa F, Vecchiet L. Effect of exogenous carnitine on muscle metabolism: a reply to Hultman et al. (1991). Eur J Appl Physiol 1992; 64: 278.
67. Roe CR, Millington DS, Maltby DA, Kahler SG, Bohan TP. L-carnitine therapy in isovaleric acidemia. J Clin Invest 1984; 74: 2290–5.
68. Siliprandi N, Siliprandi D, Ciman M. Stimulation of oxidation of mitochondrial fatty acids and acetate by acetylcarnitine. Biochem J. 1965; 96: 777–80.
69. Ciman M, Rossi CR, Siliprandi N. On the mechanism of the antiketogenic action of propionate and succinate in isolated rat liver mitochondria. FEBS Lett 1972; 22: 8–10.
70. Tassani V, Cattapan F, Magnanimi L, Peschechera A. Anaplerotic effect of propionyl carnitine in rat heart mitochondria. Biochem Biophys Res Commun 1994; 199: 949–53.
71. Di Lisa F, Menabò R, Siliprandi N. L-propionyl-carnitine protection of mitochondria in ischemic rat hearts. Mol Cell Biochem 1989; 88: 169–73.
72. Paulson DJ, Traxler J, Schimidt M, Noonan J, Shug AL. Protection of the ischaemic myocardium by L-propionylcarnitine: Effects on the recovery of cardiac output after ischaemia and reperfusion, carnitine transport, and fatty acid oxidation. Cardiovasc Res 1986; 20: 536–41.
73. Ferrari R, Ceconi C, Curello S, Pasini E, Visioli O. Protective effect of propionyl-L-carnitine against ischaemia and reperfusion-damage. Mol Cell Biochem 1989; 88: 161–8.
74. Liedtke AJ, DeMaison L, Nellis SH. Effects of L-propionylcarnitine on mechanical recovery during reflow in intact hearts. Am J Physiol 1988; 255: H169–76.
75. Packer L, Valenza M, Serbinova E, Starke-Reed P, Frost K, Kagan V. Free radical scavenging is involved in the protective effect of L-propionyl-carnitine against ischemia-reperfusion injury of the heart. Arch Biochem Biophys 1991; 288: 533–7.

76. Sassen LM, Bezstarosti K, Van der Giessen WJ, Lamers JMJ, Verdouw PD. L-propionylcarnitine increases postischemic blood flow but does not affect recovery of energy charge. Am J Physiol 1991; 261: H172–80.
77. Mortimore GE, Poso AR. Intracellular protein catabolism and its control during nutrient deprivation and supply. Annu Rev Nutr 1987; 7: 539–64.
78. Miotto G, Venerando R, Siliprandi N. Inhibitory action of isovaleryl-L-carnitine on proteolysis in perfused rat liver. Biochem Biophys Res Commun 1989; 158: 797–802.
79. Miotto G, Venerando R, Khurana KK, Siliprandi N, Mortimore GE. Control of hepatic proteolysis by leucine and isovaleryl-L-carnitine through a common locus. Evidence for a possible mechanism of recognition at the plasma membrane. J Biol Chem 1992; 267: 22066–72.
80. Pontremoli S, Melloni E. Extralysosomal protein degradation. Annu Rev Biochem 1986; 55: 455–81.
81. Pontremoli S, Melloni E, Viotti PL, Michetti M, Di Lisa F, Siliprandi N. Isovalerylcarnitine is a specific activator of the high calcium requiring calpain forms. Biochem Biophys Res Commun 1990; 167: 373–80.
82. Pontremoli S, Melloni E, Michetti M et al. Isovalerylcarnitine is a specific activator of calpain of human neutrophils. Biochem Biophys Res Commun 1987; 148: 1189–95.
83. Salamino F, Di Lisa F, Burlina A et al. Involvement of erythrocyte calpain in glycine and carnitine treated isovaleric acidemia. Pediatr Res 1994; 36: 182–86.
84. Lin A, Krockmalnic G, Penman S. Imaging cytoskeleton: mitochondrial membrane attachments by embedment-free electron microscopy of saponin-extracted cells. Proc Natl Acad Sci USA 1990; 87: 8565–9.


*Corresponding Author*: Professor Fabio Di Lisa, Università degli Studi di Padova, Dipartimento di Chimica Biologica, Via Trieste, 75, I-35121 Padova, Italy

# 4. The effects of carnitine on myocardial carbohydrate metabolism

BRETT O. SCHÖNEKESS and GARY D. LOPASCHUK

"L-propionylcarnitine administration to intact hearts increases myocardial carnitine content and results in a dramatic increase in the contribution of glucose oxidation to ATP production. Stimulation of carbohydrate oxidation may partly explain the beneficial effects of L-carnitine and L-propionylcarnitine in diabetes and hypertrophy."

## Introduction

The mammalian heart primarily meets its requirements for energy through the oxidation of fatty acids [1]. The oxidation of glucose and lactate provides most of the remaining energy needs, with glycolysis providing an additional small amount of ATP production [1, 2, 3]. An important step in the oxidation of fatty acids is the translocation of fatty acyl-CoA into the inner mitochondrial space. This is achieved by a carnitine mediated translocation involving carnitine palmitoyltransferase (CPT) I, carnitine acyltranslocase and CPT II (see Figure 1). By virtue of its role as a carrier, therefore, carnitine is essential for the oxidation of long chain fatty acids.

In addition to this critical metabolic role, carnitine can also transport acetyl groups from within the mitochondrial matrix to the cytosol [5–7]. Carnitine acetyltransferase catalyzes the transfer of acetyl groups from acetyl-CoA to carnitine forming acetylcarnitine (Figure 1). The acetylcarnitine can then be transported into the cytosol, where the acetyl groups are transferred back onto CoA. Recent interest has stemmed from this proposed role of carnitine as a modulator of the intramitochondrial acetyl-CoA/CoA ratio [4–6]. In isolated heart mitochondria, carnitine has been shown to increase free CoA levels and to reduce acetyl-CoA levels, resulting in a 10– to 20–fold decrease in the ratio of acetyl-CoA/CoA [5, 6]. In human skeletal muscle mitochondria this decrease in acetyl-CoA/CoA stimulates pyruvate oxidation, secondary to an increase in pyruvate dehydrogenase complex (PDC) activity [8]. Changes in the ratio of acetyl-CoA/CoA in the presence of L-carnitine are associated with an increased efflux of acetylcarnitine from the mitochondria

*J.W. de Jong and R. Ferrari (eds): The carnitine system*, 39–52.

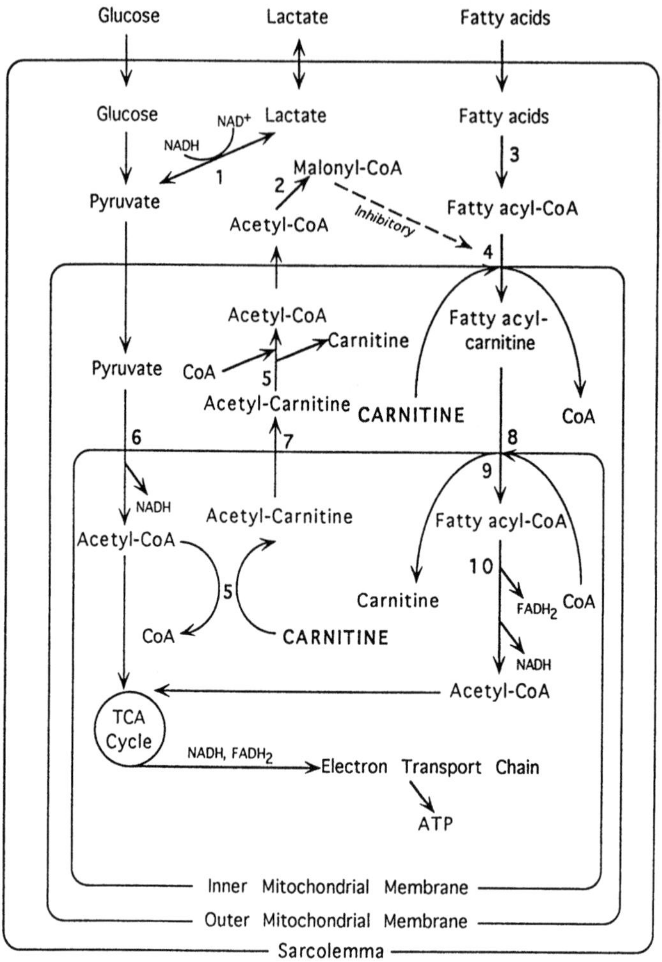


*Figure 1.* Key sites at which L-carnitine potentially modulates fatty acid and carbohydrate oxidation in the heart. Carnitine ensures an adequate supply of intramitochondrial acetyl-CoA from fatty acids at the level of 4, 8, and 9. In situations of an adequate acetyl-CoA supply from $\beta$-oxidation, carnitine can also act to lower the intramitochondrial acetyl-CoA/CoA ratio at the level of 5 and 7. This will increase pyruvate dehydrogenase complex activity, and therefore glucose oxidation. We propose that increasing the activity of 5 and 7 will also increase cytosolic acetyl-CoA levels, resulting in an increase in acetyl-CoA carboxylase activity (2). Increased malonyl-CoA production will then inhibit carnitine palmitoyltransferase 1 activity [4], resulting in a decrease in fatty acid oxidation. 1, lactate dehydrogenase complex; 2, acetyl-CoA carboxylase; 3, acyl-CoA synthetase; 4, carnitine palmitoyltransferase I; 5, carnitine acetyltransferase; 6, pyruvate dehydrogenase complex; 7, carnitine-acetylcarnitine translocase; 8, carnitine-acylcarnitine translocase; 9, carnitine palmitoyltransferase II; 10, $\beta$-oxidation. TCA = Tricarboxylic acid.

[5], which is consistent with the suggestion that carnitine increases the activity of the carnitine acetyltransferase present on mitochondrial membranes.

A role for L-carnitine in regulating the intramitochondrial acetyl-CoA/CoA ratio is supported by direct measurements of carbohydrate oxidation in the intact heart. Recently we have shown that carnitine supplementation of isolated working rat hearts will substantially increase glucose oxidation rates [4]. This increase in glucose oxidation probably occurs secondary to an increase in PDC activity, which results from the lowering of the intramitochondrial acetyl-CoA/CoA ratio (Figure 1). Of interest is the observation that this L-carnitine induced increase in glucose oxidation is accompanied by a concomitant decrease in fatty acid oxidation rates, such that overall ATP production rates remain similar [4]. While this effect of L-carnitine on fatty acid oxidation would appear paradoxical, it isn't if one considers the primary role of L-carnitine is to ensure an adequate supply of acetyl-CoA for the tricarboxylic acid (TCA) cycle. As shown in Figure 1, L-carnitine has a critical role in regulating the supply of acetyl-CoA from both PDC and from $\beta$-oxidation of fatty acids. Since the primary supply of acetyl-CoA is normally derived from fatty acid oxidation, an increase in TCA cycle activity (i.e. such as by increasing myocardial workload) will increase the supply of acetyl-CoA derived from fatty acid oxidation [1]. As will be discussed, adequate myocardial levels of carnitine are required to ensure that fatty acid oxidation is able to meet mitochondrial acetyl-CoA demand. Provided that intramitochondrial acetyl-CoA supply from fatty acid oxidation is not limited, we propose that the primary effect of L-carnitine supplementation is to regulate the supply of TCA cycle acetyl-CoA that is derived from PDC. By shuttling intramitochondrial acetyl-CoA out of the mitochondria and into the cytosol, via the carnitine acetyltransferase and carnitine acetyltranslocase pathway intramitochondrial levels of acetyl-CoA will decrease (Figure 1). The decrease acetyl-CoA/CoA ratio will result in a stimulation of PDC activity [9]. This in turn will result in increased rates of glucose oxidation [4]. Since the need for ATP at a given workload is constant, an increase in acetyl-CoA derived from PDC would be expected to result in a decrease in the requirements of acetyl-CoA derived from $\beta$-oxidation. This would explain the observed decrease in myocardial fatty acid oxidation that accompanies the increase in glucose oxidation following L-carnitine supplementation to isolated perfused hearts [4].

It is clear that the role of L-carnitine is complex in its regulation of fatty acid and carbohydrate metabolism. We believe that in severe tissue carnitine deficiencies the effects of L-carnitine supplementation on overall myocardial metabolism differs from the effects of L-carnitine supplementation when a carnitine deficiency does not exist. Whether the primary effect of L-carnitine is to stimulate fatty acid oxidation or glucose oxidation is primarily dependent on the intramitochondrial acetyl-CoA/CoA ratio.

**Potential link between the regulation of fatty acid oxidation and carbohydrate oxidation**

As discussed, L-carnitine supplementation to intact hearts increases glucose oxidation and decreases fatty acid oxidation, such that overall ATP production is maintained [4]. As shown in Figure 1, the effects of L-carnitine on glucose oxidation can be explained by a decrease in the intramitochondrial acetyl-CoA/CoA ratio. However, the effects of L-carnitine supplementation on fatty acid oxidation are less obvious, since a decrease in the acetyl-CoA/CoA ratio should also act as a stimulus to increase $\beta$-oxidation of fatty acids. However, increasing both glucose and fatty acid oxidation at a given myocardial workload would decrease myocardial efficiency since ATP production (and $O_2$ consumption) would increase in the absence of additional demands for ATP. To explain this apparent contradiction, we hypothesize that L-carnitine can inhibit fatty acid oxidation by increasing cytosolic acetyl-CoA supply to acetyl-CoA carboxylase, which by producing malonyl-CoA will inhibit CPT I activity (Figure 1).

CPT I is a key regulatory point in the oxidation of fatty acids, and is the rate-limiting step of long-chain acyl-CoA translocation into mitochondria [10]. The fact that CPT I in the heart is extremely sensitive to inhibition by malonyl-CoA ($Ki = 50$ nM) [10, 11] and that malonyl-CoA is present in measurable quantities (10–15 nmol/g dry wt) in the heart [12, 13] has led to speculation that malonyl-CoA may be an important effector for the entry of long-chain acyl-CoA's into the mitochondria and therefore a potentially important regulator of myocardial fatty acid oxidation.

Changes in the absolute levels of cytosolic malonyl-CoA may be responsible for the changes in myocardial fatty acid oxidation that occur following carnitine supplementation [4]. We hypothesize that increasing the concentration of L-carnitine in the heart will facilitate the export of intramitochondrial acetyl-CoA into the cytosol (Figure 1). The increased cytosolic levels of acetyl-CoA will then increase the activity of acetyl-CoA carboxylase (ACC). ACC catalyzes the transfer of $CO_2$ from bicarbonate to acetyl-CoA to form malonyl-CoA [14, 15], and is widely distributed in a number of different mammalian tissues, including those where fatty acid oxidation is prominent, e.g. heart, brown adipose tissue, and skeletal muscle [16–19]. While ACC in liver and white adipose tissue primarily acts as the rate limiting step in fatty acid biosynthesis, in heart it appears that ACC primarily acts to regulate fatty acid oxidation [20]. Myocardial ACC has a low affinity for acetyl-CoA [20] and cytosolic acetyl-CoA levels are very low in the heart [21]. This suggests that cytosolic acetyl-CoA levels may be an important determinant of ACC activity in the heart. We hypothesize that stimulation of carnitine acetyltransferase by L-carnitine increases cytosolic acetyl-CoA levels, increasing ACC production of malonyl-CoA (Figure 1). This would then inhibit CPT I activity, thereby decreasing fatty acid oxidation.

A role of L-carnitine and carnitine acetyltransferase as a link between glucose and fatty acid oxidation is a particularly attractive hypothesis to explain how the heart ensures an adequate supply of acetyl-CoA for the TCA cycle. When intramitochondrial acetyl-CoA demand is high both fatty acid and carbohydrate oxidation would increase to meet this demand. However, as supply exceeds demand and the acetyl-CoA/CoA ratio increases, acetyl-CoA would be shuttled out of the mitochondria via carnitine acetyltransferase and carnitine acetyltranslocase. This would increase malonyl-CoA production from ACC, resulting in a decrease in fatty acid oxidation. Carbohydrate oxidation would decrease by a direct inhibition of PDC as the levels of intramitochondrial acetyl-CoA/CoA increases.

This link between glucose and fatty acid oxidation is supported by our recent evidence demonstrating that when the supply of acetyl-CoA in the mitochondria is increased, a resultant decrease in fatty acid oxidation occurs [20]. Furthermore, we have observed a close relationship between myocardial acetyl-CoA levels and malonyl-CoA levels, with a close inverse correlation between malonyl-CoA levels and fatty acid oxidation. Evidence for the involvement of carnitine acetyltransferase in this process comes from recent studies with L-propionylcarnitine (LPC) supplementation of isolated rat hearts (Schönekess et al., unpublished). LPC administration to intact hearts increases myocardial carnitine content and results in a dramatic increase in the contribution of glucose oxidation to ATP production. This is accompanied by a decrease in the contribution of fatty acid oxidation to ATP production, and a marked increase in malonyl-CoA levels in the LPC treated hearts. This is consistent with an increase in carnitine acetyltransferase activity (Figure 1).

L-carnitine and LPC have previously been shown to improve heart function in pathological conditions such as diabetes and myocardial hypertrophy [22–26]. Because of the potential role of L-carnitine as a regulator of both carbohydrate and fatty acid oxidation we examined the effects of L-carnitine and LPC on energy metabolism in isolated fatty acid perfused hearts obtained from normal, diabetic, or aortic banded rats (Table 1). In normal hearts, L-carnitine or LPC pre-treatment results in a dramatic increase in the contribution of carbohydrate oxidation as a source of ATP production. In diabetic rat hearts, where fatty acid oxidation provides almost all of the ATP requirements [28–30], L-carnitine was also able to markedly increase glucose oxidation. Similarly, in hypertrophied hearts, LPC substantially increases carbohydrate oxidation. As a result, stimulation of carbohydrate oxidation may partly explain the beneficial effects of L-carnitine and LPC in diabetes and hypertrophy.

## Effects of myocardial carnitine deficiency on oxidative metabolism

Alterations in the metabolism of fatty acids and carbohydrates can occur when perturbations such as tissue carnitine depletion occur [2, 31]. These

*Table 1.* Acute L-carnitine or L-propionylcarnitine loading of aerobically perfused normal, diabetic or hypertrophied rat hearts: the major effects on energy substrate preference and ATP contribution.

| Perfusion condition | Energy preference and major ATP source | Reference |
|---|---|---|
| Normal rat heart | | |
| − *L-carnitine* | Fatty acid oxidation primary source of ATP<br>Carbohydrates supply 5–15% of ATP | [4,10] |
| + *L-carnitine* | Increased carbohydrate contribution to ATP production mostly through an increased PDC flux (glucose oxidation)<br>Decreased contribution of fatty acid oxidation to ATP production | |
| + *L-propionylcarnitine* | Increased carbohydrate contribution to ATP production via an increased PDC flux (glucose and lactate oxidation)<br>Decreased contribution of fatty acids to ATP production | |
| Diabetic rat heart | | |
| − *L-carnitine* | Almost all ATP from fatty acid oxidation<br>Carbohydrate metabolism almost non-existant | [27] |
| + *L-carnitine* | Dramatic increase in glucose contribution to ATP production (both glycolysis and glucose oxidation increase) | |
| Hypertrophied rat heart | | |
| − *L-propionylcarnitine* | Majority of ATP from fatty acid oxidation<br>Increased contribution of glycolysis and decreased contribution of glucose and lactate oxidation to ATP production | [10] |
| + *L-propionylcarnitine* | No change in fatty acid oxidation<br>Increased contribution of glucose and lactate oxidation to ATP production via an increased flux through PDC | |

PDC = Pyruvate dehydrogenase complex.

perturbations can lead to impairment of myocardial function [23]. Most known situations associated with myocardial carnitine deficiencies are associated with a depression in myocardial function (Table 2). Whether fatty acid oxidation rates are depressed in carnitine deficient hearts probably depends to a large extent on the severity of the carnitine deficiency, as well as the presence of circulating carbon substrates.

In primary and secondary carnitine deficiencies the depressed myocardial function is presumed to occur secondary to a depression of fatty acid oxidation. Experimentally induced carnitine deficiencies, such as following sodium pivalate treatment of rats, also results in a depression of fatty acid oxidation (DJ Paulson, personal communication). Long term treatment with sodium pivalate can result in a 50–60% reduction in myocardial carnitine

*Table 2.* Effects of carnitine deficiencies on myocardial function and energy metabolism.

| Carnitine deficient state | Effect on myocardial function | Effect on energy metabolism | Reference |
|---|---|---|---|
| Primary and secondary carnitine deficiencies | Depressed myocardial function | Presumed depression of fatty acid oxidation | [32] |
| Experimental depletion: | | | |
| *$Na^+$ pivalate* | Depressed myocardial function with extended treatment | Depressed fatty acid oxidation with enhanced glucose oxidation rates | [pers. comm.] |
| *D-carnitine supplementation* | Significant impairment of myocardial function | Probable depression of fatty acid oxidation | [22] |
| Myocardial hypertrophy | Depressed myocardial function | Fatty acid oxidation depressed<br>Glycolysis enhanced<br>Carbohydrate oxidation depressed in presence of high fat | [2, 10, 31] |
| Diabetes | Depressed myocardial function | Primary source of ATP from fatty acid oxidation<br>Depressed carbohydrate metabolism | [22, 33] |
| Reperfusion following ischemia | Depressed myocardial function | Fatty acid oxidation increased<br>Glucose oxidation depressed | [34–37] |

Personal communication, D.J. Paulson.

content. This severe carnitine deficiency results in a depression of cardiac function when the treatment is extended for periods of 24–26 weeks. These results suggest that in severe carnitine deficiencies CPT I activity is inhibited, resulting in a decrease in fatty acid oxidation.

Accompanying the decreased rates of fatty acid oxidation in sodium pivalate treated hearts is an increase in glucose oxidation rates. An increase in glucose oxidation in carnitine deficient hearts would appear to contradict the observations that L-carnitine supplementation to normal hearts also increases glucose oxidation rates [4]. However, these apparent contradictions can readily be explained by the importance of intramitochondrial acetyl-CoA/CoA in regulating glucose oxidation. In severe carnitine deficiencies, where fatty acid oxidation is inhibited, acetyl-CoA supply from $\beta$-oxidation will decrease. This will decrease the ratio of intramitochondrial acetyl-CoA/CoA, relieving the inhibition of PDC. The end result is that the activity of PDC will increase and rates of glucose oxidation will accelerate. In normal hearts where fatty acid oxidation rates are not depressed, the effects of L-carnitine on the intramitochondrial acetyl-CoA/CoA ratio would be expected to parallel what is seen in a severe carnitine deficiency, resulting a similar increase in glucose oxidation (Figure 1). As a result, the effects of a carnitine deficiency on glucose oxidation probably depends on whether the deficiency is severe enough to inhibit fatty acid oxidation.

Decreased myocardial carnitine content can also be seen in hearts obtained from diabetic animals [7, 38]. However, despite this decrease in tissue carnitine, almost all of the ATP requirements of the heart are met by the oxidation of fatty acids [28–30]. This is primarily due to the high circulating levels of fatty acids seen in uncontrolled diabetics. As a result, we believe that the mild carnitine deficiency seen in hearts from diabetic animals, when coupled to elevations in circulating fatty acids, is insufficient to cause a depression of fatty acid oxidation. Despite this observation, many studies have demonstrated that L-carnitine and LPC treatment can improve heart function in hearts from diabetic rats. We believe that the primary benefit of L-carnitine in these hearts is related to an increase in glucose oxidation (Table 1). In isolated working hearts obtained from diabetic rats, L-carnitine treatment dramatically increases glucose oxidation [27].

Decreased myocardial carnitine levels are also seen in hypertrophic hearts [39]. In severely hypertrophic hearts, carnitine depletion results in decreased rates of fatty acid oxidation [31]. In situations of mild hypertrophy, fatty acid oxidation can be depressed, but this is dependent on the perfusion conditions. In isolated working hypertrophied rat hearts, performing low work, fatty acid oxidation is depressed with a concomitant increase in glycolysis [2]. If these hearts are perfused with high levels of fatty acids, no decrease in fatty acid oxidation is observed [Schönekess et al. unpublished]. This suggests that the carnitine deficiency seen in mild hypertrophy may only limit fatty acid oxidation if other sources of energy, such as carbohydrate metabolism, are able to be used as an alternate supply of ATP. In contrast, El Alaoui-Talibi et al. [31] have found that the carnitine deficiency occurring in hypertrophic hearts does result in a decrease in fatty acid oxidation at both low and high work loads. Differences between our results and those of El Alaoui-Talibi et al. [31] probably relate to differences in the severity of hypertrophy between the two experimental models used. Our model of pressure-overload hypertrophy resulted in a 38% increase in heart size [2], whereas the latter model of volume-overload hypertrophy resulted in nearly a 100% increase in heart size [31].

## Effects of carnitine supplementation on myocardial oxidative metabolism and contractile function

The supplementation of the myocardium with carnitine or LPC results in an increased tissue carnitine content, which lessens the severity of ischemic

*Table 3.* Effects of chronic L-carnitine or L-propionylcarnitine treatment on myocardial function.

| Treatment regimen | Results on myocardial function | Reference |
|---|---|---|
| L-carnitine treatment (oral 4 g/d for 1 yr) of humans with recent myocardial infarction | Increased heart rate, systolic and diastolic pressure, lower mortality, decreased anginal attacks, and rhythm disorders | [45] |
| L-carnitine treatment of humans with cardiopathies | Improved physical performance, decreased anginal attack rate and therapeutic use of nitrates | [46] |
| L-carnitine treatment (oral 1.8 g/d for 4–8 wk) of human patients with stable angina pectoris | Improves excercise tolerance | [47] |
| L-propionylcarnitine treatment (oral 60 mg/kg for 8 wks) in aortic-banded rats | Improved cardiac work at medium and maximal workloads (isolated working hearts) | unpubl. |
| L-propionylcarnitine treatment (ia 50 mg/kg for 4 d) of aortic-banded rats | Improved cardiac function (in vivo) Lowered left ventricular end-diastolic pressure and increased relaxation rate (in vitro) | [26, 44] |
| L-propionylcarnitine treatment (ip 100 mg/kg for 8 wk) of diabetic rats | Improved post-ischemic contractile performance | [48] |

Unpublished data, Schönekess et al.

injury and improves the recovery heart function during reperfusion [24, 26, 40–44]. The effects of L-carnitine or LPC administration in a variety of pathologies is shown in Table 3. Accumulating evidence suggests that the mechanism behind the beneficial effect of L-carnitine and LPC is not always via an increased rate of fatty acid oxidation or by decreasing the levels of potentially toxic levels of long-chain acyl-CoA. Carnitine mediated increases in the rates of carbohydrate metabolism (glucose and lactate oxidation) [4, 43] provide an alternate mechanism by which L-carnitine and LPC exert their beneficial effects.

If the heart is in a carnitine deficient state it would be expected that supplementation with carnitine or carnitine derivatives should result in a normalization of carnitine levels and therefore a normalization of fatty acid oxidation. However, we have found that in situations such as mild myocardial hypertrophy, where a decrease in tissue carnitine content is seen, the primary effect of increasing tissue carnitine content is to increase carbohydrate oxidation. Following acute LPC administration to hypertrophic hearts, the major metabolic response was an increased supply of ATP from the oxidation of glucose and lactate, suggesting that PDC activity was enhanced (Schönekess et al., unpublished). In severely hypertrophic hearts, where a decrease in fatty acid oxidation occurs even in the presence of high levels of fatty acids, it is possible that increasing carnitine levels in the heart will result in an increase in fatty acid oxidation. El Alaoui-Talibi and Moravec have shown that LPC treatment will increase fatty acid oxidation rates (Z El Alaoui-Talibi, personal communication). Regardless of whether LPC acts to increase carbohydrate or fatty acid oxidation, in both studies LPC treatment improves mechanical function in the hypertrophic hearts.

Increasing myocardial tissue content can also alter heart metabolism even if a carnitine deficiency is not present. This can be seen in isolated working rat hearts which are perfused under conditions where fatty acid oxidation is the major source of ATP production. Increasing tissue carnitine content by L-carnitine pre-treatment results in a marked increase in glucose oxidation, with a parallel decrease in fatty acid oxidation [4]. Addition of LPC to hearts perfused under similar conditions also results in an increase in carbohydrate oxidation (both glucose and lactate) (Schönekess et al., unpublished). However the effects of LPC on fatty acid oxidation rates are not as dramatic as they are with L-carnitine supplementation. This may be related to the observation that an increase in mechanical function occurs following administration of LPC. However, if rates of ATP production are normalized for differences in work, LPC does cause a shift in the metabolic profile away from fatty acid oxidation and towards carbohydrate oxidation. However, even when differences in work are considered, the major effect of LPC is to increase the amount of ATP derived from the oxidation of glucose and lactate.

## Beneficial effect of L-carnitine on glucose oxidation in the reperfused ischemic heart

High levels of fatty acids have a detrimental effect on reperfusion recovery of hearts subjected to a severe episode of ischemia [30, 34]. While the exact mechanism by which fatty acid oxidation contributes to ischemic injury is not clear, our studies suggest that this may be related to their ability to inhibit glucose oxidation [34, 49–51]. High levels of fatty acid increase the intramitochondrial acetyl-CoA/CoA and NADH/$NAD^+$ ratios [9], which in turn inhibits PDC through the activation of a pyruvate dehydrogenase kinase [9, 52]. This inhibition of glucose oxidation during reperfusion can lead to a substantial imbalance between glycolysis and glucose oxidation during the actual reperfusion period [51]. This increases the production of $H^+$ ions formed by the hydrolysis of glycolytically derived ATP. It is this imbalance and exaggerated production of $H^+$ ions that we believe is mediating the detrimental effects of high levels of fatty acids on post-ischemic functional recovery. The production of $H^+$ ions during ischemia and early in reperfusion could lead to increased activity of the $Na^+/H^+$ and the $Na^+/Ca^{2+}$-exchangers and result in a potentially damaging $Ca^{2+}$ overload [53].

L-carnitine and LPC can improve functional recovery of hearts reperfused following a severe episode of ischemia [4, 43]. However, although myocardial carnitine content decreases during ischemia [54, 55], the actions of carnitine and LPC cannot be explained secondary to a stimulation of fatty acid oxidation. This is because fatty acid oxidation rates are not depressed during reperfusion of ischemic hearts. In fact, due to the high circulating levels of fatty acids normally seen during reperfusion [56], fatty acid oxidation provides over 90% of ATP production during reperfusion [34, 36, 37, 57]. In the presence of high levels of fatty acids, glucose oxidation provides only 5 to 10% of the ATP requirements. As previously mentioned, if glucose oxidation is stimulated during reperfusion it is possible to overcome the detrimental effect of high levels of fatty acids. Compounds which stimulate glucose oxidation directly by inhibiting the action of PDC kinase, such as dichloroacetate [50, 51], or indirectly such as CPT I inhibitors [34, 49] have the potential to improve post-ischemic functional recovery. Recently we have demonstrated that the beneficial effects of L-carnitine on functional recovery post-ischemia are also associated with a marked increase in glucose oxidation [43]. As a result, we hypothesize that the beneficial effects of L-carnitine and LPC in reperfused ischemic hearts occur secondary to a stimulation of glucose oxidation.

## Conclusions

Carnitine is an essential co-factor for the transportation of fatty acyl groups into the mitochondrial matrix where they undergo $\beta$-oxidation and result in the production of ATP. It is becoming evident that carnitine also has an important role in the regulation of glucose oxidation. Secondary to facilitating the intramitochondrial transfer of acetyl groups from acetyl-CoA to acetylcarnitine, L-carnitine can relieve inhibition of PDC. This role of L-carnitine may explain some of the beneficial effects associated with L-carnitine and LPC treatment in various pathological conditions. Furthermore, we believe that the well documented beneficial effects of L-carnitine and LPC in ischemic hearts are best correlated with their ability to overcome fatty acid inhibition of glucose oxidation during reperfusion.

## References


1. Neely JR, Morgan HE. Relationship between carbohydrate and lipid metabolism and the energy balance of heart muscle. Annu Rev Physiol 1974; 36: 413–59.
2. Allard MF, Schönekess BO, Henning SL, English DR, Lopaschuk GD. Contribution of oxidative metabolism and glycolysis to ATP production in hypertrophied hearts. Am J Physiol. 1994; 267: H742–H750.
3. Saddik M, Lopaschuk GD. Myocardial triglyceride turnover and contribution to energy substrate utilization in isolated working rat hearts. J Biol Chem 1991; 266: 8162–70.

4. Broderick TL, Quinney HA, Lopaschuk GD. Carnitine stimulation of glucose oxidation in the fatty acid perfused isolated working rat heart. J Biol Chem 1992; 267: 3758–63.
5. Lysiak W, Lilly K, DiLisa F, Toth PP, Bieber LL. Quantitation of the effect of L-carnitine on the levels of acid-soluble short-chain acyl-CoA and CoASH in rat heart and liver mitochondria. J Biol Chem 1988; 263: 1511–6.
6. Lysiak W, Toth PP, Suelter CH, Bieber LL. Quantitation of the efflux of acylcarnitines from rat heart, brain and liver mitochondria. J Biol Chem 1986; 261: 13698–703.
7. Pearson DJ, Tubbs PK. Carnitine and derivatives in rat tissues. Biochem J 1967; 105: 953–63.
8. Uziel G, Baravagalia B, DiDonato S. Carnitine stimulation of pyruvate dehydrogenase complex (PDHC) in isolated human skeletal muscle mitochondria. Muscle Nerve 1988; 11: 720–4.
9. Patel MS, Roche TE. Molecular biology and biochemistry of pyruvate dehydrogenase complexes. FASEB J 1990; 4: 3224–33.
10. McGarry JD, Woeltje KF, Kuwajima M, Foster DW. Regulation of ketogenesis and the renaissance of carnitine palmitoyltransferase. Diabetes Metab Rev 1989; 5: 271–84.
11. Cook GA. Differences in the sensitivity of carnitine palmitoyltransferase to inhibition by malonyl-CoA are due to differences in Ki values. J Biol Chem 1984; 259: 12030–3.
12. McGarry JD, Mills SE, Long CS, Foster DW. Observations on the affinity for carnitine, and malonyl-CoA sensitivity, of carnitine palmitoyltransferase I in animal and human tissues. Demonstration of the presence of malonyl-CoA in non-hepatic tissues of the rat. Biochem J 1983; 214: 21–8.
13. Singh B, Stakkestad JA, Bremer J, Borrebaek B. Determination of malonyl-coenzyme A in rat heart, kidney, and liver: A comparison between acetyl-coenzyme A and butyrly-coenzyme A as fatty acid synthase primers in the assay procedure. Anal Biochem 1984; 138: 107–11.
14. Mabrouk GM, Helmy IM, Thampy KG, Wakil SJ. Acute hormonal control of acetyl-CoA carboxylase: The roles of insulin, glucagon, and epinephrine. J Biol Chem 1990; 265: 6330–8.
15. Wakil SJ, Titchener EB, Gibson DM. Evidence for the participation of biotin in the enzymatic synthesis of fatty acids. Biochim Biophys Acta 1958; 29: 225–6.
16. Bianchi A, Evans JL, Iverson AJ, Nordlund AC, Watts TD, Witters LA. Identification of an isozymic form of acetyl-CoA carboxylase. J Biol Chem 1990; 265: 1502–9.
17. Thampy KG. Formation of malonyl-CoA in rat heart. Identification and purification of an isozyme of A carboxylase from rat heart. J Biol Chem 1989; 264: 17631–4.
18. Trumble GE, Smith MA, Winder WW. Evidence of a biotin dependent acetyl-coenzyme A carboxylase in rat muscle. Life Sci 1991; 49: 39–43.
19. Iverson AJ, Bianchi A, Nordlund AC, Witters LA. Immunological analysis of acetyl-CoA carboxylase mass, tissue distribution and subunit composition. Biochem J 1990; 269: 365–71.
20. Saddik M, Gamble J, Witters LA, Lopaschuk GD. Acetyl-CoA carboxylase regulation of fatty acid oxidation in the heart. J Biol Chem 1993; 268: 25836–45.
21. Idell-Wenger JA, Grotyohann LW, Neely JR. Coenzyme A and carnitine distribution in normal and ischemic hearts. J Biol Chem 1978; 253: 4310–8.
22. Paulson DJ, Schmidt MJ, Traxler JS, Ramacci MT, Shug AL. Improvement of myocardial function in diabetic rats after treatment with L-carnitine. Metabolism 1984; 33: 358–63.
23. Siliprandi N, Di Lisa F, Pivetta A, Miotto G, Siliprandi D. Transport and function of L-carnitine and L-propionylcarnitine: relevance to some cardiomyopathies and cardiac ischemia. Z Kardiol 1987; 76(Suppl 5): 34–40.
24. Paulson DJ, Traxler J, Schmidt M, Noonan J, Shug AL. Protection of the ischaemic myocardium by L-propionylcarnitine: effects on the recovery of cardiac output after ischaemia and reperfusion, carnitine transport, and fatty acid oxidation. Cardiovasc Res 1986; 20: 536–41.

25. Liedtke AJ, DeMaison L, Nellis SH. Effects of L-propionylcarnitine on mechanical recovery during reflow in intact hearts. Am J Physiol 1988; 255: H169–76.
26. Motterlini R, Samaja M, Tarantola M, Micheletti R, Bianchi G. Functional and metabolic effects of propionyl-L-carnitine in the isolated perfused hypertrophied rat heart. Mol Cell Biochem 1992; 116: 139–45.
27. Broderick TL, Quinney HA, Lopaschuk GD. L-carnitine increases glucose metabolism and mechanical function following ischemia in diabetic rat heart. Cardiovas Res (in press).
28. Garland PB, Randle PJ. Regulation of glucose uptake by muscle. 10. Effects of alloxan-diabetes, starvation, hypophysectomy and adrenalectomy and of fatty acids, ketones and pyruvate on the glycerol output and concentrations of free fatty acids, long chain fatty acyl coenzyme A, glycerol phosphate and citrate cycle intermediates in rat hearts and diaphragm muscles. Biochem J 1964; 93: 678–87.
29. Randle PJ. Fuel selection in animals. Biochem Soc Trans 1986; 14: 799–806.
30. Wall SR, Lopaschuk GD. Glucose oxidation rates in fatty acid-perfused isolated working hearts from diabetic rats. Biochim Biophys Acta 1989; 1006: 97–103.
31. El Alaoui-Talibi Z, Landormy S, Loireau A, Moravec J. Fatty acid oxidation and mechanical performance of volume-overloaded rat hearts. Am J Physiol 1992; 262: H1068–74.
32. DiDonato S, Garavaglia B, Rimoldi M, Carrara F. Introduction to clinical and biomedical aspects of carnitine deficiencies. In: Ferrari R, DiMauro S, Sherwood G, editors. L-carnitine and its role in medicine: From function to therapy. San Diego: Academic Press, 1992: 81–98.
33. Paulson DJ, Sanjak M, Shug AL. Carnitine deficiency and the diabetic heart. In: Carter AL, editor. Current concepts in carnitine research. Boca Raton: CRC Press, 1992: 215–30.
34. Lopaschuk GD, Spafford MA, Davies NJ, Wall SR. Glucose and palmitate oxidation in isolated working rat hearts reperfused after a period of transient global ischemia. Circ Res 1990; 66: 546–53.
35. Renstrom B, Nellis SH, Liedtke AJ. Metabolic oxidation of glucose during early reperfusion. Circ Res 1989; 65: 1094–101.
36. Görge G, Chatelain P, Schaper J, Lerch R. Effect of increasing degrees of ischemic injury on myocardial oxidative metabolism early after reperfusion in isolated rat hearts. Circ Res 1991; 68: 1681–92.
37. Liedtke AJ, DeMaison L, Eggleston AM, Cohen LM, Nellis SH. Changes in substrate metabolism and effects of excess fatty acids in reperfused myocardium. Circ Res 1988; 62: 535–42.
38. Vary TC, Neely JR. A mechanism for reduced myocardial carnitine content in diabetic animals. Am J Physiol 1982; 243: H154–8.
39. Reibel DK, O'Rourke B, Foster KA. Mechanisms for altered carnitine content in hypertrophied rat hearts. Am J Physiol 1987; 252: H561–5.
40. Folts JD, Shug AL, Koke JR, Bittar N. Protection of the ischemic dog myocardium with carnitine. Am J Cardiol 1978; 41: 1209–14.
41. Hülsmann WC, Dubelaar ML, Lamers JMJ, Maccari F. Protection by acyl-carnitines and phenylmethylsulfonyl fluoride in rat heart subjected to ischemia and reperfusion. Biochim Biophys Acta 1985; 847: 62–6.
42. Liedtke AJ, Nellis SH. Effects of carnitine in ischemic and fatty acid supplemented swine hearts. J Clin Invest 1979; 64: 440–7.
43. Broderick TL, Quinney HA, Barker CC, Lopaschuk GD. Beneficial effect of carnitine on mechanical recovery of rat hearts reperfused after a transient period of global ischemia is accompanied by a stimulation of glucose oxidation. Circulation 1993; 87: 972–81.
44. Yang XP, Samaja M, English E et al. Hemodynamic and metabolic activities of propionyl-L-carnitine in rats with pressure-overload cardiac hypertrophy. J Cardiovasc Pharm 1992; 20: 88–98.
45. Davini P, Bigalli A. Lamanna F, Boem A. Controlled-study on L-carnitine therapeutic efficacy in post-infarction. Drugs Exp Clin Res 1992; 18: 355–65.
46. Fernandez C, Proto C. La L-carnitina nel trattamento dell' ischemia miocardica cronica.

Analisi dei risultati di tre studi multicentrici e rassegna bibliografica. Clin Ter 1992; 140: 353–77.
47. Sotobat I, Noda S, Hayashi H et al. Clinical evaluation of the effect of levocarnitine chloride on exercise tolerance in stable angina pectoris by the serial multistage treadmill exercise testing: a muliticenter, double-blind study. Jpn Clin Pharmacol Ther 1989; 20: 607–18.
48. Paulson DJ, Shug AL, Zhao J. Protection of the ischemic diabetic heart by L-propionylcarnitine therapy. Mol Cell Biochem 1992; 116: 131–7.
49. Lopaschuk GD, Wall SR, Olley PM, Davies NJ. Etomoxir, a carnitine palmitoyltransferase I inhibitor, protects hearts form fatty acid-induced injury independent of changes in long chain acylcarnitine. Circ Res 1988; 63: 1036–43.
50. McVeigh JJ, Lopaschuk GD. Dichloroacetate stimulation of glucose oxidation improves recovery of ischemic rat hearts. Am J Physiol 1990; 259: H1079–85.
51. Lopaschuk GD, Wambolt RB, Barr RL. An imbalance between glycolysis and glucose oxidation is a possible explanation for the detrimental effects of high levels of fatty acids during aerobic perfusion of ischemic hearts. J Pharmacol Expt Ther 1993; 264: 135–44.
52. Kerbey AL, Vary TC, Randle PJ. Molecular mechanisms regulating myocardial glucose oxidation. Basic Res Cardiol 1985; 80(Suppl 2): 93–6.
53. Tani M. Mechanisms of $Ca^{2+}$ overload in reperfused ischemic myocardium. Annu Rev Physiol 1990; 52: 543–59.
54. Shug AL, Thomsen JH, Folts JD et al. Changes in tissue levels of carnitine and other metabolites during myocardial ischemia and anoxia. Arch Biochem Biophys 1978; 187: 25–33.
55. Suzuki Y, Kamikawa T, Kobayashi A, Masumura Y, Yamazaki N. Effects of L-carnitine on tissue levels of acyl carnitine, acyl coenzyme A and high energy phosphate in ischemic dog hearts. Jpn Circ J 1981; 45: 687–94.
56. Oliver MF, Kurien VA, Greenwood TW. Relation between serum-free- fatty acids and arrhythmias and death after acute myocardial infarction. Lancet 1968; 1: 710–4.
57. Saddik M, Lopaschuk GD. Myocardial triglyceride turnover during reperfusion of isolated rat hearts subjected to a transient period of global ischemia. J Biol Chem 1992; 267: 3825–31.

*Corresponding Author*: Dr Gary D. Lopaschuk, Cardiovascular Disease Research Group, 423 Heritage Medical Research Centre, University of Alberta, Edmonton, Alberta, Canada T6G 2S2

# 5. Accumulation of fatty acids and their carnitine derivatives during myocardial ischemia

GER J. VAN DER VUSSE

"The intracellular enzymatic machinery avidly converts fatty acids, keeping the cellular level as low as possible. The latter condition promotes diffusion of fatty acids from the extracellular compartment to the cytoplasm of the cardiomyocyte by creating a concentration gradient across the sarcolemma and protects cellular structures against deleterious high levels of fatty acyl moieties."

## Introduction

Fatty acids are important substrates for the heart. Under normal conditions fatty acids are continuously extracted from the extracellular space, transported through the cytoplasm to the mitochondria by fatty acid-binding proteins (FABP), and converted to fatty acyl CoA. Part of acyl CoA is used for the formation of complex lipids (triacylglycerols and phospholipids), the majority of the acyl residues is channeled across the mitochondrial inner membrane into the mitochondrial matrix with the use of a carnitine-mediated transport system [1].

Inside the mitochondria the fatty acyl residues are oxidized to $CO_2$ and $H_2O$ by the concerted action of enzymes in the $\beta$-oxidation pathway, the citric acid cycle and the respiratory chain. In this way, energy present in the fatty acyl moieties is used for the regeneration of ATP from ADP. Fatty acid oxidation is fully dependent on molecular oxygen. When oxygen supply to the heart is hampered by, for instance, obstruction of blood flow in the coronary arteries (that is, ischemia) fatty acid oxidation is inhibited. Fatty acid intermediates of the $\beta$-oxidative pathway readily accumulate in the ischemic tissue. Moreover, the tissue content of acyl carnitine and acyl CoA is found to be enhanced. Finally, fatty acids themselves accumulate in the oxygen-deprived myocardium. Increased levels of fatty acids and their CoA and carnitine derivatives might contribute to ischemia (and reperfusion) induced damage of the flow-deprived heart.

In the present overview attention will be paid to the amount of these substances present in the normoxic and ischemic heart and to the mechan-

*J.W. de Jong and R. Ferrari (eds): The carnitine system*, 53–68.

ism(s) underlying accumulation of fatty acids and their CoA and carnitine derivatives, the time course of this process and the intracellular localization during the ischemic episode.

## Fatty acid uptake and utilization in the normoxic heart

### *Supply of fatty acids to the heart*

Fatty acids are supplied to the heart either bound to plasma albumin or as triacylglycerols [2]. Triacylglycerols, forming the core of chylomicrons (released by the intestines) and very low density lipoproteins (produced by the liver), are hydrolyzed by lipoprotein lipase giving yield to fatty acids (Figure 1). Lipopotein lipase is synthetized in the cardiomyocytes and subsequently transported to the endothelial cells. Attached to the luminal side of the endothelium, the enzyme exerts its hydrolytic action on lipoprotein triacylglycerols [2]. Fatty acids complexed to the fatty acid-carrier albumin originate from adipose tissue and from hydrolyzed lipoproteins. The transport of fatty acids across the endothelial wall is most likely a passive process driven by the concentration gradient between the vascular and interstitial space. It has been suggested that interaction sites at the luminal endothelial membrane promote the release of fatty acids from albumin [1].

Fatty acids cross the sarcolemma of the cardiomyocytes either by passive diffusion or by protein-mediated transport. Inside the cardiomyocytes fatty acid-binding proteins enhance the solubility of the fatty acid molecules in the aqueous cytoplasmic environment and, hence, facilitate the bulk transport of these substrates from the sarcolemma to the mitochondrial outer membrane (Figure 1). Acyl CoA synthetase, the bulk of which is attached to the cytoplasmic side of the mitochondrial outer membrane, promotes the conversion of fatty acids to fatty acyl CoA. The latter substance is at a cross-

⟶

*Figure 1.* Schematic representation of fatty acid uptake, transport and metabolic conversion in cardiac tissue. ALB refers to albumin, VLDL to very low density lipoproteins, chylo to chylomicrons, LPL to lipoprotein lipase, TG to triacylglycerol in the core of chylomicrons and VLDL, RC to respiratory chain, FA to fatty acid or fatty acyl, FABP to fatty acid-binding protein, CoA to Coenzyme A, fp to flavoprotein, FAD to flavine adenine dinucleotide, GDP to guanosine diphosphate, GTP to guanosine triphosphate, O to molecular oxygen, numbers in brackets to number of ATP moles produced. Other numbers refer to enzymes and transport proteins; 1, fatty acyl CoA synthetase; 2, carnitine acyl transferase (CAT-1); 3, carnitine-acylcarnitine translocase; 4, carnitine acyl transferase (CAT-II); 5, fatty acyl CoA dehydrogenase; 6, enoyl CoA hydratase; 7, 3-hydroxyacyl CoA dehydrogenase; 8, 3-ketothiolase; 9, acyltransferase I and II; 10, phosphatidate phosphatase and acyl transferase III; 11, phospholipid biosynthetic pathway; 12, triacylglycerol, diacylglycerol and monoacylglycerol lipases; 13, phospholipases. Solid and broken arrow lines refer to metabolic conversions and transport routes, respectively.

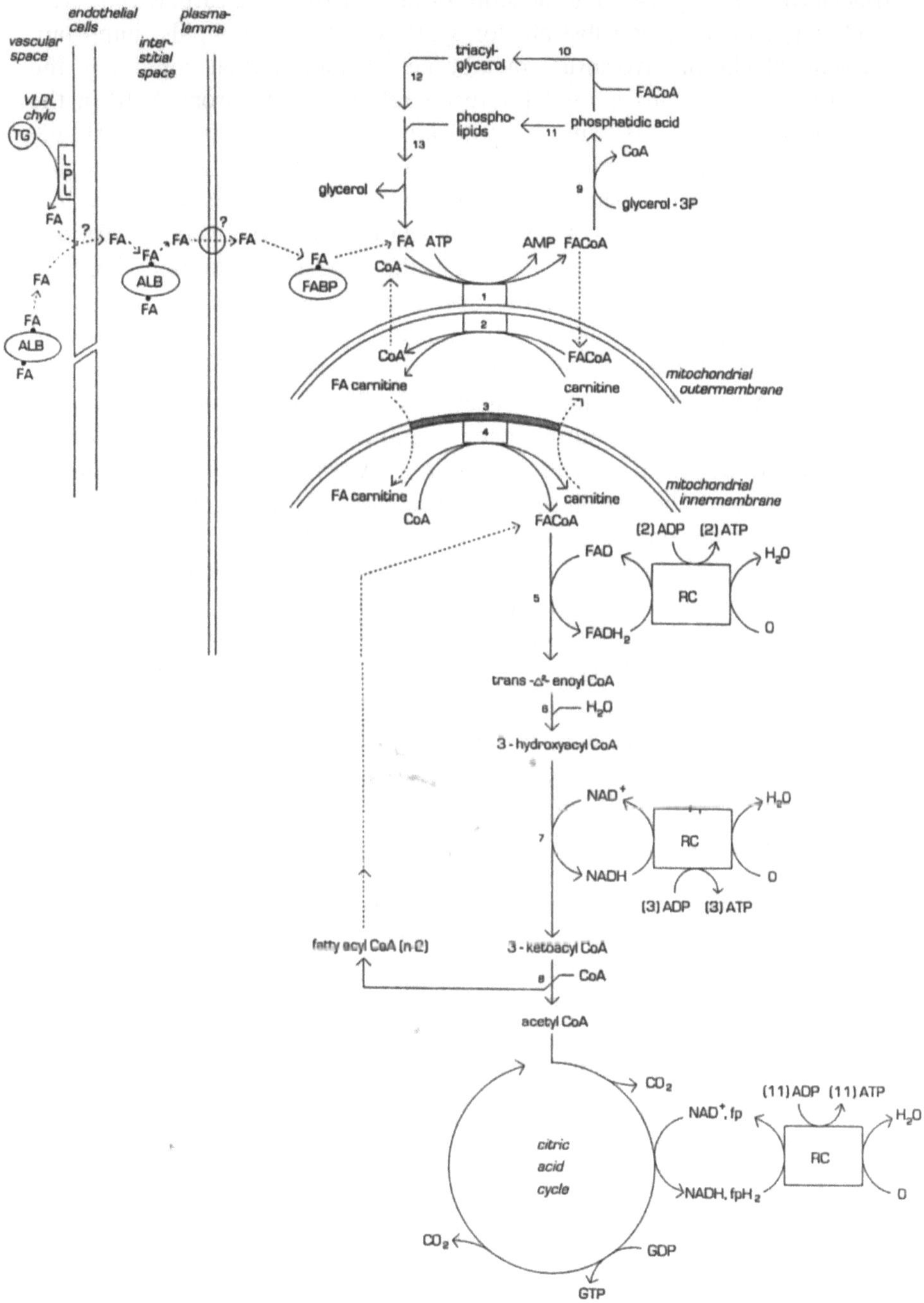
endothelial cells
plasmalemma
vascular space
interstitial space
VLDL chylo
TG
LPL
FA
ALB
FABP
triacylglycerol
phospholipids
phosphatidic acid
FACoA
CoA
glycerol
glycerol-3P
ATP
AMP
FA carnitine
carnitine
mitochondrial outermembrane
mitochondrial innermembrane
(2) ADP (2) ATP
FAD
$FADH_2$
RC
$H_2O$
O
trans-$\Delta^2$-enoyl CoA
3-hydroxyacyl CoA
$NAD^+$
NADH
(3) ADP (3) ATP
fatty acyl CoA (n-2)
3-ketoacyl CoA
acetyl CoA
$CO_2$
(11) ADP (11) ATP
$NAD^+$, fp
NADH, $fpH_2$
citric acid cycle
GDP
GTP

*Figure 1.*

road of metabolic pathways. The fatty acyl residue can be incorporated in triacylglycerols, representing the store of fatty acids in the cardiac cell. Acyl CoA can also serve as substrate for synthesis of phospholipids, important building blocks of myocardial membranes. Under normal conditions the majority of fatty acids is used for regeneration of ATP from ADP in the mitochondrial matrix. Because the mitochondrial inner membrane is virtually impermeable for long-chain fatty acyl CoA, nature has designed an elegant transport mechanism to shuttle the fatty acyl moieties into the mitochondrial matrix. To this end, the CoA residue of acyl CoA is exchanged for carnitine by carnitine acyl transferase I, located at the innerside of the mitochondrial outer membrane (Figure 1). Fatty acyl carnitine is transported across the mitochondrial inner membrane in exchange for a molecule free carnitine in a 1:1 ratio. Inside the mitochondrial matrix, fatty acyl carnitine is converted to acyl CoA by carnitine acyl transferase II. This enzyme is located at the innerside of the mitochondrial inner membrane (Figure 1). Acyl CoA is subsequently degraded to acetyl CoA in a stepwise fashion ($\beta$-oxidation). Acetyl CoA is catabolized in the citric acid cycle. Degradation of fatty acids is fully oxygen-dependent. It is uncertain which step in the overall oxidation of fatty acid is rate limiting. Under normal conditions all reaction steps seem to be very well fine-tuned by cytoplasmic and intramitochondrial levels of cofactors and intermediates, such as CoA, acyl CoA, carnitine and acyl carnitine, NADH and $NAD^+$ [3]. As a consequence of the regulatory mechanisms the enzymatic steps are in equilibrium and accumulation of substantial amounts of intermediates in the fatty acyl degradative pathway is prevented. It has been suggested that transport of fatty acyl residues across the mitochondrial inner membrane is the slowest step in cardiac fatty acid utilization when the supply of fatty acids from exogenous sources is increased.

*Rate of fatty acid oxidation in the normoxic heart*

In the normal functioning heart in situ on the order of 60 to 160 nmoles of fatty acids are consumed per gram tissue per minute [1]. The actual amount of fatty acids consumed by the heart depends on a variety of factors, including workload of the heart, blood levels of fatty acids, extracellular supply of competing substrates and the hormonal status of the animal or human person. In this regard it is noteworthy that the substrate lactate is an efficient inhibitor of cardiac fatty acid oxidation [4].

*The content of fatty acids in the normoxic heart*

The content of fatty acids, that is, the fatty acids present in the cell in the unesterified form, either free or bound to cellular proteins, has been subject of extensive studies during the past decades. Although some investigators have claimed that the level of fatty acids is high (on the order of 1000 to 25,000 nmol $\cdot$ $g^{-1}$ wet weight; summarized in [5, 6]), carefully conducted chemical analyses have revealed that the normal tissue content of fatty acids will not exceed 60 nmol · $g^{-1}$ wet weight [6–9]. This observation stresses the notion that the intracellular enzymatic machinery avidly converts fatty acids, keeping the cellular level as low as possible. The latter condition promotes diffusion of fatty acids from the extracellular compartment to the cytoplasm of the cardiomyocyte by creating a concentration gradient across the sarcolemma and protects cellular structures against deleterious effects of high levels of fatty acyl moieties [1].

*The content of CoA and acyl CoA in normoxic myocardial tissue*

During the past two decades a host of investigators has reported on the tissue content of CoA and fatty acyl CoA derivatives. In Table 1 an arbitrary selection of data available in literature is presented. In general, the total amount of CoA (free and esterified to short, medium and long-chain fatty acyl residues) is on the order of 370 to 770 nmol per gram dry weight of tissue. Species differences appear to exist (Table 1). In the normoxic heart the majority of CoA is present in its free form. In dog and swine myocardial tissue, 25 to 30% of CoA is present as long-chain fatty acyl CoA (Table 1). In rat heart a smaller proportion appears to be esterified to fatty acyl residues, that is, on the order of 5 to 20%. The actual content of fatty acyl CoA depends on the concentration of fatty acids in the perfusion medium when isolated, buffer perfused rat hearts are considered. Addition of palmitate (up to 1.2 mM) to the perfusate increases the tissue content of fatty acyl CoA by approximately 60 to 300% [10–14].

*The content of carnitine and fatty acyl carnitine in the normoxic heart*

The content of carnitine exceeds that of CoA severalfold. Table 2, representing an arbitrary selection data published on cardiac carnitine levels, shows that total carnitine is on the order of 3000 to 10,000 nmol · $g^{-1}$ dry weight in mammalian hearts. Species differences obviously exist. The data available suggest that the content in human heart is relatively high (varying from 6575 to 9600 nmol per gram dry weight of tissue), while mouse and swine hearts contain on the order of 4000 to 4500 nmol · $g^{-1}$ dry weight (Table 2). Intermediate values are reported for cat and dog heart. The bulk of carnitine is present in its free, unesterified form. The actual tissue content of fatty acyl carnitine depends on a multitude of factors, while considerable inter-laboratory differences emerge from Table 2. In general, fatty acyl carnitine levels are low in isolated, glucose perfused rat hearts. Addition of palmitate to the perfusion buffer resulted in a significant increase of tissue acyl carnitine content. In one study this increase amounted to about 5000%, when palmitate in the perfusion buffer was increased from 0 to 2.0 mM [11]. Neely and Feuvray [13] observed in isolated rat hearts an increase of the acyl carnitine content from 303 to 872 nmol · $g^{-1}$ dry weight when 1.2 mM

*Table 1.* The content of free CoA and acyl CoA in the normoxic heart.

| Species | Preparation | CoA | | | References |
|---|---|---|---|---|---|
| | | Free | Acyl | Total | |
| Rat | Unperfused[1] | 511[2] | 59 | 570 | Idell-Wenger et al. 1978 [10] |
| | Perfused, palmitate (1.2 mM) | 395[2] | 179 | 574 | |
| Rat | Perfused, glucose | 351 | 28 | 698 | Pieper et al. 1984 [11] |
| | Perfused, palmitate (2.0 mM) | 196 | 105 | 484 | |
| Rat | Unperfused[1] | 651 | 94 | 772 | Lopaschuk and Neely 1987 [12] |
| | Perfused, palmitate (1.2 mM) | 165 | 254 | 706 | |
| Rat | Perfused, glucose | | 95 | | Neely and Feuvray 1981 [13] |
| | Perfused, palmitate (0.4 mM) | | 248 | | |
| | Perfused, palmitate (1.2 mM) | | 333 | | |
| Rabbit | Unperfused[1] | | 42 | | Moore et al. 1984 [14] |
| | Perfused, glucose | | 71 | | |
| | Perfused, palmitate (0.5 mM) | | 157 | | |
| Rabbit | Unperfused[1] | 275 | 40 | 760 | Ferrari et al. 1992 [15] |
| Cat | LV, in situ | 360[2] | 80 | 440 | Reibel et al. 1983 [16] |
| Dog | LV, in situ | 130 | 85 | 425 | Shug et al. 1978 [17] |
| Swine | LV, e.c.p. | 253[2] | 118 | 372 | Liedtke and Nellis 1979 [18] |
| Swine | LV, e.c.p., FA 0.5 mM | | 143 | | Liedtke et al. 1988 [19] |
| | FA 1.5 mM | | 173 | | |

[1] Unperfused means immediately freeze-clamped after extirpation from the body. [2] Short-chain fatty acyl CoA esters are included. LV refers to left ventricular tissue; e.c.p. to extracorporal perfusion with blood; FA to fatty acids in blood. Data are expressed as nmol · $g^{-1}$ dry weight of tissue.

The conversion factor for the calculation of dry weight from wet weight was 5.0 for unperfused tissue. When the heart was perfused with a crystalloid buffer in a perfusion apparatus a factor of 6.0 was used.

*Table 2.* The content of free carnitine and acyl carnitine in the normoxic heart.

| Species | Preparation | Carnitine | | | References |
|---|---|---|---|---|---|
| | | Free | Acyl | Total | |
| Mouse | Unperfused[1] | 670 | 720 | 3960 | Stearns 1983 [20] |
| Rat | Unperfused[1] | 1940 | 660 | 5385 | Paulson et al. 1984 [21] |
| | Perfused, palmitate (1.2 mM) | 3259 | 980 | 4241 | |
| Rat | Unperfused[1] | 4835 | 1275 | 7575 | Shug et al. 1978 [17] |
| | Perfused, glucose | 4510 | 550 | 5115 | |
| Rat | Unperfused[1] | 5645[2] | 48 | 5693 | Idell-Wenger et al. 1978 [10] |
| | Perfused, palmitate (1.2 mM) | 5800[2] | 830 | 6600 | |
| Rat | Isolated cardiomyocytes | 2700[2] | 40 | 2800 | McHowat et al. 1993 [22] |
| Rat | Perfused, glucose | 3242 | 137 | 4804 | Broderick et al. 1993 [23] |
| Rat | Perfused, palmitate (1.2 mM) | 2549 | 463 | 4276 | Broderick et al. 1992 [24] |
| Rat | Perfused, glucose | 5771 | 32 | 6300 | Pieper et al. 1984 [11] |
| | Perfused, palmitate (1.2 mM) | 3071 | 131 | 4778 | |
| | Perfused, palmitate (2.0 mM) | 1392 | 1679 | 6959 | |
| Rabbit | Unperfused[1] | | 2317 | | Moore et al. 1982 [25] |
| | Perfused, glucose | | 605 | | |
| | Perfused, palmitate (0.5 mM) | | 1437 | | |
| Rabbit | Unperfused[1] | 2350 | 25 | 4575 | Ferrari et al. 1992 [15] |
| Cat | LV, in situ | 5175 | 75 | 6150 | Corr et al. 1989 [26] |
| Dog | LV, in situ | 5215 | 1070 | 8465 | Suzuki et al. 1981 [27] |
| Dog | LV, in situ | 5115 | 565 | 6575 | Shug 1979 [28] |
| Dog | LV, in situ | 7807[2] | 279 | 8101 | Vik-Mo et al. 1986 [29] |
| Swine | LV, e.c.p. | 3545 | 167 | 4346 | Molaparast-Saless et al. 1988 [30] |
| Swine | LV, e.c.p., FA 0.5 mM | 3787 | 157 | 4586 | Liedtke et al. 1988 [19] |
| | FA 1.5 mM | 3010 | 366 | 4746 | |
| Human | LV, in situ | 7000 | | 7700 | Regitz et al. 1990 [31] |
| Human | LV, in situ | 3845 | 1585 | 6575 | Masumura et al. 1990 [32] |
| Human | Right atrium in situ | 2400 | 6750 | 9600 | Böhles et al. 1986 [33] |

[1] Unperfused means immediately freeze-clamped after extirpation from the body. [2] Short-chain fatty acyl carnitine esters are included. Data are expressed as nmol · $g^{-1}$ dry weight of tissue. See also legend in Table 1.

palmitate was added to the perfusion medium. Striking inter-laboratory differences were found in rabbit hearts directly used for biochemical analysis after extirpation from the animal. Values measured varied from 25 to 2317 nmol · $g^{-1}$ dry weight of tissue [15, 25]. No satisfactory explanations can be given for the observed differences.

Very high acyl carnitine levels were found in human right atrial appendages, collected during cardiopulmonary bypass operations [33].

*Subcellular localization of fatty acids, CoA, carnitine and their derivatives in the normoxic heart*

No reliable information is available on the subcellular localization of fatty acids in normoxic myocardial tissue. Attempts to measure the content of fatty acids in subcellular fractions failed because lipolysis of endogenous lipids is stimulated during homogenization. An increase of a factor 3 in fatty acid content was observed when fresh rat cardiac tissue was homogenized in a sucrose-EDTA-Tris buffer (Van der Vusse and Roemen, unpublished results).

Earlier studies performed by Idell-Wenger and coworkers [10] have shown that over 95% of CoA is present in the mitochondrial compartment, mainly as free CoA. Only 5% of total CoA could be recovered from the cytoplasmic space. Acyl CoA was not detectable in the cytoplasm indicating that in the normoxic myocardium the majority of fatty acyl CoA is localized in the mitochondrial matrix. Later studies performed by Hütter and colleagues [34] revealed that total CoA is more equally distributed over the mitochondrial and cytoplasmic compartment (that is, 58 and 42%), respectively. On the order of 62% of long-chain acyl CoA was found to be present in the mitochondria.

According to the results of Idell-Wenger et al. [10] the majority of total carnitine is present in the cytoplasmic space. Approximately 8% of total carnitine was found to be associated with the mitochondria. No fatty acyl carnitine could be detected in the mitochondrial fraction, suggesting that almost all cellular fatty acyl carnitine molecules are present in the cytoplasm. In contrast, Hütter and colleagues [34] reported that ~35% of total carnitine could be recovered from mitochondria, while on the order of 60% of all fatty acyl carnitine molecules are present in the mitochondrial compartment. The latter authors tried to explain the striking differences between their results and those published by the group of Neely by differences in techniques used to isolate mitochondria from cardiac tissue.

Oram and coworkers [35] calculated a cytoplasmic carnitine to CoASH ratio of ~100 in normoxic rat hearts. This high ratio may function to channel fatty acids towards oxidation in the mitochondrial matrix rather than synthesis of complex lipids in the cytoplasmic compartment.

## Fatty acid oxidation in the ischemic heart

### *Uptake and utilization of fatty acids during ischemia*

Myocardial ischemia results from impeded blood flow through the coronary arteries. As a consequence supply of substrates, including fatty acids, and molecular oxygen is decreased which results in substantial alterations in energy conversion in the cardiac cells. Contractile function declines rapidly with a concomitant decrease in the energy requirement of the cardiomyocyte. Insufficient supply of oxygen leads to a shift from oxidative conversion of energy to the glycolytic production of ATP while lactate formation is increased. When ischemia is partial, that is, some blood is still flowing through the coronary vessels, glucose competes favorably with fatty acids for the residual amount of molecular oxygen [36]. Fatty acid uptake by the ischemic cells declines proportionally with the reduction in blood flow through the coronary arteries [37].

Lack of oxygen has a severe effect on fatty acid oxidation. Flavine adenine dinucleotide (FAD), the obligatory receptor of hydrogen atoms in the conversion step of fatty acyl CoA to trans-$\Delta$-$^2$ enoyl CoA (Figure 1), will accumulate in its reduced form, since the supply of molecular oxygen to convert $FADH_2$ back to FAD is insufficient. A high $FADH_2$/FAD ratio will hamper the catalytic activity of fatty acyl CoA dehydrogenase. Besides molecular oxygen is required for the regeneration of $NAD^+$ from NADH in a subsequent step in the $\beta$-oxidative pathway, that is, the conversion of 3-hydroxy acyl CoA to 3-ketoacyl CoA. Finally, oxygen is indispensable for an adequate flux of the acetyl residue of acetyl CoA in the citric acid cycle (Figure 1). From these considerations, it will be clear that lack of molecular oxygen compromises a variety of steps in the overall degradation of fatty acids in the ischemic cardiomyocytes. When the supply of fatty acids, either from extracellular or intracellular sources, exceeds the compromised capacity to dispose the fatty acyl moieties in the oxidative pathway, intermediates of the metabolic pathway will accumulate in the affected cells. Furthermore, it cannot be excluded that changes in the intracellular milieu of the ischemic cardiomyocytes have a profound effect on the activity of enzymes and transport proteins involved in fatty acid catabolism. In this respect, Pauly and coworkers [38] indicated that carnitine palmitoyl transferase I in mitochondria isolated from ischemic hearts is less sensitive to the inhibitory action of malonyl CoA than in mitochondria of normoxic cardiac tissue, which may give rise to enhanced production of fatty acyl carnitine in the flow-deprived tissue.

It should be realized that the ultimate effect of ischemia on cardiac fatty acid metabolism depends on a multitude of factors, such as the severity (no-flow vs. low-flow) and duration of ischemia, the concentration of fatty acids and the presence of competing substrates in the blood or perfusion medium flowing through the coronary vessels and the level of hormones, known to exert an effect on cardiac lipid metabolism.

*Accumulation of fatty acyl CoA and carnitine esters in the ischemic heart*

Experimental findings indicate that in isolated rabbit hearts, perfused with palmitic acid, low-flow ischemia readily results in high tissue levels of $\beta$-hydroxypalmitate and $\beta$-hydroxystearate. Moreover, accumulation of $\beta$-hydroxyacyl carnitine and $\beta$-hydroxyacyl CoA occurred in the rabbit hearts within 2–5 minutes after the onset of ischemia [25, 39]. A host of investigators has shown that also fatty acyl CoA and fatty acyl carnitine accumulate in the ischemic heart [10, 17, 26–30, 40–44]. In general, the accumulation of fatty acyl carnitine was quantitatively more pronounced than fatty acyl CoA.

Attempts to delineate the time courses of acyl CoA and acyl carnitine accumulation in flow-deprived tissue revealed that alterations in the tissue content of these fatty acyl esters occurred rapidly after the onset of ischemia. Shrago and colleagues [42] observed a doubling of the tissue content of acyl CoA within 15 minutes after ligation of the coronary artery. Comparable findings were obtained in isolated, low-flow ischemic rat hearts [21]. In the same study [21] the tissue content of acyl carnitine was found to be increased by ~150% within 30 minutes of low-flow ischemia. More detailed investigations performed by Idell-Wenger and Neely [3] indicate that in isolated rat hearts fatty acyl carnitine and fatty acyl CoA significantly accumulate already 5 minutes after the onset of ischemia. The notion that tissue levels of fatty acyl carnitine rapidly increase during ischemia is supported by findings published by Feuvray et al. [43].

Paulson and colleagues [21] have reported that the alterations in tissue levels of fatty acyl CoA and carnitine esters strongly depend on the presence of residual flow through the coronary arteries of the ischemic heart. When no-flow ischemia was applied, isolated rat hearts did not show any change in the level of fatty acyl CoA and fatty acyl carnitine in contrast to low-flow ischemic hearts. This observation is supported by earlier findings of Neely and coworkers, revealing that in autolysing rat heart tissue no significant increase in the level of fatty acyl CoA and fatty acyl carnitine occurs [45]. The observations suggest that the fatty acyl moieties of accumulating acyl CoA and acyl carnitine are mainly derived from extracellular sources, such as fatty acids complexed to albumin in the vascular and interstitial compartment [46].

*Intracellular site of accumulation of fatty acyl CoA and acyl carnitine*

Several attempts have been made to delineate the intracellular site of accumulation of acyl CoA and acyl carnitine during ischemia or hypoxia. Idell-Wenger and coworkers [10] found that increased tissue contents of fatty acyl CoA was associated with elevated mitochondrial levels of the fatty acyl ester.

These findings were corroborated by observations made by Kotaka and colleagues [40], Lochner et al. [41] and Feuvray and Plouët [46]. Detailed studies performed by Idell-Wenger and coworkers [10] and Hütter and colleagues [34] revealed that in low-flow ischemic and hypoxic rat hearts 100 and 80%, respectively, of total tissue acyl CoA was present in the mitochondrial fraction. With respect to fatty acyl carnitine deviant findings are reported in the literature. Idell-Wenger and colleagues [10] calculated that ~20% of the total fatty acyl carnitine pool of mildly ischemic rat hearts is located inside the mitochondria. Ischemia was applied for 20 minutes. In contrast, Hütter and co-investigators [34] estimated that in rat hearts made hypoxic for 90 min, over 80% of fatty acyl carnitine is confined to the mitochondrial compartment. It is uncertain whether the conflicting results regarding intracellular localization of fatty acyl carnitine can be ascribed to either differences between hypoxia and mild ischemia or to differences in duration of oxygen deprivation. In this respect it is of interest to note that McHowat and coworkers [22] observed that the proportion of fatty acyl carnitine in the mitochondrial compartment of isolated cardiomyocytes exposed to hypoxia increased from ~20% (normoxic control) to 30 and 60% when the hypoxic period lasted 10 and 20 minutes, respectively. In the latter study, the intracellular localization of fatty acyl carnitine was assessed by electron microscopical autoradiography.

Lamers and coworkers [47] used a biochemical radioisotopic technique to assess the intracellular accumulation of fatty acyl carnitine in ischemic pig hearts. They concluded that fatty acyl carnitine preferentially accumulates in the sarcolemma of flow-deprived hearts. Besides they concluded that the content of acyl carnitine present in the cell membrane was insufficiently high to account for the alterations in sarcolemmal sodium and calcium permeability that readily occur during the ischemic episode.

Recent studies performed by McHowat and coworkers [22] indicated that in isolated rat cardiomyocytes exposed to 20 minutes of hypoxia the content of fatty acyl carnitine in the sarcolemma increased ~100 fold. The content in the mitochondrial compartment doubled, while no change occurred in the cytoplasmic and nuclear space. When the enzymatic activity of carnitine acyl transferase I was blocked by sodium 2-[5-(4-chlorophenyl)-pentyl]-oxirane-2-carboxylate (abbreviated POCA), the accumulation of fatty acyl carnitine in the sarcolemma was found to be completely prevented. Interestingly, Wu and colleagues [48] observed that the hypoxia-induced increase of the sarcolemmal content of fatty acyl carnitine preferentially took place in the membranes composing the gap junctions between two adjacent cardiomyocytes. This phenomenon was associated with a significant decline in electrical cellular coupling.

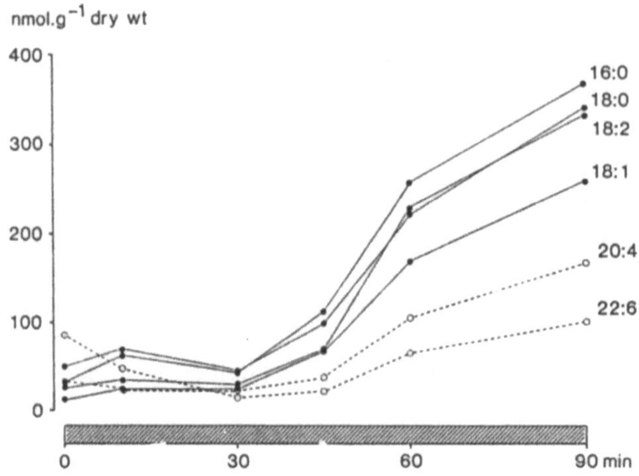


*Figure 2.* Time course of the accumulation of fatty acids in no-flow ischemic rat hearts [from ref. 58]. Hearts were obtained from male Lewis rats and perfused with a modified Krebs-Henseleit buffer containing glucose (11 mM) and pyruvate (5 mM) prior to ischemia. Hearts were freeze-clamped after 0, 10, 30, 45, 60 or 90 minutes of ischemia, respectively, and the tissue content of fatty acids was determined with a gas-chromatographic technique [54]. Fatty acids are denoted by their chemical notation.

*Accumulation of fatty acids in the ischemic heart*

In addition to fatty acyl CoA and fatty acyl carnitine, unesterified fatty acids accumulate in the ischemic heart [37, 49–57]. Unlike fatty acyl CoA and acyl carnitine fatty acid accumulation in oxygen-deprived hearts appears to be a relatively slow process. Experiments performed on regional ischemic dog hearts [49, 53] indicate that fatty acid accumulation does not occur earlier than 20 minutes after the onset of ischemia. In isolated, no-flow ischemic rat hearts fatty acids started to accumulate to a significant extent only after 30 minutes of flow cessation (Figure 2) [58].

*Relation between fatty acid accumulation and tissue fatty acyl CoA and carnitine levels in the ischemic heart*

The above mentioned findings suggest that inhibition of fatty acid oxidation in the ischemic myocardium results initially in storage of excess fatty acid moieties in the acyl carnitine and to a lesser extent the acyl CoA pool. It cannot be excluded that part of the fatty acid residues are stored in intracellular triacylglycerol due to enhanced levels of acyl CoA and glycerol-3-phosphate, the precursors of myocardial triacylglycerol synthesis [1]. Fatty acids, accumulating during prolonged myocardial ischemia, are most likely derived from the intracellular phospholipid pool, since arachidonic acid contributes substantially to the accumulating fatty acids [37, 49, 53, 54]. In normal conditions arachidonic acid is predominantly incorporated in membrane phospholipids and enhanced levels of this fatty acid in cardiac cells is commonly taken as a measure of intracellular phospholipid hydrolysis. The fact that in isolated, no-flow ischemic rat hearts, accumulation of fatty acyl carnitine is virtually absent [21, 45], while fatty acid levels are significantly enhanced [54, 57], strongly suggests that fatty acids liberated from endogenous stores have no access to the enzymatic machinery to convert fatty acids to their respective carnitine esters. Alternatively, the enzymes involved in acyl carnitine formation may become blocked under prolonged ischemic conditions. Detailed experiments are needed to elucidate the relationship between the source of fatty acids and the accumulation of acyl CoA and acyl carnitine in flow-deprived cardiac tissue as a function of time.

## Concluding remarks

Accumulation of the fatty acyl CoA and fatty acyl carnitine rapidly occurs in oxygen-deprived cardiac tissue. The cellular increase of un-esterified fatty acids is a relatively slow process indicating that excess fatty acids are preferentially stored as their CoA and carnitine esters during the acute phase of ischemia. Acyl CoA predominantly accumulates in the mitochondrial matrix, while the majority of acyl carnitine moieties preferentially accumulates in the sarcolemma of the oxygen deprived cardiomyocytes.

## Acknowledgement

The help of Claire Bollen and Emmy van Roosmalen in the preparation of the manuscript is greatly appreciated.

## References

1. Van der Vusse GJ, Glatz JFC, Stam HCG, Reneman RS. Fatty acid homeostasis in the normoxic and ischemic heart. Physiol Rev 1992; 72: 881–940.
2. Braun JEA, Severson DK. Regulation of the synthesis, processing and translocation of lipoprotein lipase. Biochem J 1992; 287: 337–47.
3. Idell-Wenger JA, Neely JR. Regulation of uptake and metabolism of fatty acids by muscles. In: Dietschy JM, Gotto AM, Ontko JA, editors. Disturbances in lipid and lipoprotein metabolism. Maryland: Am Phys Soc, 1978: 269–85.
4. Van der Vusse GJ, De Groot MJM. Interrelationship between lactate and cardiac fatty acid metabolism. Mol Cell Biochem 1992; 116: 11–7.

5. Van der Vusse GJ, Prinzen FW, Reneman RS. Are tissue non-esterified fatty acids (NEFA) involved in the impairment of biochemical and mechanical processes during acute regional ischemia in the heart? In: Ferrari R, Katz AM, Shug A, Visioli O, editors. Myocardial ischemia and lipid metabolism. New York: Plenum Press, 1984: 171–84.
6. Hunneman DH, Schweickhardt C. Mass fragmentographic determination of myocardial free fatty acids. J Mol Cell Cardiol 1982; 14: 339–51.
7. Garland PB, Randle PJ. Regulation of glucose uptake by muscle. Biochem J 1964; 93: 678–87.
8. Oram JF, Bennetch SL, Neely JR. Regulation of fatty acid utilization in isolated perfused rat hearts. J Biol Chem 1973; 248: 5299–309.
9. Van der Vusse GJ, Roemen THM, Reneman RS. Assessment of fatty acids in dog left ventricular myocardium. Biochim Biophys Acta 1980; 617: 347–52.
10. Idell-Wenger JA, Grotyohann LW, Neely JR. Coenzyme A and carnitine distribution in normal and ischemic hearts. J Biol Chem 1978; 253: 4310–8.
11. Pieper GM, Salhany JM, Murray WJ, Wu ST, Eliot RS. Lipid-mediated impairment of normal energy metabolism in the isolated perfused diabetic rat heart studied by phosphorus-31 NMR and chemical extraction. Biochim Biophys Acta 1984; 803: 229–40.
12. Lopaschuk GD, Neely JR. Stimulation of myocardial coenzyme A degradation by fatty acids. Am J Physiol 1987; 253: H41–6.
13. Neely JR, Feuvray D. Metabolic products and myocardial ischemia. Am J Pathol 1981; 102: 282–91.
14. Moore KH, Bonema JD, Solomon FJ. Long-chain acyl-CoA and acylcarnitine hydrolase activities in normal and ischemic rabbit heart. J Mol Cell Cardiol 1984; 16: 905–13.
15. Ferrari R, Di Lisa F, De Jong JW et al. Prolonged propionyl-L-carnitine pre-treatment of rabbit: Biochemical, hemodynamic and electrophysiological effects on myocardium. J Mol Cell Cardiol 1992; 24: 219–32.
16. Reibel DK, Uboh CE, Kent RL. Altered coenzyme A and carnitine metabolism in pressure-overload hypertrophied hearts. Am J Physiol 1983; 244: H839–43.
17. Shug AL, Thomson JH, Folts JD, Bittar N, Klein MI, Koke JR, Huth PJ. Changes in tissue levels of carnitine and other metabolites during myocardial ischemia and anoxia. Arch Biochem Biophys 1978; 187: 25–33.
18. Liedtke AJ, Nellis SN. Effects of carnitine in ischemic and fatty acid supplemented swine hearts. J Clin Invest 1979; 64: 440–7.
19. Liedtke AJ, DeMaison L, Eggleston AM, Cohen LM, Nellis SH. Changes in substrate metabolism and effects of excess fatty acids in reperfused myocardium. Circ Res 1988; 62: 535–42.
20. Stearns SB. Carnitine content of skeletal and cardiac muscle from genetically diabetic (db/db) and control mice. Biochem Med 1983; 29: 57–63.
21. Paulson DJ, Schmidt MJ, Romens J, Shug AL. Metabolic and physiological differences between zero-flow and low-flow myocardial ischemia: effects of L-acetylcarnitine. Basic Res Cardiol 1984; 79: 551–61.
22. McHowat J, Yamada KA, Saffitz JE, Corr PB. Subcellular distribution of endogenous long chain acyl carnitines during hypoxia in adult canine myocytes. Cardiovasc Res 1993; 27: 1327–243.
23. Broderick TL, Quinney HA, Barker CC, Lopaschuk GD. Beneficial effect of carnitine on mechanical recovery of rat hearts reperfused after a transient period of global ischemia is accompanied by a stimulation of glucose oxidation. Circulation 1993; 87: 972–81.
24. Broderick TL, Quinney HA, Lopaschuk GD. Carnitine stimulation of glucose oxidation in the fatty acid perfused isolated working rat heart. J Biol Chem 1992; 267: 3758–63.
25. Moore KH, Koen AE, Hull FE. β-Hydroxy fatty acid production by ischemic rabbit heart. Distribution and chemical states. J Clin Invest 1982; 69: 377–83.
26. Corr PB, Creer MH, Yamada KA, Saffitz JE, Sobel BE. Prophylaxis of early ventricular fibrillation by inhibition of acyl carnitine accumulation. J Clin Invest 1989; 83: 927–36.
27. Suzuki Y, Kamikawa T, Kobayashi A, Masumura Y, Yamazaki N. Effects of L-carnitine

on tissue levels of acyl carnitine, acyl coenzyme A and high energy phosphate in ischemic dog heart. Jpn Circ J 1981; 45: 687–94.
28. Shug AL. Control of carnitine-related metabolism during myocardial ischemia. Tex Rep Biol Med 1979; 39: 409–28.
29. Vik-Mo H, Mjøs OD, Neely JR, Maroko PR, Ribeiro LGT. Limitation of myocardial infarct size by metabolic interventions that reduce accumulation of fatty acid metabolites in ischemic myocardium. Am Heart J 1986; 111: 1048–54.
30. Molaparast-Saless F, Nellis SH, Liedtke AJ. The effects of propionylcarnitine taurine on cardiac performance in aerobic and ischemic myocardium. J Mol Cell Cardiol 1988; 20: 63–74.
31. Regitz V, Bossaller C, Strasser R, Müller M, Shug AL, Fleck E. Metabolic alterations in end-stage and less severe heart failure - myocardial carnitine decrease. J Clin Chem Clin Biochem 1990; 28: 611–7.
32. Masumura Y, Kobayashi A, Yamazaki N. Myocardial free carnitine and fatty acylcarnitine levels in patients with chronic heart failure. Jpn Circ J 1990; 54: 1471–6.
33. Böhles H, Noppeney Th, Akcetin Z, Rein J, Von der Emde J. The effect of preoperative L-carnitine supplementation on myocardial metabolism during aorta-coronary bypass surgery. Curr Ther Res 1986; 39: 429–35.
34. Hütter JF, Alves C, Soboll S. Effects of hypoxia and fatty acids on the distribution of metabolites in rat heart. Biochim Biophys Acta 1990; 1016: 244–52.
35. Oram JF, Wenger JI, Neely JR. Regulation of long chain fatty acid activation in heart muscle. J Biol Chem 1975; 250: 73–8.
36. Opie LH, Owen P, Riemersma RA. Relative rates of oxidation of glucose and free fatty acids by ischaemic and non-ischaemic myocardium after coronary artery ligation in the dog. Eur J Clin Invest 1973; 3: 419–35.
37. Van der Vusse GJ, Roemen THM, Prinzen FW, Coumans WA, Reneman RS. Uptake and tissue content of fatty acids in dog myocardium under normoxic and ischemic conditions. Circ Res 1982; 50: 538–46.
38. Pauly DF, Kirk KA, McMillin JB. Carnitine palmitoyltransferase in cardiac ischemia. A potential site of altered fatty acid metabolism. Circ Res 1991; 68: 1085–94.
39. Moore KH, Radloff JR, Hull FE, Sweeley CC. Incomplete fatty acid oxidation by ischemic heart: $\beta$-hydroxy fatty acid production. Am J Physiol 1980; 239: H257–65.
40. Kotaka K, Miyazaki Y, Ogawa K, Satake T, Sugiyama S, Ozawa T. Reversal of ischemia-induced mitochondrial dysfunction after coronary reperfusion. J Mol Cell Cardiol 1982; 14: 223–31.
41. Lochner A, Van Niekerk I, Kotzé JCN. Mitochondrial acyl-CoA, adenine nucleotide translocase activity and oxidative phosphorylation in myocardial ischaemia. J Mol Cell Cardiol 1981; 13: 991–7.
42. Shrago E, Shug AL, Sul H, Bittar N, Folts JD. Control of energy production in myocardial ischemia. Circ Res 1976; 38(Suppl I): 75–9.
43. Feuvray D, Idell-Wenger JA, Neely JR. Effects of ischemia on rat myocardial function and metabolism in diabetes. Circ Res 1979; 44: 322–9.
44. Hekimian G, Feuvray D. Reduction of ischemia-induced acyl carnitine accumulation by TDGA and its influence on lactate dehydrogenase release in diabetic rat hearts. Diabetes 1986; 35: 906–10.
45. Neely JR, Garber D, McDonough K, Idell-Wenger J. Relationship between ventricular function and intermediates of fatty acid metabolism during myocardial ischemia: effects of carnitine. In: Winbury MM, Abiko Y, editors. Persp Cardiovasc Res vol. 3. Ischemic myocardium and antianginal drugs. New York: Raven Press, 1979: 225–39.
46. Feuvray D, Plouët J. Relationhip between structure and fatty acid metabolism in mitochondria isolated from ischemic rat hearts. Circ Res 1981; 48: 740–7.
47. Lamers JMJ, De Jonge-Stinis JT, Verdouw PD, Hülsmann WC. On the possible role of long chain fatty acyl carnitine accumulation in producing functional and calcium permeability changes in membranes during myocardial ischaemia. Cardiovasc Res 1987; 21: 313–22.

48. Wu J, McHowat J, Saffitz JE, Yamada KA, Corr PB. Inhibition of gap junctional conductance by long-chain acylcarnitines and their preferential accumulation in junctional sarcolemma during hypoxia. Circ Res 1993; 72: 879–89.
49. Chien KR, Han A, Sen A, Buja LM, Willerson JT. Accumulation of unesterified arachidonic acid in ischemic canine myocardium. Circ Res 1984; 54: 313–22.
50. Das DK, Engelman RM, Rousou JA, Breyer RH, Otani H, Lemeshow S. Role of membrane phospholipids in myocardial injury induced by ischemia and reperfusion. Am J Physiol 1986; 251: H71–9.
51. Fox KAA, Abendschein DR, Ambos HD, Sobel BE, Bergmann SR. Efflux of metabolized and nonmetabolized fatty acid from canine myocardium. Implications for quantifying myocardial metabolism tomographically. Circ Res 1985; 57: 232–43.
52. Miura I, Hashizume H, Akutsu H, Hara Y, Abiko Y. Accumulation of nonesterified fatty acids in the dog myocardium during coronary artery occlusion determined by a method using 9–anthryldiazomethane. Heart Vessels 1987; 3: 190–4.
53. Prinzen FW, Van der Vusse GJ, Arts T, Roemen THM, Coumans WA, Reneman RS. Accumulation of nonesterified fatty acids in ischemic canine myocardium. Am J Physiol 1984; 247: H264–72.
54. Van Bilsen M, Van der Vusse GJ, Willemsen PHM, Coumans WA, Roemen THM, Reneman RS. Lipid alterations in isolated, working rat hearts during ischemia and reperfusion: Its relation to myocardial damage. Circ Res 1989; 64: 304–14.
55. Weglicki WB, Owens K, Urschel CW, Serur JR, Sonnenblick EH. Hydrolysis of myocardial lipids during acidosis and ischemia. Rec Adv Stud Cardiac Struct Metab 1973; 3: 781–93.
56. Weishaar RE, Sarma JSM, Maruyama Y, Fisher R, Bing RJ. Regional blood flow, contractility and metabolism in early myocardial infarction. Cardiology 1977; 62: 2–20.
57. De Groot MJM, Coumans WA, Willemsen PHM, Van der Vusse GJ. Substrate-induced changes in the lipid content of ischemia and reperfused myocardium. Its relation to hemodynamic recovery. Circ Res 1993; 72: 176–86.
58. Van Bilsen M. The significance of myocardial non-esterified fatty acid accumulation during ischemia and reperfusion [dissertation]. Maastricht, The Netherlands: Univ of Limburg, 1988.

*Corresponding Author*: Professor Ger J. van der Vusse, Department of Physiology, Faculty of Medicine, University of Limburg, P.O. Box 616, 6200 MD Maastricht, The Netherlands

# 6. Carnitine acylcarnitine translocase in ischemia

JEANIE B. McMILLIN

"The expression of a limitation in carnitine acylcarnitine translocase activity relevant to human ischemic heart disease may be found only in those myocardial cells which are severely compromised and thus, may only be part of the end-stage sequellae of injury and cell death."

## Introduction

### *Fatty acid oxidation in the heart: role of the carnitine translocase*

The role of fatty acid oxidation as a major source of energy to the contractile, working heart is well-established. However, the mechanism(s) by which the rates of long-chain fatty acid oxidation are controlled in the heart is (are) not as well understood as in the liver where the lipogenic substrate, malonyl-CoA, acts as a switch to partition fatty acids between synthesis and degradation [1]. Early studies in the working heart concluded that, at low levels of pressure development, rates of $\beta$-oxidation are limited by the disposal of acetyl-CoA through the citric acid cycle [2]. Recent evidence from the perfused, working rat heart [3] suggests that, similar to liver, the level of fatty acid oxidation directly reflects changes in activity of the tissue-specific acetyl-CoA carboxylase activity present in the cardiac myocyte. These results support the view that in the heart, the primary role for malonyl-CoA is the regulation of fatty acid flux through $\beta$-oxidation by inhibition of carnitine palmitoyltransferase I (CPT-I) on the outer mitochondrial membrane [4]. This situation may vary, however, depending upon the work load to the heart. When levels of cardiac work are increased, increases in fatty acylcarnitine and decreases in fatty acyl-CoA are observed concomittent with an acceleration of $\beta$-oxidation [2]. Since the long-chain acylcarnitine produced by CPT-I must be transported across the mitochondrial membrane to CPT-II in exchange for one molecule of carnitine from the matrix, these authors suggested that the increase in acylcarnitine accumulation observed at high

*J.W. de Jong and R. Ferrari (eds): The carnitine system*, 69–82.

*Table 1.* Carnitine acylcarnitine translocase activities in ischemic heart mitochondria: a comparison to aging.

| Condition | Carnitine exchange activity | Palmitoylcarnitine exchange activity | [Carnitine] |
|---|---|---|---|
| | nmol/min/mg protein | | nmol/mg |
| Control[a] | ($n = 4$) 0.54 ± 0.08 | ($n = 7$) 0.13 ± 0.03 | ($n = 12$) 1.54 ± 0.15 |
| Ischemic[a] | ($n = 9$) 0.31 ± 0.05 | ($n = 6$) 0.10 ± 0.02 | ($n = 8$) 1.28 ± 0.42 |
| Menhaden oil | | | |
| Control[b] | ($n = 8$) 0.47 ± 0.04 | ($n = 4$) 0.24 ± 0.04 | ($n = 4$) 2.27 ± 0.22 |
| Ischemic[b] | ($n = 4$) 0.35 ± 0.09 | ($n = 4$) 0.07 ± 0.02‡ | ($n = 4$) 1.02 ± 0.10‡ |
| Mature rat[c] (12 mos) | ($n = 5$) 0.88 ± 0.10 | ($n = 5$) 0.24 ± 0.01 | ($n = 7$) 1.40 ± 0.18 |
| Aged rat[c] (30 mos) | ($n = 5$) 0.49 ± 0.03* | ($n = 5$) 0.07 ± 0.01‡ | ($n = 5$) 0.96 ± 0.19* |
| Young rat[d] (6 mos) | ($n = 7$) 0.71 ± 0.03 | Not determined | ($n = 5$) 1.05 ± 0.03 |
| Aged rat[d] (24 mos) | ($n = 5$) 0.46 ± 0.04‡ | Not determined | ($n = 5$) 0.82 ± 0.02‡ |

Values represent the mean ± SE.
[a] [60]. [b] [63]. [c] [9]. [d] [8].
* Significantly different from control values $p < 0.05$.
‡ Significantly different from control values, $p < 0.001$.

work loads reflects a limitation in the carnitine-dependent pathway at the carnitine acylcarnitine translocase (CAT). The speculation that CAT may exert rate-limitation on fatty acylcarnitine oxidation under certain conditions has drawn support from observations that diabetic ketosis [5], and substrate-dependent activation [6] can up-regulate the rate of acylcarnitine translocation in liver and heart mitochondria, respectively. In the majority of these studies, it is likely that the expression of CAT activity is greatly influenced by variations in the matrix content of carnitine, the latter present at concentrations which are subsaturating under normal transport conditions [5, 7]. A decrement in matrix carnitine with aging [8,9] has been proposed to account for the decreased rates of CAT activity and palmitoylcarnitine oxidation in heart mitochondria from 24–30 month-old rats (Table 1). A related inability of palmitate to depress glucose extraction in the perfused working old rat heart suggests a direct physiological consequence of diminished acylcarnitine exchange on cardiac energy metabolism in aging [9]. The physiological importance of CAT has been further emphasized in clinical cases of genetic deficiencies reported in the translocase (see below). The functional ramifications of decreased or limiting CAT activity will be discussed below in relation to cardiac-specific effects.

*Fatty acid metabolism in ischemia*

Research on substrate metabolism in models of cardiac ischemic injury demonstrates the following changes. In substrate extraction studies following coronary artery ligation, glucose uptake and metabolism are accelerated relative to that of fatty acid, even in areas where residual oxidative metabolism continues to contribute to $^{14}CO_2$ production [10]. This pattern is consistent with a decreased aerobic production of ATP by mitochondrial oxidative phosphorylation due to lowered cellular oxygen tension, and an increase in glycolytic metabolism. As a result, tissue levels of acetyl-CoA (from fatty acids and glucose) and citrate acid cycle intermediates decrease as a result of pyruvate conversion to lactate and esterification of extracted long-chain fatty acids. The first major consequence of ischemia on metabolism in the heart is the inability of glycolysis to maintain cellular ATP at levels required for continuation of contractile function. With accumulation of lactate in the cell and increases in the myocyte $NADH/NAD^+$ ratio, glycolysis eventually becomes inhibited as a consequence of limitation at the glyceraldehyde-3-P dehydrogenase step [11]. With restoration of blood flow and oxygen to the ischemic myocardium, consistent beneficial effects of glycolytic substrates on the recovery of contractile function have been reported in a variety of models and conditions [12, 13]. In contrast, the presence of fatty acids in the serum or perfusate of ischemic heart appears to be associated with deleterious effects on cardiac contractility and rhythm.

In 1972, the British cardiologist, M.F. Oliver observed that "there is increasing evidence that high arterial blood concentrations of free fatty acids can depress myocardial performance by leading to ventricular arrhythmias and decreased contractility" [14]. Since the level of free fatty acids in the serum is important in determining the substrate selectivity of the heart, and since these fatty acids increase dramatically following ischemia, the relationship between fatty acid metabolism and the recovery of contractile function in the ischemic heart has been extensively investigated. It is established that fatty acid oxidation decreases relative to glucose utilization during ischemia [10]. As a result, the entire tissue content of CoA (95% mitochondrial) is present as long-chain acyl-CoA (299 nmol/g dry weight heart) and aclycarnitines increase five fold from 0.83 μmol/g in normoxia to 4.4 μmol/g in severe ischemia [15]. The latter levels of acylcarnitine are well within the concentration range known to affect both sarcolemmal and sarcoplasmic reticulum ion transporters, thereby potentially contributing to changes in the cellular ionic gradients during ischemia (see discussion below) [16, 17]. Free fatty acids also increase in the heart, the levels being dependent on the time of ischemia. The observation that fatty acids increase infarct size [18] has been attributed to increases in residual oxygen utilization and to ATP wasting via a proposed "triacylglycerol cycle" [19].

### Physiological role of CAT in ischemic fatty acid metabolism

Although the majority of the acylcarnitines accumulated during the ischemic interval are extramitochondrial, a net transfer of long-chain and acid-soluble carnitine into the mitochondrial matrix occurs, suggesting that a net inward translocation of acylcarnitines can take place independent of the 1:1 exchange stoichiometry of the carrier, possibly by a unidirectional mechanism of facilitated diffusion as proposed by Schulz and Racker [20] and Pande and Parvin [21]. Based on experiments with liver mitochondria from ketotic rats [5] and on the decreased tissue and mitochondrial carnitine levels found in aging rats [22, 23], it seems likely that the purpose of the unidirectional movement of carnitine is to adjust mitochondrial levels to changes in tissue carnitine concentration. Since long-chain acylcarnitines have a higher affinity for the translocase compared to free carnitine, the concentration gradient of long-chain acylcarnitine (either free or bound to the mitochondria) may appreciably influence the inward movement of carnitine and carnitine esters into the mitochondrial matrix. Thus, in ischemia, where cytosolic free carnitine concentrations decrease and acylcarnitine levels rise from 0.4 to 1.95 mM, the observed doubling of the matrix content of carnitine should reflect this inward transport of acylcarnitines into the mitochondria [15, 24]. The increased mitochondrial content of acyl-CoA and acylcarnitine is a physiological response to the pathology of ischemia where $\beta$-oxidation is slowed or inhibited. The proximity of fatty acyl groups to matrix enzymes of oxidative metabolism increases the probability of rapid catabolism upon the restoration of oxygen with reflow after ischemia (see below).

*CAT and ischemic reperfusion*

With the exception of a few studies which suggest decreased extraction and oxidation of fatty acids in the heart following ischemia using positron imaging [25] and $^{14}$C-palmitate infusion studies in humans [26], the majority of evidence from animal models of ischemia suggests that fatty acids are preferentially extracted and oxidized compared to glucose in the post-ischemic, reperfused heart [12, 27]. When both endogenous and exogenous fatty acid pools were labelled with [1-$^{14}$C]palmitate and [9,10-$^{3}$H]palmitate, respectively, measurement of fatty acid oxidation in the ischemic, reperfused hearts exceeded the steady-state, pre-ischemic rates [28]. Other investigators have also shown an increase in fatty acid oxidation during reperfusion of ischemic swine hearts [29]. Interestingly, the "burst" of endogenous fatty acid oxidation does not reflect an increased rate of endogenous tissue triacylglyceride lipolysis [28]. Although it has been suggested that the accelerated endogenous fatty acid oxidation may be explained by the presence of more than one triglyceride pool [30], it is also likely that the observed brief increase in [1-$^{14}$C]palmitate oxidation reflects, in part, immediate metabolism of the accumulated mitochondrial long-chain acyl-CoA and acylcarnitine derivatives. The sustained acceleration in the rates of oxidation of exogenous fatty acid indicates a rapid drop in cytosolic ratios of acyl-CoA/CoA and acylcarnitine/carnitine (from the elevated ischemic ratios) to facilitate fatty acid activation and $\beta$-oxidation during reperfusion [31]. It is also possible that an additional consequence of the oxidation of matrix acylcarnitine is a unidirectional efflux of free carnitine from the mitochondria to reestablish matrix:cytosol CAT equilibrium. The extent to which CAT limits $\beta$-oxidation when fatty acid metabolism is accelerated during reperfusion has not been specifically addressed. However, the small but significant elevation of long-chain acylcarnitine with reperfusion [12] is similar to the pattern observed in aerobic hearts at high work load suggesting that the rate of acylcarnitine oxidation may also become rate-limiting to acetyl-CoA production [2] under conditions of ischemic reperfusion.

*Acylcarnitine accumulation and contraction of the postischemic heart*

As discussed above, there now appears to be a consensus that fatty acid oxidation contributes significantly to energy metabolism in the *reperfused* ischemic heart [12, 27]. However, in distinction to the aerobic heart, reperfusion of the ischemic heart with fatty acids does not support the return of contractile function as well as reperfusion of hearts with glucose alone as an energy substrate. In fact, inhibition of fatty acid metabolism by specific inhibitors of CPT-I appears to be beneficial to the mechanical recovery of the ischemic heart [12, 27, 32]. These observations have been explained by the arrhythmogenic effects of membrane long-chain acylcarnitine accumulation [27, 33, 34]. The deleterious consequences of acylcarnitines on cardiac function have been proposed to result from disrupted cardiac ion homeostasis by specific inhibition of sarcolemmal $Na^+$, $K^+$-ATPase [16, 35] as well as of the $Ca^{2+}$ ATPase of the sarcoplasmic reticulum [17]. Inhibition of these membrane pumps has been proposed to account for the rise in cellular $Ca^{2+}$ documented in ischemic and postischemic heart [36, 37]. Increased cellular $Ca^{2+}$ occurs as a result of a stimulation of $Na^+$-$Ca^{2+}$ exchange activity subsequent to $Na^+$ pump inhibition [37], and increased $Ca^{2+}$ efflux from the sarcoplasmic reticulum [17]. In a cultured myocyte model of hypoxia, long-chain acylcarnitine concentrations increase 70-fold in the sarcolemma with concomittant depression of cell membrane potentials [33]. These effects were prevented by incubation of the cells with the CPT-I inhibitor, phenylalkyloxirane carboxylic acid (POCA). The effects of palmitoylcarnitine may also be explained by its inhibitory action on cytosolic lysophospholipase [38] thereby permitting lysophosphatidylcholine to accumulate and exert additive electrophysiological effects [38]. Meszaros and Pappano have speculated that palmitoylcarnitine may also affect the voltage and time-dependent $Ca^{2+}$ currents by lengthening the action potential duration and eventually producing cellular $Ca^{2+}$ overload [39]. In addition, the ability of palmitoylcarnitine to reduce the negative surface charge of the sarcolemma appears to shift the voltage-dependent activation of the calcium current to less negative potentials [40]. Alterations in cellular $Ca^{2+}$ metabolism in cardiac ischemia [37] and the role of metabolic amphiphiles in this process have been reviewed in detail by Katz and Messineo [41]. Although the accumulation of long-chain acylcarnitines appears to be contributory to the onset of arrhythmias, other workers have dissociated the cellular accumulation of these compounds from the overall depression in mechanical performance ("stunning") which occurs during ischemic reperfusion [42, 43]. Other studies also in the isolated, perfused rat heart employed a presumptive inhibitor of CAT which prevents mitochondrial, but not cytosolic acylcarnitine accumulation. In the presence of this inhibitor, the accelerating effects of fatty acids on cardiac injury were not observed [44]. Despite an elevation of cytosolic palmitoylcarnitine, no causal effects between this accumulation and ischemic cell injury could be shown.

*Clinical findings in carnitine translocase deficiency*

Conditions which mimic the interrupted fatty acid metabolism of myocardial ischemia are genetic deficiencies in carnitine palmitoyltransferase II and in the carnitine acylcarnitine translocase. Loss in the expression of either of these activities would result in acylcarnitine accumulation, either by an inability to transesterify acylcarnitine to coenzyme A via CPT-II, or by the inhibition of acylcarnitine transfer to the mitochondrial matrix on the translocase, respectively. In the first example, CPT-II deficiency was suspected following the inability of a 3 month-old boy to produce ketone bodies following a long-chain triacylglyceride loading. A greater than 90% CPT-II deficiency, diagnosed in fibroblasts from the patient was associated with elevated plasma creatine kinase, indicating muscle cell damage [45]. Consistent with the experimental findings of electrophysiological abnormalities accompanied by long-chain acylcarnitine accumulation, the affected child demonstrated "premature ventricular complexes, auriculoventricular block, and ventricular tachycardia". Sudden death followed an overnight fast.

The capability of measuring mitochondrial CAT activity in fibroblasts and in whole muscle biopsies from patients was developed by Murthy et al. [46]. By monitoring [$^{14}$C]acetylcarnitine synthesis from [2-$^{14}$C]pyruvate, it is possible to detect possible deficiencies in CAT activity in small samples without the need to isolate mitochondria. Using this methodology, a virtually complete blockage of $\beta$-oxidation in an infant was localized to the relative absence of CAT, but not CPT-I or CPT-II, in the patient's fibroblasts [47]. At early onset of the disorder, the patient exhibited recurrent premature ventricular contractions, tachycardia and hypotension. Before death at 3 years of age, plasma creatine kinase was elevated and mild ventricular hypertrophy was noted. In a related report on a male infant who died 8 days after birth, the deficiency in CAT content was total [48]. Again, oxidation of long-chain fatty acids was almost absent although both CPT-I and CPT-II levels were normal as described in the preceding case. Although not measured, a marked build-up of long-chain acylcarnitines is to be expected. Consistent with this prediction, cardiac abnormalities consisted of a first degree auriculoventricular block and left bundle branch block. It follows that a deficiency in the cardiac isoform of CPT-I should not be associated with electrophysiological abnormalities. Such a specific lesion localized specifically to muscle has not yet been reported. However, a defect in CPT-I expression in the liver is not associated with palmitoylcarnitine accumulation and cardiac involvement is lacking [49, 50]. Therefore, drawing from the body of knowledge concerning the relationship between accumulation of cellular amphiphiles in ischemia, and associated arrhythmias, it has been suggested that neonates and infants presenting with cardiovascular symptoms and cardiac arrest should be considered as candidates for potential genetic defects in long-chain fatty acid oxidation [48].

## Physiological factors as modifiers of CAT activity

Apart from genetic mechanisms which act to alter CAT levels and activity in the mitochondria, cellular factors may also play a role in carrier function. Mitochondria isolated from ischemic heart show markedly depressed oxidation of palmitic acid [51] as well as palmitoylcarnitine [52]. While decreased $\beta$-oxidation in ischemia can be attributed to decreased oxygen tension in the affected tissue, mitochondria are still unable to oxidize long-chain fatty acids at control rates when they are subsequently isolated and exposed to oxygenated medium. Other studies on mitochondria isolated from chronically ischemic dog heart demonstrate unaltered oxidation of medium-chain fatty acid substrates, suggesting a potential defect in the carnitine palmitoyltransferase-translocase system [53]. Changes in the kinetics of both CPT-I and CPT-II have been detected in mitochondria from ischemic heart [54]; however, only a few studies have examined the activity of CAT as a consequence of myocardial ischemia. The observation that the rate constant for liver CAT exceeds that of cardiac mitochondria by 15-fold [21] suggests that any further decrement of CAT activity in heart could potentially limit fatty acid oxidation when fatty acid pressure is high, e.g. during initial stages of ischemic reperfusion.

### *Thiol reduction*

An early characterization of the properties of the translocase demonstrated that the protein activity is very sensitive to thiol modification [55]. Approximately 4 thiol residues appear to be intimately associated with the carnitine binding site since inhibition by the reversible SH inhibitor, mersalyl, is competitive with respect to carnitine [55]. For optimal activity, it appears that CAT must retain these thiols in a reduced state. In the physiological setting, the high intracellular concentration of glutathione (GSH) appears to act as the major intracellular protein disulfide reductant [56]. Although large changes in cellular redox do not occur normally in most tissues, a proposed role for oxidative stress during ischemia and ischemic-reperfusion injury [57] is consistent with the lowered ischemic tissue levels of glutathione [58, 59]. A decrease in mitochondrial matrix glutathione (GSH) correlates with decreased CAT activity following one hour of total circumflex artery occlusion

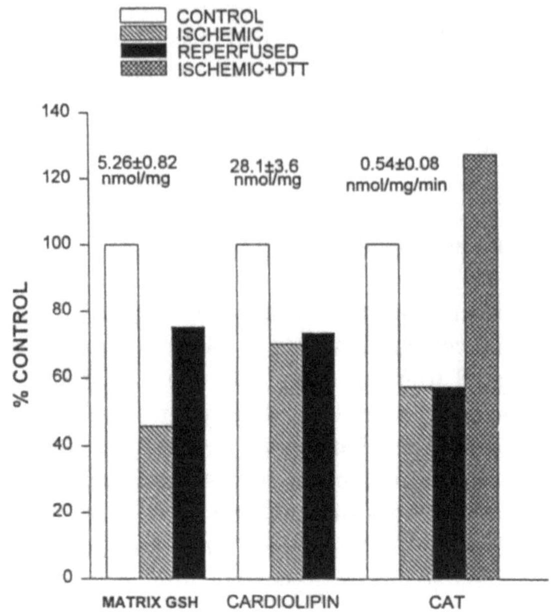


*Figure 1.* The relationship between ischemia-associated changes in heart mitochondria and carnitine acylcarnitine translocase activity. The quantitative values for the control parameters are found on the appropriate control (open) bars. All results are the mean ± SE. Subsequent changes in matrix reduced glutathione (GSH), mitochondrial membrane cardiolipin, and carnitine acylcarnitine translocase (CAT) are expressed as % Control following either sixty minutes left circumflex occlusion (striped bar), occlusion followed by twenty minutes reperfusion (closed bar), or mitochondria isolated from sixty-minute ischemic canine heart and incubated with the thiol reductant, dithiothreitol (DTT, cross-hatched bar). For the GSH measurements, the number of animals are: Control, N = 6; Ischemic, N = 4; Reperfused, N = 5. For cardiolipin determinations: Control, N = 7; Ischemic, N = 6; Reperfused, N = 6, and for carnitine aclycarnitine translocase activities (CAT), the numbers of animals are: Control, N = 4; Ischemic, N =9; and Reperfused, N = 2.

[60] (Figure 1). Protein thiols from mitochondria isolated from ischemic hearts are also significantly decreased from the preocclusion value [54]. Furthermore, incubation of ischemic mitochondria with either GSH or the chemical reducing agent, dithiothreitol, reverses the loss in activity measured (Figure 1). Although similar reducing protocols were not carried out on mitochondrial CAT from reperfused hearts, it was anticipated that the partial restoration of matrix GSH observed on reperfusion (Figure 1) would increase CAT activity, at least slightly. However, CAT activity remains depressed after 20 minutes of reflow [60]. Since the high redox states representative of

tissue GSH/GSSG are more than adequate to maintain a range of enzyme activities in vitro, it is presumed that the normal matrix GSH/GSSG redox also reflects the CAT protein's sulhydryl redox. On the other hand, since cytosolic GSH/GSSG appears to be oxidized further with reperfusion in some studies [58], the extramitochondrial pool of glutathione may play a role in the sustained activity decrease.

*Relevance of membrane phospholipid composition to CAT activity*

Another important cellular factor essential to CAT expression is the phospholipid milieu of the protein. A mixture of phosphatidylcholine, phosphatidylethanolamine and cardiolipin is most effective for reconstitution of octylglucoside-solubilized protein [61]. Cardiolipin itself was shown to be essential for reconstitution of activity as well as for the activity of CAT in intact mitochondria [61]. That myocardial membrane structure and phospholipid composition are altered in ischemia is supported by several reports. First, total concentrations of tissue free linoleic and arachidonic acids in ischemic heart are thought to reflect membrane phospholipase $A_2$ activation (for review, see Van der Vusse et al. [19]). Changes in myocardial $Ca^{2+}$ levels with ischemia as well as peroxidation of polyunsaturated fatty acids by ischemic generation of oxygen free radicals are likely forerunners to the phospholipase $A_2$-mediated excision of the affected fatty acids and consequent membrane phospholipid depletion [19]. Mitochondrial membrane phospholipid composition is also specifically affected by ischemia, presumably from a combination of oxidative stress as exemplified by the significant fall in the GSH/GSSG ratio, and by subsequent mitochondrial $Ca^{2+}$ uptake due to cellular $Ca^{2+}$ overload [62]. In mitochondria isolated from ischemic and ischemic/reperfused hearts, the total phospholipid content is significantly decreased with ischemia and reperfusion from 298.2 ± 15.0 nmol/mg protein in control hearts to 265.4 ± 27.9 and 236.8 ± 16.0 nmol/mg in ischemia and reperfusion, respectively. Of that decrease, a 30% decrease was observed in the mitochondrial content of cardiolipin (Figure 1). No change in the membrane content of phosphatidylcholine and only a 15% decrease in phosphatidylethanolamine was seen with ischemia, suggesting that the lack of change in these phospholipids could be dissociated from effects of ischemia on CAT activity. The lack of return of CAT activity with reperfusion mirrors the decreased cardiolipin content during both ischemia and reperfusion (Figure 1). A similar correlation between decreased CAT activity and mitochondrial cardiolipin content is observed in the aging Fischer rat (30 months) [9]. However, in contrast to ischemic mitochondria where carnitine content is not significantly altered from control, carnitine content in mitochondria from the old rats is diminished by 30% (Table 1 and [9]). This decreased pool of exchangeable carnitine is believed to be the primary cause of decreased exchange activity in the aging rat [8]. Thus, it is difficult to assign any contributory role of cardiolipin to altered CAT activity in this setting. Further studies designed to understand the possible role of phospholipid composition on CAT have been carried out in dogs fed a diet that has been supplemented with menhaden oil [63]. The mitochondrial membrane phospholipids demonstrate a decrease in the mole percent of arachidonic acid with a concomittant increase in eicosapentaenoic and docosahexaenoic acids, increasing the n-3 to n-6 ratio of fatty acids over 3.5-fold with no alteration in the unsaturation index. In this case the only significant change in the mitochondrial membrane phospholipid classes is a 50% *increase* in cardiolipin. Neither palmitoylcarnitine oxidation nor CAT activity (carnitine-carnitine and palmitoylcarnitine-carnitine exchange were measured, Table 1) is affected by this altered lipid environment. However, in mitochondria isolated from ischemic hearts of menhaden oil-fed dogs, the matrix levels of carnitine are decreased by 50%, in contrast to a maintained carnitine content in mitochondria isolated from normal dog hearts made ischemic for a similar period of time (Table 1 and [60]). Because of the decreased carnitine in the mitochondrial matrix of the menhaden oil-fed dogs and the apparent susceptibility of the same membranes to permeability alterations in the presence of palmitoylcarnitine, it was concluded that mitochondrial membranes enriched in n-3 fatty acids may become more labile following ischemia. In this case, the decrease in mitochondrial oxygen uptake with palmitoylcarnitine in ischemia appears to reflect a smaller pool of matrix carnitine available for exchange.

The question still is unsettled as to whether the modifications observed in CAT activity in vitro have a physiological impact on fatty acid oxidation in ischemia. Reports that palmitic acid is the preferred substrate with ischemic reperfusion, supplying over 90% of the energy to the isolated, perfused rat heart do not support the concept that there are any meaningful alterations in CPT or CAT activity in this model [12]. A decrease in CAT activity in the aging rat heart, however, does correlate with an inability of perfusate fatty acid to suppress glucose oxidation. These findings suggest that the senescent heart prefers glucose as an energy substrate, and that this preference might be explained mechanistically by lowered carnitine exchange across the mitochondrial inner membrane. Thus, protocols to augment carnitine content in hearts from old animals would be a valuable approach for protection against ischemic injury in the aging population. There are few relevant studies on human which address this issue. However, in heart muscle biopsies from patients with congestive heart failure and in cardiomyopathic hamster heart, the levels of cardiac free carnitine are decreased [64]. Significant elevations in long-chain acylcarnitines are suggestive of a defect in $\beta$-oxidation distal to CPT-I. Although the diminished CAT activity in mitochondria from ischemic canine heart is not as dramatic as reported in the aging rat model, the rates are sufficiently diminished in vitro when palmitoylcarnitine concentrations that are consistent with the ischemic milieu are added to stimulate exchange activity [54]. This decrease in CAT can account fully for the depressed oxygen consumption in isolated mitochondria with palmitoylcarnitine as substrate. In summary, it is clear that certain pathophysiological conditions can impact on CAT activity in the heart. However, studies using either the rat or swine models of ischemia and reperfusion suggest that under the specific experimental conditions used [12, 27], neither CAT nor any component of the carnitine-mediated pathway appears to exert inhibitory effects on fatty acid flux through $\beta$-oxidation. Therefore, the expression of a limitation in CAT activity relevant to human ischemic heart disease may be found only in those myocardial cells which are severely compromised and thus, may only be part of the end-stage sequellae of injury and cell death.

**Acknowledgement**

The work on the carnitine translocase in ischemia from the author's laboratory is supported by funding from the National Heart, Lung and Blood Institute, HL-RO1 38863 to J.B.M.

**References**

1. McGarry JD, Woeltje KF, Kuwajima M, Foster DW. Regulation of ketogenesis and the renaissance of carnitine palmitoyltransferase. Diabetes/Metab Rev 1989; 5: 271–84.
2. Oram JF, Bennetch SL, Neely JR. Regulation of fatty acid utilization in isolated perfused rat hearts. J Biol Chem 1973; 248: 5299–309.
3. Saddik M, Gamble J, Witters LA, Lopaschuk GD. Acetyl-CoA carboxylase regulation of fatty acid oxidation in the heart. J Biol Chem 1993; 268: 25836–45.
4. Murthy MSR, Pande SV. Characterization of a solubilized malonyl-CoA-sensitive carnitine palmitoyltransferase from the mitochondrial outer membrane as a protein distinct from the malonyl-CoA-insensitive carnitine palmitoyltransferase of the inner membrane. Biochem J 1990; 268: 599–604.
5. Parvin R, Pande SV. Enhancement of mitochondrial carnitine and carnitine acylcarnitine translocase-mediated transport of fatty acids into liver mitochondria under ketogenic conditions. J Biol Chem 1979; 254: 5423–9.
6. Wolkowicz PE, Pauly DF, VanWinkle WB, McMillin JB. Chymotrypsin activates cardiac mitochondrial carnitine-acylcarnitine translocase. Biochem J 1989; 261: 363–70.
7. Idell-Wenger JA. Carnitine: acylcarnitine translocase of rat heart mitochondria. J Biol Chem 1981; 256: 5597–603.
8. Hansford RG. Lipid oxidation by heart mitochondria from young adult and senescent rats. Biochem J 1978; 170: 285–95.
9. McMillin JB, Taffet GE, Taegtmeyer H, Hudson EK, Tate CA. Mitochondrial metabolism and substrate competition in the aging Fischer rat heart. Cardiovasc Res 1993; 27: 2222–8.
10. Opie LH, Owen P, Riemersma RA. Relative rates of oxidation of glucose and free fatty acids by ischaemic and non-ischaemic myocardium after coronary artery ligation in the dog. Eur J Clin Invest 1973; 3: 419–35.
11. Rovetto MJ, Lamberton WF, Neely JR. Mechanisms of glycolytic inhibition in ischemic rat hearts. Circ Res 1975; 37: 742–51.
12. Lopaschuk GD, Spafford MA, Davies NJ, Wall SR. Glucose and palmitate oxidation in isolated working rat hearts reperfused after a period of transient global ischemia. Circ Res 1990; 66: 546–53.
13. Liedtke AJ, Nellis SH, Neely JR, Hughes HC. Effects of treatment with pyruvate and tromethamine in experimental myocardial ischemia. Circ Res 1976; 39: 378–87.

14. Oliver MF. Metabolic response during impending myocardial infarction. II. Clinical implications. Circulation 1972; 45: 491–500.
15. Idell-Wenger JA, Grotyohann LW, Neely JR. Coenzyme A and carnitine distribution in normal and ischemic hearts. J Biol Chem 1978; 253: 4310–8.
16. Wood (McMillin) J, Bush B, Pitts BJR, Schwartz A. Inhibition of bovine $Na^+$, $K^+$-ATPase by palmitylcarnitine and palmityl-CoA. Biochem Biophys Res Commun 1977; 74: 677–84.
17. Pitts BJR, Tate CA, VanWinkle WB, Wood (McMillin) J, Entman ML. Palmitylcarnitine inhibition of the calcium pump in cardiac sarcoplasmic reticulum: A possible role in myocardial ischemia. Life Sci 1978; 23: 391–402.
18. Stanley Jr AW, Moraski RE, Russell Jr RO et al. Effects of glucose-insulin-potassium on myocardial substrate availability and utilization in stable coronary artery disease. Am J Cardiol 1975; 36: 929–37.
19. Van Der Vusse GJ, Glatz JFC, Stam HCG, Reneman RS. Fatty acid homeostasis in the normoxic and ischemic heart. Physiol Rev 1992; 72: 881–940.
20. Schulz H, Racker E. Carnitine transport in submitochondrial particles and reconstituted proteoliposomes. Biochem Biophys Res Commun 1979; 89: 134–40.
21. Pande SV, Parvin R. Carnitine-acylcarnitine translocase catalyzes an equilibrating unidirectional transport as well. J Biol Chem 1980; 255: 2994–3001.
22. Abu-Erreish GM, Neely JR, Whitmer JT, Whitman V, Sanadi DR. Fatty acid oxidation by isolated perfused working hearts of aged rats. Am J Physiol 1977; 232: E258–62.
23. Hansford R, Castro F. Age-linked changes in the activity of enzymes of the tricarboxylic acid cycle and lipid oxidation, and of carnitine content, in muscles of the rat. Mech Aging Dev 1982; 19: 191–201.
24. Idell-Wenger JA, Neely JR. Regulation of uptake and metabolism of fatty acids by muscle. In: Dietschy JM, Gotto AM, Ontko JA, editors. Disturbances in lipid and lipoprotein metabolism. Bethesda: Am Physiol Soc, 1978: 2669–84.
25. Schwaiger M, Schelbert HR, Keen R et al. Retention and clearance of C-11 palmitic acid in ischemic and reperfused canine myocardium. J Am Coll Cardiol 1985; 6: 311–20.
26. Teoh KH, Mickle DAG, Weisel RD et al. Decreased postoperative myocardial fatty acid oxidation. J Surg Res 1988; 44: 36–44.
27. Renstrom B, Nellis SH, Liedke AJ. Metabolic oxidation of glucose during early myocardial reperfusion. Circ Res 1989; 65: 1094–101.
28. Saddik M, Lopaschuk GD. Myocardial triglyceride turnover during reperfusion of isolated rat hearts subjected to a transient period of global ischemia. J Biol Chem 1992; 267: 3825–31.
29. Liedtke AJ, DeMaison L, Eggleston AM, Cohen LM, Nellis SH. Changes in substrate metabolism and effects of excess fatty acids in reperfused myocardium. Circ Res 1988; 62: 535–42.
30. Paulson DJ, Crass III MF. Endogenous triacylglycerol metabolism in diabetic heart. Am J Physiol 1982; 242: H1084–94.
31. Idell-Wenger JA, Grotyohann LW, Neely JR. Regulation of fatty acid utilization in heart. Role of the carnitine-acetyl-CoA transferase and carnitine-acetyl carnitine translocase system. J Mol Cell Cardiol 1982; 14: 413–7.
32. Liedtke AJ, Nellis SH, Mjøs OD. Effects of reducing fatty acid metabolism on mechanical function in regionally ischemic hearts. Am J Physiol 1984; 247: H387–94.
33. Knabb MT, Saffitz JE, Corr PB, Sobel BE. The dependence of electrophysiological derangements on accumulation of endogenous long-chain acyl carnitine in hypoxic neonatal myocytes. Circ Res 1986; 58: 230–40.
34. DaTorre SD, Creer MH, Pogwizd SM, Corr PB. Amphipathic lipid metabolites and their relation to arrhythmogenesis in the ischemic heart. J Mol Cell Cardiol 1991; 23(Suppl 1): 11–22.
35. Adams RJ, Cohen DW, Gupta S et al. In vitro effects of palmitylcarnitine on cardiac plasma membrane Na, K-ATPase, and sarcoplasmic reticulum $Ca^{2+}$-ATPase and $Ca^{2+}$ transport. J Biol Chem 1979; 254: 12404–10.

36. Shen AC, Jennings RB. Kinetics of calcium accumulation in acute myocardial cell injury. Am J Path 1972; 67: 441–52.
37. Morris AC, Hagler HK, Willerson JT, Buja LM. Relationship between calcium loading and impaired energy metabolism during $Na^+$, $K^+$ pump inhibition and metabolic inhibition in cultured neonatal rat cardiac myocytes. J Clin Invest 1989; 83: 1876–87.
38. Corr PB, Snyder DW, Cain ME, Crafford Jr WA, Gross RW, Sobel BE. Electrophysiological effects of amphiphiles on canine Purkinje fibers: Implications for dysrhythmia secondary to ischemia. Circ Res 1981; 49: 354–63.
39. Meszaros J, Pappano AJ. Electrophysiological effects of L-palmitoylcarnitine in single ventricular myocytes. Am J Physiol 1990; 258: H931–8.
40. Inoue D, Pappano AJ. L-Palmitylcarnitine and calcium ions act similarly on excitatory ionic currents in avian ventricular muscle. Circ Res 1983; 52: 625–34.
41. Katz AM, Messineo FC. Lipid-membrane interactions and the pathogenesis of ischemic damage in the myocardium. Circ Res 1981; 48: 1–16.
42. Lopaschuk GD, Wall SR, Olley PM, Davies NJ. Etomoxir, a carnitine palmitoyltransferase I inhibitor, protects hearts from fatty acid-induced ischemic injury independent of changes in long chain acylcarnitine. Circ Res 1988; 63: 1036–43.
43. Ichihara K, Neely JR. Recovery of ventricular function in reperfused ischemic rat hearts exposed to fatty acids. Am J Physiol 1985; 249: H492–7.
44. Hutter JF, Alves C, Soboll S. Effects of hypoxia and fatty acids on the distribution of metabolites in rat heart. Biochim Biophys Acta 1990; 1016: 244–52.
45. DeMaugre F, Bonnefont JP, Colonna M, Cepanec C, Leroux J-P, Saudubray J-M. Infantile form of carnitine palmitoyltransferase II deficiency with hepatomuscular symptoms and sudden death. Physiopathological approach to carnitine palmitoyltransferase II deficiencies. J Clin Invest 1991; 87: 859–64.
46. Murthy MSR, Kamanna VS, Pande SV. A carnitine/acylcarnitine translocase assay applicable to biopsied muscle specimens without requiring mitochondrial isolation. Biochem J 1986; 236: 143–8.
47. Stanley CA, Hale DE, Berry GT, Deleeuw S, Boxer J, Bonnefont J-P. Brief report: A deficiency of carnitine-acylcarnitine translocase in the inner mitochondrial membrane. N Engl J Med 1992; 327: 19–23.
48. Pande SV, Brivet M, Slama A, DeMaugre F, Aufrant C, Saudubray J-M. Carnitine-acylcarnitine translocase deficiency with severe hypoglycemia and auriculo ventricular block. Translocase assay in permeabilized fibroblasts. J Clin Invest 1993; 91: 1247–52.
49. Haworth JC, DeMaugre F, Booth FA et al. Atypical features of the hepatic form of carnitine palmitoyltransferase deficiency in a Hutterite family. J Pediatr 1992; 121: 553–7.
50. Vianey-Saban C, Mousson B, Bertrand C et al. Carnitine palmitoyl transferase I deficiency presenting as a Reye-like syndrome without hypoglycemia. Eur J Pediatr 1993; 152: 334–8.
51. Wood (McMillin) J, Hanley H, Entman ML et al. Biochemical and morphologic correlates of acute experimental myocardial ischemia. III. Energy producing mechanisms during very early ischemia. Circ Res 1979; 44: 52–61.
52. Yoon SB, McMillin-Wood JB, Michael LH, Lewis RM, Entman ML. Protection of canine cardiac mitochondrial function by verapamil-cardioplegia during ischemic arrest. Circ Res 1985; 56: 704–8.
53. Schwartz A, (McMillin) Wood J, Allen JC et al. Biochemical and morphologic correlates of cardiac ischemia. I. Membrane systems. Am J Cardiol 1973; 32: 46–61.
54. Pauly DF, Kirk KA, McMillin JB. Carnitine palmitoyltransferase in cardiac ischemia: A potential site for altered fatty acid metabolism. Circ Res 1991; 68: 1085–94.
55. Pande SV, Parvin R. Characterization of carnitine acylcarnitine translocase system of heart mitochondria. J Biol Chem 1976; 251: 6683–91.
56. Ziegler DM. Role of reversible oxidation-reduction of enzyme thiols-disulfides in metabolic regulation. Annu Rev Biochem 1985; 54: 305–29.

57. Myers ML, Bolli R, Lekich RF, Hartley CS, Roberts R. Enhancement of recovery of myocardial function by oxygen free-radical scavengers after reversible regional ischemia. Circulation 1985; 72: 915–21.
58. Curello S, Ceconi C, Bigoli C, Ferrari R, Albertino R, Guarnieri C. Changes in the cardiac glutathione status after ischemia and reperfusion. Experientia 1985; 41: 42–3.
59. Janssen M, Koster JF, Bos E, De Jong JW. Malondialdehyde and glutathione production in isolated perfused human and rat hearts. Circ Res 1993; 73: 681–8.
60. Pauly DF, Yoon SB, McMillin JB. Carnitine-acylcarnitine translocase in ischemia: Evidence for sulfhydryl group modification. Am J Physiol 1987; 253: H1557–65.
61. Noel H, Pande SV. An essential requirement of cardiolipin for mitochondrial carnitine acylcarnitine translocase activity. Eur J Biochem 1986; 155: 99–102.
62. Beatrice MC, Stiers DL, Pfeiffer DR. Increased permeability of mitochondria during $Ca^{2+}$ release induced by *t*-butylhydroperoxide or oxaloacetate. The effect of ruthenium red. J Biol Chem 1982; 257: 7161–71.
63. McMillin JB, Bick RJ, Benedict CR. Influence of dietary fish oil on mitochondrial function and response to ischemia. Am J Physiol 1992; 263: H1479–85.
64. Kobayashi A, Masumura Y, Yamazaki N. L-Carnitine treatment for congestive heart failure – experimental and clinical study. Jpn Circ J 1992; 56: 86–94.

*Corresponding Author*: Dr Jeanie B. McMillin, Department of Pathology & Laboratory Medicine, University of Texas Medical School, Health Science Center at Houston, 6431 Fannin, Houston, TX 77030, USA

# 7. Amphiphilic interactions of long-chain fatty acylcarnitines with membranes: potential involvement in ischemic injury

JOS M.J. LAMERS

"Altogether, definitive evidence for long chain fatty acylcarnitine as a mediator of impaired contractile function is still lacking, but experimental evidence for arrhythmogenicity of accumulated acylcarnitine seems rather convincing."

## Introduction

Under physiological conditions, the heart preferentially utilizes fatty acids to meet its energy requirements. Non-esterified free fatty acids (NEFA) in plasma are bound to albumin with only small quantities free in solution, in equilibrium with the albumin-bound NEFA [1]. A significant portion of NEFA utilized by the cardiomyocyte originates from hydrolysis by endothelial surface-bound lipoprotein lipase of triacylglycerols of the circulating lipoproteins [2]. Unbound NEFA transverses the sarcolemma (SL) either passively or facilitated by specific membrane proteins such as SL fatty acid binding protein [3]. Like plasma NEFA, cytosolic unbound NEFA exist in equilibrium with a larger quantity of intracellular NEFA bound to fatty acid binding protein [3]. Metabolism of cytosolic NEFA proceeds initially by thioesterification into fatty acyl-CoA esters catalyzed by acyl-CoA synthetase that is localized predominantly on the outer mitochondrial membrane (Figure 1, normoxia). Due to the limited availability of CoA, most of which is contained in the mitochondrial matrix, the fatty acyl-CoA synthetase activity is substrate dependent. Long chain fatty acyl-CoA cannot pass the inner mitochondrial membrane to become degraded by the β-oxidation. The cytosolic and matrix pools of CoA are also strictly separated and the cytosolic long chain fatty acyl-CoA is first converted to long chain fatty acylcarnitine (LCAC) by carnitine fatty acyltransferase I, localized at the inner surface of the mitochondrial outer membrane [4]. LCAC thus generated is translocated across the mitochondrial inner membrane via the specific acylcarnitine-carnitine antiporter (translocase) [5]. Inside the mitochondrial matrix the LCAC is transesterified to yield free carnitine and fatty acyl-CoA by carnitine

*J.W. de Jong and R. Ferrari (eds): The carnitine system*, 83–100.

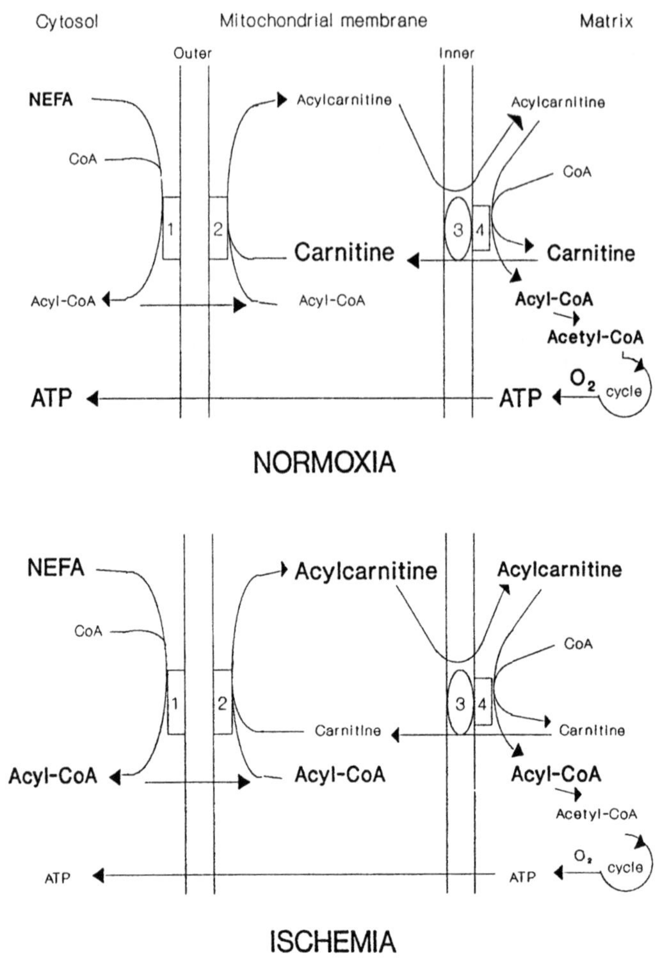


*Figure 1.* Schematical representation of carnitine-linked non-esterified fatty acid oxidation in normoxic and ischemic heart. The size of the letters is meant to symbolize the relative concentration of the metabolites which changes dramatically during myocardial ischemia. The numbers refer to the following enzymes/carriers involved: 1) fatty acyl-CoA synthetase; 2) carnitine fatty acyltransferase I; 3) carnitine acylcarnitinetranslocase; 4) carnitine fatty acyltransferase II. Abbreviations: cycle = tricarboxylic acid cycle, which operates in conjunction with the $O_2$ requiring oxidative phosphorylation to yield ATP from oxidation of acetyl-CoA; NEFA = non-esterified fatty acids.

*Table 1.* Factors determining the extent of accumulation of non-esterified fatty acids (NEFA) and long-chain fatty acylcarnitines (LCAC) in the ischemic heart.

| Factor | Pathogenic mechanism involved |
|---|---|
| Elevated plasma NEFA | Stress-induced fatty acid mobilization from adipose tissue |
| | Impaired cardiac tissue perfusion |
| Elevated intracellular formation of NEFA | Activation of cardiac phospholipases |
| | Activation of cardiac lipases |
| Blocked NEFA utilization | $O_2$ deficiency causing inhibition of cardiac $\beta$-oxidation |

acyltransferase II localized at the inner surface of the mitochondrial inner membrane [3]. Total carnitine content (free and esterified) on each side of the inner mitochondrial membrane remains relatively constant. Complete $\beta$-oxidation of fatty acyl-CoA results in the production of acetyl-CoA fragments with concomitant reduction of $NAD^+$ and FAD. Acetyl-CoA enters the citric acid cycle with production of additional NADH and $FADH_2$. The reducing equivalents formed yield ATP in the mitochondrial electron transport chain tightly coupled to the ATP synthetase.

In the normoxic myocardium NEFA can be partially stored in tissue triacylglycerols by esterification reactions [3]. It is evident that the NEFA fraction that is incorporated in the cardiomyocyte into triacylglycerol and membrane phospholipids, or is oxidized to $CO_2$ changes in response to myocardial energy requirements and availability of circulating NEFA.

When oxygen is limited during myocardial ischemia or hypoxia, cellular NADH and $FADH_2$ accumulate, inhibiting the stepwise degradation of fatty acid by $\beta$-oxidation and diminishing the rate of formation of acetyl-CoA. As a consequence accumulation of intermediates of the pathway of fatty acid oxidation, such as fatty acyl-CoA and LCAC, readily occurs (Table 1; Figure 1, ischemia) [6]. This accumulation of lipid intermediates is further augmented in the presence of excess NEFA which occur in ischemic myocardium, in part by increased intracellular phospholipid and neutral lipid breakdown [3] and in part by increase in serum NEFA which are released by hormonal effects on the peripheral adipose tissue stores (Table 1) [7]. It is the $Ca^{2+}$ dependent phospholipase $A_2$ that is generally believed to partially contribute to the NEFA accumulation during ischemia [3]. There is now abundant evidence that the cellular accumulation of the amphiphilic lipid intermediates NEFA, long chain fatty acyl-CoA, LCAC and lysophospholipid that occur in patients with ischemic heart disease, produce significant abnormalities in cardiac function by modifying the structure and function of membranes. In the last decade several excellent reviews have appeared on the effects of lipid burden on ischemic myocardium [3, 8–13]. The aim of this report will be to focus particularly on membrane disturbing actions of amphiphilic LCAC with emphasis on recent advances in this field.

## Membrane functions, critically involved in myocardial excitability and excitation-contraction coupling

For understanding the potential hazards of interference by amphiphilic LCAC on cardiac membranes, it is important first to consider the critical role of membrane systems in excitability and excitation-contraction coupling of myocardium. In the absence of excitation, cardiac myocytes maintain a resting membrane potential of −80 to −95 mV [14]. The resting potential ($V_r$) arises as a result of the selective permeability of the SL to $K^+$ and the transmembrane $K^+$ gradient, which in turn is maintained by the SL $Na^+$, $K^+$-pump. When an excitatory stimulus depolarizes the membrane beyond the threshold potential (about −70 mV) an action potential is generated which successively causes opening of $Na^+$- and L-type $Ca^{2+}$ channels with concomitant further depolarization of the membrane [14]. Following the peak of the upstroke of the action potential, cardiomyocytes undergo a phase of repolarization to $V_r$ during which $Na^+$ and $Ca^{2+}$ channels close and $K^+$ channels reopen. The $Na^+/K^+$-pump contributes also to repolarization by restoring the $Na^+$ gradient.

Depolarization of the $V_r$ from the normal level by accumulation of extracellular $K^+$ appears to be a prominent feature of ischemia associated with cardiac arrhythmias. The ischemic heart is electrically unstable and more prone to arrhythmias and disturbances in electrical conduction [14–17]. Changes in membrane excitability and cell-to-cell uncoupling through membrane gapjunctions are likely to represent the basic electrophysiological alterations leading to arrhythmias and disturbances of propagation. Moreover, the increase in $\alpha_1$-adrenergic responsiveness has been implicated in the electrical instability of ischemic myocardium as blockade of the $\alpha_1$-adrenoceptors was found to be anti-arrhythmic [18].

Normal systolic and diastolic function of myocardium requires the release and reuptake of $Ca^{2+}$. There are two major $Ca^{2+}$ dependent mechanisms that alter the contractile state of the heart: changing the availability of $Ca^{2+}$ to the contractile proteins and changing the responsiveness of the contractile proteins to activation by intracellular $Ca^{2+}$ [19]. The availability of intracellular $Ca^{2+}$ is regulated by the SL and sarcoplasmic reticulum (SR) membrane systems and the $Ca^{2+}$ responsiveness is controlled by the myofilaments and regulatory troponin-tropomyosin complex. Excitation-contraction coupling is initiated when L-type $Ca^{2+}$ channels are opened and the $Na^+/Ca^{2+}$ exchanger operates in a reversed manner by depolarization of SL, permitting $Ca^{2+}$ to enter the cytosol. This small amount of $Ca^{2+}$ entering induces a release of a much larger quantity of activated $Ca^{2+}$ from the intracellular stores in the SR [20]. The released $Ca^{2+}$ interacts with troponin C, which is a part of the regulatory complex of the myofilaments thereby initiating cardiac contraction. Relaxation starts when $Ca^{2+}$ is sequestered by the SR $Ca^{2+}$ pump, so that $Ca^{2+}$ dissociates from the contractile apparatus. Some $Ca^{2+}$ is also extruded from the cardiomyocyte by the SL $Na^+/Ca^{2+}$ exchanger and the $Ca^{2+}$ pump.

Abnormal handling of intracellular $Ca^{2+}$ at any of the afore-mentioned steps of the excitation-contraction coupling will cause contractile dysfunction. This will occur for instance, when a membrane-active compound slightly increases the passive $Ca^{2+}$ permeability of the membrane. The signalling function of $Ca^{2+}$ demands a very low ionic concentration (during the resting state 10,000 fold lower than outside) of $Ca^{2+}$ inside the myocardial cell. It is only a very small amount of $Ca^{2+}$ entering the cell during each depolarization that needs to be extruded to prevent $Ca^{2+}$ overloading of the myocytes. It is the bulk of $Ca^{2+}$ released from the SR that must be taken up again to its site of origin, the binding sites at calsequestrin inside the SR, in order to be released in the next cardiac cycle. On an integrated level, the SR can be considered as the transport system presiding over the rapid and fine regulation of intracellular $Ca^{2+}$ linked to the contraction/relaxation cycle [21]. For instance, a depression of the SR $Ca^{2+}$ pump will primarily cause relaxation abnormalities but it will secondarily decrease the availability of activator $Ca^{2+}$, resulting in contractile dysfunction.

**In vitro effects of long-chain fatty acylcarnitine on cardiac membranes**

A large body of evidence has indicated that LCAC can alter functional properties of myocardial membranes in vitro (reviewed in [8–12]). Examples of important enzymatic interactions include inhibition of the SL $Na^+/K^+$-pumping ATPase [22, 23] and $Na^+/Ca^{2+}$ antiporter [23, 24] and the SL [23] and SR $Ca^{2+}$-pumping ATPase [24, 25]. Inhibition of $Na^+$, $K^+$-ATPase was observed when LCAC was added to detergent-(pre)treated SL membranes whereas no effect was seen with intact SL membrane vesicles [22, 23, 26]. These findings may indicate that $Na^+$/ $K^+$-ATPase is not a target for LCAC interaction in vivo. In this regard the recent work on voltage-clamped intact single guinea pig ventricular myocytes should obtain attention. The specific $Na^+/K^+$-pump inhibitor ouabain and 1–5 μM L-palmitoylcarnitine reversibly suppressed and reduced the rate constant for the decline of a transient outward $Na^+/K^+$ antiport current [27]. This result is consistent with the observed suppression by L-palmitoylcarnitine of $Na^+/K^+$-ATPase in SL vesicles isolated from cardiac muscle [22, 23, 26]. Moreover, we and others showed that palmitoylcarnitine at relatively low concentrations also enhances the passive $Ca^{2+}$ and $Na^+$ permeability of SL and SR membrane vesicles [23, 25]. Although LCAC has been found to be a potent inhibitor of $Na^+/K^+$-ATPase in detergent-treated SL vesicles, the micromolar concentrations required to inhibit the enzyme appeared to be higher than those needed to alter the passive $Ca^{2+}$ and $Na^+$ permeability of the intact SL membrane vesicles (Figure 2) [23]. We could show by studying the latency of major

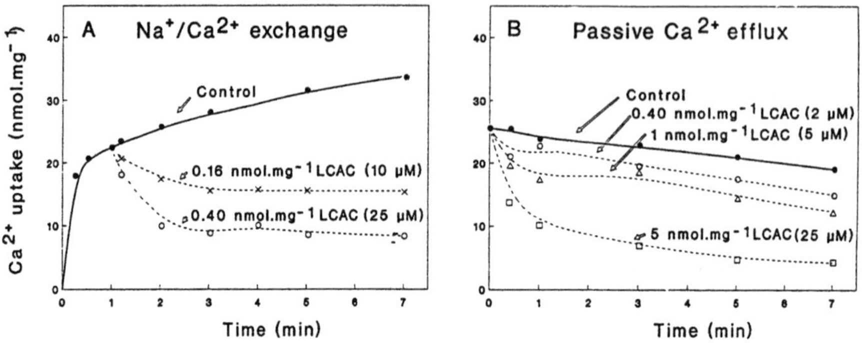


*Figure 2.* Palmitoylcarnitine (LCAC) effects on the $Na^+/Ca^{2+}$ exchanger activity and the passive $Ca^{2+}$ permeability of isolated cardiac sarcolemmal vesicles. Sarcolemmal (SL) vesicles were isolated from pig heart by the procedure described in [23]. Prior to the experiment the SL vesicles were loaded with 160 mM NaCl in 20 mM 4–morpholinepropanesulfonic acid (Mops), pH = 7.4. For measuring $Na^+/Ca^{2+}$ exchange or antiport activity (*A*) a 10 μl aliquot (12.5 μg SL protein) was rapidly diluted into 200 μl 160 mM KCl in 20 mM Mops (pH = 7.4) containing 50 μM $^{45}Ca$ (0.1 Ci/mmol) to which various concentrations of L-palmitoylcarnitine (10 and 25 μM LCAC) were added. The 21–fold downhill $Na^+$ gradient drives the uphill $Ca^{2+}$ transport into the SL vesicles via the $Na^+/Ca^{2+}$ antiporter. For measuring the passive $Ca^{2+}$ transport (*B*), a 50 μl aliquot (6.25 μg) of sarcolemmal $Na^+/Ca^{2+}$ antiport reaction after 2 min (see above) was rapidly diluted into 1200 μl 160 mM KCl, 100 μM EGTA, 20 mM Mops (pH = 7.4) and various concentrations of L-palmitoylcarnitine (2, 5 and 25 μM LCAC). Note that the concentration of membrane protein in the $Ca^{2+}$ uptake reaction mixtures used in A was 12.5 higher than in passive $Ca^{2+}$ efflux reaction mixtures used in B. It means that, in spite of its presence at the same absolute concentration (μM), L-palmitoylcarnitine (LCAC) concentration expressed in nmol · $mg^{-1}$ protein differed by a factor 12.5 comparing reactions A and B. This difference could not be avoided due to the necessary extra dilution step (into KCl and EGTA containing medium) of SL vesicles to assure the pure measurement of passive $Ca^{2+}$ efflux.

cyclic AMP-dependent protein phosphorylation sites (e.g. 15 and 9 kDa proteins) of SL that palmitoylcarnitine did increase the SL vesicle permeability for $Na^+$ and $Ca^{2+}$ ions but not for small molecules such as ATP and heat-stable inhibitor of cyclic AMP-dependent protein kinase [23]. The observed inhibition of $Na^+/Ca^{2+}$ exchange by LCAC may be indirectly caused by increases in membrane $Ca^{2+}$ and $Na^+$ ion permeabilities [23, 24]. This conclusion is based on the results on SL vesicles isolated from porcine heart as illustrated in Figure 2. $Ca^{2+}$ accumulated by $Na^+/Ca^{2+}$ exchange activity was released by L-palmitoylcarnitine (LCAC) depending on its concentration (10 and 25 μM). These results demonstrate that the mechanism of action of LCAC cannot only be ascribed to inhibition of the $Na^+/Ca^{2+}$ antiporter. Apparently, there is also a marked effect of LCAC on the $Ca^{2+}$

permeability. The conditions of no external free $Ca^{2+}$ level, created by the presence of 100 μM ethylene glycol bis (β-aminoethylether)-N,N,N′,N′-tetraacetic acid (EGTA), and of the absence of external $Na^+$, excluded the contribution of the $Na^+/Ca^{2+}$ antiporter activity to the measured $Ca^{2+}$ efflux. This $Ca^{2+}$ efflux can therefore be seen as a pure passive unidirectional process (Figure 2B). Under this condition already the lowest concentration of LCAC (2 μM) increased the rate of $Ca^{2+}$ efflux during the first 2 min. Others [25] demonstrated that the extent of inhibition of the $Ca^{2+}$ pump in isolated SR vesicles mostly depended on the ratio of LCAC to SR membrane protein. It was also shown that almost 90% of the LCAC added to an SL vesicle suspension was bound to an LCAC concentration of 200 to 1500 nmol.$mg^{-1}$ SL protein [26]. Therefore, in the experiments described and illustrated in Figure 2 not only the absolute concentrations of LCAC but also the membrane-associated LCAC · $mg^{-1}$ protein should be taken into account in comparing LCAC effects of $Na^+/Ca^{2+}$ antiporter activity and passive $Ca^{2+}$ efflux rate. If this consideration is correct, it means that the inhibitory effect of LCAC on $Na^+/Ca^{2+}$ exchanger cannot be completely ascribed to an increase of SL $Ca^{2+}$ permeability. In [23] we demonstrated that LCAC also increased $Na^+$ permeability of SL vesicles which might explain why the $Na^+/Ca^{2+}$ antiporter is blocked by LCAC. The observed effects of LCAC on $Ca^{2+}$ and $Na^+$ permeability of SL membrane raise the question by which mechanism(s) transmembrane passive ion diffusion is facilitated by LCAC. Likely targets for LCAC action are ion-specific channels. L-Palmitoylcarnitine affected the fast $Na^+$ channel and slow $Ca^{2+}$ channel in avian ventricular muscle strips and single ventricular myocytes [28, 29]. Concentrations of 30–300 μM L-palmitoylcarnitine influenced $Na^+$ and $Ca^{2+}$ channel operation by an effect on membrane surface charge recorded by action potential measurements. Both L-and D-palmitoylcarnitine can mimic the actions of the specific slow $Ca^{2+}$ channel activator, Bay K8644, and can counteract the effects of $Ca^{2+}$ entry blockers on chick embryonic myocytes [30]. It is unlikely that the LCAC effects on passive $Na^+$ and $Ca^{2+}$ efflux, observed in isolated SL vesicles, are related to the LCAC effects seen on action potentials or L-type $Ca^{2+}$ channel activity in intact cells. For instance, voltage-operated $Ca^{2+}$ channels are kept in the inactivated closed state [31, 32] in isolated SL vesicles. Depolarization can be elicited in vitro in SL vesicles inducing the uptake of $Ca^{2+}$, but the characteristics of this $Ca^{2+}$ entry (e.g. sensitivity to inhibition by $Ca^{2+}$ entry blockers) are not alike those of $Ca^{2+}$ entry through L-type $Ca^{2+}$ channels [31,32]. Moreover, recently it was shown in whole cell voltage-clamp procedures in isolated adult guinea pig ventricular myocytes that LCAC [1–25 μM] suppresses rather than stimulates the voltage-dependent L-type $Ca^{2+}$ channel current [16]. In these experiments it was shown that LCAC induces afterdepolarizations and triggered activity which could contribute to the arrhythmogenic effects during ischemia [16]. Insertion of various amphiphilic compounds in the outer leaflet of the SL lipid bilayer produces shifts in current-voltage relationship and inactivation curves of L-type $Ca^{2+}$ channel current. Some of these effects can be explained by changes in the actual $V_r$ of the membrane [33].

Low concentrations of LCAC stimulate the SL and SR $Ca^{2+}$-ATPase, whereas higher concentrations inhibit [23,25]. Lower concentrations of palmitoylcarnitine stimulate $Ca^{2+}$-ATPases likely by uncoupling the active $Ca^{2+}$ uptake process from the ATP hydrolysis through an increase of the passive $Ca^{2+}$ permeability of the membrane vesicles. At present it is still unknown if the micromolar concentration of long chain acylcarnitines tested in the in vitro experiments are indeed reached during ischemia in the heart. From values reported by others for the intact heart it was calculated that the cytoplasmic content of LCAC in normal and ischemic heart is 150 and 780 nmol/g wet weight, respectively [6]. It is important to recognize that the affinity of LCAC for the membrane bilayer is such that there will be very little free of this acylcarnitine either within the cell or in in vitro SL vesicle suspensions. The ratio of palmitoylcarnitine to SR protein was calculated by assuming the presence of about 3 mg SR/g wet weight of heart and a contribution of 60% of the total cellular membrane area by the network SR [25]. Based upon these assumptions membrane concentrations of 30 and 160 nmol LCAC per mg SR protein in normal and ischemic hearts were calculated, respectively. If these estimations are true for SR and SL it is evident from the previous in vitro studies on isolated membrane vesicles [22–25] that LCAC concentrations in the ischemic heart become high enough to affect significantly the passive $Ca^{2+}$ and $Na^+$ permeability of SL and SR (compare also the effective LCAC concentrations in the in vitro studies on isolated SL vesicles as illustrated in Figure 2). Alternatively, if the total cellular phospholipid pool is taken as the main dissolving compartment into which LCAC distributes (see also later), it can be calculated that accumulation of 780 nmol LCAC/g wet weight equals to 1 nmol LCAC/25 nmol phospholipid. This is close to the molar ratios of LCAC per membrane phospholipid that are found to be effective in the in vitro studies on isolated membrane vesicle [22–25].

The question arises by which mechanisms LCAC alters the passive ion permeability, apart from its possible direct effects on $Na^+$ and $Ca^{2+}$ channels, and ion pump activities of cardiac membranes. At relatively low concentrations, below the cricital micelle concentration (13 $\mu$M according to [34]), LCAC exist in solution as monomers that can be inserted into the hydrophobic environment of the lipid membrane. It has been suggested that insertion of "wedge shaped" molecules can result in an abnormal shape or curvature of the membrane and thereby alter membrane fluidity [10]. These "wedge shaped" structures can also disrupt SL by their effect on overall membrane topography [8]. LCAC-induced fluidity changes in canine myocardial SL [35, 36] and human erythrocyte membranes [37] have been demonstrated using the spin-label technique. Calcium and L-palmitoylcarnitine reduce erythrocyte electrophoretic mobility which was ascribed to an alteration in membrane surface charge [38]. Another team of investigators has elucidated the transbilayer reorientation (flip-flop) of LCAC in erythrocytes after addition of 3 nmol labelled LCAC/1000 nmol membrane phospholipid [39]. These authors showed that LCAC is incorporated into the membrane in alignement with the head groups and the acyl chains of the membrane phospholipids, and reorient from one leaflet of the bilayer to the other by a slow process (2.6 h at 37°C), termed "flip rate" [39]. This means that during cytosolic accumulation of LCAC during ischemia in vivo the inner leaflet of the SL will be overloaded with LCAC molecules which slowly "flip" to the outer bilayer leaflet. The incorporation of amphiphilic molecules into the membrane may not only modify membrane functions by fluidity change but also directly influence integral membrane protein functions by altering the composition of the boundary lipids (or lipid annulus) of the particular membrane-bound enzyme, G-protein modulated ion channel, membrane receptor protein or transport carrier protein [8]. At a high concentration of LCAC (>13 μM), the monomers aggregate into micelles in which the hydrophilic region of the molecule remains in contact with the aqueous medium and the lipophilic portions are clustered into the hydrophobic core. On the one hand these micelles may act as a reservoir continuously providing monomers for insertion [10]. These LCAC micelles also have the ability to incorporate membrane phospholipid into their structure, thereby forming mixed micelles [8]. The latter LCAC effect will lead to partial disruption of the membrane bilayer, liberating endogenous membrane phospholipids. In fact, it may represent a detergent-like action destroying the integrity of the membrane by mixed micelle formation. The detergent-like action, of LCAC could be responsible for the changes in passive $Ca^{2+}$ and $Na^{+}$ ion permeability, as we and others observed them [23, 25].

## Other useful criteria for evaluating the potential role of long-chain fatty acylcarnitine in ischemia

The data on isolated membranes reviewed above indicate that LCAC-induced loss of the ability of myocardial membranes to act as a permeability barrier to $Ca^{2+}$ and $Na^{+}$ plays a significant role in the pathogenesis of cell death in the ischemic or hypoxic heart. However, definite conclusions cannot be drawn from results of these in vitro membrane preparations. In evaluating the potential importance of LCAC in producing abnormalities of ischemic myocardium, the following criteria, as previously outlined for candidate metabolites in general [35], become particularly useful:

1. Increase or decrease of LCAC accumulation by an exogenously administered compound with respective parallel exacerbating or ameliorating loss of function, arrhythmias, increase of $\alpha_1$-adrenergic responsiveness or

lactate dehydrogenase (LDH) release despite induction of comparably severe ischemia.
2. Demonstration of a time-course relationship between accumulation of LCAC and the development of abnormal function or tissue damage (e.g. contractile dysfunction, arrhythmias, $\alpha_1$-adrenergic responsiveness and enzyme leakage).
3. Localization of the accumulated LCAC to a cellular site where it most likely interacts with certain membrane(proteins) during ischema.
4. Induction of biological effects comparable to those seen during ischemia by exposure of normal myocardium to exogenous LCAC at concentrations comparable to those encountered in the vicinity of sarcolemma in vivo.

During the past decade numerous studies on the possible role of LCAC during ischemia have been carried out, in each of which the evaluation of the effects of LCAC was based upon one of the criteria 1, 3 or 4. Below we will summarize these experimental data. It is curious to see that no investigations on the potential role of LCAC during ischemia are available that were based on the second criterium. This is likely due to the fact that the myocardial LCAC concentration increases immediately after the onset of ischemia, rapidly reaching a steady state level. At least this is what we observed in porcine myocardium during coronary ligation [40]. The level of LCAC increases with myocardial ischemia and remains elevated for hours even after reperfusion of the ischemic heart [41–43]. It is postulated that a decrease in sensitivity of carnitine fatty acyl-CoA transferase I to malonyl-CoA inhibition would blunt the regulatory role of malonyl-CoA and result in increased quantities of LCAC synthetized from any given concentration of malonyl-CoA [44]. Therefore elevation in the activity of carnitine fatty acyl-CoA transferase I or alteration in its regulation by malonyl-CoA may keep the LCAC levels high. Criterium 2 seems not applicable to define the influence of accumulated LCAC on cardiac performance. Apparently, a certain period of time is required for the amphiphilic interactions of LCAC with membranes to become deteriorating. In the previous in vitro studies on isolated SL membranes [22–26] or intact single cardiomyocytes [27–30], however, no indications were given for any need of long-lasting preincubation of the membrane- or cell preparation with exogenous LCAC.

## Inhibitors of specific steps in fatty acid metabolism as tools to investigate the role of long-chain fatty acylcarnitine

One of the strategies to investigate the potential role of LCAC accumulation in ischemia was the use of several naturally occurring and synthetic compounds which selectively block either carnitine acyltransferase I and II on the outer- and inner surfaces of the mitochondrial membrane, respectively (Table 2). Representatives include the carnitine acyltransferase I inhibitors,

*Table 2.* Inhibitors of enzymatic reactions involved in fatty acyl-CoA metabolism.

| Enzymatic reaction | Inhibitors [References] | Likely consequence during ischemia |
|---|---|---|
| Carnitine fatty acyltransferase I | POCA* [17, 45, 50, 56, 57, 59]<br>Etomoxir* [46]<br>Oxfenicine* [47–49, 51]<br>TDGA* [47–49] | LCAC accumulation is decreased |
| Carnitine fatty acyltransferase II | Aminocarnitine [54]<br>D-Octanoylcarnitine [53] | LCAC accumulation is increased |

* Abbreviations: POCA = sodium 2-(5-(4-chlorophenyl)pentyl)-oxirane-2-carboxylate); etomoxir = ethyl-2-[6-(4-chlorophenoxy)hexyl] oxirane-2-carboxylate; oxfenicine = 2-tetradecylglycidic acid; TDGA = tetradecylglycidic acid.

POCA, etomoxir, oxfenicine and TDGA. POCA and the even more potent analogue of POCA, etomoxir, are esterified to CoA derivatives in the cytosol. The thioesters strongly bind to carnitine fatty acyltransferase I, thereby inhibiting the enzyme [45, 46]. As a consequence, the ischemia-induced accumulation of LCAC will be reduced. Oxfenicine is transaminated to 4-hydroxyphenylglyoxylate and this metabolite inhibits fatty acid oxidation at the level of carnitine fatty acyltransferase I [47]. Several studies have appeared which demonstrate that agents blocking the carnitine fatty acyltransferase I can protect ischemic myocardium, using recovery of mechanical function during reperfusion as a parameter of beneficial action. Oxfenicine and TDGA were shown to selectively reduce the LCAC level in intact myocardium. The trend was noted toward improved mechanical performance in aerobic and ischemic perfused heart muscle [47–49]. Others demonstrated protective effects of POCA [50] and oxfenicine [51] in in situ pig hearts. In contrast, yet other groups [45, 46, 52] failed to demonstrate a protective effect of carnitine acyltransferase I inhibitors. In one of the latter studies the investigators looked at the effect of a low ($10^{-9}$ M) and high ($10^{-6}$ M) dose of etomoxir on fatty acid-induced ischemic injury [46]. It was demonstrated that etomoxir indeed can prevent the fatty acid-induced failure of ischemic hearts. However, unlike the other studies recovery of heart contractility was not correlated with the attenuation of accumulation of LCAC. Thus a dissociation was found between LCAC accumulation and post-ischemic recovery. A decrease in oxygen consumption per unit heartwork, due to an increase in glucose utilization, was postulated to account for the beneficial action of etomoxir [46]. More recently it was confirmed that administration of POCA does not affect functional recovery and LDH release, but results in about 2–fold increase of NEFA levels upon reperfusion as compared to glucose-perfused hearts [45]. The accumulation of NEFA could be explained by assuming that the oxidation of NEFA released from endogenous lipid pools is inhibited by POCA [45]. In the latter study the tissue levels of LCAC were not measured; hence here the role of LCAC in ischemic injury cannot be evaluated. In conclusion, further studies are required to unravel the role of these inhibitors in myocardial ischemia.

Carnitine fatty acyl-CoA transferase II inhibitors D-octanoyl- [53] and aminocarnitine [54] might also be interesting tools to investigate the role of LCAC accumulation. During normoxic Langendorff perfusion of rat heart with aminocarnitine, LCAC accumulates in heart cells, from which it is excreted. Heart function remaines intact during that process [54]. It appeared that LCAC had no membrane toxicity when acidosis was absent and coronary flow ensured continuous removal of LCAC. This need not be so when LCAC accumulates during ischemia when the coronary flow is zero and acidosis is present due to lactate accumulation [54]. At present no studies are available on the influence of carnitine fatty acyl-CoA transferase II inhibitors during ischemia.

**Subcellular accumulation sites of long-chain fatty acylcarnitine indicative for the mechanism of injurious action**

Another approach for studying the role of LCAC is to characterize the subcellular distribution of LCAC under normoxic and ischemic conditions. This can reveal the preferential accumulation sites and so the likely loci of their deteriorating actions during myocardial ischemia. One group of investigators produced a series of reports describing the subcellular distribution of [$^3$H]-carnitine(derivatives) in monolayer cultures of neonatal rat cardiomyocytes and collagenase-dissociated adult canine cardiomyocytes [17, 18, 55–59]. Homogeneous cell preparations were used to overcome the difficulty of discrimination between LCAC present in myocardial and non-myocardial cellular elements. In one of their first reports, processing for electronmicroscopy of neonatal rat cardiomyocytes, prelabelled with [$^3$H]-carnitine, is described by a procedure specifically developed for selective extraction of endogenous tritiated short-chain and free carnitine but retention of endogenous tritiated LCAC [56]. (The authors used the term "endogenous tritiated LCAC" to describe tracer [$^3$H]-carnitine incorporated into cellular membranes, "exogenous" for unincorporated [$^3$H]-carnitine.) In normoxic-perfused cells, [$^3$H]-LCAC was concentrated in mitochondria and cytoplasmic membranous components. Only very small amounts were present in the SL. Hypoxia increased mitochondrial [$^3$H]-LCAC 10-fold and SL [$^3$H]-LCAC 70-fold. After 60 min of hypoxia, SL contained even $1.4 \times 10^7$ LCAC molecules/$\mu m^3$ of membrane volume, a value corresponding to approximately 3.5% of total SL phospholipid [56]. In this regard, it is interesting to note that these amounts are close to the molar ratios of LCAC per membrane phospholipid that were found to be effective in the in vitro studies on isolated membrane vesicles [22–26]. Hypoxia of the rat cardiomyocytes also produced electrophysiological abnormalities, such as decreased maximum diastolic potential, action potential amplitude and maximum upstroke velocity of phase zero [56]. Furthermore, the carnitine fatty acyltransferase I inhibitor POCA inhibited accumulation of [$^3$H]-LCAC in each subcellular compartment and prevented the depression of electrophysiological function induced by hypoxia. Recently, the same group has confirmed that a subcellular distribution pattern of accumulated LCAC similar to that in neonatal rat cardiomyocytes was found in adult canine cardiomyocytes exposed to 10–20 min hypoxia [58]. $\alpha_1$-Adrenergic receptor number has been shown to increase markedly with ischemia in cat, dog and guinea pig hearts. The increased $\alpha_1$-adrenergic responsivity is generally believed to be causally associated with the high incidence of ventricular fibrillation and frequence of premature ventricular complexes during coronary artery occlusion [18, 57]. A procedure was developed for measuring $\alpha_1$-adrenergic receptors in isolated, $Ca^{2+}$-tolerant adult canine cardiomyocytes. Myocytes exposed to only 10 min hypoxia at 37°C exhibited a two- to threefold increase in $\alpha_1$-adrenoceptor number; this correlated well with a threefold increase of LCAC (normoxia *vs* ischemia: 21 *vs* 66 pmol·mg$^{-1}$ protein, see [18]). Again, pretreatment with POCA completely abolished not only the increase of LCAC induced by hypoxia, but also prevented the increase in $\alpha_1$-adrenergic receptor number. Moreover, exogenous palmitoylcarnitine (1 μM), in the presence or absence of POCA (10 μM), led to a significant increase in $\alpha_1$-adrenergic receptor density. Accumulation of LCAC in SL membrane with subsequent alterations of SL fluidity and the microenvironment surrounding the $\alpha_1$-adrenoceptors was postulated as the likely mechanism for the increased exposure of the receptors, by unmasking latent receptors closely associated with the sarcolemma [18, 57]. Protein synthesis of new receptors and subsequent transfer and externalization to the surface of SL could be excluded as these obviously require an extended period of time, at least substantially more than 10 min duration of hypoxia or exposure to exogenous LCAC [57]. Two recent studies demonstrated that SL accumulation of LCAC during ischemia might be responsible for the inhibition of gapjunctional conductance leading to cellular uncoupling [17, 59]. A change in cell-cell coupling likely represents the basic electrophysiological alteration leading to disturbances in propagation. Also here, POCA (10 μM) as well as oxfenicine (100 μM) pretreatment did not only prevent the LCAC accumulation but also markedly delayed the secondary decrease in conduction velocity normally observed in response to ischemia [17].

We used a radioisotope procedure to determine LCAC concentrations in subcellular fractions of porcine myocardium that had been subjected to different periods of ischemia (0, 1, 2 and 3 h) [40]. In myocardial tissue from non-ligated hearts LCAC concentrations were 0.3 and 1.5 nmol · mg$^{-1}$ protein in homogenate and SL enriched membrane fractions respectively, findings in agreement with a preferential membrane localization of LCAC observed by electronmicoscopy/autoradiography before [17, 18, 55–59]. Both the total and SL localized LCAC were increased twofold after 2 h of ischemia. However, the accumulated LCAC was not correlated temporally with the decrease in the $Ca^{2+}$ pumping activity, increase of $Na^+/Ca^{2+}$ antiporter activity and decrease of $Ca^{2+}$ permeability measured in the isolated membrane preparations [40]. In the same study it was shown that in vitro incubation of isolated SL membranes with [$^{14}$C]-palmitoylcarnitine, labeled molecules (clearly identified as [$^{14}$C]-palmitoylcarnitine) remained associated with the SL membranes even after repeated washing. The absolute amounts (up to about $6\,nmol \cdot mg^{-1}$ SL protein) of incorporated [$^{14}$C]-palmitoylcarnitine were in the same range as endogenously incorporated LCAC in SL that was isolated from ischemic porcine myocardium [40]. However, no changes were observed in the intrinsic $Na^+/Ca^{2+}$ antiporter activity of the SL vesicles when in vitro 6 nmol [$^{14}$C]-palmitoylcarnitine per mg protein was bound to or incorporated into the membrane lipid phase. At any rate, these results suggested that the intracellular increase in LCAC during almost zero myocardial flow is not critical to sarcolemmal $Na^+$ and $Ca^{2+}$ permeability and $Ca^{2+}$ pumping activity.

## Effects of exposure of normoxic myocardium to exogenous long-chain fatty acylcarnitine

If LCAC accumulation alone is responsible for injurious action on membranes during ischemia, it should be possible to evoke these effects by exogenous LCAC in normoxic myocardium. One investigation on the hypoxia-induced increase in $\alpha_1$-adrenergic receptor number, proving this principle, has already been discussed [57].

In another study rat hearts were perfused by the Langendorff technique; cellular release of myoglobin (as an index of SL damage) was induced in a dose-dependent manner by exogenous palmitoylcarnitine in concentrations exceeding 1.6 μM in the perfusion solution [60]. The presence of albumin in the perfusion solution could prevent this LCAC effect. A concentration threshold between 1 and 5 μM exogenous palmitoylcarnitine for inducing hypercontracture was observed before by us using a similar model [61]. It should, however, be noted that a membrane stabilizing effect was found by perfusing rat hearts with 1 μM palmitoylcarnitine [61]. In summary, relatively low concentrations are necessary to cause SL disruption when the cells are allowed to accumulate LCAC from an external supply [60].

## Concluding remarks

This comprehensive analysis of literature data supporting or rejecting the potential role of amphiphilic interactions of accumulated LCAC during myocardial ischemia could help redirect many scientists who have carried out experimental studies in this field. Altogether, definitive evidence for LCAC as a mediator of impaired contractile function is still lacking, but experimental evidence for arrhythmogenicity of accumulated LCAC seems rather convincing. Most of the studies focussed on the function of the cardiomyocyte(membranes), perhaps in vain. Ischemia-induced dysfunction of vascular endothelium causes loss of the ability to regulate vasomotor tone in the coronary vasculature and deterioration of myocardial damage in the ischemic area. In this context, very recent work showing accumulated LCAC to be a mediator of impaired vascular endothelial function, should receive attention [62]. The information in this review might also be of help in the search for therapeutical interventions aimed at attenuation of LCAC accumulation to ameliorate ischemic injury.

## References


1. Spector AA, Fletcher JE, Ashbrook JD. Analysis of long-chain free fatty acid binding to bovine serum albumin by determination of stepwise equilibrium constants. Biochemistry 1971; 10: 3229–32.
2. Stam HCG, Schoonderwoerd K, Breeman W, Hülsmann WC. Effects of hormones, fasting and diabetes on triglyceride lipase activities in rat heart and liver. Horm Metab Res 1984; 16: 293–7.
3. Van der Vusse GJ, Glatz JFC, Stam HCG, Reneman RS. Fatty acid homeostasis in the normoxic and ischemic heart. Physiol Rev 1992; 72: 881–935.
4. Murthy MSE, Pande SV. Malonyl-CoA binding site and the overt carnitine palmitoyltransferase activity reside on the opposite sides of the outer mitochondrial membrane. Proc Natl Acad Sci USA 1987; 84: 378–82.
5. Wolkowicz PE, Pownall HJ, McMillin-Wood JB. (1–Pyrenbutyryl)carnitine and 1-pyrenebutyryl coenzym A: Fluorescent probes for lipid metabolite studies in artificial and natural membranes. Biochemistry 1982; 21: 2990–6.
6. Whitmer JT, Idell-Wenger JA, Rovetto MJ, Neely JR. Control of fatty acid metabolism in ischemic and hypoxic hearts. J Biol Chem 1978; 253: 4305–9.
7. Oliver MF, Kurien VA, Greenwood TW. Relation between serum-free-fatty-acids and arrhythmias in death after acute myocardial infarction. Lancet 1968; 1: 710–4.
8. Katz AM, Messineo FC. Lipid-membrane interactions and the pathogenesis of ischemic damage in the myocardium. Circ Res 1981; 48: 1–16.
9. Lamers JMJ, Stinis JT, Montfoort A, Hülsmann WC. Modulation of membrane function by lipid intermediates: a possible role in myocardial ischemia. In: Ferrari R, Katz AM, Shug AL, Visioli O, editors. Myocardial ischemia and lipid metabolism. New York: Plenum Publishing Corporation, 1984: 107–25.
10. Corr PB, Gross RW, Sobel BE. Amphipathic metabolites and membrane dysfunction in ischemic myocardium. Circ Res 1984; 55: 135–54.
11. Van der Vusse GJ, Prinzen FW, Van Bilsen, Engels W, Reneman RS. Accumulation of lipids and lipid-intermediates in the heart during ischaemia. Basic Res Cardiol 1987; 82(Suppl 1): 157–67.
12. Liedtke AJ. Lipid burden in ischemic myocardium. J Mol Cell Cardiol 1988; 20(Suppl II): 65–74.
13. Fritz IB, Arrigoni-Martelli E. Sites of action of carnitine and its derivatives on the cardiovascular system: interactions with membranes. Trends Pharmacol Sci 1993; 141: 355–60.
14. Ten Eick RE, Whalley DW, Rasmussen HH. Connections: heart disease, cellular electrophysiology and ion channels. FASEB J 1992; 6: 2568–80.
15. DaTorre SD, Creer MH, Pogwizd SM, Corr PB. Amphipathic lipid metabolites and their

relation to arrhythmogenesis in the ischemic heart. J Mol Cell Cardiol 1991; 23(Suppl I): 11–22.
16. Wu J, Corr PB. Influence of long-chain acylcarnitines on voltage-dependent calcium current in adult ventricular myocytes. Am J Physiol 1992; 263: H410–7.
17. Yamada KA, McHowat J, Yan G-X, Donahue K, Peirick J, Kléber, Corr PB. Cellular uncoupling induced by accumulation of long-chain acylcarnitine during ischemia. Circ Res 1994; 74: 83–95.
18. Corr PB, Yamada KA, DaTorre SD. Modulation of $\alpha$-adrenergic receptors and their intracellular coupling in the ischemic heart. Basic Res Cardiol 1990; 85(Suppl 1): 31–45.
19. Morgan JP. Abnormal intracellular modulation of calcium as a major cause of cardiac contractile dysfunction. N Engl J Med 1991; 325: 625–31.
20. Stern MD, Lakatta EG. Excitation-contraction coupling in the heart: state of the question. FASEB J 1992; 6: 3092–100.
21. Carafoli E. The homeostasis of calcium in heart cells. J Mol Cell Cardiol 1985; 17: 203–12.
22. McMillin Wood J, Bush B, Pitts BJR, Schwartz A. Inhibition of bovine heart $Na^+,K^+$-ATPase by palmitylcarnitine and palmityl-CoA. Biochem Biophys Res Commun 1977; 74: 677–84.
23. Lamers JMJ, Stinis JT, Montfoort A, Hülsmann WC. The effect of lipid intermediates on $Ca^{2+}$ and $Na^+$ permeability and $(Na^+ + K^+)$-ATPase of cardiac sarcolemma. A possible role in myocardial ischemia. Biochim Biophys Acta 1984; 774: 127–37.
24. Adams RJ, Cohen DW, Gupte S, Johnson JD, Wallick ET, Wang T, Schwartz A. In vitro effects of palmitylcarnitine on cardiac plasma membrane Na,K-ATPase, and sarcoplasmic reticulum $Ca^{2+}$-ATPase and $Ca^{2+}$ transport. J Biol Chem 1979; 254: 12404–10.
25. Pitts BJR, Tate CA, Van Winkle B, Wood JM, Entman ML. Palmitylcarnitine inhibition of the calcium pump in cardiac sarcoplasmic reticulum: a possible role in myocardial ischemia. Life Sci 1978; 23: 391–402.
26. Owens K, Kennett FF, Weglicki WB. Effects of fatty acid intermediates in $Na^+,K^+$-ATPase activity. Am J Physiol 1982; 242: H456–61.
27. Tanaka M, Gilbert J, Pappano AJ. Inhibition of sodium pump by L-palmitoylcarnitine in single guinea pig ventricular myocytes. J Mol Cell Cardiol 1992; 24: 711–20.
28. Inoue D, Pappano AJ. L-Palmitoylcarnitine and calcium ions act similarly on excitatory ionic currents in avian ventricular muscle. Circ Res 1983; 52: 625–34.
29. Meszaros J, Pappano AJ. Electrophysiological effects of L-palmitoylcarnitine in single ventricular myocytes. Am J Physiol 1990; 258: H931–8.
30. Patmore L, Duncan GP, Spedding M. Interaction of palmitoyl carnitine with calcium antagonists in myocytes. Br J Pharmacol 1989; 97: 443–50.
31. Philipson KD. "Calciductin" and the voltage-sensitive calcium uptake. J Mol Cell Cardiol 1983; 15: 867–9.
32. Lamers JMJ. Calcium transport systems in cardiac sarcolemma and their regulation by the second messengers cyclic AMP and calcium-calmodulin. Gen Physiol Biophys 1985; 4: 143–54.
33. Post JA, Kenneth SJ, Langer GA. Effects of charged amphiphiles on cardiac cell contractility are mediated via effects on $Ca^{2+}$ current. Am J Physiol 1991; 260: H759–69.
34. Piper MH, Sezer O, Schwartz P, Hütter JF, Schweickhardt C, Spieckermann PG. Acylcarnitine effects on isolated cardiac mitochondria and erythrocytes. Basic Res Cardiol 1984; 79: 186–98.
35. Corr PB, Gross RW, Sobel BE. Arrhythmogenic amphiphilic lipids and the myocardial cell membrane. J Mol Cell Cardiol 1982; 14: 619–26.
36. Fink KL, Gross RW. Modulation of canine myocardial sarcolemmal membrane fluidity by amphiphilic compounds. Circ Res 1984; 55: 585–94.
37. Kobayashi A, Watanabe H, Fujisawa S, Yamamoto T, Yamazaki N. Effects of L-carnitine and palmitoylcarnitine on membrane fluidity of human erythrocytes. Biochim Biophys Acta 1989; 986: 83–8.
38. Meszaros J, Villanova L, Pappano AJ. Calcium ions and L-palmitoyl carnitine reduce

erythrocyte electrophoretic mobility: test of a surface charge hypothesis. J Mol Cell Cardiol 1988; 20: 481–92.
39. Classen J, Deuticke B, Haest CWM. Nonmediated flip-flop of phospholipid analogues in the erythrocyte membrane as probed by palmitoylcarnitine: basic properties and influence of membrane modification. J Membr Biol 1989; 111: 169–78.
40. Lamers JMJ, De Jonge-Stinis JT, Verdouw PD, Hülsmann WC. On the possible role of long chain fatty acylcarnitine accumulation in producing functional and calcium permeability changes in membranes during myocardial ischaemia. Cardiovasc Res 1987; 21: 313–22.
41. Ichihara K, Neely JR. Recovery of ventricular function in reperfused ischemic rat hearts exposed to fatty acids. Am J Physiol 1985; 249: H492–7.
42. Liedtke AJ, DeMaison L, Eggleston A, Cohen LM, Nellis SH. Changes in substrate metabolism and effects of excess fatty acids in reperfused myocardium. Circ Res 1988; 62: 535–42.
43. Reeves RC, Evanochko WT, Canby RC, McMillin JB, Pohost GM. Demonstration of increased myocardial lipid with post-ischemic dysfunction ("myocardial stunning") by proton nuclear magnetic resonance spectroscopy. J Am Coll Cardiol 1989; 13: 739–44.
44. Pauly DF, Kirk KA, McMillin JB. Carnitine palmitoyltransferase in cardiac ischemia. A potential site for altered fatty acid metabolism. Circ Res 1991; 68: 1085–94.
45. Van Bilsen M, Van der Vusse GJ, Willemsen PHM, Coumans WA, Reneman RS. Fatty acid accumulation during ischemia and reperfusion: Effects of pyruvate and POCA, a carnitine palmitoyltransferase I inhibitor. J Mol Cell Cardiol 1991; 23: 1437–47.
46. Lopaschuk GD, Wall SR, Olley PM, Davies NJ. Etomoxir, a carnitine palmitoyltransferase I inhibitor, protects hearts from fatty acid-induced ischemic injury independent of changes in long chain acylcarnitine. Circ Res 1988; 63: 1036–43.
47. Liedtke AJ, Nellis SH, Mjos OD. Effects of reducing fatty acid metabolism on mechanical function in regionally ischemic hearts. Am J Physiol 1984; 247: H387–94.
48. Liedtke AJ, Nellis SH, Neely JR. Effects of excess free fatty acids on mechanical and metabolic function in normal and ischemic myocardium in swine. Circ Res 1978; 43: 652–61.
49. Miller WP, Liedtke AJ, Nellis SH. Effects of 2–tetradecylglycidic acid on myocardial function in swine hearts. Am J Physiol 1986; 251: H547–52.
50. Paulson DJ, Noonan JJ, Ward KM, Stanley H, Sherratt A, Shug AL. Effects of POCA on metabolism and function in ischemic rat heart. Basis Res Cardiol 1986; 81: 180–7.
51. Molaparast-Saless F, Liedtke AJ, Nellis SH. Effects of fatty acid blocking agents, oxfenicine and 4–bromocrotonic acid, on performance in aerobic and ischemic myocardium. J Mol Cell Cardiol 1987; 19: 509–20.
52. Seitelberger R, Kraupp O, Winkler M, Brugger G, Raberger G. Effect of the acylcarnitine-transferase blocking agent sodium 2[5-(4-chlorophenyl)pentyl]-oxirane-2-carboxylate (POCA) on metabolism and regional function in the underperfused myocardium. J Cardiovasc Pharmacol 1985; 7: 273–80.
53. McGarry JD, Foster DW. Studies with (+)-octanoylcarnitine in experimental diabetic ketoacidosis. Diabetes 1974; 23: 485–93.
54. Hülsmann WC, Schneijdenberg CTWM, Verkleij AJ. Accumulation and excretion of long-chain acylcarnitine by rat hearts; studies with aminocarnitine. Biochim Biophys Acta 1991; 1097: 263–9.
55. Knabb MT, Ahumada CG, Sobel BE, Saffitz JE. A fixation procedure suitable for autoradiography of endogenous long-chain acyl carnitine. J Histochem Cytochem 1985; 33: 744–8.
56. Knabb MT, Saffit JE, Corr PB, Sobel BE. The dependence of electrophysiological derangements on accumulation of endogenous long-chain acyl carnitine in hypoxic neonatal rat myocytes. Circ Res 1986; 58: 230–40.
57. Heathers GP, Yamada KA, Kanter EM, Corr PB. Long-chain acylcarnitines mediate the hypoxia-induced increase in $\alpha_1$-adrenergic receptors on adult canine myocytes. Circ Res 1987; 61: 735–46.
58. McHowat J, Yamada KA, Saffitz JE, Corr PB. Subcellular distribution of endogenous long

chain acylcarnitines during hypoxia in adult canine myocytes. Cardiovasc Res 1993; 27: 1237–43.
59. Wu J, McHowat J, Saffitz JE, Yamada KA, Corr PB. Inhibition of gap junctional conductance by long-chain acylcarnitines and their preferential accumulation in junctional sarcolemma during hypoxia. Circ Res 1993; 72: 879–89.
60. Busselen P, Sercu D, Verdonck F. Exogenous palmitoyl carnitine and membrane damage in rat hearts. J Mol Cell Cardiol 1986; 20: 905–16.
61. Hülsmann WC, Dubelaar M-L, Lamers JMJ, Maccari F. Protection by acyl-carnitines and phenylmethylsulfonyl fluoride of rat heart subjected to ischemia and reperfusion. Biochim Biophys Acta 1985; 847: 62–6.
62. Inoue N, Hirata K-I, Akita H, Yokoyama M. Palmitoyl-L-carnitine modifies the function of vascular endothelium. Cardiovasc Res 1994; 28: 129–34.

*Corresponding Author:* Professor Jos M.J. Lamers, Department of Biochemistry I, Cardiovascular Research Institute COEUR, Erasmus University P.O. Box 1738, 3000 DR Rotterdam, The Netherlands

# 8. Mitochondrial injury in the ischemic-reoxygenated cardiomyocyte: the role of lipids and other pathogenic factors

HANS MICHAEL PIPER, THOMAS NOLL
and BERTHOLD SIEGMUND

"It seems questionable whether the interference of amphiphilic long-chain acyl derivatives with mitochondrial functions plays an important causal role for postischemic dysfunction or the onset of irreversibility in ischemic-reperfused myocardium."

## Introduction

In human pathology, the most frequent cause for myocardial oxygen deficiency is regional ischemia, caused by a partial or complete occlusion of the supplying coronary artery. In the past 15 years four theories have been predominantly discussed that relate the onset of irreversibility to changes in the cells' biochemical state: (i) a critical energy loss; (ii) a critical accumulation of cellular calcium; (iii) deleterious effects of accumulation of long-chain acyl compounds; or (iv) the effects of free radical formation. These theories do not represent mutually exclusive alternatives, each of them may accentuate a certain part of the pathophysiological process leading to irreversible myocardial injury. Here, selected aspects of the second and third theory are discussed.

In hypoxic and ischemic myocardial tissue, long-chain fatty acyl-CoA and carnitine esters were found to quickly accumulate even in the absence of exogenous fatty acids [1]. Most of the accumulating long-chain acyl derivatives are hydrolyzed from triglycerides and from membrane phospholipids [2]. Exogenous fatty acids, being the main fuel for the myocardium under aerobic conditions, are disadvantageous under oxygen deprivation since their presence further augments the concentration of long-chain acyl esters in the myocardial cell [3, 4] (Figure 1). In addition to these esters a number of other amphiphilic long-chain acyl derivatives also accumulate in ischemic myocardium [5, 6]. The accumulation of lipids as free amphiphiles and the degradation of lipids from cellular membranes may contribute to the progression of injury. Both the detergent effect of the amphiphilic compounds and the loss of constituent phospholipids can alter the barrier and

*J.W. de Jong and R. Ferrari (eds): The carnitine system*, 101–121.

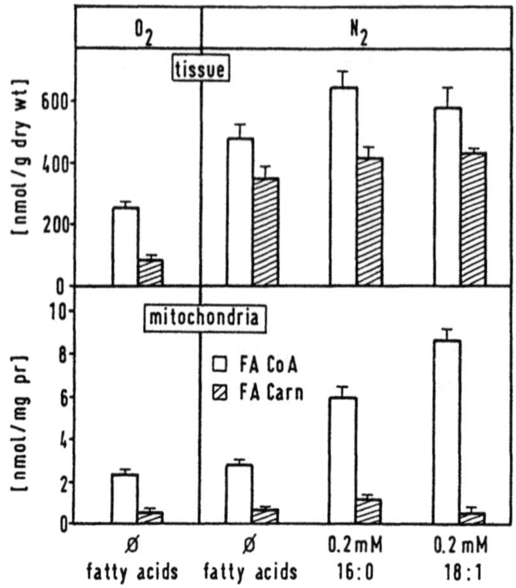


*Figure 1.* Tissue and mitochondrial contents of long-chain acyl coenzyme A (FA CoA) and carnitine (FA Carn) esters in ischemic hearts. Langendorff-hearts (guinea pig) were perfused for 60 min either with an $O_2$-equilibrated saline perfusion fluid (flow 10 ml/g wet weight per min) or an $N_2$-equilibrated saline perfusion fluid (flow 0.3 ml/g wet weight per min). Mitochondria were isolated at the end of perfusion. The anoxic perfusion fluid contained no fatty acid, palmitic (16:0) or oleic acid (18:1) (complexed in 1:5 molar ratio to albumin). Means ± SD, n = 5. pr = protein.

transport functions of cellular membranes and lead, ultimately, to their physical destruction (reviews [7, 8]). The potential harm of the amphiphiles relates to their chemical nature, e.g. the type of the long-chain acyl moiety [3, 4, 9, 10].

Because of the central role of mitochondria in myocardial energy metabolism, numerous studies have addressed their pathophysiological role in ischemia-reperfusion and other conditions of oxygen depletion and reoxygenation. This review is focused on the injury of mitochondria in oxygen depleted and reoxygenated myocardium. Parts of this paper were published recently in another form [11]. Various defects have been found in mitochondria isolated from ischemic or hypoxic myocardial tissue. Such defects, if not reversible, may limit the recovery of mitochondrial function upon reoxygenation. The present knowledge of mitochondrial pathophysiology is mainly based on three types of experimental approaches: Experiments in which mitochondrial functions are inferred indirectly from the metabolism of whole myocardium; experiments in which mitochondrial functions are investigated in single cardiomyocytes; and experiments in which the functional properties of mitochondria are investigated when isolated from tissue or cells under predetermined experimental conditions. In the latter experiments mitochondrial functions are analyzed in greatest detail. It is however, unclear to what extent mitochondria after isolation continue to represent their previous functional state in the living myocardial cell.

The effects of intracellularly accumulating amphiphilic lipids on the development of mitochondrial functional defects may be direct or indirect. If the noxious agents are harmful for the cell as a whole this can cause indirect effects also on mitochondrial integrity, e.g. as a consequence of an increase in cytosolic $Ca^{2+}$ load. It must be noted that it is in general difficult to decide whether mitochondrial damage developing in tissue concomitantly with an increase in free lipids is influenced by these in a direct or indirect way.

## Mitochondrial functional failure in the oxygen deprived cell

Due to variations of collateral flow, the residual oxygen supply and consequently the residual oxygen concentration varies in ischemic myocardium. The metabolic response of oxygen-depleted myocardium as a whole represents only a statistical average. In the core of severely ischemic tissue the concentration of oxygen may fall to very low levels. As yet it is generally not known how rapidly the residual oxygen content of ischemic tissue further declines once it has reached a partial pressure of about 1 mm Hg, the practical limit of most techniques to measure oxygen in tissue [12]. A number of factors may oppose a rapid progression of oxygen depletion beyond this level: First, even in regionally ischemic tissue with a complete occlusion of the supplying vessel oxygen supply is not zero, due to some collateral flow and the mere diffusion of oxygen from outside into the ischemic core. Second, the more the oxygen tension falls the more the oxygen consumption of myocardial cells declines, due to the Michaelis-Menten-type kinetics of mitochondrial respiration. From the studies on isolated cardiomyocytes it can be extrapolated that at $pO_2 \leqslant 0.1$ mm Hg the oxygen consumption is reduced $\geqslant$100-fold compared to the normoxic control levels [13]. The combination of these factors should prevent the oxygen concentration in regionally ischemic tissue from falling virtually indefinitely. In the ischemic myocardium this may have the implication that the oxygen supply is too low to satisfy the cellular demand for oxidative phosphorylation but still sufficient for a low rate of electron flux which preserves the polarized state of the inner mitochondrial membrane.

When myocardial cells become deprived of oxygen due to a stop of perfusion (ischemia) or to lack of oxygen in the perfusate (hypoxic perfusion), the respiratory generation of ATP is slowed down until it (virtually) ceases. Even though myocardial cells respond to the lack of oxygen with a pronounced acceleration of glycolytic flux, this anaerobic mechanism of energy production usually remains insufficient to fully compensate for the loss of respiratory ATP. The balance of energy therefore soon becomes negative and the cellular stores of high-energy phosphates are progressively depleted.

The mitochondrial $F_1$,$F_0$-proton ATPase is a multicomponent enzyme complex which reversibly catalyzes synthesis and hydrolysis of ATP. Under normal respiratory conditions the transport of electrons within the inner mitochondrial membrane builds up a membrane potential and a gradient of protons ("proton electrochemical gradient"). The energy ("protonmotive force", pmf) stored in the transmembrane distribution of charges and protons is used to drive the synthesis of ATP from ADP and $P_i$, which represents an "uphill" reaction in terms of energy. This reaction is catalyzed by the $F_1$ ATPase/synthase part of the $F_1$,$F_0$-proton ATPase. In depolarized mitochondria as well as in non-polarized submitochondrial membrane fractions the pmf is zero. This enzyme complex then catalyzes the net hydrolysis of ATP. In terms of energy, net ATP hydrolysis may occur whenever the free energy of ATP hydrolysis in the matrix space, $\Delta G_{ATP}$, exceeds the free energy stored in the electrochemical gradient, $\Delta G_{pmf}$. The reversal of net ATP synthesis, i.e. ATP hydrolysis, can nevertheless be activated whenever the electrochemical gradient is somewhat reduced [14]. This may occur in ischemic or hypoxic myocardial cells when the electron flow approaches zero.

The maximal rate of mitochondrial ATP hydrolysis is so large that, theoretically, it could degrade all ATP contained in a myocardial cell with a half-time of about 100 ms [15]. Accordingly, an extremely rapid depletion of cellular ATP reserves can be observed when mitochondria are depolarized by use of very effective chemical uncouplers, i.e. agents increasing the permeability of the inner mitochondrial membrane for protons. Oligomycin is a specific inhibitor of the mitochondrial ATPase/synthase. With the use of this inhibitor it has been demonstrated [16, 17] that in the ischemic dog heart mitochondrial ATP hydrolysis contributes to the progression of energy loss, but at a much slower rate than theoretically possible. This has been attributed to the beneficial action of the natural inhibitor protein $IF_1$ contained within the matrix space, which binds to the $F_1$ component of the ATPase complex when mitochondria become depolarized [18–20]. Acidotic conditions, as found in ischemia, favor the binding of the inhibitor protein and thus reduce the activity of mitochondrial ATP hydrolysis [16]. The binding of this protein is reversible. When mitochondria are re-polarized it can dissociate from its binding site [19] and may therefore not inhibit net ATP synthesis in re-oxygenated and re-polarized mitochondria. It is not known how rapid the process of dissociation is in the whole cell; it may occur with some delay [21, 22]. It has also been suggested that under normoxic conditions a variable portion of the ATPase/synthase units are inhibited by binding of the inhibitor protein, in correspondence to the energy demand of the myocardial cell [23].

Rouslin et al. [17] found that activation of mitochondrial ATP hydrolysis can contribute to energy depletion in ischemic myocardium (from dog). They hypothesized the following mechanism of activation [17, 24]: When the blood flow to the cardiac muscle is interrupted, the tissue is rapidly depleted of oxygen. Cessation of mitochondrial electron flow then leads to dissipation of the mitochondrial transmembrane electrochemical gradient and activation of net ATP hydrolysis within the mitochondria. This hypothesis is supported by the finding in isolated mitochondria that the forward reaction of mitochondrial ATP synthase is reduced whereas its backward reaction (hydrolysis) is greatly stimulated when electron flux is low [14]. As argued above, it is an open question however, whether cessation or pronounced depression of electron flux is sufficient to cause net ATP hydrolysis in the living ischemic cell. Activation of glycolysis and the net production of lactate is often considered to indicate the stop of mitochondrial respiration. It only indicates, however, that oxidative phosphorylation has decreased to an extent insufficient to meet the cellular energy demand. In a study by our own group using isolated cardiomyocytes it was demonstrated [13] that the myocardial cell may fully activate glycolysis and yet retain a polarized mitochondrial state when the ambient oxygen pressure is reduced one-thousandfold below the normal arterial level, i.e. $\leqslant$0.1 mm Hg. Even at this very low level of ambient oxygen some net synthesis of ATP took place. The activation of mitochondrial net ATP hydrolysis observed in ischemic tissue may alternatively be due to the accumulation of chemical factors such as lipids (and others, see below) which alter the permeability of the inner mitochondrial membrane. When the rate of proton pumping is low a relatively subtle increase in permeability can cause a substantial mitochondrial depolarization. In vitro long-chain acyl compounds have been found to inhibit and ultimately uncouple mitochondrial respiration if present at high concentrations [9, 10]. In the oxygen-deprived cells with minimal electron flux much lower concentrations may be sufficient to induce mitochondrial depolarization. The experiments shown in Figure 2 demonstrate that in deep hypoxia depolarization of mitochondrial membranes can cause a substantial difference in the rate of ATP loss.

## Structural injury of mitochondria isolated from ischemic and reperfused tissue

In ischemic myocardium, mitochondria as well as other cell structures become progressively injured and, under certain conditions, reoxygenation may even aggravate this damage. Since mitochondrial functions cannot be investigated in detail in experiments with whole cells or tissue, many studies have investigated in vitro the properties of mitochondria isolated from myocardium in various pathophysiological states. Results from such studies are reviewed in this section.

In mitochondria isolated from ischemic tissue, components of the respiratory chain are progressively lost or denatured (Figure 3). Among the cyto-

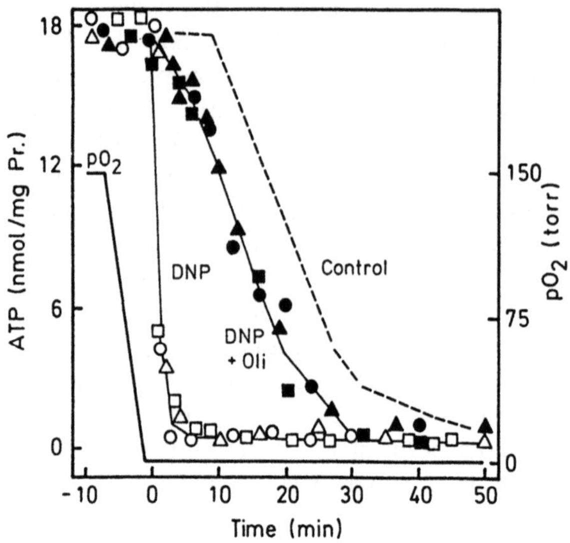


*Figure 2.* Effect of mitochondrial depolarization on the rate of ATP depletion in cardiomyocytes (from adult rat) under deep hypoxia. Cardiomyocytes were incubated in substrate-free saline medium in which the $PO_2$ was reduced to $\leq 0.1$ torr. The time course of ATP loss under hypoxic control conditions is indicated by the broken line. When the depolarizing agent 2,4-dinitrophenol (DNP; 50 μM) was added at time 0, i.e. 1 min after $PO_2 \leq 0.1$ torr had been reached, the ATP loss was greatly accelerated (open symbols). This acceleration could be antagonized by simultaneous addition of oligomycin (20 μM), a specific inhibitor of the $F_1$ ATPase (closed symbols). Data of 3 experiments (symbolized by squares, circles and triangles). Pr = protein. (From [13], with permission.)

chromes [25], the loss is most pronounced for cytochrome c, known to be only loosely bound at the outer surface of the inner mitochondrial membrane, But integral components of the membrane structure, such as cytochromes $aa_3$, are also extensively lost. In low-flow ischemic hearts perfused with fatty acids, a characteristic loss of cytochrome b was observed [26].

NADH reductase (complex I of the respiratory chain) is impaired early in hypoxic and ischemic myocardium [25–30]. Since it is particularly sensitive to acidosis [28], the oxidation of NAD-dependent substrates is reduced most distinctly in mitochondria from ischemic tissue. The presence of exogenous fatty acids in the ischemic myocardium aggravates the loss of function (Figure 4). This early impairment, however, is still a reversible phenomenon [25]. Subsequently, the functions of coenzyme Q-reductase (complex III) and of succinate dehydrogenase (complex II) also become affected so that both the oxidation of NAD- and FAD-dependent substrates are reduced [25]. The activity of the mitochondrial ATP synthase is found most reduced in ischemia

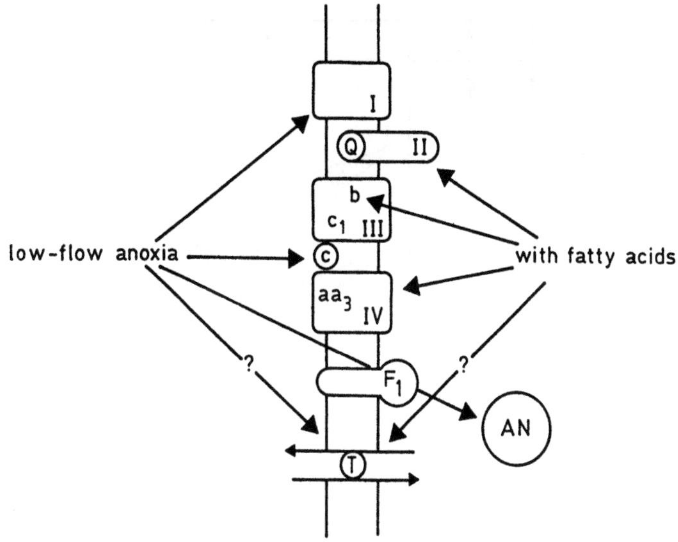


*Figure 3.* Schematic representation of the components of the respiratory chain and the major points of injury when hearts are perfused in low flow anoxia (5% of normal flow) with addition of fatty acids. Abbreviations: I to IV, complexes I to IV; b, $c_1$, c, $aa_3$, cytochromes; Q, coenzyme Q; $F_1$, $F_1$ ATPase/synthase; T, adenine nucleotide translocator; AN, mitochondrial content of adenine nucleotides. (From [3], with permission.)

with pronounced acidosis [18], due to an enhanced binding of the inhibitor binding protein (see above). Other parts of the respiratory apparatus are only slightly affected, e.g. the activity of cytochrome c oxidase (complex IV) [25, 29].

Parallel to the loss of adenine nucleotides from the whole tissue, a loss of adenine nucleotide contents can also be determined in mitochondria isolated from ischemic tissue [25, 31, 32]. The mechanism of this loss is not clear. The mitochondrial matrix is devoid of the enzymes catalyzing the breakdown of adenine nucleotides to nucleosides. In normal polarized mitochondria the mitochondrial pool of adenine nucleotides remains constant, since export of ATP and import of ADP mediated by the adenine nucleotide translocator strictly follows a 1:1 stoichiometry [33, 34]. However, under certain conditions a net release of adenine nucleotides can occur, e.g. when the extramitochondrial phosphate concentration increases [35]. The release seems to be the consequence of a permeability transition of the inner mitochondrial membrane (see below).

Synthesis of ATP by the mitochondria is of use to the reoxygenated cell only if the ATP produced can also be exported to the cytosol, i.e. if the

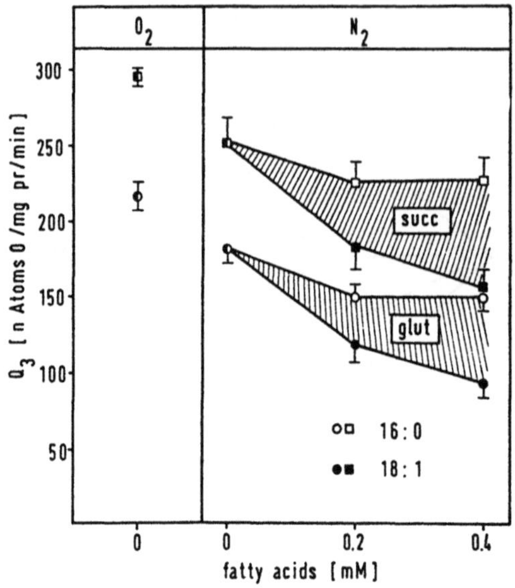


*Figure 4.* Presence of fatty acids in ischemic hearts accelerates the decline in NAD-dependent (substrate glutamate) and FAD-dependent (substrate succinate) phosphorylating state 3 respiration. Mitochondria were isolated from oxygenated or anoxic perfused Langendorff-hearts (guinea pig). Conditions as for Figure 1. The anoxic perfusate contained no fatty acid, palmitic acid (16:0) or oleic acid (18:1) (complexed in 1:5 molar ratio to albumin). Means ± SD, n = 5. glut = glutamate; pr = protein; succ = succinate.

adenine nucleotide translocator is still functioning. In mitochondria isolated from ischemic myocardium the activity of this carrier was found impaired [25–38]. This may be due to the reduction in intramitochondrial adenine nucleotide concentrations, falling below the $K_m$ of the inner binding site for adenine nucleotides of the carrier [32]. It has also been hypothesized that increased levels of long-chain coenzyme A (CoA) esters in the ischemic tissue are responsible for the reduction of activity [36, 39–41]. This hypothesis is based on the finding that long-chain acyl-CoA accumulates in the mitochondrial matrix when $\beta$-oxidation stops [1]. Using inner mitochondrial membranes turned inside-out (submitochondrial particles) it has been demonstrated that long-chain acyl-CoA indeed inhibits the translocase from its matrix side [42]. It is however, unclear whether free concentrations of these compounds ever reach inhibitory levels in mitochondria of ischemic cells. This is because the intracellular distribution of these amphiphilic compounds cannot be exactly determined. In the cell, long-chain acyl compounds bind not only to mitochondrial membranes but also to other cell structures and soluble proteins, and there is no established way to preserve this distribution in cell fractionation [43, 44].

In mitochondria from ischemic hearts not only the function, but also the number of adenine nucleotide translocator proteins decreases, either by loss from the membrane or by denaturation. A close correspondence was found between the recovery of activity and the remaining amount of non-denatured protein of the adenine nucleotide translocator [25]. Thus in ischemia the mitochondrial content of adenine nucleotides and the activity of the adenine nucleotide translocator decrease. This may not be very important for the myocardial cells as long as oxygen remains absent, but it can limit the ability of the mitochondria to produce and export ATP when they become reoxygenated.

## Hyperpermeability of the inner mitochondrial membrane

The inner membrane of energized cardiac mitochondria is normally permeable to small solutes only in a very selective manner. A number of conditions [15, 45–47] that cause a large increase in permeability of the inner membrane are however known. It then becomes permeable in a non-selective manner to solutes up to a molecular weight of about 1000, including adenine and pyridine nucleotides. Hyperpermeability of the inner membrane leads to rapid dissipation of the electrochemical potential and activation of mitochondrial net ATP hydrolysis if the latter is not prevented by the binding of the inhibitor protein (see above).

In vitro, mitochondrial hyperpermeability can be induced by a large number of conditions (many are listed in [46]). For example, the exposure of respiring mitochondria to high extra-mitochondrial concentrations of $Ca^{2+}$ plus inorganic phosphate [15, 46] can induce mitochondrial hyperpermeability. It has been demonstrated that the rise in the concentration of ionized $Ca^{2+}$ within the matrix space is the decisive factor. A rise of $Ca^{2+}$ alone is however not sufficient; additional factors are needed to induce hyperpermeability.

Besides a high concentration of inorganic phosphate, several other factors or complex conditions can also induce hyperpermeability in the presence of a high extramitochondrial $Ca^{2+}$ concentration [44]. Among these "inducing factors" are: a depletion of mitochondrial adenine nucleotides [37–50] (see above); atractyloside blocking the nucleotide binding side of the adenine nucleotide translocator in outward orientation [51, 52]; long-chain acyl-CoA [53, 54]; lysophospholipids [49, 55]; pro-oxidants [52, 55–57]; and oxidation and hydrolysis of the mitochondrial pyridine nucleotides [57, 58].

The mechanism(s) causing the permeability transition is (are) known only partially. It may be due to the opening of a defined, non-selective proteinaceous "pore" (estimated diameter approximately 2 nm) or "megachannel" [59] in the inner mitochondrial membrane, a hypothesis first proposed by

Haworth and Hunter [45]. The concept of a pre-existent structure of the inner mitochondrial membrane which can open and close is supported by the rapidity of induction and reversibility of the hyperpermeable state. Thus the effect of a rise of extramitochondrial $Ca^{2+}$ (in the presence of phosphate) is rapidly reversed by addition of $Ca^{2+}$ chelators [60]. The fact that hyperpermeability can be blocked by a single agent, cyclosporin A [61], also argues for this concept. The fact that atractyloside, a blocker of the outward-oriented binding side of the adenine nucleotide translocator, can modulate a permeability transition [51] has led to speculations that the "pore" may represent the translocator in a special conformational state. The molecular nature of the hypothetical "pore" has not yet been identified.

The group of Pfeiffer [46] has favored the concept that a lipid phase alternation causes the permeability transition. The main support for this hypothesis comes from the finding that conditions favoring a degradation of membrane phospholipids are also inducers of hyperpermeability whereas conditions stabilizing phospholipids, e.g. inhibitors of the mitochondrial $Ca^{2+}$-sensitive phospholipase $A_2$, antagonize the induction of hyperpermeability [46]. At present it seems unclear whether there are two different mechanisms leading to a similar end-result of permeability transition. In the following we will discuss the permeability transition in terms of the "pore" theory only. Since "pores" have not been identified at the molecular level, this is only a semantic decision in the context of this review.

For a pathophysiological interpretation of the phenomenon of the permeability transition, the roles of $Ca^{2+}$, $P_i$, adenine nucleotides and pH are worth considering. For the transition to occur, a combination of high $Ca^{2+}$ ($>1$ μM), low ATP ($<1$ mM) and high $P_i$ ($>10$ mM) concentrations in the extramitochondrial milieu is very effective [62]. These are conditions found in the cytosol of the cardiomyocytes after a prolonged time of oxygen depletion. Since the permeability transition requires accumulation of $Ca^{2+}$ within the matrix space, early dissipation of the electrochemical potential and presence of factors interfering with the mitochondrial mechanism of $Ca^{2+}$ uptake, e.g. a high cytosolic concentration of $Mg^{2+}$ [63], may lead to early (re)sealing of the mitochondria [64]. The opening of "pores" is inhibited by an extramitochondrial concentration of ATP > 1 mM [65] and also by a low pH [66]. Reversal of the opening is favored by elevated concentrations of ADP ($K_{0.5}$ 30 μM) [65] and a reduced state of pyridine nucleotides [54]. Ischemic conditions contain therefore both factors favoring and factors inhibiting "pore" opening. Since "pore" opening in a population of equally treated mitochondria seems to be statistically distributed [15], the combined action of these factors may lead to a gradual variation of the number of hyperpermeable mitochondria in ischemic and reoxygenated myocardial cells and tissue. Consistent with a deleterious role of "pore" opening in the energy-depleted cardiomyocyte is the recent finding that a blocker of the "pore", cyclosporin A, can retard severe cell damage in anoxically incubated isolated cardiomyocytes [62]. Protection against ischemia-reperfusion by cyclosporin A has

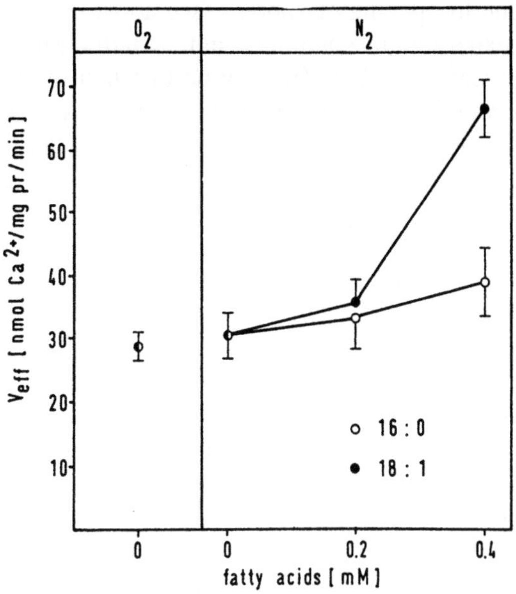


*Figure 5.* Presence of fatty acids in ischemic hearts increases mitochondrial leakiness for $Ca^{2+}$, an indicator for "pore opening". Rate of ruthenium-red induced $Ca^{2+}$ efflux from mitochondria preloaded with 100 nmol $Ca^{2+}$/mg protein. Mitochondria were isolated from oxygenated or anoxic perfused Langendorff-hearts (guinea pig). Conditions as for Figure 1. The anoxic perfusate containing no fatty acid, palmitic acid (16:0) or oleic acid (18:1) (complexed in 1:5 molar ratio to albumin). Means ± SD, n = 5.

also been achieved in other tissues [66–69]. More studies are needed before a judgement on the potential role of "pore opening" or "permeability transition" in the pathophysiology of hypoxia-reoxygenation or ischemia-reperfusion of the myocardial cell can be made.

Induction of hyperpermeability of the inner mitochondrial membrane may thus be disadvantageous in the ischemic or hypoxic cell since it causes activation of mitochondrial net ATP hydrolysis (Figure 5). Amphiphilic long-chain acyl compounds are possible inducers of this state of hyperpermeability.

## Mitochondrial $Ca^{2+}$ overload

The accumulation of $Ca^{2+}$ within the mitochondria proceeds at the expense of the electrochemical potential. In order to keep this constant, respiratory energy must by diverted from oxidative phosphorylation. Mitochondria forced to accumulate $Ca^{2+}$ do therefore produce less ATP [70] and may eventually depolarize due to opening of "pores" (see above).

Under certain experimental conditions, mitochondria can accumulate enormous amounts of $Ca^{2+}$ [70, 71]. This can occur when the $Ca^{2+}$ taken up is precipitated in a complex with phosphate (hydroxyapatite) which appears as crystalline electron-dense material in electron microscopy. Presence of adenine nucleotides is required for this process [70]. In contrast to a normal $Ca^{2+}$ load of 1–2 nmol $Ca^{2+}$/mg mitochondrial protein, loads up to 3000 nmol $Ca^{2+}$/mg mitochondrial protein can be obtained, and after loading the mitochondria can remain functionally competent. Preservation of functional competence is apparently possible [70, 71] when the concentration of ionized $Ca^{2+}$ in the matrix space is low after the loading process. Mitochondria swell during the loading process, indicative of "pore" opening. It has been suggested that the role of ATP or ADP during the loading process consists in a facilitation of formation of the precipitate. It seems also possible that ADP (generated by hydrolysis of ATP) antagonizes "pore" opening to some extent to prevent an extensive loss of membrane polarization during the loading process [46].

Mitochondria with a manifest $Ca^{2+}$ overload, isolated from ischemic-reperfused myocardium were usually found to be functionally impaired. A higher total $Ca^{2+}$ load was associated with larger reductions in the capability for oxidative phosphorylation [72, 73]. The failure could be partially reversed when the mitochondria were first exposed to an exogenous $Ca^{2+}$ chelating agent [72]. As this treatment is now known to induce closure of hyperpermeable "pores" (see above), it seems possible that the functional failure of these $Ca^{2+}$-loaded mitochondrial preparations was due to a permeability transition. These studies on isolated mitochondria have, however, to be interpreted with some caution. If the massive overload of the isolated mitochondria with $Ca^{2+}$ originated from their history in the tissue, they could not have been hyperpermeable at the same time. It must therefore be considered that the isolation process had artificially aggravated the degree of mitochondrial injury. Because of the methodological difficulties in evaluating the functions of mitochondria in tissue the pathophysiological significance of mitochondrial $Ca^{2+}$ overload remains a contentious issue.

It is clear from experiments using isolated mitochondria that mitochondria deteriorate when excess $Ca^{2+}$ taken up cannot be deposited in the form of insoluble precipitates but when it elevates the intramitochondrial concentration of ionized $Ca^{2+}$. The causes for such damage may be manyfold; one is probably the activation of intramitochondrial phospholipase $A_2$ [46, 74]. It must be emphasized that in ischemia-reperfusion this "suicidal" mechanism can occur only when the cells are reoxygenated because the accumulation of $Ca^{2+}$ in the mitochondrial matrix requires respiratory energy. This has been confirmed by measurements of intracellular distribution of total $Ca^{2+}$ in situ using electron probe microanalysis. Myocardial cells severely injured by ischemia-reperfusion developed a pronounced mitochondrial overload with calcium only upon reperfusion [75, 76]. The ultrastructural equivalent of this overload is the appearance of characteristic granular densities in mitochondrial matrices, documented by transmission electron microscopy by many investigators (e.g. [77]).

Deterioration of mitochondrial functional competence may be preventable when mitochondrial $Ca^{2+}$ uptake is blocked at the onset of reoxygenation, e.g. with ruthenium red. Less respiratory energy should then be diverted from the oxidative production of ATP. Since this energy is needed for a reactivation of the pumps clearing the cytosol from excess $Ca^{2+}$ this strategy should also remove the initial cause of mitochondrial $Ca^{2+}$ accumulation, the $Ca^{2+}$ overload of the cytosol. When this has been achieved, a further blockade of mitochondrial $Ca^{2+}$ uptake becomes unnecessary.

## Mitochondrial contribution to stunning and the onset of irreversible injury in ischemic-reperfused myocardium

In the previous section we discussed the results of studies investigating functions of mitochondria isolated from injured myocardium. These studies indicate that mitochondria are progressively damaged. It cannot however, be directly inferred from these studies that a deficit of recovery in reoxygenated myocardium is due to mitochondrial malfunction. Difficulties of interpreting these data arise for the following reasons: First, the isolation procedure may add something to or substract something from the functional state in vivo. Second, it is conceivable that injury affecting initially only part of the mitochondria may remain undetected, if the isolation procedure is selective. Differences in selectivity may explain why in some studies ischemia seemed to cause distinct mitochondrial changes whereas in others, applying similar protocols of ischemia, mitochondrial functions seemed unaltered. Third, the reduction of mitochondrial function in an in-vitro assay, designed to measure a functional maximum, may not represent a relevant impairment within the living cell. One of the reasons is that the experimental conditions may not reflect the cytosolic milieu properly (in most studies, mitochondrial functions were tested in the absence of $Ca^{2+}$, $Mg^{2+}$ and $Na^{+}$). Another is that the functional maximum may not normally be required for the myocardial cell. For these reasons, it is necessary to evaluate the recovery of mitochondrial function in intact myocardial cells and tissue (Figure 6).

Prolonged but transient post-ischemic myocardial dysfunction has been termed "stunning" [78]. It characterizes a condition of the myocardial tissue which (i) has suffered from ischemia long enough to prevent an immediate full recovery but less than needed for manifest irreversible damage, (ii) has been reperfused adequately in relation to its actual energy demand, and (iii) has not yet recovered a normal contractile function. It has been reported in many studies that mitochondria isolated from ischemic or ischemic-reperfused myocardium prone to exhibit "stunning" are functionally impaired. As

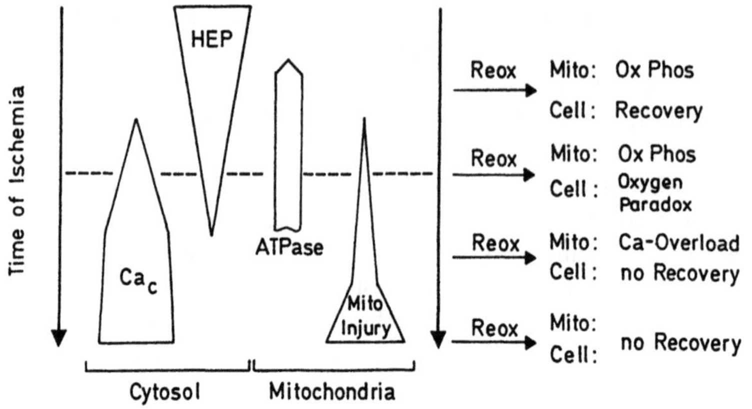


*Figure 6.* Hypothesis on the sequence of events leading to irreversible mitochondrial (Mito) and cell injury in ischemia/hypoxia and reoxygenation of myocardium. With time of ischemia, reserves of high-energy phosphates (HEP) are depleted, partly due to activation of mitochondrial net ATP hydrolysis. Cytosolic $Ca^{2+}$ ($Ca_c$) rises in the energy-depleted cells. Development of mitochondrial structural injury progresses, reaching late a state of irreversibility. When ischemic myocardial cells are early reoxygenated, oxidative phosphorylation (Ox Phos) is re-initiated and the cells recover. Later, when a cytosolic $Ca^{2+}$ overload has developed, reoxygenation provokes the deleterious "oxygen paradox" injury. This defines the usual practical limit of reversibility (dashed line). At a next stage, when other mechanisms for clearing excess $Ca^{2+}$ from the cytosol are impaired, reoxygenated mitochondria are forced to actively accumulate large amounts of $Ca^{2+}$. This accumulation causes a further increase of their injury. Finally, mitochondria have reached a state of irreversible injury already before reoxygenation. (From [11], with permission.)

an example, mentioned above, a very early sign of mitochondrial ischemic injury is a complex I defect, reducing the rate of oxidation of NAD-dependent substrates. Asimakis et al. [32] investigated several parameters of mitochondrial integrity in mitochondria isolated from "stunned" myocardium. The reduction of translocator activity and of the respiratory activity under phosphorylating conditions (state 3) were found rapidly reversible, as also reported by others [25]. The extent of mitochondrial dysfunction did not correlate with the recovery of function, indicating that an insufficient rate of ATP generation is not responsible for the failure of function in "stunning". This conclusion is in accordance with measurements of ATP production in vivo using NMR techniques [79]. The key observation proving functional competence of mitochondria in reperfused "stunned myocardium" is that the energy can be provided for an increase in contractile performance. With inotropic stimulation, a normalization of function can be achieved without increasing the energy deficit [80–83]. The changes in the pattern of substrate oxidation found in "stunned myocardium" [84–86] cannot be the cause, therefore, for its mechanical dysfunction. In the case of "stunned myocardium", therefore, post-ischemic dysfunction seems not causally related to mitochondrial injury, probably because it is not severe enough and therefore does not become the limiting factor for contractile activation.

When energy-depletion in ischemia or hypoxia is extended in time beyond the stage where reoxygenation still leads to spontaneous recovery, reoxygenation may have a "paradoxical" result on myocardial cells. It can lead to an abrupt aggravation of tissue injury ("oxygen paradox") characterized by a sudden onset of contracture and massive enzyme release [87, 88]. Provocation of this lethal reoxygenation injury normally represents the practical limit for reversibility [88]. This form of acute and severe tissue injury is dependent on the ability of mitochondria to resume oxidative energy production; it is attenuated or does not occur when this ability is impaired or abolished, e.g. when tissue reoxygenation is performed in the presence of uncouplers or inhibitors of mitochondrial respiration [88, 89]. When the oxygen paradox develops in reoxygenated myocardial tissue, a disruption of the sarcolemma and a consecutive massive influx of $Ca^{2+}$ into disrupted myocardial cells soon terminate mitochondrial ATP production. The explanation of the oxygen paradox seems to be the following: Under prolonged conditions of energy depletion cardiomyocytes develop a pronounced cytosolic $Ca^{2+}$ overload [90–96]. This happens also in the ischemic heart [97–102]. At the myofibrils, the $Ca^{2+}$ overload creates a state of potential activation, which remains without mechanical consequences as long as the energy of ATP hydrolysis is too low to drive the cross bridge cycle. When oxidative phosphorylation is re-initiated upon re-supply of oxygen, the energy for cross bridge cycling becomes available [103]. This causes a very forceful, uncontrolled contraction, leading to disruptions within the cytoskeleton, i.e. the structural damage of "hypercontracture". In tissue, hypercontracture of adjacent cardiomyocytes leads to their mutual disruption, release of intracellular constituents and to a massive secondary influx of $Ca^{2+}$ into the broken cells [104] which then determines the morphological picture of irreversibly damaged myocardium [77]. There is no indication that accumulation of fatty acid derivatives plays a crucial role in this sequence of events. It may nevertheless act indirectly by causing acceleration of energy loss, e.g. by inducing early mitochondrial depolarisation in the ischemic or hypoxic cell.

At still later stages of ischemic or hypoxic cell injury the process of structural disintegration has progressed so far that mitochondrial and other cell functions are irreversibly impaired already prior to the onset of reoxygenation. Such a state is reached, for example, if in ischemic tissue the sarcolemma disintegrates. It seems possible that incorporation of amphiphilic lipids into cell membranes speeds up the process of disintegration by a detergent-like effect. As the practical limit of reversibility was reached before, however, these autolytic phenomena are of only minor pathophysiological interest.

## Conclusions

In ischemic myocardium, mitochondria lose progressively their normal functional competence. Simultaneously amphiphilic long-chain acyl compounds with potentially deleterious actions on cell membranes accumulate. To date it has not been possible to determine the free concentrations of the relevant lipid moieties in the cytosol and it has therefore remained uncertain whether the concentrations found deleterious in vitro are comparable to those found in ischemic myocardial cells. Consequently a causal relationship between mitochondrial injury and accumulation of long-chain acyl compounds is unproven.

The pathophysiological significance of the defects found in mitochondria isolated from ischemic or hypoxic myocardium is also unclear. There is no convincing evidence that mitochondrial injury developed during energy-depletion represents the decisive cause for reversible contractile dysfunction ("stunning"). A regular mitochondrial action in conjunction with unfavorable other circumstances is responsible for the deleterious result of "oxygen paradox" injury. An increase in amphiphilic lipids is not part of these circumstances.

Presence of elevated levels of amphiphiles in the cytosol may accelerate the process of energy depletion in the oxygen-deprived myocardial cell as mitochondrial net ATP hydrolysis may become activated as a consequence of mitochondrial depolarisation. To initiate depolarisation a small increase in membrane permeability might be sufficient, e.g. caused by the incorporation of amphiphilic long-chain acyl derivatives into the inner membrane of the mitochondria.

In the reoxygenated, $Ca^{2+}$-overloaded cell respiring mitochondria can be forced to take up large amounts of $Ca^{2+}$ which may terminate their metabolic survival together with that of the entire cell. A prerequisite for this accumulation is mitochondrial polarization. Since lipids at high concentration tend to depolarize mitochondria, there is no specific role for lipids in this pathomechanism. The primary cause for irreversibility, a cellular state in which mitochondrial $Ca^{2+}$ overload occurs, seems outside these organelles as persistently high cytosolic $Ca^{2+}$ concentrations indicate severe dysfunction of the sarcolemma.

In summary, it seems questionable whether the interference of amphiphilic long-chain acyl derivatives with mitochondrial functions plays an important causal role for postischemic dysfunction or the onset of irreversibility in ischemic-reperfused myocardium.

## Acknowledgement

This study was supported by the Deutsche Forschungsgemeinschaft, grant C6 of SFB 242, and a BIOMED-1 grant of the European Union.

## References


1. Idell-Wenger JA, Grotyohann LW, Neely JR. Coenzyme A and carnitine distribution in normal and ischemic hearts. J Biol Chem 1978; 253: 4310–8.
2. Van der Vusse GJ, Stam H. Accumulation of lipids and lipid-intermediates in the heart during ischaemia. Basic Res Cardiol 1987; 82(Suppl 1): 157–67.
3. Piper HM, Das A. Detrimental actions of endogenous fatty acids and their derivatives. A study of ischaemic mitochondrial injury. Basic Res Cardiol 1987; 82(Suppl 1): 187–96.
4. Piper HM, Das A. The role of fatty acids in ischemic tissue injury: difference between oleic and palmitic acid. Basic Res Cardiol 1986; 81: 373–83.
5. Van Bilsen M, Van der Vusse GJ, Willemsen PHM, Coumans WA, Roemen THM, Reneman RS. Lipid alterations in isolated, working rat hearts during ischemia and reperfusion: Its relation to myocardial damage. Circ Res 1989; 64: 304–14.
6. Davies NJ, Schulz R, Olley PM, Strynadka KD, Panas DL, Lopaschuk GD. Lysoplasmenylethanolamine accumulation in ischemic/reperfused isolated fatty acid-perfused hearts. Circ Res 1992; 70: 1161–8.
7. Katz AM, Messineo FC. Lipid-membrane interactions and the pathogenesis of ischemic damage in the myocardium. Circ Res 1981; 48: 1–16.
8. Corr PB, Gross RW, Sobel BE. Amphipathic metabolites and membrane dysfunction in ischemic myocardium. Circ Res 1984; 55: 135–54.
9. Piper HM, Sezer O, Schwartz P, Hütter JF, Spieckermann PG. Fatty acid-membrane interactions in isolated cardiac mitochondria and erythrocytes. Biochim Biophys Acta 1983; 732: 193–203.
10. Piper HM, Sezer O, Schwartz P, Hütter JF, Schweickhardt C, Spiekermann PG, Acylcarnitine effects on isolated cardiac mitochondria and erythrocytes. Basic Res Cardiol 1984; 79: 186–98.
11. Piper HM, Noll T, Siegmund B. Mitochondrial function in the oxygen depleted and reoxygenated myocardial cell. Cardiovasc Res 1994; 28: 1–15.
12. Vanderkooi JM, Erecinska M, Silver IA. Oxygen in mammalian tissue: methods of measurement and affinities of various reactions. Am J Physiol 1990; 260: C1131–50.
13. Noll T, Koop A, Piper HM. Mitochondrial ATP-synthase activity in cardiomyocytes after aerobic-anaerobic metabolic transition. Am J Physiol 1992; 262: C1297–303.
14. LaNoue KF, Jeffries FMH, Radda GK. Kinetic control of mitochondrial ATP synthesis. Biochemistry 1986; 25: 7667–75.
15. Crompton M. The role of $Ca^{2+}$ in the function and dysfunction of heart mitochondria. In: Langer GA, editor. Calcium and the heart. New York: Raven Press, 1990: 167–98.
16. Rouslin W. Protonic inhibition of the mitochondrial oligo-mycin-sensitive adenosine 5′-triphosphatase in ischemic and autolyzing cardiac muscle. J Biol Chem 1983; 258: 9657–61.
17. Rouslin W, Erickson JL, Solaro RJ. Effects of oligomycin and acidosis on rates of ATP depletion in ischemic heart muscle. Am J Physiol 1986; 250: H503–8.
18. Pullman ME, Monroy GC. A naturally occuring inhibitor of mitochondrial adenosine triphosphatase. J Biol Chem 1963; 238: 3862–9.
19. Schwerzmann K, Pedersen P. Regulation of the mitochondrial ATP synthase/ATPase complex. Arch Biochem Biophys 1986; 250: 1–18.
20. Rouslin W. Regulation of the mitochondrial ATPase in situ in cardiac muscle: role of the inhibitor unit. J Bioenerg Biomembr 1991; 23: 873–88.
21. Mittnacht S, Sherman SC, Farber JL. Reversal of ischemic mitochondrial dysfunction. J Biol Chem 1979; 254: 9871–8.
22. Aw TY, Andersson BS, Jones DP. Suppression of mitochondrial respiratory function after short-term anoxia. Am J Physiol 1987; 252: C362–8.
23. Das AM, Harris DA. Control of mitochondrial ATP synthase in heart cells: inactive to active transitions caused by beating or positive inotropic agents. Cardiovasc Res 1990; 24: 411–7.

24. Rouslin W, Broge CW, Grupp IL. ATP depletion and functional loss during ischemia in slow and fast heart-rate hearts. Am J Physiol 1990; 259: H1759–66.
25. Piper HM, Sezer O, Schleyer M, Schwartz P, Hütter JF, Spieckermann PG. Development of ischemia-induced damage in defined mitochondrial subpopulations. J Mol Cell Cardiol 1985; 17: 186–98.
26. Piper HM, Das A. Detrimental actions of endogenous fatty acids and their derivatives. A study of ischaemic mitochondrial injury. Basic Res Cardiol 1987; 82(Suppl 1): 187–96.
27. Lochner A, van Niederkerk I, Whitesell LF. Mitochondrial acyl-CoA, adenine nucleotide translocase activity and oxidative phosphorylation in myocardial ischemia. J Mol Cell Cardiol 1981; 13: 991–7.
28. Kotaka K, Miyazaki Y, Ogawa K, Satake T, Sugiyama S, Ozawa T. Reversal of ischemia-induced mitochondrial dysfunction after coronary reperfusion. J Mol Cell Cardiol 1982; 14: 223–31.
29. Rouslin W. Mitochondrial complexes I, II, III, IV, and V in myocardial ischemia and autolysis. Am J Physiol 1983; 244: H743–8.
30. Hardy L, Clark JB, Darley-Usmar VM, Stone D. Reoxygenation-dependent decrease in mitochondrial NADH:CoQ reductase (Complex I) activity in the hypoxic/reoxygenated rat heart. Biochem J 1991; 274: 133–7.
31. Asimakis GK, Conti VR. Myocardial ischemia: correlation of mitochondrial adenine nucleotide and respiratory function. J Mol Cell Cardiol 1984; 16: 439–48.
32. Asimakis GK, Zwischenberger JB, Inners-McBride K, Sordahl LA, Conti VR. Postischemic recovery of mitochondrial adenine nucleotides in the heart. Circularion 1992; 85: 2212–20.
33. Klingenberg M. The ADP-ATP translocation in mitochondria. A membrane potential controlled transport. J Membr Biol 1980; 56: 97–105.
34. Klingenberg M. Molecular aspects of the adenine nucleotide carrier from mitochondria. Arch Biochem Biophys 1989; 270: 1–14.
35. Asimakis GK, Conti VR. Phophate-induced efflux of adenine nucleotides from heart mitochondria. Am J Physiol 1985; 249: H1009–16.
36. Shug AL, Shrago E, Bittar N, Folts JD, Roke JR. Acyl CoA inhibition of adenine nucleotide translocation in ischemic myocardium. Am J Physiol 1975; 228: 689–92.
37. Regitz V, Paulson DJ, Hodach RJ, Little SE, Schaper W, Shug AL. Mitochondrial damage during myocardial ischemia. Basic Res Cardiol 1984; 79: 207–17.
38. Duan J, Karmazyn M. Relationship between oxidative phosphorylation and adenine nucleotide translocase activity of two populations of cardiac mitochondria and mechanical recovery of ischemic hearts following reperfusion. Can J Physiol Pharmacol 1989; 67: 704–9.
39. Pande SV, Blanchaer MC. Reversible inhibition of mitochondrial adenosine diphosphate phosphorylation by long chain acyl coenzyme A esters. J Biol Chem 1971; 246: 402–11.
40. Shug AL, Lerner C, Elson O, Shrago E. The inhibition of adenine nucleotide translocase by oleoyl CoA and its reversal in rat liver mitochondria. Biochem Biophys Res Commun 1971; 43: 557–63.
41. Shrago E, Shug AL, Sul H, Bittar N, Folts JD. Control of energy production in myocardial ischemia. Circ Res 1976; 38:(Suppl I): 75–9.
42. Woldegiorgis G, Shrago E. The recognition of two specific binding sites of the adenine nucleotide translocase by palmityl CoA in bovine heart mitochondria and submitochondrial particles. Biochem Biophys Res Commun 1979; 89: 837–44.
43. Liedtke AJ, Shrago E. Detrimental effects of fatty acids and their derivatives in ischemic and reperfused myocardium. In: Piper HM, editor. Pathophysiology of severe ischemic myocardial injury. Dordrecht: Kluwer Academic Publishers, 1990: 149–66.
44. Van der Vusse GJ, Van Bilsen M, Sonderkamp T, Reneman RS. Hydrolysis of phospholipids and cellular integrity. In: Piper HM, editor. Pathophysiology of severe ischemic myocardial injury. Dordrecht: Kluwer Academic Publishers, 1990: 167–94.
45. Haworth RA, Hunter DR. Allosteric inhibition of the $Ca^{2+}$ activated hydrophilic channel of the mitochondrial inner membrane by nucleotides. J Membr Biol 1980; 54: 231–6.

46. Gunter TE, Pfeiffer DR. Mechanisms by which mitochondria transport calcium. Am J Physiol 1990; 258: C755–86.
47. Crompton M, Andreeva L. On the involvement of a mitochondrial pore in reperfusion injury. Basic Res Cardiol 1993; 88: 513–23.
48. Harris EJ. Modulation of $Ca^{2+}$ efflux from heart mitochondria. Biochem J 1979; 178: 673–80.
49. Harris EJ, Cooper MB. Calcium and magnesium losses in response to stimulants of efflux applied to heart, liver and kidney mitochondria. Biochem Biophys Res Commun 1981; 103: 788–96.
50. Sordahl LA, Stewart ML. Mechanism(s) of altered mitochondrial calcium transport in acutely ischemic canine hearts. Circ Res 1980; 47: 814–20.
51. Le Quoc K, Le Quoc D. Involvement of the ATP/ADP carrier in calcium-induced perturbations of the mitochondrial inner membrane permeability: Importance of the orientation of the nucleotide bindng site. Arch Biochem Biophys 1988; 265: 249–57.
52. Carbonera D, Azzone GF. Permeability of inner mitochondrial membrane and oxidative stress. Biochim Biophys Acta 1988; 943: 245–55.
53. Palmer JW, Pfeiffer DR. The control of $Ca^{2+}$ release from heart mitochondria. J Biol Chem 1981; 256: 6742–50.
54. De Villiers M, Lochner A. Mitochondrial $Ca^{2+}$ fluxes: role of free fatty acids, acyl-CoA and acylcarnitine. Biochim Biophys Acta 1986; 876: 309–17.
55. Beatrice MC, Stiers O, Pfeiffer DR. Increased permeability of mitochondria during $Ca^{2+}$ release induced by t-butyl hydroperoxide or oxalacetate. J Biol Chem 1982; 357: 7161–71.
56. Crompton M, Costi A. Kinetic evidence for a heart mitochondrial pore activated by $Ca^{2+}$, inorganic phosphate and oxidative stress. Eur J Biochem 1988; 178: 489–501.
57. Richter C, Frei B. $Ca^{2+}$ release from mitochondria induced by prooxidants. Free Rad Biol Med 1988; 4: 365–75.
58. Lehninger AL, Vercesi A, Bababunmi EA. Regulation of $Ca^{2+}$ release from mitochondria by the oxidation-reduction state of pyridine nucleotides. Proc Natl Acad Sci USA 1978; 75: 1690–4.
59. Szabo I, Zoratti M. The mitochondrial megachannel is the permeability pore. J Bioenerg Biomembr 1992; 24: 111–7.
60. Al-Nasser I, Crompton M. The entrapment of the $Ca^{2+}$ indicator arsenazo III in the matrix space of rat liver mitochondria by permeabilisation and resealing. Biochem J 1986; 239: 31–40.
61. Crompton M, Ellinger H, Costi A. Inhibition of cyclosporin A of a $Ca^{2+}$-dependent pore in heart mitochondria activated by inorganic phosphate and oxidative stress. Biochem J 1988; 255: 357–60.
62. Nazareth W, Yafei N, Crompton M. Inhibition of anoxia-induced injury in heart myocytes by cyclosporin A. J Mol Cell Cardiol 1991; 23: 1351–4.
63. McCormack JG, Halestrap AP, Denton RM. Role of calcium ions in regulation of mammalian intramitochondrial metabolism. Physiol Rev 1990; 70: 391–425.
64. Petrollini V, Cola C, Bernardi P. Modulation of the mitochondrial cyclosporin A-sensitive permeability transition pore. II. The minimal requirements for pore induction underscore a key role for transmembrane electrical potential, matrix pH, and matrix $Ca^{2+}$. J Biol Chem 1993; 268: 1011–6.
65. McGuiness O, Crompton M. Cyclosporin A and mitochondrial dysfunction. Biochem Soc Transact 1990; 18: 883–4.
66. Halestrap AP. Calcium-dependent opening of a non-specific pore in the mitochondrial inner membrane is inhibited at pH values below 7. Implications for the protective effect of low pH against chemical and hypoxic cell damage. Biochem J 1991; 278: 715–9.
67. Crompton M, Costi A. A heart mitochondrial $Ca^{2+}$-dependent pore of possible relevance to reperfusion-induced injury. Evidence that ADP facilitates pore interconversion between the closed and open states. Biochem J 1990; 266: 33–9.

68. Kurokawa T, Kobayashi H, Nonami T et al. Beneficial effects of cyclosporine on postischemic liver injury in rats. Transplantation 1992; 53: 308–11.
69. Shiga Y, Onodera H, Matsuo Y, Kogure K. Cyclosporin A protects against ischemia-reperfusion injury in the brain. Brain Res 1992; 595: 145–52.
70. Lehninger AL. Mitochondria and calcium transport. Biochem J 1970; 119: 129–38.
71. Carafoli E. Intracellular calcium homeostasis. Annu Rev Biochem 1987; 56: 395–433.
72. Peng CF, Kane JJ, Murphy ML, Straub KD, Abnormal mitochondrial oxidative phosphorylation of ischemic myocardium reversed by $Ca^{2+}$-chelating agents. J Mol Cell Cardiol 1977; 9: 897–908.
73. Ferrari R, Williams A, Di Lisa F. The role of mitochondrial function in the ischemic and reperfused myocardium. In: Caldarera CM, Harris P, editors. Advances in studies on heart metabolism. Bologna: CLUEB, 1988: 245–55.
74. Broekemeier KM, Schmid PC, Schmid HHO, Pfeiffer DR. Effects of phospholipase $A_2$ inhibitors on the permeability of the liver mitochondrial inner membrane. J Biol Chem 1986; 26: 14018–24.
75. Walsh LG, Tormay JMcD. Subcellular electrolyte shifts during in vitro myocardial ischemia and reperfusion. Am J Physiol 1988; 255: H917–28.
76. Walsh LG, Tormey JMcD. Cellular compartmentation in ischemic myocardium: indirect analysis by electron probe. Am J Physiol 1988; 255: H929–36.
77. Reimer KA, Jennings RB. Myocardial ischemia, hypoxia, and infarction. In: Fozzard HA, Haber E, Jennings RB, Katz AM, editors. The heart and cardiovascular system, scientific foundations, 2nd ed. New York: Raven Press, 1990: 1875–974.
78. Braunwald E, Kloner RA. The stunned myocardium: prolonged postischemic ventricular dysfunction. Circulation 1982; 66: 1146–9.
79. Sako EY, Kingsley-Hickman PB, From AHL, Foker JE, Ugurbil K. ARP synthesis kinetics and mitochondrial function in the postischemic myocardium as studied by $^{31}P$ NMR. J Biol Chem 1988; 263: 10600–7.
80. Mercier JC, Lando U, Kanmatsuse K et al. Divergent effects of inotropic stimulation on the ischemic and severely depressed reperfused myocardium. Circulation 1982; 66: 397–400.
81. Ellis SG, Wyne J. Braunwald E, Henschke CI, Sandor T, Kloner RA. Response of reperfusion-salvaged, stunned myocardium to inotropic stimulation. Am Heart J 1984; 107: 13–9.
82. Becker LC, Levine JH, DiPaula AF, Guarnieri T, Aversano T. Reversal of dysfunction in postishcemic stunned myocardium by epinephrine and postextasystolic potentiation. J Am Coll Cardiol 1986; 7: 580–9.
83. Ambrosio G, Jacobus WE, Bergman CA, Weisman HF, Becker LC. Preserved high-energy phosphate metabolic reserve in globally "stunned" hearts despite reduction of basal ATP content and contractility. J Mol Cell Cardiol 1987; 19: 953–64.
84. Görge G, Chatelain P, Schaper J, Lerch R. Effects of increasing degrees of ischemic injury on myocardial oxidative metabolism early after reperfusion in isolated rat hearts. Circ Res 1991; 68: 1681–92.
85. Nellis SH, Liedtke AJ, Renstrom B. Distribution of carbon flux witin fatty acid utilization during myocardial ischemia and reperfusion. Circ Res 1991; 69: 779–90.
86. Lerch R. Oxidative substrate metabolism during postischemic reperfusion. Basic Res Cardiol 1993; 88: 525–44.
87. Hearse DJ, Humphrey SM, Chain EB. Abrupt reoxygenation of the anoxic potassium-arrested perfused rat heart: A study of myocardial enzyme release. J Mol Cell Cardiol 1973; 5: 395–407.
88. Ganote CE. Contraction band necrosis and irreversible myocardial injury. J Mol Cell Cardiol 1983; 15: 67–73.
89. Vander Heide RS, Angelo JP, Altschuld RA, Ganote CE. Energy dependence of contraction band formation in perfused hearts and isolated adult cardiomyocytes. Am J Pathol 1986; 125: 55–68.

90. Siegmund B, Zude R, Piper HM. Recovery of anoxic-reoxygenated cardiomyocytes from severe $Ca^{2+}$-overload. Am J Physiol 1992; 263: H1262–9.
91. Siegmund B, Klietz T, Schwartz P, Piper HM. Temporary contractile blockade prevents hypercontracture in anoxic-reoxygenated cardiomyocytes. Am J Physiol 1991; 260: H426–35.
92. Siegmund B, Schlüter KD, Piper HM. Calcium and the oxygen paradox. Cardiovasc Res 1993; 27: 1778–83.
93. Siegmund B, Ladilov Yu, Piper HM. Importance of sodium for recovery of calcium control in reoxygenated cardiomyocytes. Am J Physiol. 1994; 267: H506–13.
94. Allshire A, Piper HM, Cuthbertson KSR, Cobbold PH. Cytosolic free $Ca^{2+}$ in single rat heart cells during anoxia and reoxygenation. Biochem J 1987; 244: 381–5.
95. Li Q, Altschuld RA, Stokes BT. Myocyte deenergization and intracellular free calcium dynamics. Am J Physiol 1988; 255: C162–8.
96. Miyata H, Lakatta EG, Stern MD, Silverman HS. Relation of mitochondrial and cytosolic free calcium to cardiac myocyte recovery after exposure to anoxia. Circ Res 1992; 17: 605–13.
97. Marban E, Kitakaze M, Kusuoka H, Porterfield JK, Yue DT, Chacko VP. Intracellular free calcium concentration measured with $^{19}F$ NMR spectroscopy in intact ferret hearts. Proc Natl Acad Sci USA 1987; 84: 6005–9.
98. Lee HC, Mohabir R, Smith N, Franz MR, Clusin WT. Effect of ischemia on calcium-dependent fluorescence transients in rabbit hearts containing indo-1. Circulation 1988; 78: 1047–59.
99. Steenbergen C, Murphy E, Levy L, London RE. Elevation in cytosolic free calcium concentration early in myocardial ischemia in perfused rat heart. Circ Res 1987; 60: 700–7.
100. Kihara Y, Grossmann W, Morgan JP. Direct mesurement of changes in intracellular calcium transients during hypoxia, ischemia, and reperfusion of the intact mammalian heart. Circ Res 1989; 65: 1029–44.
101. Marban E. Pathogenetic role for calcium in stunning. Cardiovasc Drugs Ther 1991; 5: 891–4.
102. Steenbergen C, Murphy E, Watts JE, London RE. Correlation between cytosolic free calcium, contracture, ATP, and irreversible injury in perfused rat heart. Circ Res 1990; 66: 135–46.
103. Piper HM. Energy deficiency, calcium overload or oxidative stress: Possible causes of irreversible ischemic myocardial injury. Klin Wochenschr 1989; 67: 465–76.
104. Elz J, Nayler WG. Contractile activity and reperfusion-induced calcium gain after ischemia in the isolated rat heart. Lab Invest 1988; 58: 653–9.

*Corresponding Author*: Professor Hans Michael Piper, Physiological Institute, Justus-Liebig-University, Aulweg 129, D-35392 Giessen, Germany

# 9. Free radical-mediated damage and carnitine esters

JOHAN FOKKE KOSTER

"The administration of propionylcarnitine leads to less hydrogen peroxide formation and less available free iron, resulting in attenuation of the free radical-mediated damage. It is, however, not unlikely that propionylcarnitine directly inhibits the $O_2^-$ generation, perhaps by stabilizing the membrane."

## Introduction

It is now well recognized that oxygen free radicals are involved in the tissue damage occurring upon reperfusion after an ischemic period. Although it is quite obvious that oxygen supply must be restored as soon as possible, this beneficial effect is counteracted by the generation of free oxygen radicals.

A free radical is by definition an atom or a molecule with one or more impaired electrons in its outer orbital. Molecular oxygen is a biradical by itself because it has two impaired electrons in its outer orbitals. These two impaired electrons have parallel spins. Therefore all reactions in which molecular oxygen is involved are spin restricted, resulting in rather slow reaction rates.

Normally in biological oxygenation, oxygen is tetravalently reduced to water in the mitochondria by the enzyme cytochrome *c* oxidase. It is, however, assumed that about 1% of the oxygen consumption by the cell is not tetravalently reduced, but gives rise to oxygen free radicals generation. If oxygen is reduced in univalent steps, three types of reactive intermediates can be formed:

$$O_2 + e^- \rightarrow O_2^- \qquad \text{(superoxide)}$$

$$O_2^- + e^- + 2H^+ \rightarrow H_2O_2 \qquad \text{(hydrogen peroxide)}$$

$$H_2O_2 + e^- + H^+ \rightarrow H_2O + OH \qquad \text{(hydroxyl radical)}$$

*J.W. de Jong and R. Ferrari (eds): The carnitine system*, 123–132.

The addition of a single electron results in the formation of superoxide anion ($O_2^-$), divalent reduction results in the formation of hydrogen peroxide ($H_2O_2$) and a trivalent reduction in the formation of hydroxyl radicals ($OH^\cdot$). Superoxide anion and hydroxyl radicals are free radicals, but $H_2O_2$ although a strong oxidant, is not a free radical because all its electrons are paired.

In biological systems, e.g. heart tissue, the toxicity of superoxide is not regarded as very high [1]. Nevertheless, the protonated form of $O_2^-$, the hydroperoxyl radical, is known to be more aggressive. It can cause the peroxidation of lipids. Hydrogen peroxide is a stable compound of limited reactivity and no further reduction takes place in the absence of a suitable catalyst.

## Protection against free radicals

As already stated about 1% of the oxygen consumed is not tetravalently reduced to water, but to intermediate oxygen free radicals. So during aerobic metabolism there is a continuous generation of $O_2^-$ and $H_2O_2$. These are even used in biosynthesis, intracellular signalling and in the defence against invading microorganisms [2, 3]. To cope with this generation of free oxygen radicals, and to prevent damage from them, the cell has several lines of defence, an enzymatical and a non-enzymatical system [4]. The detoxification of oxygen radicals by the enzymatic system occurs through an array of various enzymes (Figure 1). The dismutation reaction of $O_2^-$ is speeded up 10,000 times by the enzyme superoxide dismutase to yield $H_2O_2$ and $O_2$. The $H_2O_2$ is degraded to $H_2O$ and $O_2$ by catalase. Physiologically more important is the glutathione peroxidase, which removes $H_2O_2$ at the expense of reduced glutathione (GSH). This enzyme can also detoxify lipid hydroperoxide in a similar reaction, but not lipid hydroperoxides in phospholipids. The latter compounds are detoxified by a membrane bound version of glutathione peroxidase [5]. It is clear that these lines of defence are strongly dependent on the availability of GSH. This makes the system dependent of the re-reduction of oxidized glutathione (GSSG). The reduction of GSSG is catalyzed by glutathione reductase at the expense of NADPH. The latter is provided by carbohydrate degradation through the pentose phosphate shunt.

Figure 1 also shows a part of the non-enzymatic defence, namely alpha-tocopherol (vitamine E) and ascorbate (vitamine C) as direct scavengers and chain breakers. The product itself is a radical, however less reactive. But it has to be removed. It has been proposed that vitamine E radical reacts with vitamine C yielding dehydroascorbic acid [4]. These are very interesting reactions because they have to occur on the water-lipid surfaces in the membrane.

Recently Winterbourne [6] proposed a cycle in which radicals react directly with oxygen to yield superoxide or through GSH to yield superoxide and GSSG. The hypothesis was put forward that in the absence of superoxide

Superoxide Dismutase

$$O_2^{\cdot-} + HO_2^{\cdot} \xrightarrow{H^+} H_2O_2 + O_2$$

Catalase

$$H_2O_2 + H_2O_2 \longrightarrow 2\,H_2O + O_2$$

Glutathione Peroxidases

$$ROOH + 2GSH \longrightarrow ROH + GSSG + H_2O$$

Scavenging by Vitamins

$$ROO^{\cdot} + \text{vitE-OH} \longrightarrow ROOH + \text{vitE-O}^{\cdot}$$

$$\text{vitE-O}^{\cdot} + \text{vitC-H} \longrightarrow \text{vitE-OH} + \text{vitC-}^{\cdot}$$

$$\text{vitC-}^{\cdot} + 2\,H^+ \longrightarrow \text{vitC-H} + DHA$$

*Figure 1.* Cellular defence against free radicals. Reactions catalyzed by superoxide dismutase, catalase and glutathione peroxidases. Non enzymatic removal occurs through a number of chain breaking scavengers. Here the concerted action of vitamin E (vitE) and vitamin C (vitC) is presented. The end product of the ROO radicals is the non toxic dehydroascorbic acid (DHA). GSH and GSSG, reduced and oxidized glutathione, respectively.

dismutase, superoxide would continue the cycle with GSH to yield hydrogen peroxide and GSSG. Removal of superoxide by superoxide dismutase would reduce the production of hydrogen peroxide and GSSG. In this way only one enzyme would be necessary to remove a whole range of radicals.

Superoxide and hydrogen peroxide are not so toxic as is often claimed [1, 7]. It is the availability of free transition metals which makes these agents so toxic, due to the formation of the highly aggressive hydroxyl radical ($OH^{\cdot}$), according the following reactions:

$$O_2^- + O_2^- + 2\,H^+ \rightarrow O_2 + H_2O_2 \tag{1}$$

$$O_2^- + Fe^{3+} \rightarrow Fe^{2+} + O_2 \tag{2}$$

$$H_2O_2 + Fe^{2+} \rightarrow Fe^{3+} + OH^- + OH^{\cdot} \tag{3}$$

$$\text{red} + Fe^{3+} \rightarrow Fe^{2+} + \text{ox} \tag{4}$$

in which red and ox denote reductant and oxidant, respectively.

The first reaction is catalyzed by superoxide dismutase. Reactions 2 and 3 are known as the Haber-Weiss reaction. The reactions are written down for iron but also take place with copper. Ferritin, transferrin and lactoferrin are often mentioned as radical scavengers, but this is based on their iron chelating capacities which prevent the formation of hydroxyl radicals. Due to the extremely low solubility of ferric hydroxide, the $Fe^3+$ ion concentration will be very low under normal physiological pH. Therefore the Haber-Weiss (reactions 2 and 3) will not proceed [8, 9]. Only in the presence of a suitable chelator that keeps iron in solution, it can participate in these reactions. The chelator must allow one electron transfer so at least one coordination site must be occupied by an easy displaceable ligand [10]. Therefore, the nature of the complex determines why some iron chelators enhance and others completely prevent the iron driven radical reactions [11]. It is quite clear that under normal physiological conditions the amount of free iron is very low and almost all of the cellular iron is located in ferritin. Iron stored in ferritin is unable to initiate lipid peroxidation, but under pathological conditions, as e.g. inflammation, iron can be released from ferritin by superoxide [12]. It has also been shown that other reducing equivalents (e.g. NADPH) can release iron from ferritin in order to initiate lipid peroxidation of microsomes [13].

## The role of iron in ischemia-reperfusion syndrome

A wide variety of experiments has substantiated the role of iron as catalytic transition metal in post-ischemic free radical mediated damage [8]. However, the role of iron was indirectly shown with the aid of iron chelators. These iron chelators, which also inhibit the lipid peroxidation process, were shown to attenuate the free radical mediated damage [14, 15]. The iron chelator desferal (desferoxamine) has been shown to decrease hydroxyl radical production [16] and preserve membrane phospholipid in post-ischemic rat hearts [17]. In vivo studies using dogs [18] and pigs [19] have confirmed the beneficial effect in whole animals. Smith et al. [20] have shown that the protective effect of desferal is due to its chelating capacities and not to a radical scavenging effect, since the beneficial effect is abolished by using the iron containing counterpart of desferal (ferrioxamine). More direct evidence for the important role of iron comes from iron loading of endothelial cells [21] and from the increased sensitivity towards mild anoxic insults of heart from iron overloaded rats [22].

Although the role of iron during post-ischemia from radical mediated damage is well established, it is unclear where the catalytic iron originates from or in which form it is present. As mentioned above only a very small amount of iron has to be present in the socalled low molecular weight pool (LMWP) [23, 24] in which the iron will be chelated to small molecules as ATP and/or AMP [24]. This form has been shown to catalyze hydroxyl radical formation and lipid peroxidation [25]. Also evidence was presented that iron is delocalized in this pool during ischemia in dog heart [26], gerbil brain [27] and rat kidneys [28]. Except for the last study [28] in the other studies [26, 27] it was not realized that the iron could have been released from ferritin during the procedures used. Healing et al. [28] have realized this possibility but argue that since the iron measured was chelatable, it must have been loosely bound and therefore potentially catalytic. Modifying the method of Gower et al. [29], we have recently developed a method to measure the LMW iron pool directly [30]. With this method it was clearly shown that during ischemia there is a drastic increment of the LMW iron pool, even to a concentration of 40 μM [30]. This concentration is high enough to initate the lipid peroxidation reaction.

**Source of oxygen radicals**

All aerobic life forms use oxygen through a stepwise enzymatic reduction of the molecule using transition metals to overcome the spin restrictions [31]. This causes a constant generation of superoxide and hydrogen peroxide from different sources, which may be accidently through leakage, or for the production of substrate for further reactions [32]. Some of these metabolic pathways contribute to radical generation during reperfusion.

Mitochondrial respiration is one of the main sources of reactive oxygen in normal metabolism [33]. The last step in the mitochondrial respiratory chain is the tetravalent reduction of molecular oxygen in one step. Separate from leakage from respiring mitochondria, superoxide is generated through an NADH oxidase that may exist on the mitochondrial outermembrane. It is absent in liver mitochondria [34, 35]. The mitochondrial generation of reactive oxygen is enhanced by ischemia and reperfusion [35, 36]. The peroxisomal ß oxidation accounts for an important part of the fatty acid oxidation and is therefore a constant source of hydrogen peroxide [37]. In the formation of prostaglandins and leukotrienes through cyclooxygenase and lipoxygenase, respectively, lipid peroxides as intermediates are involved. These pathways are stimulated by either free arachidonic acid, superoxide, hydrogen peroxide and lipid peroxide [38], and have been shown to generate reactive oxygen species [39]. Furthermore, ischemia leads to an elevation of free arachidonic acid [40], making these pathways a likely source of oxygen radicals during reperfusion [39]. The release of noradrenaline during reperfusion may contribute to the radical formation. Autooxidation of catecholamines generates superoxide [41]. The endothelium derived releasing factor nitric oxide (NO) is also a radical and reacts with superoxide. A balanced production of both radicals by the endothelium is a way to regulate the vascular tone [42]. It is also prognosed that the reaction between nitric oxide and superoxide leads to the formation of peroxynitrite, which is converted to a hydroxyl-like radical which could induce endothelial damage [43]. In vivo, the activation of granulocytes leads to a sudden increase in oxygen consumption. Ninety percent of this oxygen consumption (the so-called "respiratory burst") is converted to superoxide through the activated NAD(P)H oxidase. This superoxide is in turn converted to hydrogen peroxide. The latter is used for the formation of hypochlorous acid, which is toxic and kills the bacteria.

Granulocytes accumulate in the infarcted area after in vivo ischemia and can certainly contribute to the generation of superoxide upon reperfusion [44].

It has been described that during ischemia xanthine dehydrogenase is converted to xanthine oxidase by ischemia [45]. However this possibility is strongly debated [46]. For the human heart is is quite evident that this tissue has a low activity of xanthine oxidase [47]. So the contribution of free radicals by xanthine oxidase in the human heart will be negligible.

## The effect of carnitine ester on free radical-mediated tissue damage

Tong Mak et al. [48] have demonstrated that the lipid peroxidation of sarcolemma is enhanced by the addition of palmitoyl-CoA, palmitoyl-L-carnitine and lysophosphatidylcholine, while free fatty acid, coenzyme A and L-carnitine have no effect. Palmitoyl-CoA ester has a greater promoting effect on the lipid peroxidation than palmitoyl-L-carnitine ester. It is interesting to note that both acyl esters increase during ischemia. In contrast to the long-chain acyl esters, acetyl-L-carnitine seems to have a protective effect on the NADPH induced lipid peroxidation of heart microsomes [49]. This effect on lipid peroxidation is measured on the chemiluminiscence signal. It has also been shown that acetyl-L-carnitine does not inhibit the $O_2^-$ production from granulocytes activated by phorbolester.

Provoking mitochondrial lipid peroxidation with ferrous ions as catalyst, Ferrari et al. [50] showed the L-carnitine and acetyl-L-carnitine failed to prevent mitochondrial damage. However, propionyl-L-carnitine improved the mitochondrial functions significantly, but this was not reflected in a decrease of malondialdehyde formation. This can indicate that inhibition of lipid peroxidation is not involved but that propionyl-L-carnitine inhibits other radical mediated damage, e.g. on proteins. Shug et al. [51] claimed that acetyl-L-carnitine as well as propionyl-L-carnitine inhibit the mitochondrial $O_2^-$ production while L-carnitine does not affect it. These authors also report a beneficial effect of acetyl-L-carnitine and propionyl-L-carnitine on the cardiac functions upon reperfusion after a 90 min low flow ischemia.

More experimental evidence for the beneficial effects of propionyl-L-carnitine is provided by the group of Packer [52, 53]. They show that propionyl-L-carnitine does prevent the formation of carbonyls in the protein, which strengthened the findings of Ferrari et al. [50] who found an improvement of mitochondrial functions without a decrease of malondialdehyde, as mentioned before. However, in contrast to Shug et al. [51], the Packer group failed to find any inhibition of $O_2^-$ production by submitochondrial particles. In a model system of hydrogen peroxide with $FeSO_4$ with measurement of the luminal chemiluminiscence, propionyl-L-carnitine and propionyl-D-carnitine equally inhibit the chemiluminiscence response. It should be realized that high concentrations of either carnitine ester are necessary (75 mM) to obtain this inhibitory effect. To inhibit the ascorbate oxidation by iron, high concentrations of the carnitine esters are needed. The reaction of hydrogen peroxide and ferrous ion reveals radicals which can be detected by electron spin resonance. The radical is stabilized by and trapped with 5,5′-dimethyl-l-pyroline-N-oxide (DMPO). High concentrations (75 mM) of either propionyl-L-carnitine or the D-form are necessary to block the DMPO-OH signal. It should be considered that they used adventitious iron or added 75 μM $FeSO_4$. The more surprising is their finding that for the generation of the DPMO spin adducts in the heart perfusate after 40 min of ischemia, only 10 mM propionyl-L-carnitine is needed [52, 53]. Voogd et al. [30] have found that the LMW iron is then about 40 μM. Based on these findings we suggested that a part of the beneficial effect of propionylcarnitine is due to an iron-chelating capacity. However, if this is the case one would expect that instead of the DMPO-OH spin adduct the DMPO-OOH spin adduct is seen, which is not the case. The latter is found if $O_2^-$ is generated. It is therefore more likely that in some way propionyl-L-carnitine blocks the formation of $O_2^-$.

It is known that ischemic hearts are more susceptible to hydrogen peroxide than normoxic hearts [54]. Voogd et al. [55] have shown that this increased susceptibility is due to the increase of LMW iron. It is estimated [56] that the relative contribution of peroxisomes to the total oxidation capacity is 10–30% for common fatty acids and about 45% for fatty acids with a chain length of more than 22 C. Oxidation through the peroxisomes leads to the formation of hydrogen peroxide. Acyl-CoA esters do enhance the lipid peroxidation more than acylcarnitine [48], raising the possibility that the addition of propionyl-L-carnitine increases the L-carnitine level. This can result in the conversion of acyl-CoA to acylcarnitine, which also decreases the substrate supply in the cytosol for peroxisomal oxidation. The latter results also in less hydrogen peroxide formation. As mentioned before, during ischemia also the LMW iron is increased. Propionyl-L-carnitine may chelate a part of this iron pool. Altogether the administration of propionyl-carnitine leads to less hydrogen peroxide formation and less available free iron, resulting in an attenuation of the free radical-mediated damage. It is, however, not unlikely that propionylcarnitine directly inhibits the $O_2^-$ generation, perhaps by stabilizing the membrane.

## References

1. Singh A. Chemical and biochemical aspects of superoxide radicals and related species of activated oxygen. Can J Physiol Pharmacol 1982; 60: 1330–42.
2. Barja G. Oxygen radicals, a failure or a success of evolution? Free Rad Res Commun 1993; 18: 63–70.
3. Koster JF, Sluiter W. Physiological relevance of free radicals and their relation to iron. In: Nohl H, Esterbauer H, Rice-Evans C (eds), Free Radical Series Vol VIII. London: Richelieu Press, 1994: 409–27.

4. Sies H. Oxidative stress: From basic research to clinical application. Am J Med 1991; 91: 31S–8S.
5. Thomas JP, Maiorino M, Ursini F, Girotti AW. Protective action of phospholipid hydroperoxide glutathione peroxidase against membrane-damaging lipid peroxidation. In situ reduction of phospholipid and cholesterol hydroperoxides. J Biol Chem 1990; 265: 454–61.
6. Winterbourne CC. Superoxide as an intracellular radical sink. Free Rad Med Biol 1993; 14: 85–90.
7. Sawyer DF, Valentine JS. How super is superoxide? Acc Chem Res 1981; 14: 393–400.
8. Halliwell B, Gutteridge JMC. Role of free radicals and catalytic metal ions in human disease: an overview. Methods Enzymol 1990; 186: 1–85.
9. Dunford HB. Free radicals in iron-containing systems. Free Rad Biol Med 1987; 3: 405–22.
10. Graf E, Mahoney JR, Bryant RG, Eaton JW. Iron-catalyzed hydroxyl radical formation. Stringent requirement for free iron coordination site. J Biol Chem 1984; 259: 3620–4.
11. Mostert LJ, Van Dorst JALM, Koster JF, Van Eijk HG, Konthoghiorghes GJ. Free radicals and cytotoxic effect of chelators and their iron complexes in the hepatocyte. Free Rad Res Commun 1987; 3: 379–88.
12. Biemond P, Van Eijk HG, Swaak AJG, Koster JF. Iron mobilization from ferritin by superoxide derived from stimulated polymorphonuclear leukocytes. Possible mechanism in inflammation diseases. J Clin Invest 1984; 73: 1576–9.
13. Koster JF, Slee RG. Ferritin, a physiological iron donor for microsomal lipid peroxidation. FEBS Lett 1986; 199: 85–8.
14. Opie LH. Reperfusion of the ischemic myocardium. J Mol Cell Cardiol 1977; 9: 605–16.
15. Van der Kraaij AMM, Van Eijk HG, Koster JF. Prevention of postischemic cardiac injury by the orally active iron chelator 1,2-dimethyl-3-hydroxy-4-pyridone (L1) and the antioxidant (+)-cyanidanol-3. Circulation 1989; 80: 158–64.
16. Takemura G, Onodera T, Ashraf M. Quantification of hydroxyl radical and its lack of evidence to myocardial injury during early reperfusion after graded ischemia in rat hearts. Circ Res 1992; 71: 96–105.
17. Liu X, Prasad R, Engelsman RM, Jones RM, Das DK. Role of iron on membrane phospholipid breakdown in ischemic-perfused rat heart. Am J Physiol 1990; 259: H1101–7.
18. Conte Jr JV, Katz NM, Foegh ML, Wallace RB, Ramwell PW. Iron chelation therapy and lung transplantation. Effects of deferoxamine on lung preservation in canine single lung transplantation. J Thorac Cardiovasc Surg 1991; 101: 1024–9.
19. Mergner GW, Weglicki WB, Kramer JH. Postischemic free radical production in the venous blood of the regionally ischemic swine heart. Effect of deferoxamine. Circulation 1991; 84: 2079–90.
20. Smith JK, Carden DL, Grisham MB, Granger DN, Korthuis RJ. Role of iron in post-ischemic microvascular injury. Am J Physiol 1989; 256: H1472–7.
21. Balla G, Vercellotti GM, Eaton JW, Jacob HS. Iron loading of endothelial cells augments oxidant damage. J Lab Clin Med 1990; 116: 546–54.
22. Van der Kraaij AMM, Mostert LJ, Van Eijk HG, Koster JF. Iron-load increases the susceptibility of rat hearts to oxygen reperfusion damage. Protection by the antioxidant (+)-cyanidanol-3 and deferoxamine. Circulation 1988; 78: 442–9.
23. Fontecave H, Pierre JL. Iron metabolism: The low-molecular mass iron pool. Biol Metals 1991; 4: 133–5.
24. Weaver J, Pollack S. Low-Mr iron isolated from guinea pig reticulocytes as AMP-Fe and ATP-Fe complexes. Biochem J 1989; 261: 787–92.
25. Rush JD, Maskos Z, Koppenol WH. Reactions of iron (II) nucleotide complexes with hydrogen peroxide. FEBS Lett 1990; 261: 121–3.
26. Holt S, Gunderson M, Joyce K et al. Myocardial tissue iron delocalization and evidence for lipid peroxidation after two hours of ischemia. Ann Emerg Med 1986; 15: 1155–9.
27. Komara JS, Nayini NR, Bialick HA et al. Brain iron delocalization and lipid peroxidation following cardiac arrest. Ann Emerg Med 1986; 15: 384–9.

28. Healing G, Gower J, Fuller B, Green C. Intracellular iron redistribution. An important determinant of reperfusion damage to rabbit kidneys. Biochem Pharmacol 1990; 39: 1239–45.
29. Gower JD, Healing G, Green CJ. Determination of desferrioxamine-available iron in biological tissues by high-pressure liquid chromatography. Anal Biochem 1989; 180: 126–30.
30. Voogd A, Sluiter W, Van Eijk HG, Koster JF. Low molecular weight iron and the oxygen paradox in isolated rat hearts. J Clin Invest 1992; 90: 2050–5.
31. Hill HAO. Oxygen, oxygenases and the essential trace metals. Philos Trans R Soc London Ser B 1981; 294: 119–28.
32. Freeman BH, Crapo JD. Biology of disease: Free radicals and tissue injury. Lab Invest 1982; 47: 412–26.
33. Boveris A. Mitochondrial production of superoxide radical and hydrogen peroxide. Adv Exp Med Biol 1977; 78: 67–82.
34. Nohl H. Demonstration of the existence of an organo-specific NADH dehydrogenase in heart mitochondria. Eur J Biochem 1987; 169: 585–91.
35. VandePlassche G, Hermans C, Thone F, Borgers M. Mitochondrial hydrogen peroxide generation by NADH oxidase activity following regional myocardial ischemia in the dog. J Mol Cell Cardiol 1989; 21: 383–92.
36. Otani H, Tanaka H, Inoue T et al. In vitro study on contribution of oxidative metabolism of isolated rabbit heart mitochondria to myocardial reperfusion injury. Circ Res 1984; 55: 168–75.
37. Hull FE, Radloff JF, Sweeley CC. Fatty acid oxidation by ischemic myocardium. Rec Adv Stud Card Struct Metab 1975; 8: 153–65.
38. Lands WEM. Interaction of lipid hydroperoxides with eicosanoid biosynthesis. J Free Rad Biol Med 1985; 1: 97–101.
39. Kukreja RC, Kontos HA, Hess ML, Ellis EF. PGH synthase and lipoxygenase generate superoxide in the presence of NADH or NADPH. Circ Res 1986; 59: 612–9.
40. Karmazyn M, Moffat MP. Toxic properties of arachidonic acid on normal ischemic and reperfused hearts. Indirect evidence for free radical involvement. Prostag Leukotr Med 1985; 17: 251–64.
41. Singal PK, Kapur N, Dhillon KS, Beamish RE, Dhalla NS. Role of free radicals in catecholamine-induced cardiomyopathy. Can J Physiol Pharmacol 1982; 60: 1390–7.
42. Rubanyi GM, Vanhoutte PM. Oxygen-derived free radicals, endothelium, and responsiveness of vascular smooth muscle. Am J Physiol 1986; 250: H815–21.
43. Beckman JS, Beckman TW, Chen J, Marshall PA, Freeman BA. Apparent hydroxyl radical production by peroxynitrite: Implications for endothelial injury from nitric oxide and superoxide. Proc Natl Acad Sci USA 1990; 87: 1620–4.
44. Lucchesi BR, Mullane KM. Leukocytes and ischemia-induced myocardial injury. Ann Rev Pharmacol Toxicol 1986; 26: 201–24.
45. McCord JM. Oxygen-derived free radicals in post-ischemic tissue injury. N Engl J Med 1985; 312: 159–63.
46. Downey JM, Miura T, Eddy LJ, Chamers DE et al. Xanthine oxidase is not a source of free radicals in the ischemic rabbit heart. J Mol Cell Cardiol 1987; 19: 1053–60.
47. Janssen M, Van der Meer P, De Jong JW. Antioxidant defences in rat, pig, guinea pig, and human hearts: comparison with xanthine oxidoreductase activity. Cardiovasc Res 1993; 27: 2052–7.
48. Tong Mak I, Kramer JH, Weglicki WB. Potentiation of free radical-induced lipid peroxidative injury to sarcolemmal membranes by lipid amphiphiles. J Biol Chem 1986; 261: 1153–7.
49. Schinetti ML, Rossini D, Greco A, Bertelli A. Protective action of acetylcarnitine on NADPH-induced lipid peroxidation of cardiac microsomes. Drugs Expl Clin Res 1987; 13: 509–15.

50. Ferrari R, Ciampalini G, Agnoletti G, Cargnoni A, Ceconi C, Visioli O. Effect of L-carnitine derivatives on heart mitochondrial damage induced by lipid peroxidation. Pharmacol Res Commun 1988; 20: 125–32.
51. Shug A, Paulson D, Subramanian R, Regitz V. Protective effects of propionyl-L-carnitine during ischemia and reperfusion. Cardiovasc Drugs Ther 1991; 5(Suppl 1): 77–84.
52. Packer L, Valenza M, Serbinova E, Starke-Reed P, Frost K, Kagan V. Free radical scavenging is involved in the protective effect of L-propionyl-carnitine against ischemia-reperfusion injury of the heart. Arch Biochem Biophys 1991; 288: 533–7.
53. Reznick AZ, Kagan VE, Ramsey R et al. Antiradical effects of L-propionyl carnitine protection of the heart against ischemia-reperfusion injury: The possible role of iron chelation. Arch Biochem Biophys 1992; 296: 394–401.
54. Shattock MJ, Manning AS, Hearse DJ. Effects of hydrogen peroxide on cardiac function and post-ischemic function recovery in isolated working heart. Pharmacology 1982; 24: 118–32.
55. Voogd A, Sluiter W, Koster JF. The increased susceptibility to hydrogen peroxide of the (post-)ischemic rat heart is associated with the magnitude of the low molecular weight iron pool. Free Rad Biol Med 1994; 16: 453–8.
56. Van der Vusse GJ, Glatz JFC, Stam HCG, Reneman RS. Fatty acid homeostasis in the normoxic and ischemic heart. Physiol Rev 1992; 72: 881–940.

*Corresponding Author*: Professor Johan Fokke Koster, Department of Biochemistry, Ee 642, Faculty of Medicine and Health Science, Erasmus University Rotterdam, P.O. Box 1738, 3000 DR Rotterdam, The Netherlands

# 10. Carnitine transport in volume-overloaded rat hearts

JOSEF MORAVEC, ZAINAB EL ALAOUI-TABLIBI
and CHRISTIAN BRUNOLD

"An alteration of active carnitine transport did occur during the development of cardiac hypertrophy: both total and carrier-mediated carnitine transport were significantly depressed. The alterations of carrier-mediated transport might be related to a decreased affinity of membrane carrier for L-carnitine (higher apparent $K_M$ for carnitine) rather than to a decreased number of carriers (Vmax unchanged)."

## Introduction

Carnitine concentration in tissue is generally related to mitochondrial volume-density and ability to oxidize fatty acids. The highest tissue carnitine has been detected in ventricular myocardium which, compared to other tissues, presents elevated rates of oxidative phosphorylation [1]. The ability of cardiac mitochondria to oxidize long chain fatty acids is also much higher when compared to skeletal muscle or liver sarcosomes (Table 1). Paradoxically enough, it has been known for many years [3–5] that the heart is missing $\gamma$-butyrobetaine hydroxylase [6, 7], the last enzyme of carnitine synthesizing pathway, and that in the myocardium of different species including man, the carnitine synthesis stops at the level of deoxycarnitine ($\gamma$-butyrobetaine). The tissue presenting the highest carnitine concentrations must therefore take up this essential co-factor of lipid metabolism from the blood where it is supplied by liver and, in some species, by kidney [6, 8].

## Carnitine transport across cardiac cell membrane

Since carnitine concentration in cardiac muscle is considerably higher than in the blood (Table 1), carnitine transport against a large concentration gradient is necessary [9]. The mechanism of carnitine transport has been studied in a variety of experimental systems: human fetal heart cells in culture [10, 11], adult rat cardiomyocytes [12], isolated skeletal muscle [13], heart

*J.W. de Jong and R. Ferrari (eds): The carnitine system*, 133–144.

*Table 1.* Total tissue carnitine and rate of fatty acid oxidation by isolated mitochondria from different tissues of the rat.

| Tissue | Carnitine (μmol/g wet wt) | State 3 $QO_2$ (μmol/min/mg protein) |
|---|---|---|
| Heart | 1.00–1.40 | 180 |
| Muscle | 0.50–1.00 | 80 |
| Liver | 0.20–0.40 | 60 |
| Blood | 0.05–0.08 | – |

From Siliprandi et al. [1] and Bode & Klingenberg [2].

slices [14] and perfused rat hearts [9]. Despite heterogeneity of these preparations, the existence of a saturable carrier-mediated transport could always be established. This latter process is optimally active at physiological concentrations of extracellular carnitine: at 50 μM extracellular carnitine about 80% of uptake occurs by carrier-mediated system. At higher concentrations (100–1000 μM), free diffusion of L-carnitine can significantly contribute to carnitine uptake by tissue [9, 15]. At 200 μM extracellular carnitine, about half of the L-carnitine transported is taken up by the saturable carrier-mediated process, the remaining 50% cross the membrane by passive diffusion [9, 12].

Cardiac carrier-mediated carnitine transport was initially characterized on an established line (CCL-27) of fetal cardiac cells in culture [10] and on isolated adult rat heart myocytes [12, 16]. According to these studies, the carnitine transport was found three- to ten-fold faster in heart cells than in fibroblasts. γ-Butyrobetaine was taken up to the same extent as carnitine [10], and other compounds that contained trimethylamino- and carboxylic groups reduced carnitine uptake. The substances without quaternary ammonia were found without effect. The membrane carrier had about 25-fold higher affinity for L-carnitine than for the D-isomer [10]. A similar situation also occurred in the intact rat heart [9] where, on the other hand, the membrane carrier proved to have higher affinity for short-chain acylcarnitine than for L-carnitine itself [17]. This may explain why upon a single injection of propionyl-L-carnitine in vivo the myocardial carnitine levels attain higher levels than after the administration of L-carnitine per se (Moravec et al., unpublished observation).

The kinetic parameters for the carrier-mediated carnitine transport differ according to the preparation used [8, 9]. The apparent $K_M$ for carnitine is about 5 μM in Girardi heart cells [10]. It is close to 60 μM for isolated myocytes from adult rats [16] and, in isolated rat hearts, it ranges from about 25 μM [9] to 83 μM [18]. Recently, it has been suggested that, at least in human cultured myoblasts [15], two distinct components of carrier-mediated saturable transport can be distinguished: a high affinity uptake occurs between 0.5 and 10 μM carnitine, and a low affinity uptake of carnitine operates at carnitine concentrations between 25 and 200 μM. The high affinity uptake ($K_M$ 4.17–5.50 μM, Vmax 11.80 –19.60 pmol/h/mg protein) does not change during maturation of cells in culture. On the other hand, the low affinity uptake is influenced by muscle maturation in vitro. The decrease in Vmax (from 140 to 39 pmol/h/mg protein) was interpreted to suggest a reduction in total number of carriers available. The concomitant decrease of apparent $K_M$ (from 160 to 14 μM) compensated for this former change by increased specificity. The authors concluded that the maturation of muscle cultures may be accompanied by a dedifferentiation of carnitine binding protein of Cantrell & Borum [19]. They also suggested that a similar defect of low affinity carnitine uptake may be the cause of the primary carnitine deficiency syndromes and lipid storage myopathies. In this respect it is of interest to note that alterations of carnitine binding protein may occur in hereditary cardiomyopathies [20] and that prednisolone and other glucocorticoids proved to have therapeutic effects on different carnitine deficiency syndromes [21]. When $10^{-8}$ to $10^{-5}$ M prednisolone was added into the established cell line in culture, it actually increased the carnitine uptake and intracellular carnitine concentration to about 160 and 120%, respectively [21].

The results obtained by different authors are consistent with the idea that the uphill transport of L-carnitine by muscle cells needs energy. However, a careful review of the literature reveals that carnitine transport, as studied on different muscle cell preparations, does not require a direct mobilization of energy production. For instance, in the experiments of Molstad et al. [11], inhibitors of glycolysis or oxidative phosphorylation did not consistently inhibit the uphill transport of carnitine in human cultured cells. In the experiments of Rebouche [13] on isolated rat muscle, $CN^-$ had only a slight inhibitory effect. Similarly, severe anoxia did not affect the carnitine transport by isolated rat hearts [9]. All these experiments seem to suggest that the energy source required for carnitine transport is not ATP directly. Alternatively, the uptake could occur by a functional trapping of newly transported intracellular carnitine through acylation [9] or by carnitine/deoxycarnitine exchange [22]. The fact that total tissue carnitine did not change significantly during carnitine loading experiments [13] argues against an excessive carnitine trapping by acylation [23]. Therefore, the existence of an antiport insensitive to metabolic inhibitors but sensitive to sulfydryl group blocking agents was suggested [14, 22]. According to Siliprandi et al. [1] who worked on rat heart slices, the exchange between internal deoxycarnitine and external carnitine is a saturable process with an apparent $K_M$ (carnitine) of 23 μM. The temperature dependence of the exchange does not fit with a simple diffusion mechanism, but suggests a facilitated transport [22]. This concept may be integrated in a more general schema of carnitine/deoxycarnitine exchange between different organs such as liver (kidney) on one hand and muscle and heart on the other hand [6, 8].

However, in kidney [24], brain [25] and skeletal muscle [13], carnitine transport has been suggested to occur down the $Na^+$ concentration gradient, perhaps involving a co-transport system similar to that of amino acids [26]. According to Vary and Neely [27] this could also be the case in cardiac muscle. These authors suggested that carnitine transport into the intact perfused rat hearts was influenced strongly by the membrane electrochemical potential: lowering of extracellular $Na^+$ (from 140 to 25 mM) led to a decrease of carnitine transport whereas reduction of extracellular $K^+$ (from 5.9 to 0.6 mM) stimulated the carnitine transport by 35%. These effects were visible over a range of extracellular carnitine concentrations examined (15 to 100 μM) and resulted primarily from changes of maximal rate of transport (Vmax) and not the apparent $K_M$ for carnitine [27]. To determine whether the effects of $Na^+$ and $K^+$ were due to changes in the $Na^+$ flux across the sarcolemmel membrane, Vary and Neely [26] also tested various compounds known to alter $Na^+$ movements. Ouabain ($10^{-3}$ M) did not reduce carnitine uptake, suggesting that carnitine transport was independent of the $Na^+/K^+$-ATPase activity. Tetrodotoxin ($10^{-5}$ M), a fast channel blocker, induced cardiac arrest and stimulated carnitine transport by about 40%. This indicated that inward carnitine transport was not dependent on the fast $Na^+$ channel and that it could be dissociated from membrane depolarization. Gramicidin ($5 \cdot 10^{-6}$ M) inhibited slightly carnitine transport as it reduced $Na^+$ transmembrane gradient. All these experiments are in qualitative agreement that ventricular myocardium, like the brain [25], kidney [24] and skeletal muscle [13], may accumulate L-carnitine and its analogs without direct hydrolysis of ATP. According to this hypothesis, the energy for carnitine transport against a large concentration gradient is derived from the movement of $Na^+$ down its electrochemical gradient.

## Carnitine deficiency cardiomyopathies

One of the principal consequences of well-regulated carnitine uptake by the heart consists in relatively constant tissue levels of L-carnitine [9]. This is necessary in order to ensure an appropriate control of lipid [28–30] and carbohydrate [31] metabolism (see also Chapter 4). In fact, carnitine concentrations were shown to decrease under a variety of experimental conditions that are also characterized by altered myocardial fatty acid utilization. In rats, carnitine content of the heart is relatively low at birth [32]. It increases rapidly during postnatal life and decreases again in old animals [33]. In addition to these physiological fluctuations, tissue L-carnitine content has been found to decrease under different pathological conditions [34, 35]. In some cases, myocardial carnitine deficiency results from a hereditary defect of carnitine synthesis in the liver which leads to a dramatic decrease of circulating blood carnitine (primarily systemic carnitine deficiency syndromes). In other conditions, the defect of carnitine liberation from the liver may be acquired during late postnatal life (secondary systemic carnitine deficiency syndromes). This seems to be the case of experimental diabetes characterized by a retention of carnitine in the liver and by decreased circulating levels of free carnitine [34]. According to Vary and Neely [23], reduced levels of carnitine in hearts of alloxan-treated rats may result from a lower transport rate due to decreased serum carnitine concentrations (below the apparent $K_M$ for carnitine). The kinetics of carrier-mediated carnitine transport per se is supposed to be unchanged.

In other conditions, a selective muscle carnitine deficiency is attributed to intrinsic defects of carrier-mediated carnitine transport occurring in the absence of any alteration of carnitine synthesis in the liver [35]. Carnitine distribution in the body is also unaffected (circulating carnitine concentrations are in the range of physiological values). There are several case reports of human carnitine deficiency syndromes [35, 36] which enter into this special category (primary muscle carnitine deficiencies). As concerns the cardiac muscle, it may become carnitine deficient under a variety of pathological conditions. Syrian cardiomyopathic hamsters which develop cardiac hypertrophy and failure [29, 37] were shown to have reduced cardiac carnitine concentrations that probably resulted from impaired transport of carnitine into the myocytes [20]. A similar situation may also occur in hearts of other species in response to a chronic mechanical overloading [28, 38–41]. However, systematic studies of the kinetics of carrier-mediated carnitine transport in hypertrophied and dilated hearts still remain quite scarce [18, 41, 42].

After the demonstration that banding of abdominal aorta results in a significant decrease of total tissue L-carnitine [39], Reibel and co-workers have studied carnitine transport in the hearts of rats with 4-wk-old aortic stenosis [42]. According to this work, there was a 20% reduction of carrier-mediated transport of carnitine at all concentrations examined, i.e. 45, 100 and 200 μM, whereas there was no change in carnitine uptake by diffusion. Total tissue carnitine was decreased by about 20%, while there was no change in serum carnitine concentrations. A similar situation also prevailed in hearts of rats with chronic volume-overload related to a surgically created aorto-caval communication [18]. The principal results of this series of experiments are described in the following section.

## Carnitine transport in volume-overloaded rat hearts

A chronic volume-overload was induced in 2-mo-old Wistar rats by a surgical opening of the aorto-caval fistula [43]. Sham-operated animals from the same litters were used as controls. The animals were sacrificed 3 months after surgery; their hearts were used for in vitro experiments [18]. After the initial 10-min period, which is necessary to remove blood and stabilize the heart rate at about 240 beats/min, Langendorff-perfused hearts were recirculated for variable periods (10, 15, 20, 30, 45 min) with 200 ml of Krebs-Henseleit buffer containing 11 mM glucose and 50 μM L-$^{14}$C-carnitine (specific activity about 450 cpm/nmol). In experiments designed to study the kinetic properties of carrier-mediated carnitine transport, the hearts were perfused for 30 min with the same buffer containing variable concentrations of L-$^{14}$C- carnitine

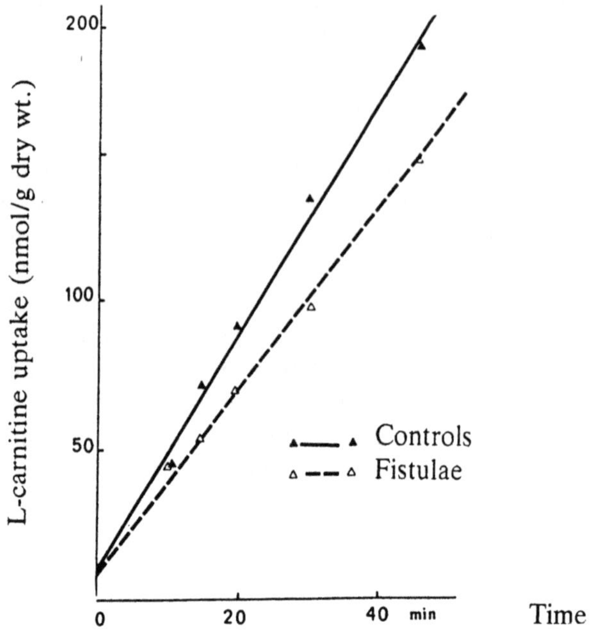


*Figure 1.* L-$^{14}$C-Carnitine uptake as function of time. Hearts were perfused with Krebs-Henseleit bicarbonate buffer containing 11 mM glucose and 50 μM L-carnitine (specific activity 450 counts $\cdot$ min$^{-1}$. nmol$^{-1}$) for various periods of time (10–45 min). After this loading period, the hearts were perfused for an additional 6 min with a buffer that did not contain carnitine. They were then frozen for analysis of tissue radioactivity. Carnitine uptake rate was calculated from tissue radioactivity and perfusate specific activity. Each point represents 5–10 observations. Note decreased slope of carnitine uptake rate in volume-overloaded hearts.

(10, 20, 50, 70, 100, 200 and 250 μM). At the end of the perfusion with labelled carnitine the hearts were perfused for an additional 6 min with a carnitine-free buffer. This proved to be necessary for the complete elimination of radioactive label from the extracellular spaces [9]. Beyond this point, tissue radioactivity in both control and volume-overloaded hearts remained steady for a further 10 min, suggesting that no major leaks of the intracellular L-$^{14}$C-carnitine occurred during the washout period. Following this washout perfusion, the hearts were rapidly frozen by cool Wollenberger clamps and stored in liquid nitrogen. The tissue was then lyophilized and aliquots of about 100 mg dry wt were introduced into the polycarbonate capsules. The samples were mineralized in a catalytic oven and $^{14}CO_2$ quantitatively trapped in a scintillation liquid containing phenylethylamine [18]. The amount of L-carnitine taken by the hearts was expressed in nmol $\cdot$ g$^{-1}$ dry wt $\cdot$ h$^{-1}$, after the radioactivity of the tissue and the specific activity of

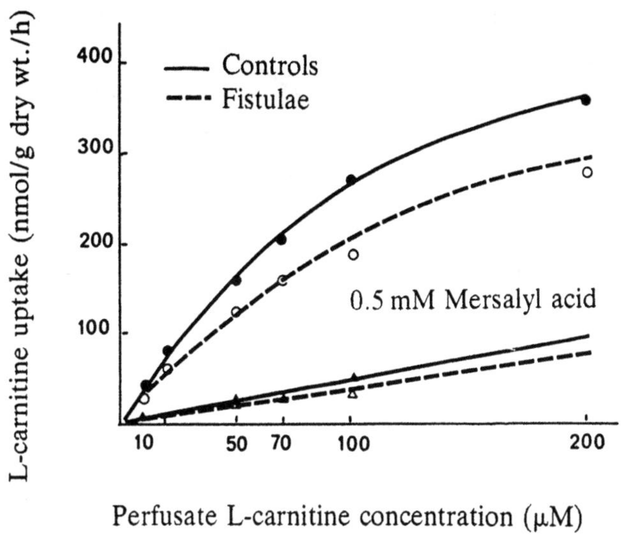


*Figure 2.* Uptake of L-$^{14}$C-carnitine by control and volume-overloaded hearts as function of perfusate L-carnitine concentrations. Hearts were recirculated for 20 min with a Krebs-Henseleit buffer containing 11 mM glucose and variable concentrations of L-$^{14}$C-carnitine. In some experiments, 0.05 mM mersalyl acid was added to the perfusion medium in order to assess passive carnitine transport. Mean values for 5–10 measurements are indicated.

L-carnitine in the perfusion medium (cpm · nmole$^{-1}$) were determined. In some experiments, 0.05 mM mersalyl acid was added to the perfusate in order to subtract the passive component from total carnitine transport to the heart [9]. The analysis of the saturation curves thus obtained allowed to compare the respective kinetics of carrier-mediated transport in control and overloaded hearts. The Vmax and apparent $K_M$ of saturable carnitine transport were determined from the Lineweaver-Burk reciprocal plot.

The heart weights of rats exposed to a chronic aorto-caval fistula were increased by about 85% when compared to those of sham-operated controls. Serum carnitine concentrations and tissue contents of total tissue L-carnitine are given in Table 2. It can be seen that chronic volume-overload resulted in similar changes as banding of abdominal aorta [42]: a 30% depletion of tissue carnitine occurred in the absence of any alteration of circulating L-carnitine suggesting thus an impaired carnitine transport to tissue. The measurements of carnitine uptake rate by control and volume-overload hearts confirmed that this was really the case (Figure 1). When the control hearts were perfused with 50 μM L-$^{14}$C- carnitine, their tissue radioactivity progressively increased. A similar linear relationship also occurred in mechanically

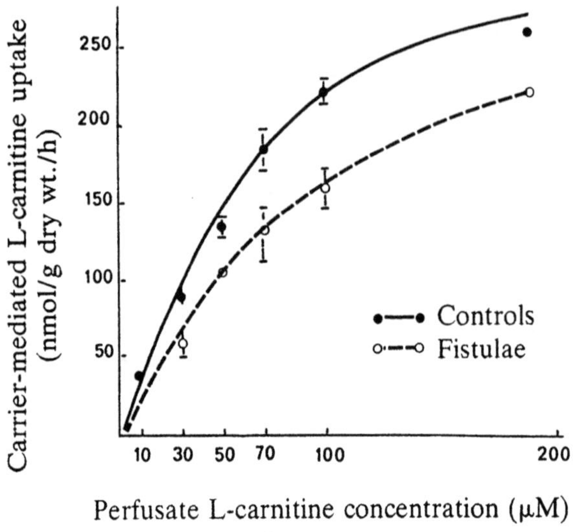


*Figure 3.* Saturation curves of carrier-mediated carnitine transport as observed in control and volume-overloaded rat hearts. Curves generated from data given in Figure 2 by substraction of diffusion component from total carnitine uptake. Mean values ± SEM (n = 5–10).

*Table 2.* Myocardial and blood plasma carnitine in control and volume-overloaded rats

| Condition | Total tissue carnitine (nmol · $g^{-1}$ dry wt) | Blood plasma carnitine (μM) |
|---|---|---|
| Controls | 5550 ± 150 | 45.8 ± 1.2 |
| Fistulae | 3864 ± 175* | 44.0 ± 2.4 |

The values are mean ± SEM for n = 10; * $p < 0.001$.

overloaded hearts. However, in this case, the slope of L-carnitine accumulation was considerably depressed (Figure 1). This decrease in L-carnitine uptake did not seem to be related to changes in coronary flow rate which was quite comparable in control and overloaded hearts. Figures 2 and 3 show clearly that an alteration of active carnitine transport did occur during the development of cardiac hypertrophy: both total and carrier-mediated carnitine transport were significantly depressed over the range of exogenous carnitine concentrations examined (10–200 μM). On the other hand, the passive diffusion of L-carnitine was unchanged. According to Lineweaver-Burk analysis of the respective saturation curves (Figure 4), the alterations of carrier-mediated transport might be related to a decreased affinity of membrane carrier for L-carnitine (higher apparent $K_M$ for carnitine) rather than to a decreased number of carriers (Vmax unchanged). These data argue

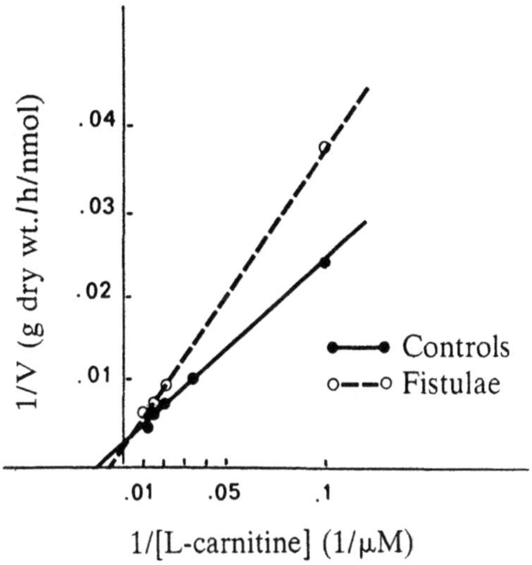


*Figure 4.* Kinetic analysis of saturable carnitine transport by reciprocal plots of Lineweaver and Burk. Note increase of apparent $K_M$ in volume-overloaded hearts (125 instead of 83 μM). On the other hand, Vmax of carnitine transport is unchanged (380 nmol · $g^{-1}$ dry wt per h). These alterations suggest that a decreased affinity of membrane carrier for L-carnitine occurs during the development of cardiac hypertrophy.

against any major accumulation of intracellular $Na^+$ (extracellular sodium concentration was kept constant by use of a freshly prepared medium containing 138 mM $Na^+$): the decrease of the $Na^+$ electrochemical gradient was shown to affect primarily the maximal rate of transport and not the apparent Michaelis constant for carnitine [27]. In fact, to date, there is no direct evidence indicating that the $Na^+$ electrochemical gradient is altered in hypertrophied hearts [41, 42].

Recently, the carnitine-binding protein prepared from hearts of cardiomyopathic hamsters has been shown to have a lower maximal carnitine binding and an increased dissociation constant when compared with the protein prepared from hearts of normal hamsters [20]. It is not excluded that similar alterations of the cardiac carnitine-binding protein occur in other models of hypertrophic and dilated cardiomyopathies. They could then contribute to the reduced affinity of carnitine carrier for L-carnitine that we described in chronically volume-overloaded hearts. However, the alterations of membrane lipids that also interfere with carrier-mediated transport processes [44] should also be considered. In this respect, the report of Reibel et al. [45] is of interest. These authors have found that the fatty acyl composition of total membrane phospholipids changes during the development of pressure-induced cardiac hypertrophy. Three weeks after the constriction of abdominal aorta in rats, the content of myocardial phosphatidylcholine, phosphatidylinositol and sphingomyelin increased by 10–20%. The essential fatty acid, linoleic acid, was markedly reduced in major classes of cardiac phospholipids, while arachidonic acid tended to increase [45]. The authors concluded that these alterations may influence the function of membrane-bound enzymes and mainly that of carnitine carrier. However, systematic studies of sarcolemmal membranes are still needed in order to verify whether the alterations of phospholipid and essential fatty acid composition affect this particular membrane and whether they really interfere with carnitine transport.

In conclusion, the data presented in this work confirm the contention of other authors [28, 29, 38, 39] concerning the mechanism of decreased carnitine content in mechanically overloaded hearts. In hearts of rats with aortic banding [42] as well as in those with an aorto-caval fistula [18, 41], decreased tissue L-carnitine relies on altered carrier-mediated carnitine uptake by the myocardium. In vivo, cardiac carnitine depletion occurs in the presence of normal circulating levels of L-carnitine. This suggests that hypertrophied and failing hearts may develop a secondary carnitine deficiency resulting from a dysfunction of carnitine-binding protein [19]. One of the principal consequences of decreased tissue levels of L-carnitine consists in impaired long-chain fatty acid utilization [28, 43]. The latter can be partly corrected by the administration of exogenous L-carnitine [28] or of its short-chain acyl derivatives such as propionyl-L-carnitine [46, 47].

## References


1. Siliprandi N, Di Lisa F, Pivetta A, Miotto G, Siliprandi D. Transport and function of L-carnitine and L-propionylcarnitine: relevance to some cardiomyopathies and cardiac ischemia. Z Kardiol 1987; 76(Suppl 5): 34–40.
2. Bode Ch, Klingenberg M. Die Veratmung von Fettsaüren in isolierten Mitochondrien. Biochem Z 1965; 341: 271–99.
3. Frenkel RA, McGarry JD, editors. Carnitine biosynthesis, metabolism and functions. New York: Academic Press, 1980.
4. Cox RA, Hoppel CL. Carnitine and trimethylaminobutyrate synthesis in the rat tissues. Biochem J 1974; 142: 699–701.
5. Sachan DS, Hoppel CL. Carnitine biosynthesis. Hydroxylation of N6–trimethyl-lysine to 3–hydroxy-N6–trimethyl-lysine. Biochem J 1980; 188: 529–34.
6. Rebouche CJ. Comparative aspects of carnitine biosynthesis. In: Frenkel RA, McGarry JD, editors. Carnitine biosynthesis, metabolism and function. New York: Academic Press, 1980: 50–110.
7. Engel AG. Possible causes and effects of carnitine deficiency in man. In: Frenkel RA, McGarry JD, editors. Carnitine biosynthesis, metabolism and function. New York: Academic Press, 1980: 271–84.
8. Bohmer T, Molstad P. Carnitine transport across the plasma membrane. Frenkel RA, McGarry JD, editors. Carnitine biosynthesis, metabolism and function. New York: Academic Press, 1980: 73–88.

9. Vary TC, Neely JR. Characterization of carnitine transport in isolated perfused adult rat hearts. Am J Physiol 1982; 242: H585–92.
10. Bohmer T, Eiklid K, Jonsen J. Carnitine transport into human heart cells in culture. Biochim Biophys Acta 1977; 465: 627–33.
11. Molstad P, Bohmer T, Eiklid K. Specificity and characteristics of the carnitine transport in human heart cells (CCL 27) in culture. Biochim Biophys Acta 1977; 471: 296–304.
12. Bahl JJ, Navin TR, Bressler R. Carnitine uptake and stimulation of carnitine uptake in isolated beating adult rat heart myocytes. In: Frenkel RA, McGarry JD, editors. Carnitine biosynthesis, metabolism and function. New York: Academic Press, 1980: 91–112.
13. Rebouche CJ. Carnitine movement across muscle cell membranes. Studies in isolated rat muscle. Biochim Biophys Acta 1977; 471: 145–55.
14. Sartorelli L, Ciman M, Siliprandi N. Carnitine transport in rat heart slices: I. The action of thiol reagents on the acetylcarnitine/carnitine exchange. Ital J Biochem 1985; 34: 275–81.
15. Martinuzzi A, Vergani L, Rosa M, Angelini C. L-carnitine uptake in differentiating human cultured muscle. Biochim Biophys Acta 1991; 1095: 217–22.
16. Bahl J, Navin T, Manian AA, Bressler R. Carnitine transport in isolated adult rat heart cardiomyocytes and the effect of 7,8–diOH chlorpromazine. Circ Res 1981; 48: 378–85.
17. Paulson DJ, Traxler J, Schmidt M, Noonan J, Shug AL. Protection of the ischaemic myocardium by L-propionylcarnitine: effects on the recovery of cardiac output after ischaemia and reperfusion, carnitine transport, and fatty acid oxidation. Cardiovasc Res 1986; 20: 536–41.
18. El Alaoui-Talibi Z, Moravec J. Carnitine transport and exogenous palmitate oxidation in chronically volume-overloaded rat hearts. Biochim Biophys Acta 1989; 1003: 109–14.
19. Cantrell CR, Borum PR. Identification of a cardiac carnitine binding protein. J Biol Chem 1982; 257: 10599–604.
20. York CM, Cantrell CB, Borum PR. Cardiac carnitine deficiency and altered carnitine transport in cardiomyopathic hamsters. Arch Biochem Biophys 1983; 221: 526–33.
21. Molstad P, Bohmer T. The effect of diphteria toxin on the cellular uptake and efflux of L-carnitine. Evidence for a protective effect of prednisolone. Biochim Biophys Acta 1981; 641: 71–8.
22. Sartorelli L, Ciman M, Mantovani G, Siliprandi N. Carnitine transport in rat heart slices: II. The carnitine/deoxycarnitine antiport. Biochim Biophys Acta 1985; 34: 282–7.
23. Vary TC, Neely JR. A mechanism for reduced myocardial carnitine levels in diabetic animals. Am J Physiol 1982; 243: H154–8.
24. Huth PJ, Shug AL. Properties of carnitine transport in rat kidney cortex slices. Biochim Biophys Acta 1980; 602: 621–34.
25. Huth PJ, Schmidt MJ, Hall PV, Fariello RG, Shug AL. The uptake of carnitine by slices of rat cerebral cortex. J Neurochem 1981; 36: 715–23.
26. Shultz SG, Curran PF. Coupled transport of sodium and organic solutes. Physiol Rev 1970; 50: 636–718.
27. Vary TC, Neely JR. Sodium dependence of carnitine transport in isolated perfused adult rat hearts. Am J Physiol 1983; 244: H247–52.
28. Wittels B, Spann Jr JFJ. Defective lipid metabolism in the failing heart. J Clin Invest 1968; 47: 1787–93.
29. Whitmer JT. Energy metabolism and mechanical function in perfused hearts of Syrian hamsters with dilated or hypertrophic cardiomyopathy. J Mol Cell Cardiol 1986; 18: 307–17.
30. Vary TC, Reibel DK, Neely JR. Control of energy metabolism of heart muscle. Annu Rev Physiol 1981; 43: 419–30.
31. Broderick TL, Quinney HA, Barker CC, Lopaschuk GD. Beneficial effect of carnitine on mechanical recovery of rat hearts reperfused after a transient period of global ischemia is accompanied by a stimulation of glucose oxidation. Circulation 1993; 87: 972–81.

32. Wittels B, Bressler P. Lipid metabolism in the newborn hearts. J Clin Invest 1965; 44: 1639–46.
33. Abu-Erreish GM, Neely JR, Whitmer JT, Whitman V, Sanadi DR. Fatty acid oxidation by isolated perfused working hearts of aged rats. Am J Physiol 1977; 232: E258–62.
34. Borum PR. Carnitine. Ann Rev Nutr 1983; 3: 233–59.
35. Rebouche CJ, Engel AG. Kinetic compartmental analysis of carnitine metabolism in the human carnitine deficiency syndromes. Evidence for alterations in tissue carnitine transport. J Clin Invest 1984; 73: 857–67.
36. VanDyke DH, Griggs RC, Markesbery W, Dimauro S. Hereditary carnitine deficiency of muscle. Neurology 1975; 25: 154–9.
37. Whitmer JT. L-carnitine treatment improves cardiac performance and restores high-energy phosphate pools in cardiomyopathic Syrian hamster. Circ Res 1987; 61: 396–408.
38. Revis NW, Cameron AJV. Metabolism of lipids in experimental hypertrophic hearts of rabbits. Metabolism 1979; 28: 601–13.
39. Reibel DK, Uboh CE, Kent RL. Altered coenzyme A and carnitine metabolism in pressure-overload hypertrophied rat hearts. Am J Physiol 1983; 244: H839–43.
40. Bowé C, Nzonzi J, Corsin A, Moravec J, Feuvray D. Lipid intermediates in chronically volume-overloaded rat hearts. Effect of diffuse ischemia. Pflügers Arch 1984; 402: 317–20.
41. El Alaoui-Talibi Z, Moravec J. Decreased L-carnitine transport in mechanically overloaded rat hearts. In: Jacob R, Just H, Holubarsch, editors. Cardiac energetics. New York: Springer Verlag, 1987: 223–32.
42. Reibel DK, O'Rourke B, Foster KA. Mechanisms for altered carnitine content in hypertrophied rat hearts. Am J Physiol 1987; 252: H561–5.
43. El Alaoui-Talibi Z, Landormy S, Loireau A, Moravec J. Fatty acid oxidation and mechanical performance of volume-overloaded rat hearts. Am J Physiol 1992; 262: H1068–74.
44. Spector AA, Yorek MA. Membrane lipid composition and cellular function. J Lipid Res 1985; 26: 1015–35.
45. Reibel DK, O'Rourke B, Foster KA, Hutchinson H, Uboh CE, Kent RL. Altered phospholipid metabolism in pressure-overload hypertrophied hearts. Am J Physiol 1986; 250: H1–6.
46. El Alaoui-Talibi Z, Bouhaddioni N, Moravec J. Assessment of the cardiostimulant action of propionyl-L-carnitine on chronically volume-overloaded rat hearts. Cardiovasc Drugs Ther 1993; 7: 357–63.
47. Torielli L, Conti F, Cinato E, Ceppi E, Anversa P, Bianchi G, Ferrari P. Alterations in energy metabolism of hypertrophied rat cardiomyocytes: influence of propionyl-L-carnitine. Am J Physiol. In press.

*Corresponding Author*: Dr Josef Moravec, Lab. d'Energetique et de Cardiologie Cellulaire, INSERM, Faculté de Pharmacie, Université de Bourgogne, 7, Bvd Jeanne d'Arc, F-21033 Dijon, France; ***present address:*** Dept. of Physiology, Bât 404, Claude Bernard University, 43 Bd du 11 Novembre 1918, 69622 Villeurbanne, France

# 11. Myocardial carnitine deficiency in human cardiomyopathy

VERA REGITZ-ZAGROSEK and ECKART FLECK

"In heart failure, the myocardial carnitine level can decrease to concentrations in the range of the $K_M$ of carnitine palmityltransferase for free carnitine. Thus, in these cases a reduced availability of free carnitine may limit the transferase reaction and thereby fatty acid oxidation."

## Introduction

In the literature an important role for carnitine has been discussed in myocardial ischemia and several non-ischemic heart diseases. The presumably broad spectrum of carnitine effects in the heart is based on the central role of this compound in fatty acid oxidation and in the control of intermediary metabolism [1–3]. Long-chain acyl-CoA esters may only penetrate into the mitochondrial matrix in the form of their carnitine esters, and intracellular concentrations of long-chain acyl-CoA and long-chain acylcarnitine as well as free CoA depend on the availability of free carnitine. Thus, this availability controls basic cellular functions such as energy production and energy transport from the mitochondria into cytoplasm. The interaction between carnitine and potentially toxic products of fatty acid metabolism, as long-chain acyl-CoA esters, may explain the protection of mitochondrial function and the modulation of the adenine nucleotide translocator activity by carnitine as well as its effects on the integrity and fluidity of sarcolemmal membranes.

Systemic carnitine deficiency with low plasma and tissue carnitine levels represents an undisputed indication for the therapeutic use of carnitine, independent of the cause of carnitine deficiency. In inborn organ-specific carnitine deficiency with elevated plasma and reduced tissue carnitine levels, the therapeutic use of carnitine is still questionable. Secondary acquired myocardial carnitine deficiency may occur in ischemia, in diphtheria, untreated diabetes mellitus and adriamycin toxicity and responds to carnitine substitution in some animal models [4–6]. Patients in the end-stage of heart

*J.W. de Jong and R. Ferrari (eds): The carnitine system*, 145–166.

failure also exhibit disturbances of carnitine metabolism; elevated serum carnitine and low myocardial carnitine levels have been documented [7, 8].

The purpose of this chapter is to review the different forms of carnitine deficiency that are associated with impaired cardiac function or the clinical syndrome of heart failure. Inborn systemic and organ-specific carnitine deficiency in humans and animal models will be discussed as well as acquired forms of myocardial carnitine deficiency in adult heart failure.

## Human primary systemic carnitine deficiency

In the 1970s and 1980s a number of cases of systemic carnitine deficiency have been described [9–14]. The case presented by our own group in 1982 fulfills all the classic criteria for systemic carnitine deficiency [15]. Four out of 5 children from healthy, non-related parents developed dilated cardiomyopathy. The two oldest children died in respiratory failure at the age of 15 and 22 months. The third child showed cardiomyopathy at the age of 11 months, progressing to open heart failure [New York Heart Association (NYHA) class III] with exercise tolerance far below the 10% percentile at 8 years. A skeletal muscle biopsy was taken and the carnitine concentrations were determined in plasma and skeletal muscle. As carnitine was low in both compartments, the patient received carnitine substitution. The 4th child was healthy. In the 5th child respiratory infection led to death at the age of 2 years. Plasma carnitine levels were low normal in their parents and severely decreased in the 3rd and 5th child. Tissue carnitine concentrations in patient 5, where samples were obtained at autopsy, were 1% below normal in heart and liver. Light and electron microscopy of muscle biopsies showed excessive accumulation of lipids together with anomalous mitochondria in patients 3 and 5 (Figure 1).

Oral substitution with L-carnitine in patient 3 led to a significant clinical improvement within two weeks, with NYHA class III going to IIA. Mitral valve regurgitation disappeared, left atrial and ventricular sizes decreased and the ejection fraction increased. In a repeated skeletal muscle biopsy after 6 months of carnitine substitution, vacuolation disappeared, mitochondria normalized, and the amount of accumulated lipids was reduced (Figure 1). Correspondingly, the carnitine content of the skeletal muscle rose from 30 nmol/g before therapy to 300 nmol/g, which is still below normal. Low tissue carnitine levels in the presence of elevated plasma carnitine concentrations support the idea of a transport defect, particularly a defect in the accumulation of carnitine in muscles. Such a defect has indeed been found in fibroblasts from carnitine-deficient children [16].

Comparable patients with severely reduced systemic carnitine levels have been described by Chapoy et al. [11] and Waber et al. [17]. All symptoms in these cases can be explained by the low carnitine levels and the impossibility to utilize long-chain fatty acids. Classical symptoms are hypotension,

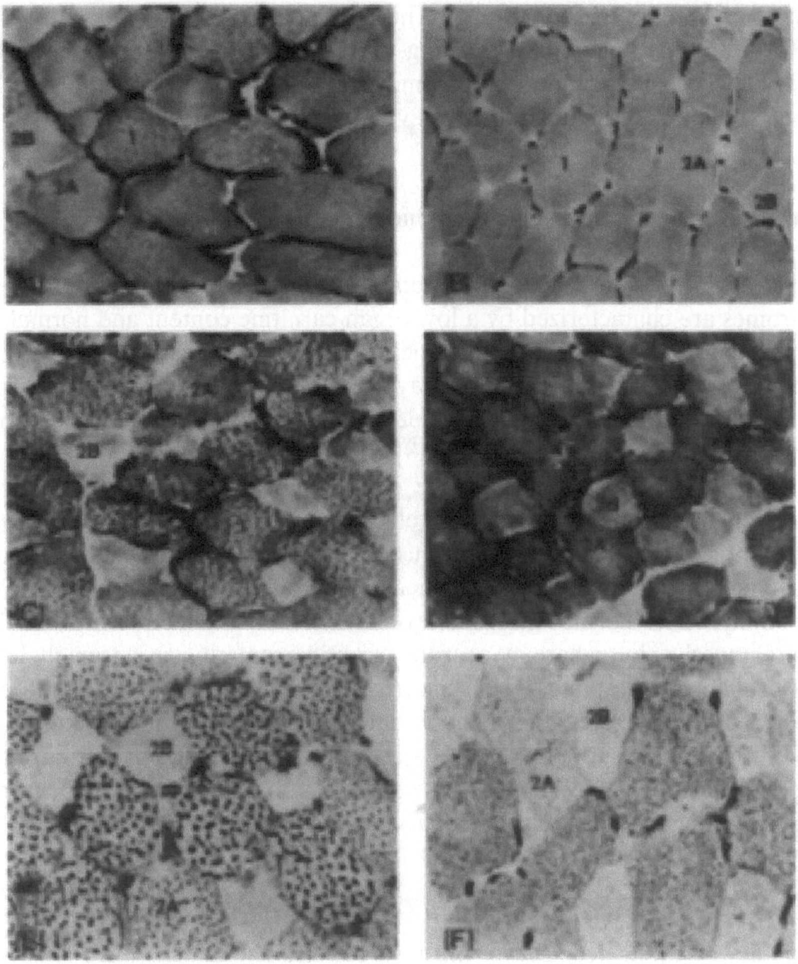

*Figure 1.* Skeletal muscle biopsy in a carnitine-deficient patient (child 3) before carnitine therapy showed variations in the fiber diameter (hematoxilin eosin stain) [A], and a large number of vacuoles, identified by the NADH stain as mitochondria [C]. Staining with Sudan red gave evidence of the accumulation of lipids in type 1 fibers [E]. 2A and 2B refer to subtypes of type 2 muscle fibers; the latter were atrophic. After 6 months of carnitine therapy, fiber diameters normalized [B], and vacuoles [D] and lipid accumulation [F] were reversed.

muscle weakness, and excessive accumulation of lipids in the skeletal muscles. Hypoglycemia occurs frequently, as energy utilization depends almost completely on carbohydrates. The participation of the myocardium in the form of a cardiomyopathy leading to heart failure determines the prognosis. Diagnosis is easy to obtain by measurements of carnitine levels in the blood, and improvement can be achieved by carnitine substitution.

## Human primary organ-specific carnitine-deficiency syndromes

In contrast to systemic carnitine deficiency, organ-specific carnitine-deficiency syndromes are characterized by a low organ carnitine content and normal or elevated plasma levels. The pathogenesis is not clear; organ-specific defects in the transport system for carnitine have been discussed. Normally, carnitine concentrations in the myocytes of skeletal muscle or heart are 20 to 50-fold above the blood concentrations [18, 19]. Increased leakage of carnitine from the respective cells can cause tissue carnitine deficiency. The best documentation has been obtained for skeletal muscle carnitine-deficiency syndromes [9, 20]. Heart and skeletal muscles are frequently affected together, but isolated myocardial carnitine deficiency has also been described [21–23] in a family where normal carnitine levels were found in the plasma and the skeletal muscles, but where myocardial carnitine content was extremely low. This probably corresponds to an inborn isolated myocardial carnitine deficiency. Different expression of the carnitine carrier in different organs can explain the clinical and laboratory features in the mentioned patient groups. Inborn defects of palmityl carnitine transferase may manifest with features similar to organ-specific carnitine deficiency, such as muscle weakness, renal failure, and elevated serum lipids [24, 25].

## Secondary carnitine-deficiency syndromes and animal models

Acquired carnitine-deficiency syndromes in children are frequently caused by antiepileptic treatment or defects in branched-chain amino acid metabolism [26–28]. In adults, ischemia, thermal injury, and shock are frequent causes of carnitine deficiency [6, 29]. In addition, a 50% to 75% reduction in myocardial carnitine levels was found in heart- muscle biopsies of patients treated with adriamycin [AL Shug, unpublished data]. Pathophysiological aspects of adriamycin-induced cardiomyopathy have been studied in a rat model [30]. Rats were treated for 6 weeks with adriamycin and carnitine, adriamycin and placebo, or carnitine and placebo. After 6 weeks of treatment hearts were studied in an isolated working heart system. Adriamycin-treated hearts showed reduced cardiac output and left ventricular systolic pressure compared to controls or to adriamycin- and carnitine-treated hearts. The myocardial carnitine content in the non-perfused hearts was not influenced by adriamycin therapy, and muscle, kidney, and liver carnitine levels were unchanged. However, total plasma carnitine in the adriamycin group was significantly elevated, based on increased carnitine esters. This suggests that adriamycin induced mitochondrial damage and interferes with fatty acid oxidation. The mechanism by which carnitine protects the heart against adriamycin-induced cardiomyopathy is unclear, but light- and electron-microscopic histology indicates that carnitine prevents mitochondrial damage. Adriamycin probably forms a complex at the inner mitochondrial membrane and disturbs mitochondrial calcium uptake. Free radicals are probably involved in this process, and it is speculated that L-carnitine prevents mitochondrial damage by adriamycin by preventing free radical formation.

A different pharmacological way to cause myocardial carnitine deficiency and functional impairment is the induction of diabetes by streptozotocin in rats. Streptozotocin-treated diabetic rats develop myocardial carnitine deficiency, associated with functional impairment, as can be measured in an isolated heart model [31, 32]. Parallel to functional impairment and the myocardial carnitine loss, long-chain acyl-CoA esters increase significantly in the myocardium. When myocardial carnitine levels are reduced to only about 74% of normal, long-chain acyl-CoA esters rise to 150% of controls. Thus, small changes in myocardial carnitine concentrations are associated with other significant biochemical defects in this model. Substitution of carnitine improves myocardial function as well as tolerance to ischemia. The concentration of long-chain acyl carnitine increases, whereas that of long-chain acyl-CoA decreases with carnitine treatment.

In contrast to these models with acquired carnitine deficiency, other animal models exist where carnitine deficiency is inherited. Spontaneous dilated cardiomyopathy has been reported recently in dogs [33]. One canine family showed characteristic signs of severe congestive heart failure: peripheral and pulmonary edema, atrial fibrillation and reduced fractional shortening in echocardiography. The myocardial carnitine content was severely reduced in comparison with normal controls. Treatment with high doses of L-carnitine led to an increase of myocardial L-carnitine concentrations together with an improvement in clinical status and myocardial function. Withdrawal of L-carnitine caused myocardial dysfunction and clinical signs of heart failure to recur. Thus myocardial function in this dog model of familial cardiomyopathy was parallelled by the myocardial carnitine content.

Syrian hamsters with inborn hypertrophic and dilated cardiomyopathy develop myocardial carnitine deficiency early in their lifes. Reduced myocardial carnitine content is associated with significant alterations in myocardial energy metabolism. Carnitine substitution restores myocardial carnitine concentrations to normal, improves myocardial function, and restores myocardial ATP-concentrations [34].

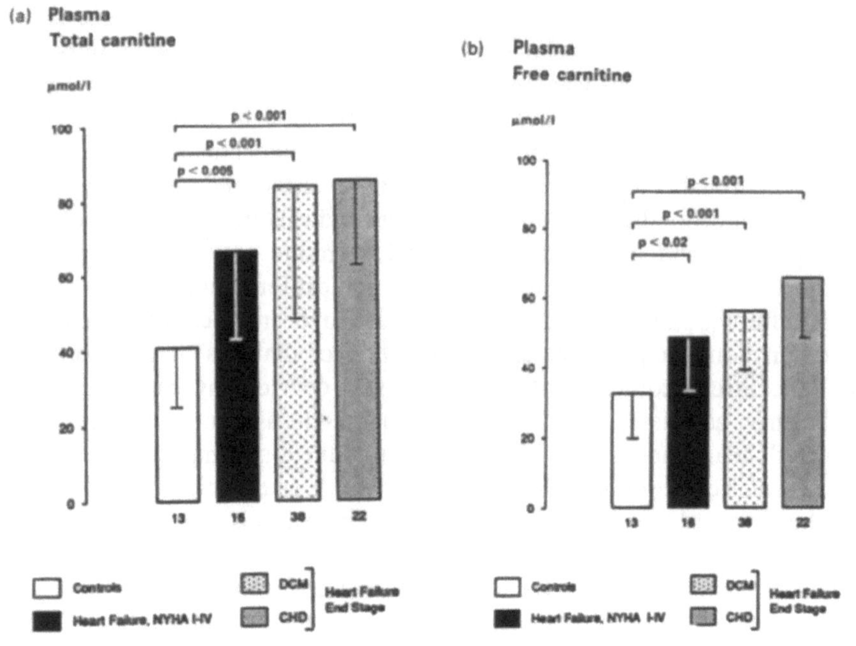


*Figure 2.* Plasma total (a) and free (b) carnitine concentrations in controls and in patients with moderate and end-stage heart failure. CHD = coronary heart disease; DCM = dilated cardiomyopathy; NYHA = New York Heart Association class. Mean ± SD, with n below bars.

## Plasma carnitine in human cardiovascular diseases

Based on the experience that systemic carnitine deficiency can cause the clinical and hemodynamic features of cardiomyopathy in children, several investigators studied the carnitine concentrations in adult dilated cardiomyopathy, expecting that at least in some of the patients systemic carnitine deficiency might cause heart failure. However, changes in plasma carnitine were found in most patients, but were in the opposite direction than expected. Plasma carnitine concentrations in adults with dilated cardiomyopathy were increased in patients with heart failure [35–37]. Conte et al. [36] even claimed an association between the increase in plasma carnitine levels and prognostic impairment in human heart failure. In our own study, plasma carnitine in heart failure was increased with the highest concentrations occurring in end-stage heart failure [7, 35]. No difference was found between different origins of heart failure, i.e. cardiomyopathy or coronary heart disease (Figure 2). Free carnitine in the patients with moderate heart failure or end-stage heart failure was also significantly elevated in comparison with controls. The ratio of free carnitine to total carnitine in the heart-failure patients (0.71 ± 0.03) was slightly lower than in the control group (0.81 ± 0.05, ns). Thus, there was a tendency towards an increased percentage of carnitine esters, which could be induced by the accumulation of long-chain acyl-CoA esters, but this tendency was not significant. The rise in plasma carnitine occurred independently of serum creatinine. Increased plasma carnitine levels (>mean ± 1 SD, corresponding to 52 μM) were found in 12 patients. Of these 12, 8 had normal serum creatinine levels (<1.2 mg/dl). No significant correlation was found between the plasma carnitine content and serum creatinine levels.

The reasons for the increased plasma carnitine levels in heart failure are not yet clear. Impaired renal function, which is frequently found in such patients, could certainly lead to increased plasma carnitine concentrations [38]. In this case, however, a correlation between an increase in creatinine levels and an increase in plasma carnitine levels would have been expected. The lack of such a correlation and the fact that most of our patients had normal or only moderately impaired renal function argues against a significant contribution of decreased renal elimination to increased plasma carnitine levels in heart failure. An increase in carnitine synthesis in the liver could also increase plasma carnitine levels; however, no experimental evidence has been obtained for such a mechanism. Finally, leakage from tissues with high carnitine concentrations should be considered. Usually, carnitine, after synthesis in the liver, is taken up from the plasma by the myocardium and skeletal muscles with the help of specialized transport systems, and it is accumulated in these tissues to concentrations up to 20 to 50 times the plasma levels. High myocardial carnitine concentrations are maintained by the action of a specific carnitine carrier and the relative impermeability of the myocyte membrane for carnitine [1, 3, 9, 19, 39, 40]. Defects in the carrier system, as well as non-specific membrane damage, for example ischemia, can both lead to reduced myocardial carnitine concentrations and probably a number of different toxic mechanisms can affect carnitine transport in a comparable way [6, 41, 42]. Thus, the observed increase in plasma carnitine in adult heart failure could result from a leakage of carnitine from the heart or skeletal muscle. To ascertain such a hypothesis, measurements of carnitine concentrations in tissue are necessary.

## Measurement of carnitine concentrations in human myocardium

To prove the hypothesis of a carnitine leakage from the muscles into the plasma in heart failure, tissue concentrations must be determined. Care must be taken to obtain adequate diseased and control tissues. Explanted hearts are frequently used as a model for diseased tissues. This is valid, if the time between explantation of the heart and freezing of the tissue samples remains short. A delay of only a few minutes induces ischemia and will change the ratio between free carnitine and esterified carnitine to a significant degree. Long-chain acyl-CoA esters accumulate during ischemia and are transferred to carnitine forming long-chain acylcarnitines, thus confounding the in-vivo concentrations of free and esterified carnitine. Carnitine itself is only slowly broken down enzymatically and if no washout of the tissue occurs, it can be measured safely even 1 or 2 h after explantation of the heart.

Our own studies in hearts obtained at autopsy, however, indicate that longer delays of about 12 or 24 h are not tolerable, because degradation does occur [43]. Thus, the measurement of valid carnitine levels in end-stage failing hearts, obtained at transplantation, appears safe, but the determination of normal control levels from autopsy hearts is obsolete and therefore, the determination of normal values remains a problem. The so-called "healthy donor hearts" are also subjected to a number of medical procedures and therapies and are usually not completely healthy. Further, there is often a considerable delay for medical reasons before freezing of the hearts is possible. Explanted hearts are particularly useful to study the regional distribution of carnitine. We found the highest concentrations in the left ventricle and the lowest in the right atrium. Within the left ventricle, there was a gradient with highest carnitine concentrations at the base and lowest at the apex. The differences between local carnitine concentrations in explanted human hearts were small as carnitine in all areas ranged between 7 and 5 nmol/mg non-collagen protein (NCP, Figure 3) [35, 43].

**Establishment of normal human myocardial carnitine concentrations**

As pointed out, normal values for free and total carnitine cannot be obtained at autopsy, and as hearts undergoing surgery are usually not healthy, control values are best obtained by endomyocardial biopsy. The assessment of carnitine concentrations in endomyocardial biopsies requires, in addition to validated micromethods for the determination of carnitine, the search for an adequate reference system [44]. Relating measured metabolites to wet weight leads to artefacts caused by varying water and fibrous tissue content of the samples and blood contamination. Using total protein will lead to errors based on varying fibrous tissue content. The introduction of tissue creatinine or mitochondrial marker enzymes did not improve the precision of measurements. Different investigators used non-collagen protein as one of the best compromises [34, 45, 46]. This reduced the variability considerably. In our experience, the coefficient of variance in a group of normal controls was huge (70%) if wet weight was used as a reference system, whereas the variance was significantly decreased if non-collagen protein was used (30%) (Figure 4). Using non-collagen protein as a reference system, we determined the content of total and free carnitine in a group of patients undergoing endomyocardial biopsy for suspected myocarditis. In these patients, left

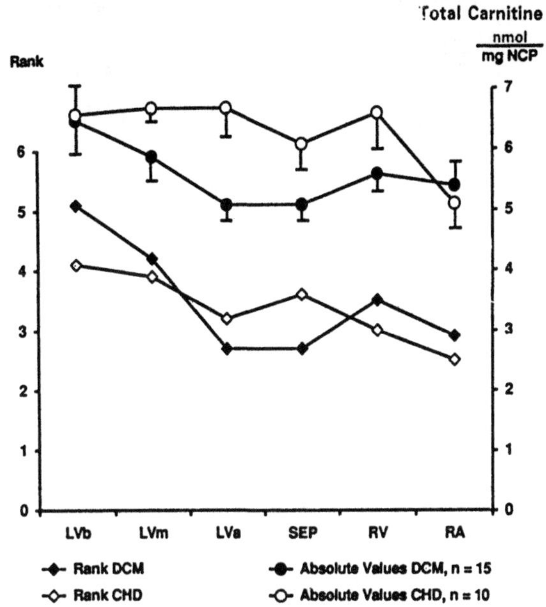


*Figure 3.* Regional distribution of carnitine in explanted human hearts with end-stage failure due to dilated cardiomyopathy (DCM, filled symbols) or coronary heart disease (CHD, open symbols). Carnitine concentrations and respective rank orders are shown for the base (b), mid area (m) and apex (a) of the left ventricle (LV), septum (SEP), right ventricle (RV) and right atrium (RA). Mean ± SD.

ventricular function and coronary arteries were normal and the suspected myocarditis was excluded histologically. Therefore, they may be regarded as healthy controls.

## Myocardial carnitine concentrations in patients with heart failure

Suzuki et al. [47] measured myocardial carnitine concentrations in papillary muscle obtained at surgery from 16 patients with mitral valve disease. Total myocardial carnitine was found to be normal, whereas free carnitine was decreased. Although, these patients were supposed to have congestive heart failure, myocardial failure may not have been present, because mitral valve disease usually leads to heart failure independent of myocardial function. Thus, carnitine concentrations in mitral valve patients do not necessarily reflect those in failing myocardium. Spagnoli et al. [48] measured carnitine concentrations at autopsy in patients with myocardial infarction. They found

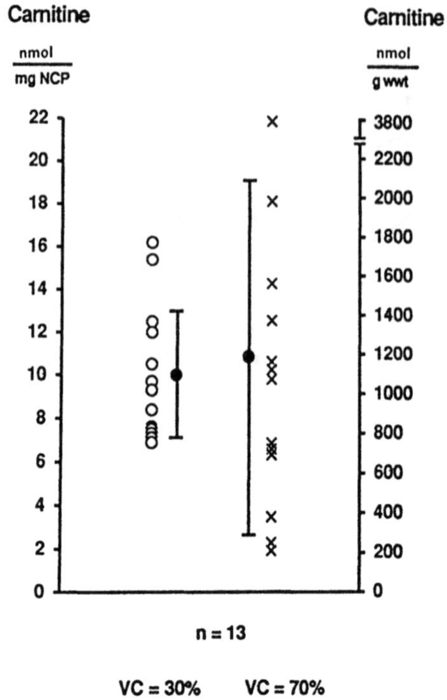


*Figure 4.* Variance of carnitine determinations in endomyocardial biopsies depending on the reference system. The variance for non-collagen-protein (NCP) related data (left) is far below that for wet-weight (wwt) related values (right). Mean ± SD. VC = variation coefficient.

low free carnitine concentrations in comparison with healthy myocardium. In these cases, ischemia may have significantly contributed to the loss of carnitine from the border zone. Pierpoint et al. [8] investigated patients with dilated cardiomyopathy, coronary artery disease and myocarditis undergoing cardiac transplantation. Low total myocardial carnitine was found in the left ventricle; the reduction of carnitine was more pronounced in the left than in the right ventricle. Free carnitine was not measured, plasma carnitine was high, and no correlation with hemodynamic parameters was found. In this study, the difference between controls and heart failure was only significant in 7 of 51 patients. The reason for this may have been that the control levels were obtained at autopsy and were rather low (left ventricle 5.7 ± 1.0, right ventricle 6.2 ± 1.8 nmol/mg NCP). Neither the underlying disease nor the interval between death and autopsy were mentioned. As several systemic diseases are associated with altered carnitine handling and as a degradation of carnitine occurs after 12 to 24 h, the use of autopsy material may have led to a systematic underestimation of normal carnitine levels and therefore to an underestimation of the differences between controls and heart failure. Patel et al. [49] investigated 10 patients with cardiomyopathy and 8 with heart failure of different origin. Endomyocardial biopsy was used to study right ventricular septum. Total as well as free myocardial carnitine were reduced in comparison with controls. No correlation was found between myocardial carnitine content and the degree of left ventricular dysfunction. In summary, the 4 studies argue in favor of a reduction of myocardial carnitine concentrations in human heart failure. However, the statistical significance remains weak. The use of improper controls is probably the reason for the failure of the studies mentioned to show decreased myocardial carnitine concentrations in human heart failure.

Our own studies, where endomyocardial biopsies from 28 patients with dilated cardiomyopathy and 9 with coronary artery disease, valvular heart disease or hypertension were investigated, showed a decrease of myocardial carnitine concentrations in patients with heart failure in comparison with controls (HF $6.2 \pm 2.4$, controls $10.1 \pm 3.1$ nmol/mg NCP, $p < 0.001$) [7]. The loss of myocardial carnitine in patients with left ventricular ejection fractions below 30% was bigger than that in patients with left-ventricular ejection fractions from 30% to 55% (Figure 5a). Patients with heart failure due to dilated cardiomyopathy and heart failure due to other origins showed a comparable loss of myocardial carnitine (Figure 5b). In addition, free myocardial carnitine in these patients was studied. Patients with heart failure ($n = 22$) had significantly lower free myocardial carnitine concentrations than 11 controls (heart failure $4.6 \pm 1.4$, controls $9.7 \pm 2.7$ nmol/mg NCP, $p < 0.0001$). There was no difference in free myocardial carnitine content in patients with heart failure due to dilated cardiomyopathy or coronary or valvular heart disease. A significant non-linear correlation was shown to exist between the myocardial free carnitine content and the left-ventricular ejection fraction (Figure 6). The ratio of free to total carnitine in the myocardium was $0.81 \pm 0.05$ in the heart failure group in comparison with $0.91 \pm 0.03$ in the control group. Although this difference was not statistically significant, it may reflect an interesting tendency.

Measurements in explanted hearts by our group yielded comparable results. A significant loss of myocardial carnitine concentrations was found in patients with dilated cardiomyopathy or coronary heart disease undergoing heart transplantation for end-stage failure (Figure 7) [35, 43]. This loss of myocardial carnitine was comparable in patients with dilated cardiomyopathy and coronary heart disease. In both groups, the loss of myocardial carnitine affected all areas of the failing hearts. Only total carnitine has been studied in these explanted hearts, because, as discussed previously, esterification of free carnitine may take place very quickly and therefore ratios of free to esterified carnitine may not be usable.

The ratio of free carnitine to total carnitine was decreased in the plasma as well as in the tissue of heart-failure patients in comparison with controls.

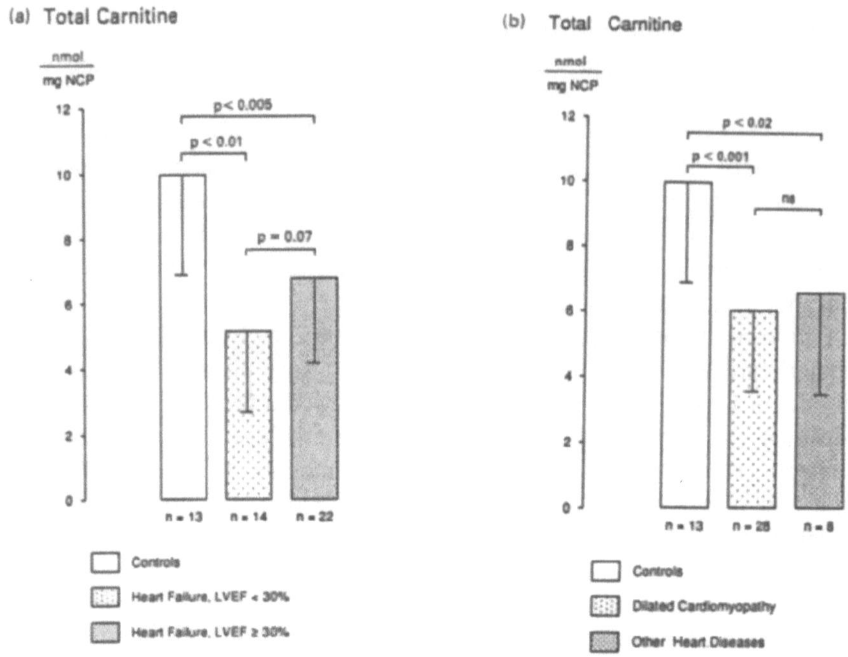


*Figure 5.* Loss of total myocardial carnitine in patients with heart failure and left-ventricular ejection fraction (LVEF) < 30% or LVEF 30–55% (a) and in patients with dilated cardiomyopathy and other heart diseases (b) in comparison with controls. Mean ± SD.

Although these differences were not significant, the tendency that increased amounts of long-chain acylcarnitine esters are produced in the diseased patients should at least be taken into account. The free carnitine that is available for reaction with CoA-esters decreases.

## Compartmentalization of carnitine in the myocardium

The loss of carnitine that is measured in tissue homogenates does not allow any conclusions regarding the compartments that are mainly affected: cytosol, membranes or mitochondria. From animal experiments it is, however, known that more than 90% of total carnitine in the non-ischemic heart is contained in the cytosol and less than 10% in the mitochondria [50]. Therefore, the loss of myocardial carnitine in heart failure probably affects mainly the cytosol. It cannot be explained by a reduction of mitochondrial volume

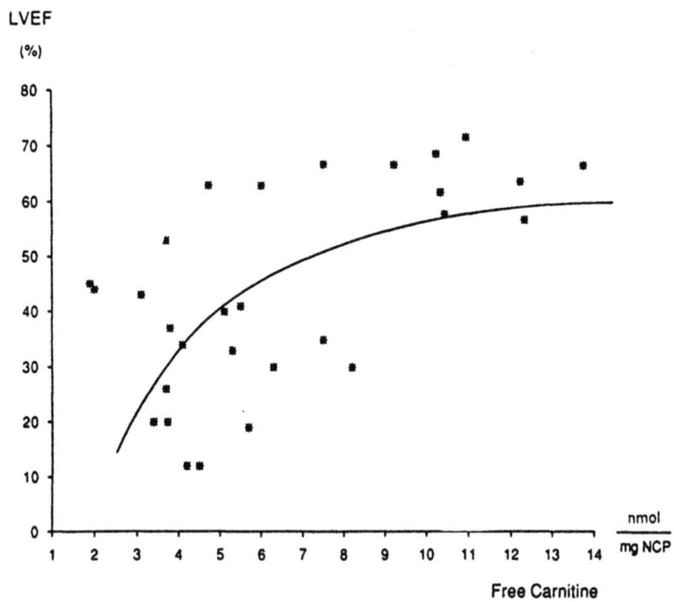


*Figure 6.* Non-linear correlation between free myocardial carnitine and left-ventricular ejection fraction (r = 0.443).

that is shown to occur in heart failure. Thus, in patients with heart failure, a reduction of the effective carnitine concentrations in the cytoplasm must be considered.

## Carnitine palmityltransferase activity in heart failure

Carnitine palmityltransferase (CPT, EC 2.3.1.21) plays a key role in myocardial fatty acid oxidation. The enzyme catalyses the reaction of long-chain acyl-CoA esters and free carnitine to form long-chain acylcarnitine esters, entering the mitochondria and undergoing fatty acid oxidation [3]. The enzyme has several allosteric regulators. Usually, free carnitine is not rate-limiting [2]. Only if free carnitine concentrations decrease below the Michaelis-Menton constant ($K_M$) of carnitine palmityltransferase for free carnitine, is the reaction velocity controlled by the availability of carnitine. The $K_M$ of carnitine palmityltransferase for free carnitine is significantly below the physiological concentration of human myocardial free carnitine. However, in heart failure, the myocardial carnitine level can decrease to concentrations in the range of the $K_M$ of carnitine palmityltransferase for

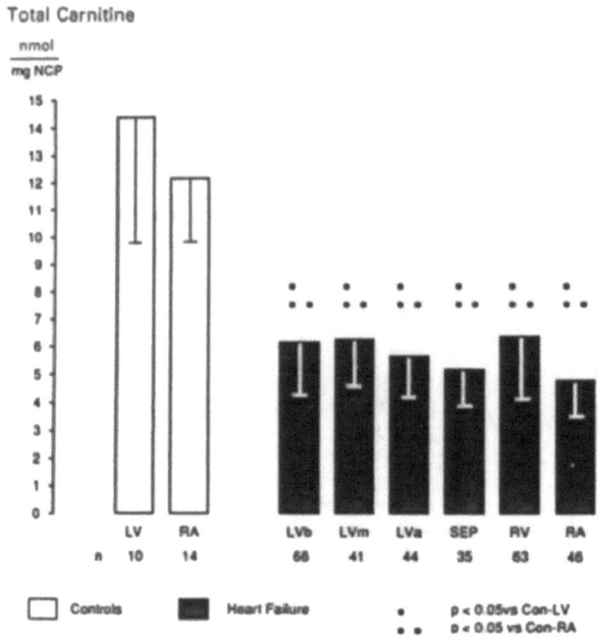


*Figure 7.* Loss of total myocardial carnitine in explanted hearts from patients with end-stage dilated cardiomyopathy or coronary heart disease in comparison with controls. For abbreviations, see legend to Figure 3. Means ± SD.

free carnitine. Thus, in these cases a reduced availability of free carnitine may limit the transferase reaction and thereby fatty acid oxidation.

## Long-chain acylcarnitines and arrhythmia

Spontaneous ventricular arrhythmia represent a major problem in human heart failure. Altered membrane function probably facilitates arrhythmogenesis by reentry loops and from autonomous foci [51]. Free and esterified fatty acids and acylcarnitines have been shown to alter membrane function by a number of different mechanisms. The compounds exist at relatively low concentrations as monomers and can insert into the inner portion of the lipid bilayer, thereby altering the physical properties and functions of the membranes. At higher concentrations, the compounds can aggregate into micelles, and these can also insert into the lipid bilayer or can damage the membranes by their detergent effects. By these mechanisms, conformation and function of membrane proteins may be modified substantially. Further, insertion of amphiphiles into membranes may displace calcium from negatively charged sites on membrane phospholipids, because of a change in orientation of the polar head group [52,53]. Long-chain acylcarnitines can alter membrane function by "wedging" into the membrane. Their lytic activity is determined by the length of the hydrophobic acyl chain. Monomer insertion into the membranes affects ionic movement and enzyme activity of integral membrane proteins.

The effect of long-chain acylcarnitines on membrane function and arrhythmia has been studied in several animal models. In adult canine myocytes the expression of alpha-1-adrenergic receptors was increased significantly without change in receptor affinity by exposure to ischemia [54]. The concentration of long-chain acylcarnitine in these myocytes also increased in parallel. Inhibition of carnitine acyltransferase-I abolished the accumulation of long-chain acylcarnitine and the increase in alpha-1-adrenergic receptor number induced by 30 min of hypoxia. Incubation of normoxic cells with exogenous palmitylcarnitine also increased alpha-1-adrenergic receptor number in the presence or absence of acyltransferase inhibitors. Thus, hypoxia results in an increase in alpha-1-adrenergic receptors based on an increase in endogenous long-chain acylcarnitines, and the inhibition of carnitine acyltransferase prevents both events [54]. Comparable effects were obtained in rat myocytes [55]. The administration of a carnitine acyltransferase inhibitor inhibited the accumulation of long-chain acylcarnitine induced by hypoxia and prevented the depression of electrophysiological function by hypoxia. In a cat model with occlusion of the left anterior coronary artery, the inhibition of carnitine acyltransferase-I reduced the incidence of lethal arrhythmia induced after the onset of ischemia. It also prevented the increase of both long-chain acylcarnitine and lysophosphatidylcholin.

Multiple lines of evidence support the capability of long-chain acylcarnitines to cause arrhythmia in animal models. It may be concluded that long-chain acylcarnitines may also contribute to arrhythmia in human diseases. Long-chain acylcarnitines can be eliminated from the cell via the carnitine transporter in the sarcolemmal membrane in exchange with free carnitine on the outside. Therefore, the increase in extracellular free carnitine available for exchange with intracellular long-chain acylcarnitine esters might be tested in order to reduce arrhythmia in human heart failure.

## Biochemical consequences of reduced myocardial carnitine content

Altered myocardial lipid metabolism has already been described in human heart failure [56]. We have found some evidence that reduced myocardial carnitine concentration might limit fatty acid utilization. Fatty acids provide about 70% of the energy needs of the human myocardium whereas the remaining 30% are covered by glucose and lactate. An increase in glucose

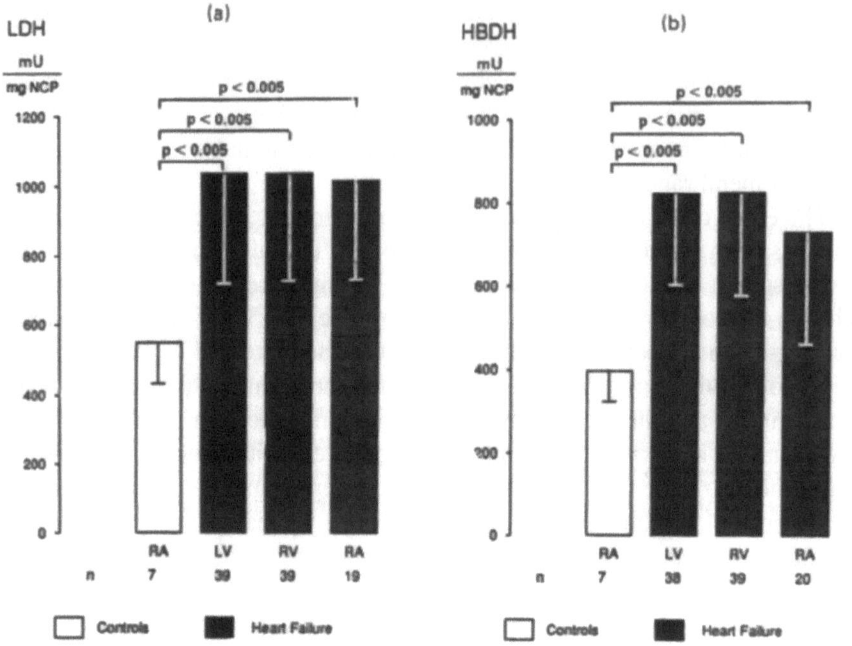


*Figure 8.* Myocardial lactate dehydrogenase (LDH) activity (a) and β-hydroxybutyrate dehydrogenase (HBDH) activity (b) in explanted human hearts with end-stage heart failure. For abbreviations, see legend to Figure 3. Means ± SD.

and lactate utilization may compensate for impaired fatty acid oxidation [57, 58]. However, an increase in anerobic glycolysis – i.e. lactate production from glucose – can only cover about 5% of the total myocardial energy demand [59]. Oxidative utilization of glucose and lactate via pyruvate dehydrogenase are much more efficient. The equilibrium between glucose and lactate is determined by lactate dehydrogenase (LDH) isoenzyme activity. Isoenzyme 1 and 2 β-hydroxybutyrate dehydrogenase (HBDH) favor the formation of pyruvate from lactate [60].

The activity of glucose and lactate utilization have been measured in explanted hearts as well as in endomyocardial biopsies from patients with moderate heart failure [43]. Total LDH activity in the explanted hearts was significantly increased in comparison with controls. No significant difference between patients with dilated cardiomyopathy and coronary heart disease was found (Figure 8a). In biopsies from patients with moderate heart failure, a significant increase of LDH activity was also found in comparison with normal controls. LDH activities in the atria from healthy donor hearts and in the right ventricles from patients with normal left ventricular function were within the same range, and so were the LDH activities in the atria and ventricles of the explanted hearts. A significant increase of HBDH was also found in heart failure patients in comparison with controls. Again, there was no significant difference between cardiomyopathy and coronary artery disease (Figure 8b). The ratio of HBDH to LDH was comparable in the atria of controls and heart failure patients (0.72). In both ventricles of heart failure patients, the ratio HBDH/LDH was increased to 0.80. Thus, an increase in total glucose utilization and probably a more pronounced increase in oxidative glucose utilization occurs in human-heart failure.

As lipids provide the main substrate for energy utilization in the normal heart and as there is evidence for a shift from lipid to glucose utilization in heart failure, it has been speculated that this may induce energy deficiency in heart failure. We therefore determined the myocardial adenine-nucleotide content in heart-failure patients in comparison with controls [61]. In 18 patients with heart failure due to dilated cardiomyopathy, the myocardial adenosine triphosphate (ATP) concentrations and total adenine nucleotides (ATP + ADP + AMP) in comparison with controls were not significantly different (Figure 9). Patients with dilated cardiomyopathy and a left ventricular ejection fraction below 30% also showed normal adenine-nucleotide concentrations in the myocardium, as did patients with cardiomyopathy and increased left-ventricular end-diastolic pressures. In patients with heart failure due to other origins, myocardial ATP concentrations were also not significantly different from controls. As normal myocardial adenine-nucleotide concentrations are maintained in patients with severely depressed left-ventricular ejection fraction and severely reduced myocardial carnitine concentrations, the alterations in carnitine concentrations are probably not the cause of impaired energy metabolism. Probably, increased glucose utilization is able to compensate for a potential decrease in lipid utilization in energy metabolism.

## Summary and conclusions

The association between carnitine metabolism and myocardial function is based on the central role of carnitine in lipid utilization and membrane function and has been documented in several human diseases and animal models. Systemic carnitine deficiency is characterized by muscle weakness, accumulation of lipids, hypoglycemia, and cardiomyopathy. It is probably caused by a transport defect and is rapidly improved by carnitine substitution. Transport defects can probably also concern single organs, like skeletal muscle or heart, and lead to organ-specific carnitine-deficiency. A number of secondary or organ-specific carnitine-deficiency syndromes has been described in humans, such as treatment with valproic acid and defects in branched-chain amino-acid metabolism, and in animal models, where a loss

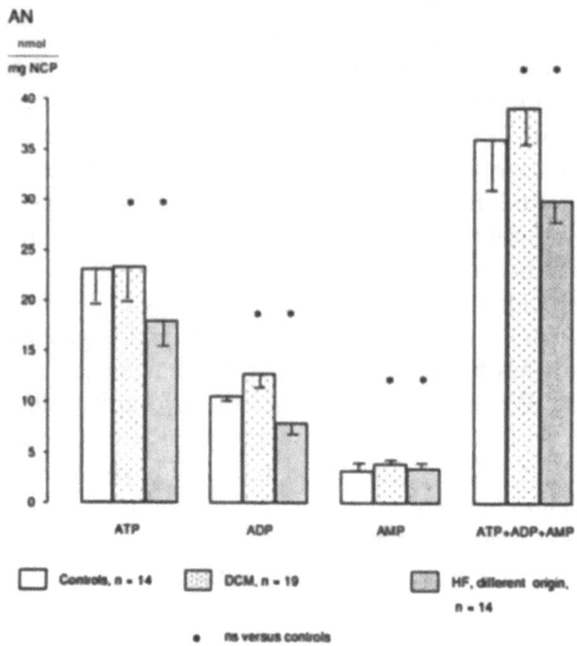


*Figure 9.* Unchanged myocardial adenine-nucleotide concentrations (AN: ATP, ADP and AMP) in patients with heart failure due to dilated or ischemic cardiomyopathy in comparison with controls. DCM = dilated cardiomyopathy; HF = heart failure; NCP = non-collagen protein. Means ± SD.

of carnitine from the myocardium associated with functional impairment was documented for adriamycin-induced cardiomyopathy, streptozotocin-induced diabetes and for the inherited cardiomyopathy of Syrian hamster tribes. Substitution of carnitine in these syndromes usually improves function, but does not normalize all physiological and biochemical changes. In human heart failure, plasma free and total carnitine levels are usually increased. The percentage of carnitine esters is slightly elevated, possibly indicating an increased production of carnitine esters. As increased leakage from the myocardium or skeletal muscles in heart failure has been suspected to cause the increased plasma carnitine concentrations, myocardial carnitine levels have been determined in endomyocardial biopsies as well as in larger samples from explanted hearts. The establishment of valid normal values for these studies is of major importance. The failure of several groups to document reduced myocardial carnitine levels in heart failure and to observe only trends towards decreased myocardial carnitine content, although a significant decrease was probably present, is due to improper controls. Indeed, clear decrease in myocardial carnitine concentrations occurs in human heart failure of different origins. A non-linear correlation exists between myocardial carnitine concentrations and left-ventricular function. Free carnitine is reduced as well as total carnitine, and the percentage of carnitine esters tends to be increased in human heart failure. The decrease in myocardial carnitine in human heart failure might limit the velocity of the carnitine palmityltransferase reaction and thus of lipid oxidation. Long-chain acylcarnitines may play a major role in arrhythmogenesis, as they may decrease membrane stability and may increase the automaticity of arrhythmogenic foci. Biochemical consequences of reduced myocardial carnitine content have been sought. Decreased utilization of lipids is probably compensated for by increased utilization of glucose, as indicated by an increase in myocardial LDH-activity. However, myocardial adenine-nucleotide concentrations are not altered in human heart failure, indicating that the compensation is effective and the loss of myocardial carnitine does not limit energy metabolism and does not cause energy deficiency as a mechanism of heart failure.

## References


1. Fritz IB. Carnitine and its role in fatty acid metabolism. In: Paoletti R, Kritchevsky D, editors. Advances in lipid research. New York: Academic Press, 1963: 285–334.
2. Bremer J. Carnitine – metabolism and functions. Physiol Rev 1983; 63: 1420–81.
3. Bremer J. In: Gran FC, editor. Cellular compartmentalization and control of fatty acid metabolism. New York: Academic Press, 1968: 65–8.
4. Challoner DR, Prols HG. Free fatty oxid oxidation and carnitine levels in diphteritic guinea pig myocardium. J Clin Invest 1972; 51: 2071–6.
5. Wittels B, Bressler RJ. Biochemical lesion of diphteria toxin in the heart. J Clin Invest 1965; 44: 1639–44.
6. Shug AL, Thomsen JH, Folts JD et al. Changes in tissue levels of carnitine and other metabolites during myocardial ischemia and anoxia. Arch Biochem Biophys 1978; 187: 25–33.
7. Regitz V, Shug AL, Fleck E. Defective myocardial carnitine metabolism in congestive heart failure secondary to dilated cardiomyopathy and to coronary, hypertensive and valvular heart diseases. Am J Cardiol 1990; 65: 755–60.
8. Pierpont MEM, Judd D, Goldenberg IF, Rings WS, Olivari MT, Pierpont GL. Myocardial carnitine in end-stage congestive heart failure. Am J Cardiol 1989; 64: 56–60.
9. Engel AG. Possible causes and effects of carnitine deficiency in man. In: Frenkel RA, McGarry JD, editors. Carnitine biosynthesis, metabolism and functions. New York: Academic Press, 1980: 271–86.
10. Boudin G, Mikol J, Guillard A, Engel AG. Fatal systemic carnitine deficiency with lipid storage in skeletal muscle, heart, liver and kidney. J Neurol Sci 1976; 30: 313–25.
11. Chapoy PR, Angelini C, Brown WJ, Stiff JE, Shug AL, Cederbaum SD. Systemic carnitine deficiency: a treatable inherited lipid-storage disease presenting as Reye's syndrome. N Engl J Med 1980; 303: 1389–94.
12. Cornelio F, Di Donato S, Peluchetti D et al. Fatal cases of lipid storage myopathy with carnitine deficiency. J Neurol Neurosurg Psych 1977; 40: 170–8.
13. Karpati G, Carpenter S, Engel AG et al. The syndrome of systemic carnitine deficiency.

Clinical, morphologic, biochemical, and pathophysiologic features. Neurology 1975; 25: 16–24.
14. Tripp ME, Katcher ML, Peters HA et al. Systemic carnitine deficiency presenting as familial endocardial fibroelastosis: a treatable cardiomyopathy. N Engl J Med 1981; 305: 385–90.
15. Regitz V, Hodach RJ, Shug AL. Carnitin-Mangel: Eine behandelbare Ursache kindlicher Kardiomyopathien. Klin Wochenschr 1982; 60: 393–400.
16. Treem WR, Stanley CA, Finegold DN, Hale DE, Coates PM. Primary carnitine deficiency due to a failure of carnitine transport in kidney, muscle, and fibroblasts. N Engl J Med 1988; 319: 1331–6.
17. Waber LJ, Valle D, Neill C, DiMauro S, Shug A. Carnitine deficiency presenting as familial cardiomyopathy: A treatable defect in carnitine transport. J Pediatr 1982; 101: 700–5.
18. Borum PR, Park JH, Law PK, Roelops RJ. Altered tissue carnitine levels in animals with hereditary muscular dystrophy. J Neurol Sci 1978; 38: 113–21.
19. Böhmer T, Rydning A, Solberg HE. Carnitine levels in human serum in health and disease. Clin Chim Acta 1974; 57: 55–61.
20. Angelini C, Lücke S, Cantarutti F. Carnitine deficiency of skeletal muscle: Report of a treated case. Neurology 1976; 26: 633–7.
21. Hart ZH, Chang CH, Di Mauro S, Farooki Q, Ayyar R. Muscle carnitine deficiency and fatal cardiomyopathy. Neurology 1978; 28: 147–51.
22. VanDyke DH, Griggs RC, Markesberry W, DiMauro S. Hereditary carnitine deficiency of muscle. Neurology 1975; 25: 154–9.
23. Deacon JSR, Gilbert EF, Viseskul C, Herrmann J, Angevine JM, Albert AE. Familial cardiac lipidosis. Can Med Ass J 1979; 120: 181–94.
24. DiMauro S, DiMauro PM. Muscle carnitine palmityltransferase deficiency and myoglobinuria. Science 1973; 182: 929–31.
25. Hostetler KY, Hoppel CL, Romine JS, Sipe JC, Gross SR, Higginbottom PA. Partial deficiency of muscle carnitine palmitoyltransferase with normal ketone production. N Engl J Med 1978; 298: 553–7.
26. Roe CR, Millington SD, Maltby DA, Kahler SG, Bohan TP. L-Carnitine therapy in isovaleric acidemia. J Clin Invest 1988; 74: 2290–5.
27. Hale DE, Batshaw ML, Coates PM. Long-chain acyl coenzyme A dehydrogenase deficiency: an inherited cause of nonketonic hypoglycemia. Pediatr Res 1985; 19: 666–71.
28. Winter SC, Szabo-Aczel S, Curry CJR, Hutchinson HT, Hogue R, Shug A. Plasma carnitine deficiency. Clinical observations in 51 pediatric patients. Am J Dis Child 1987; 141: 660–5.
29. Shug AL, Hayes B, Huth PJ et al. Changes in carnitine-linked metabolism during ischemia, thermal injury, and shock. In: Frenkel RA, McGarry JD, editors. Carnitine biosynthesis, metabolism, and functions. New York: Academic Press, 1980: 321–39.
30. Shug AL. Protection from adriamycin-induced cardiomyopathy in rats. Z Kardiol 1987; 76(Suppl 5): 46–52.
31. Paulson DJ, Schmidt MJ, Traxler JS, Ramacci MT, Shug AL. Improvement of myocardial function in diabetic rats after treatment with L-carnitine. Metabolism 1984; 33: 358–63.
32. Paulson DJ, Shug AL. Effects of myocardial ischemia and long chain acyl CoA on mitochondrial adenine nucleotide translocator. In: Ferrari R, Katz A, Shug S, editors. Myocardial ischemia and lipid metabolism. New York: Plenum Press, 1984: 185–202.
33. Keene BW, Panciera DL, Atkins CE, Regitz V, Schmidt MJ, Shug AL. Myocardial L-carnitine deficiency in a family of dogs with dilated cardiomyopathy. J Am Vet Med Assoc 1991; 198: 647–50.
34. Whitmer JT. L-Carnitine treatment improves cardiac performance and restores high-energy phosphate pools in cardiomyopathic Syrian hamster. Circ Res 1987; 61: 396–408.
35. Regitz V, Bossaller C, Strasser R, Müller M, Shug AL, Fleck E. Metabolic alterations in end-stage and less severe heart failure – myocardial carnitine decrease. J Clin Chem Clin Biochem 1990; 28: 611–7.

36. Conte A, Hess OM, Maire R et al. Klinische Bedeutung des Serumcarnitins für den Verlauf und die Prognose der dilatativen Kardiomyopathie. Z Kardiol 1987; 76: 15–24.
37. Tripp ME, Shug AL. Plasma carnitine concentrations in cardiomyopathy patients. Biochem Med 1984; 32: 199–206.
38. Chen SH, Lincoln SD. Increased serum carnitine concentration in renal insufficiency. Clin Chem 1977; 23: 278–80.
39. Fritz IB.: Gran C, editor. Cellular compartmentalization and control of fatty acid metabolism. New York: Academic Press, 1968: 39–63.
40. Vary TC, Neely JR. Characterization of carnitine transport in isolated perfused adult rat hearts. Am J Physiol 1982; 244: H585–92.
41. Bressler R, Wittels B. The effect of diphtheria toxin on carnitine metabolism in the heart. Biochim Biophys Acta 1965; 104: 39–45.
42. Molstadt P, Bohmer T, Eiklid K. Specificity and characteristics of the carnitine transport in human heart cells (CCL 27) in culture. Biochim Biophys Acta 1977; 471: 296–304.
43. Regitz V, Müller M, Strasser R et al. Untersuchung des Myokardstoffwechsels bei Herzinsuffizienz – Variabilität von Metabolitkonzentrationen und Enzymaktivitäten im Myokard. Z Herz Thorax Gefäßchir 1989; 3: 107–14.
44. Parvin R, Pande SV. Microdetermination of (−)carnitine and carnitine acetyltransferase activity. Anal Biochem 1977; 79: 190–201.
45. Pierpont GL, Francis GS, DeMaster EG, Levine TB, Bolman RM, Cohn JN. Elevated left ventricular myocardial dopamine in preterminal idiopathic dilated cardiomyopathy. Am J Cardiol 1983; 52: 1033–5.
46. Bittl JA, Weisfeldt ML, Jacobus WE. Creatine kinase of heart mitochondria. The progressive loss of enzyme activity during in vivo ischemia and its correlation to depressed myocardial function. J Biol Chem 1985; 260: 208–14.
47. Suzuki Y, Masumura Y, Kobayashi A et al. Myocardial carnitine deficiency in chronic heart failure. Lancet 1982; 1: 116 (Lett to the Ed).
48. Spagnoli LG, Corsi M, Villaschi S, Palmieri G, Maccari F. Myocardial carnitine deficiency in acute myocardial infarction. Lancet 1982; 1: 1419–20 (Lett to the Ed).
49. Patel AK, Thomsen JH, Kosolcharoen PK, Shug AL. Myocardial carnitine status: Clinical, prognostic and therapeutic significance. In: Ferrari R, DiMauro S, Sherwood G, editors. L-Carnitine and its role in medicine: From function to therapy. London: Academic Press, 1992: 325–35.
50. Idell-Wenger JA, Grotyohann LW, Neely JR. Coenzyme A and carnitine distribution in normal and ischemic hearts. J Biol Chem 1978; 253: 4310–8.
51. Pogwizd SM, Corr PB. Reentrant and nonreentrant mechanisms contribute to arrhythmogenesis during early myocardial ischemia: Results using three-dimensional mapping. Circ Res 1987; 61: 352–71.
52. Corr PB, Gross RW, Sobel BE. Amphipathic metabolites and membrane dysfunction in ischemic myocardium. Circ Res 1984; 55: 135–51.
53. Corr PB, Creer MH, Yamada KA, Saffitz JE, Sobel BE. Prophylaxis of early ventricular fibrillation by inhibition of acylcarnitine accumulation. J Clin Invest 1989; 83: 927–36.
54. Heathers GP, Yamada KA, Kanter EM, Corr PB. Long-chain acylcarnitines mediate the hypoxia-induced increase in alpha 1-adrenergic receptors on adult canine myocytes. Circ Res 1987; 61: 735–46.
55. Knabb MT, Saffitz JE, Corr PB, Sobel BE. The dependence of electrophysiological derangements on accumulation of endogenous long-chain acyl carnitine in hypoxic neonatal rat myocytes. Circ Res 1986; 58: 230–40.
56. Wittels B, Spann Jr JF. Defective lipid metabolism in the failing heart. J Clin Invest 1968; 47: 1787–94.
57. Nägele S, Hockerts T, Bögelmann G. Untersuchungen zum Stoffwechsel des Herzmuskels bei Ischämie. Klin Wochenschr 1963; 41: 1020–30.
58. Neely JR, Morgan HE. Relationship between carbohydrate and lipid metabolism and the energy balance of heart muscle. Annu Rev Physiol 1974; 36: 413–59.

59. Neely JR, Denton RM, England PJ, Randle PJ. The effects of increased heart work on the tricarboxylate cycle and its interactions with glycolysis in the perfused rat heart. Biochem J 1972; 128: 147–59.
60. Cahn RD, Kaplan NO, Levine L, Zwilling E. Nature and development of lactic dehydrogenases. Science 1962; 136: 962–9.
61. Regitz V, Fleck E. Myocardial adenine nucleotide concentrations and myocardial norepinephrine content in patients with heart failure secondary to idiopathic dilated or ischemic cardiomyopathy. Am J Cardiol 1992; 69: 1574–80.

*Corresponding Author*: Dr Vera Regitz-Zagrosek, Department of Cardiology, German Heart Institute Berlin, Augustenburger Platz 1, P.O. Box 650505, D-13305 Berlin, Germany

PART TWO

# Therapeutic efficacy of L-carnitine at the myocardial level

# 12. Is the carnitine system part of the heart antioxidant network?

ARDUINO ARDUINI, SECONDO DOTTORI, FRANCESCO MOLAJONI, RUTH KIRK and EDOARDO ARRIGONI-MARTELLI

"Carnitine and carnitine palmitoyltransferase can be considered integral components of membrane phospholipid fatty acid turnover in human erythrocytes and neuronal rat cells. Since this pathway is essentially related to the secondary antioxidant response to oxidatively damaged membrane phospholipids, one may envisage that the carnitine system takes part in the heart antioxidant network as a member of the secondary defence line."

## Introduction

Free radical reactions are considered to be an important pathophysiologic determinant of a broad range of inflammatory and ischemic diseases. The latter disease state, in particular, is still a very active area of free radical research since the discovery of the so-called *oxygen paradox* [1, 2]. Indeed, the sudden reoxygenation of the myocardium after a transient period of global ischemia results in cellular necrosis and intracellular calcium overload. Current hypotheses on the involvement of oxygen-derived free radicals in the course of the reperfusion event are essentially based on animal studies, in which the inclusion of free radical scavengers in the perfusion medium significantly improved hemodynamic and biochemical parameters [3–6]. In addition, electron spin resonance, low level chemiluminescence and reflectance studies on perfused rat heart suggest that free radical production is already present in the ischemic phase and that shortly after the onset of the oxygen readmission, a burst of free radical generation occurs [7–9].

The heart contains a well-integrated enzymatic and non-enzymatic antioxidant defence system, which in principle should prevent the noxious effect of a free radical attack. The antioxidant enzymatic system, which includes the Cu-Zn- and Mn-containing superoxide dismutases, selenium-dependent and non-selenium-dependent glutathione peroxidases, and catalase, is localized in different cellular compartments [10,11]. The non-enzymatic defence system encompasses an array of molecules able to counteract oxygen-derived free radicals in both lipophilic (vitamins E and A) and hydrophilic

*J.W. de Jong and R. Ferrari (eds): The carnitine system*, 169–181.

(ascorbic acid, reduced glutathione, uric acid, etc.) molecular environments [12]. According to the above postulated antioxidant network, it would seem unlikely that a free radical reaction is capable of exerting its deleterious action on heart cells. However, several lines of evidence indicate that in the course of the ischemic/reperfusion injury, oxidative changes toward lipid and protein components of the cell still take place [11, 13–16]. In other words, the primary antioxidant network does not seem to fully protect the cell from a free radical attack. Under these circumstances, a system capable of eventually removing and possibly repairing aberrant products of the oxidative insult would be further beneficial for cellular survival. It is becoming more and more evident that such a system, commonly regarded as the secondary antioxidant defence line, is operative towards damaged proteins, lipids, and DNA. Thus, enzymes promptly remove oxidatively damaged proteins [17–19], repair peroxidized phospholipids [20–22], and repair oxidatively damaged DNA [23–25].

Several groups have recently addressed the possibility that carnitine and its acyl-esters may act as antioxidants, and that this action may partly explain the therapeutical efficacy of these compounds in the ischemic heart [26–32]. Although the concept of a potential involvement of carnitine and/or its congeners in the antioxidant network may be appealing, it certainly raises a number of intriguing questions. For example, is there any chemical rationale that allows the consideration of these compounds as true primary antioxidants? On the other hand, is there direct experimental evidence that clearly shows that the carnitines possess a scavenging action toward free radicals? Is it instead possible that some of the anti-radical findings of carnitine are simply related to the proposed metabolic improvement of the heart bioenergetics in the course of ischemic disease [33, 34]? In fact, such an amelioration would most likely make the heart cell more resistant to a free radical attack and/or reduce free radical production.

We have recently shown that carnitine and carnitine palmitoyltransferase can be considered integral components of membrane phospholipid fatty acid turnover in human erythrocytes and neuronal rat cells [35–37]. Since this pathway is essentially related to the secondary antioxidant response to oxidatively damaged membrane phospholipids, one may envisage that the carnitine system takes part in the heart antioxidant network as a member of the secondary defence line. This chapter will try to dissect in some detail the possible location of the carnitine system in the antioxidant network, and to hopefully offer answers and new elements of discussion.

**Is the carnitine family a member of the primary antioxidant system?**

By definition, an antioxidant is a compound capable of efficiently preventing or delaying the oxidation of another compound. The oxidation of biologically relevant molecules is commonly thought to be caused by oxygen-derived free radicals. There are various mechanisms by which an antioxidant may exert its action [12]. An important aspect of the antioxidant action is related to the concentration at which the molecule is active: the antioxidant level should be low, at least as low as the oxidizable target. Of course, thermodynamic and kinetic properties of the reaction between an antioxidant and a free radical essentially dictate such a quantitative feature. Another indispensable requirement for those antioxidants which are consumed in the course of their action is that the product is either a less reactive free radical species or a non-radical species unable to cause damage.

Among the oxidative effects caused by oxygen-derived free radicals toward cellular components, lipid peroxidation is probably the most popular and well studied example. Consequently, most of the non-enzymatic antioxidants are best known as agents able to significantly quench the peroxidation of polyunsatured fatty acids. According to Figure 1, an antioxidant may prevent the *first-chain initiation* of lipid peroxidation by removing the initiator radical (i.e. hydroxyl radical scavenger), by chelating transition metal ions (iron or copper), to prevent Fenton-type reactions responsible for the formation of initiating species and/or by decomposition of lipid peroxide to peroxy or alcoxy radicals [12, 38]. The decomposition of lipid peroxide to the corresponding unreactive lipid alcohol, instead, is another mechanism of antioxidant activity carried out by enzymes such as glutathione-dependent peroxidases [39, 40]. Finally, chain-breaking antioxidants interrupt the peroxidative sequence by scavenging lipid peroxy or alkoxy radicals able to propagate the peroxidation of polyunsatured fatty acids [41, 42].

From the chemical point of view, both monosubstituted and polysubstituted phenols with electron-releasing groups are among the best non-enzymatic antioxidants used in biological systems. Carnitine and its short- and long-chain esters do not belong to this category of molecules, and it does not seem likely that the chemical functional groups present in the carnitines have an antioxidant action. However, several studies suggest that carnitine and some of its short-chain derivatives, namely acetyl-L-carnitine and propionyl-L-carnitine, could act as agents able to slow down lipid peroxidative processes [26, 28, 29, 32, 43]. In addition, protein carbonyl levels, an index of protein oxidative injury, were significantly reduced by treating ischemic-reperfused rat heart with either the D- or L-stereoisomer of propionylcarnitine in comparison to untreated ischemic-reperfused heart [16, 31]. An important common denominator in most of these studies is that to observe any significant protecting effect of L-carnitine and its congeners toward oxidative phenomena the concentration of these molecules has to be in a high millimolar range (10–100 mM). On the other hand, Ferrari et al. have shown that the iron-induced in vitro peroxidation of cardiac mitochondria was not affected by the inclusion in the reaction mixture of 1 mM propionyl-L-carnitine [44]. It is clear that the physiological or pharmacological value of these antioxidant effects is dubious, since the same effect can be obtained with either naturally-occurring antioxidants or synthetic ones at much lower con-

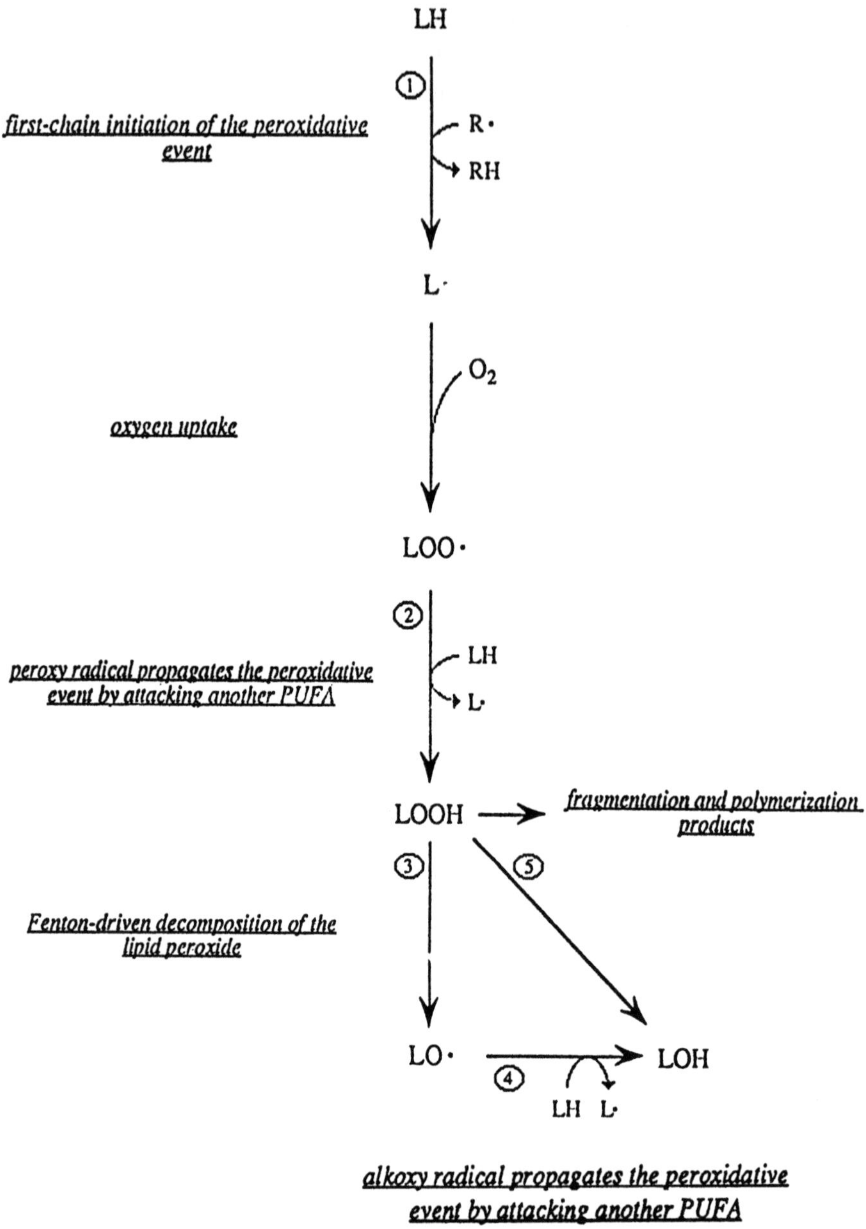


*Figure 1.* Lipid peroxidation of polyunsaturated fatty acid (PUFA). The scheme is mainly intended to show the primary site of action of antioxidants, which are indicated by circled numbers. 1) compounds able to block the initiation of the peroxidation process, such as hydroxyl radical scavengers; 2) & 4) chain-breaking antioxidants, such as monosubstituted and polysubstituted phenols; 3) iron-chelating agents able to block the Fenton-driven decomposition of lipid peroxide; 5) enzymatic reduction of lipid peroxide (LOOH) to the corresponding lipid alcohol (LOH) catalyzed by GSH-peroxidases. In the first peroxidative reaction the initial hydrogen abstraction is shown. Other abbreviations used are: R˙, free radical; L˙, carbon radical present in the PUFA; LOO˙, peroxy radical; LO˙, alkoxy radical.

centration (usually in the micromolar range), with a broader spectrum of action [12]. With respect to a specific scavenging action of the carnitines toward oxygen-derived free radicals, there is unequivocal evidence that such compounds are not effective. We have studied the reactivity of both hydroxyl radicals and superoxide anions with regard to L-carnitine and some of its short-chain esters [43]. In this report, hydroxyl radicals and superoxide anions were produced through the radiolysis of water: by irradiating aqueous samples with X-rays generated by a linear accelerator. The antioxidant activity of L-carnitine, acetyl-L-carnitine and propionyl-L-carnitine was evaluated by following the fragmentation of bovine serum albumin, the free radical target, present in the irradiated aqueous samples. Any scavenger able to remove oxygen free radicals would significantly decrease the fragmentation of the bovine serum albumin. The presence of L-carnitine, acetyl-L-carnitine or propionyl-L-carnitine at concentrations up to 10 mM did not affect the fragmentation. In addition, by using two different sources of superoxide anion, it has been shown that there is no decrease in superoxide anion production in the presence of millimolar concentration of carnitine or its short-chain esters [16, 45]. However, acetyl-L-carnitine was capable to partially inhibit the activity of xanthine oxidase, an enzyme often used as a superoxide anion-generating system, though at concentrations higher than 40 mM [32]. Also, these molecules did not act as chain-breaking antioxidants, when, in a phycoerythrin-fluorescence based assay, peroxyl radicals were generated in an incubation mixture containing carnitine or propionyl-L-carnitine [16]. It is worth noting that chain-breaking antioxidants represent in the lipid peroxidation chemistry the most common and efficient scavengers of lipid radicals.

Nevertheless, several pieces of experimental evidence suggest that both L-carnitine and propionyl-L-carnitine are effective in protecting the heart against the ischemia-reperfusion injury, even though the ester derivative provided better results. Although the mechanism of action is not entirely clear, a metabolic effect devoted to the improvement of myocyte bioenergetics is the most credited hypothesis [33, 34, see also this book]. If this is the case, one may also expect a reduced susceptibility of the heart to the free radical insult in the course of ischemia-reperfusion event. This would also explain why in the propionyl-L-carnitine perfused heart, the myocardial levels of lipid peroxidation by-products are lower than in the untreated sample at the end of the post-ischemic reperfusion.

An interesting aspect of the proposed antiradical effect of propionyl-L-carnitine in the ischemia-reperfusion injury of the heart is the potential role in iron chelation. As mentioned above, when iron is complexed with certain chelating agents, inorganic or organic hydroperoxide cannot be decomposed to the respective harmful hydroxyl or peroxyl radicals. Thus, the inclusion of desferrioxamine, an iron-chelating drug successfully used in the therapy of thalassemic patients, completely blocks lipid peroxidation in an in vitro system where the oxidative insult is triggered by iron salts [46]. Reznick et al. have shown that the propionate ester of carnitine, either the L- or D-stereoisomer, was active in decreasing hydroxyl radical formation in two different Fenton systems [16]. In addition, these compounds inhibited the iron-induced ascorbate oxidation, an indirect estimate for iron chelation. Interestingly, L-carnitine resulted essentially inactive in both the experimental models of iron chelation, whereas, from a chemical standpoint, it is not clear why the presence of the propionate ester moiety is critical for carnitine to manifest such iron-chelating activity. When the propionyl-L-carnitine data are compared with those obtained in the same experimental models with desferrioxamine, however, the latter compound was always more efficient in inhibiting the iron-driven Fenton reaction, even at concentrations one thousand times lower than the carnitine derivatives.

### Membrane phospholipid repair process: A potential link between the carnitine system and the heart antioxidant network

The oxidative deterioration of membrane phospholipids may compromise a variety of cellular functions, which normally rely on the integrity of the chemico-physical status of the membrane bilayer. This is chemically manifested by the peroxidation of polyunsatured fatty acids esterified on the glycerol backbone of membrane phospholipids. A biochemically relevant feature observed in the course of membrane phospholipid peroxidation is the concurrent increased activity of different phospholipases [21, 22, 47–56]. In particular, phospholipase $A_2$ activity seems to be mainly involved in the excision of the fatty acid hydroperoxide, when tissues are subjected to oxidative stress. This enzymatic process is considered the first response of the secondary antioxidant defence line against the oxidative insult of membrane lipids. It has been also shown that the action of glutathione peroxidase, a component of the primary defence system, is greatly enhanced by phospholipase $A_2$ [39]. This concerted enzymatic action is based on the fact that glutathione peroxidase enzymes (either selenium- or non-selenium-dependent) are capable of reducing lipid hydroperoxide to the corresponding non-reactive lipid alcohol only if the peroxide group is present in the unesterified fatty acid. The ability of phospholipase $A_2$ to remove the peroxidized fatty acid present in phospholipids has been shown to occur in purified lipid mixtures [50, 52, 55]. In these kinds of studies it was possible to demonstrate that the alteration of lipid fluidity induced by the peroxidation of liposomes composed of unsaturated phospholipids is associated with an enhanced phospholipase $A_2$ attack, suggesting a possible structural link between peroxidized lipid and activation of the hydrolytic enzyme [52, 55]. In addition, once the peroxidative event triggered the phospholipase $A_2$ activity, peroxidized phospholipids were hydrolyzed at higher rates and peroxidation of host phospholipid did not affect the hydrolytic rate of non-oxidized phospholipids [55].

If the removal of the peroxidized fatty acid represents an important stage in the membrane phospholipid repair process, a further step is required to fulfil the repair action: the reacylation of the lysophospholipid. This step is accomplished by the action of the enzyme lysophospholipid acyl-CoA transferase (LAT), which catalyzes the transfer of the fatty acid from CoA to lysophospholipid. The overall process of membrane phospholipid repair activity is essentially a deacylation-reacylation cycle, a metabolic pathway devoted to tailoring the fatty acid composition of membrane phospholipids [57]. It should be taken into account that the reacylation step requires an adequate supply of acyl-CoA, which in turn is generated by the ATP-dependent enzyme acyl-CoA synthetase (ACS).

An experimental model successfully used to study the membrane phospholipid repair process is the erythrocyte. Lubin et al. provided the first experimental evidence that membrane phospholipid reacylation was greatly enhanced in intact vitamin E-deficient human red cells exposed to hydrogen peroxide [20]. The enhanced radioactive fatty acid incorporation was higher in membrane phosphatidylethanolamine (PE) than phosphatidylcholine (PC). Dise and Goodman have shown that the exposure of human red cells to t-butylhydroperoxide (t-BOOH) caused an increased incorporation of radioactive oleate or palmitate into membrane PE, although, the reacylation of membrane PC in t-BOOH-treated red cells was partially depressed [58]. We have observed that the incorporation of saturated and monounsatured fatty acids into membrane phospholipids of oxidatively challenged red cells was greatly enhanced in comparison to control cells [59]. On the other hand, the incorporation rates of polyunsatured fatty acids in such cells were lower than in control ones (Arduini et al., submitted). In support of the notion that during oxidative challenge, polyunsaturated fatty acids are not good reacylating substrates, Allen et al. found that the incorporation of arachidonic acid into membrane PC and PE of red cells treated with different oxidizing agents was inhibited [60]. The inhibition of polyunsaturated fatty acid incorporation into membrane phospholipids was also seen in alveolar macrophages and brain synaptosomes exposed to hydroperoxides [61, 62]. The reasons for a reduced ability of these different biological systems to repair membrane phospholipids with polyunsatured fatty acids are diverse, though it is not possible to recognize a common underlying mechanism.

A closer look at the proposed membrane phospholipid repair mechanism poses, in our opinion, some relevant questions on the biochemical requirements for the physiological expression of the reacylation step. For example, the increased demand of activated acyl-units to reacylate lysolipids generated by the phospholipase $A_2$ action on oxidatively challenged phospholipids may exceed the intracellular availability of acyl-CoA. This suggests that the activity of the reacylating enzyme has to be finely assisted in terms of activated acyl-unit supply. Since ACS is the enzyme devoted to the production of activated acyl-units, one would predict that either ACS generates large amounts of acyl-CoA to buffer any potential LAT substrate request, or ACS

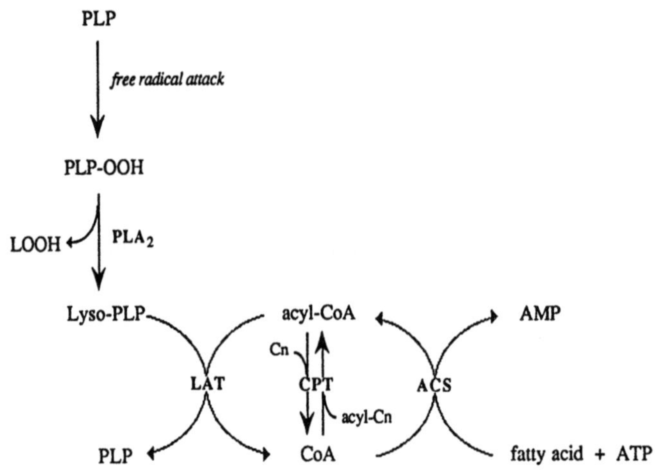


*Figure 2.* The carnitine system and membrane phospholipid repair reactions. Abbreviations used are: PLP, phospholipids; PLP-OOH, phospholipid hydroperoxide; $PLA_2$, phospholipase $A_2$; LAT, lysophospholipid acyl-CoA transferase; Cn, carnitine; acyl-Cn, acylcarnitine; CPT, carnitine palmitoyltransferase; ACS, acyl-CoA synthetase.

and LAT rates are somehow coupled to guarantee the phospholipid repair process can respond adequately to oxidative challenge. The first prediction is difficult to sustain, for the simple reason that the cellular free CoA levels are usually low and compartmentalized (i.e. in the heart 95% of total CoA is present in the mitochondrial matrix) [63]. In addition, an extensive free CoA acylation may impair several metabolic pathways. The second suggestion presents some kinetic impediment, as the specific activities of ACS and LAT are in several cases very diverse for a given fatty acid and its corresponding CoA ester [36]. Rate differences of these two enzymes also imply an alteration of the acyl-CoA/free CoA ratio, which may be an additional deterrent for the membrane phospholipid reacylation [64]. LAT and glycerophosphate acyl-CoA transferase activities can be inhibited in vitro by relatively high levels of free CoA [35, 64]. It is worth noting that the oxidative insult occurring in ischemic disease is also associated with a drop of intracellular ATP levels, which in turn is expected to limit the activity of ACS. In addition, the increased breakdown of ATP gives rise to AMP and adenosine, compounds able to inhibit ACS [65].

A possible way to surmount such a metabolic stall is to introduce carnitine and carnitine palmitoyltransferase (CPT) in the membrane phospholipid repair scenario (Figure 2). This enzyme, known for its role played in mitochondrial fatty acid oxidation, catalyzes the reversible transfer of long-chain fatty acids from CoA to carnitine [66]. Given the sensitivity of CPT to the mass action ratio of the substrates, any variation of the acyl-CoA/free CoA ratio can be compensated by this enzyme. CPT action would be twofold: provide acyl-units at no ATP-cost and buffer the harmful elevation of free CoA.

We have shown that CPT can be considered as an integral component of the pathway for membrane phospholipid and triglyceride fatty acid turnover in both human erythrocytes and neuronal cells [36, 37]. In these biological models, the inhibition of CPT caused a depression of the reacylating capability of membrane complex lipids. It was also demonstrated that the long-chain acylcarnitine pool is an important reservoir of acyl-units, which are actively utilized when the cell is in a deenergized state. Furthermore, CPT buffered the excess of acyl-CoA produced by ACS, to prevent free CoA depletion and to maintain an optimal acyl-CoA/free CoA ratio for the membrane phospholipid and triglyceride fatty acid turnover. Since this pathway represents the physiological expression of the reacylation process, one may extend these concepts to the membrane phospholipid repair activity. Thus, human red cells, treated with either reversible or irreversible CPT inhibitors and exposed to t-BOOH, incorporated significantly less amounts of radioactive fatty acids into membrane PE and PC than untreated cells exposed to t-BOOH (Arduini et al., submitted). Pulse-chase experiments with t-BOOH-treated red cells showed an increased acyl-trafficking between the acylcarnitine pool and the reacylation step of membrane phospholipids, which could be abolished by inhibiting CPT [67]. Also, when ATP-depleted red cells were incubated with radioactive fatty acid complexed to bovive serum albumin, a reduced radioactive acylcarnitine formation was observed [36]. On the other hand, ATP-depleted red cells incubated with radiolabeled palmitoyl-L-carnitine incorporated more radioactive palmitate into membrane PC and PE than control cells [36]. Finally, preliminary data from our laboratory suggest that the membrane phospholipid repair process is operative in oxidatively challenged endothelial cells, and that the inhibition of CPT activity is accompanied by a decreased ability of these cells to properly repair membrane phospholipids.

In conclusion, the data and concepts illustrated above, and the biochemical analogies existing between ischemic disease and oxidative challenge, should more easily orient future work to ascertain involvement of the carnitine system in the secondary antioxidant network of the heart.

## References

1. Hearse DJ, Humphrey SM, Nayler WG, Slade A, Border D. Ultrastructural damage associated with reoxygenation of the anoxic myocardium. J Mol Cell Cardiol 1975; 7: 315–24.
2. McCord JM. Oxygen-derived free radicals in postischemic tissue injury. N Engl J Med 1985; 312: 159–63.
3. Guarnieri C, Ferrari R, Visioli O, Caldarera CM, Nayler WG. Effect of α-tocopherol on

hypoxic-perfused and reoxygenated rabbit heart muscle. J Mol Cell Cardiol 1978; 10: 893–906.
4. Jolly SR, Kane WJ, Bailie MB, Abrams GD, Lucchesi BR. Canine myocardial reperfusion injury. Its reduction by the combined administration of superoxide dismutase and catalase. Circ Res 1984; 54: 277–85.
5. Ambrosio G, Becker LC, Hutchins GM, Weisman HF, Weisfeldt ML. Reduction in experimental infarct size by recombinant human superoxide dismutase: insights into the pathophysiology of reperfusion injury. Circulation 1986; 74: 1424–33.
6. Chi L, Tamura Y, Hoff PT et al. Effect of superoxide dismutase on myocardial infarct size in the canine heart after six hours of regional ischemia and reperfusion: a demonstration of myocardial salvage. Circ Res 1989; 64: 665–75.
7. Zweier JL, Flaherty JT, Weisfeldt ML. Direct measurement of free radical generation following reperfusion of ischemic myocardium. Proc Natl Acad Sci USA 1987; 84: 1404–7.
8. Arduini A, Eddy L, Hochstein P. Detection of ferryl myoglobin in the isolated ischemic rat heart. Free Rad Biol Med 1990; 9: 511–3.
9. Kumar C, Okuda M, Ikai I, Chance B. Luminol enhanced chemiluminescence of the perfused rat heart during ischemia and reperfusion. FEBS Lett 1990; 272: 121–4.
10. Ferrari R, Ceconi C, Curello S et al. Oxygen-mediated myocardial damage during ischemia and reperfusion:role of the cellular defences against oxygen toxicity. J Mol Cell Cardiol 1985; 17: 937–45.
11. Arduini A, Mezzetti A, Porreca E et al. Effect of ischemia and reperfusion on antioxidant enzymes and mitochondrial inner membrane proteins in perfused rat heart. Biochim Biophys Acta 1988; 970: 113–21.
12. Halliwell B, Gutteridge JMC. Free radicals in biology and medicine. 2nd rev ed. Oxford: Clarendon Press, 1989: 86–276.
13. Romaschin AD, Rebeyka I, Wilson GJ, Mickle DAG. Conjugated dienes in ischemic and reperfused myocardium: an in vivo chemical signature of oxygen free radical mediated injury. J Mol Cell Cardiol 1987; 19: 289–302.
14. Otani H, Prasad MR, Jones RM, Das DK. Mechanism of membrane phospholipid degradation in ischemic-reperfused rat hearts. Am J Physiol 1989; 257: H252–8.
15. Park Y, Kanekal S, Kehrer JP. Oxidative changes in hypoxic rat heart tissue. Am J Physiol 1991; 260: H1395–405.
16. Reznick AZ, Kagan VE, Ramsay R et al. Antiradical effects in L-propionylcarnitine protection of the heart against ischemia-reperfusion injury: the possible role of iron chelation. Arch Biochem Biophys 1992; 296: 394–401.
17. Levine RL. Preferential degradation of the oxidatively modified form of glutamine synthetase by intracellular mammalian proteases. J Biol Chem 1985; 260: 300–5.
18. Davies KJA. Protein damage and degradation by oxygen radicals. J Biol Chem 1987; 262: 9895–901.
19. Davies KJA. Proteolytic systems as secondary antioxidant defenses. In: Chow CK, editor. Cellular antioxidant defense mechanisms. Boca Raton: CRC Press, 1988: 25–67.
20. Lubin BH, Shohet SB, Nathan DG. Changes in fatty acid metabolism after erythrocyte peroxidation: stimulation of a membrane repair process. J Clin Invest 1972; 51: 338–44.
21. Sevanian A, Stein RA, Mead JF. Metabolism of epoxidized phosphatidylcholine by phospholipase $A_2$ and epoxide hydrolase. Lipids 1981; 16: 781–9.
22. Van Kuijk FVG, Sevanian A, Handelman GJ, Dratz EA. A new role for phospholipase $A_2$: protection of membrane from lipid peroxidation damage. Trends Biochem Sci 1987; 12: 31–4.
23. Doetsch PW, Henner WD, Cunnigham RP, Toney JH, Helland DE. A highly conserved endonuclease activity present in Escherichia coli, bovine and human cells recognizes oxidative DNA damage at sites of pyrimidines. Mol Cell Biol 1987; 7: 26–31.
24. Ramotar D, Auchincloss AH, Fraser MJ. Nuclear endo-exonuclease of Neurospora crassa. J Biol Chem 1987; 262: 425–9.
25. Demple B, Greenberg JT, Johnson A, Levin JD. Cellular defenses against oxidative dam-

age: roles of inducible resistance in bacteria and a widespread DNA repair system for oxidative strand breaks. In: Friedberg EC, Hanawalt PC, editors. Mechanisms and consequences of DNA damage processing. New York: Alan R Liss, 1988: 159–72.
26. Schinetti ML, Rossini D, Greco R, Bertelli A. Protective action of acetylcarnitine on NADPH-induced lipid peroxidation of cardiac microsomes. Drugs Exp Clin Res 1987; 8: 509–15.
27. Geremia E, Santoro C, Baratta D, Scalia M, Sichel G. Antioxidant action of acetyl-L-carnitine: in vitro study. Med Sci Res 1988; 16: 699–700.
28. Kumari SS, Menon VP. Effect of carnitine administration on levels of lipid peroxides and activities of superoxide dismutase and catalase in isoproterenol-induced myocardial infarction in rats. J Biosci 1988; 13: 257–62.
29. Masini E, Bianchi S, Gambassi F et al. Ischemia reperfusion injury and histamine release in isolated and perfused guinea-pig heart: pharmacological interventions. Agents Actions 1990; 30: 198–200.
30. Leipala JA, Bhatnagar R, Pineda E, Najibi S, Massoumi K, Packer L. Protection of the reperfused heart by L-propionylcarnitine. J Appl Physiol 1991; 71: 1518–22.
31. Packer L, Valenza M, Serbinova E, Starke-Reed P, Frost K, Kagan V. Free radical scavenging is involved in the protective effect of propionyl-L-carnitine against ischemia-reperfusion injury of the heart. Arch Biochem Biophys 1991; 288: 533–7.
32. Di Giacomo C, Latteri F, Fichera C et al. Effect of acetyl-L-carnitine on lipid peroxidation and xanthine oxidase activity in rat skeletal muscle. Neurochem Res 1993; 18: 1157–62.
33. Di Lisa F, Menabo R, Siliprandi N. L-Propionylcarnitine protection of mitochondria in ischemic rat hearts. Mol Cell Biochem 1989; 88: 169–73.
34. Broderick TL, Quinney HA, Barker CC, Lopaschuk GD. Beneficial effect of carnitine on mechanical recovery of rat hearts reperfused after a transient period of global ischemia is accompanied by a stimulation of glucose oxidation. Circulation 1993; 87: 972–81.
35. Arduini A, Mancinelli G, Ramsay RR. Palmitoyl-L-carnitine, a metabolic intermediate of the fatty acid incorporation pathway in erythrocyte membrane phospholipids. Biochem Biophys Res Commun 1990; 173: 212–7.
36. Arduini A, Mancinelli G, Radatti GL, Dottori S, Molajoni F, Ramsay RR. Role of carnitine and carnitine palmitoyltransferase as integral components of the pathway for membrane phospholipid fatty acid turnover in intact human erythrocytes. J Biol Chem 1992; 267: 12673–81.
37. Arduini A, Denisova N, Virmani A, Avrova N, Federici G, Arrigoni-Martelli E. Evidence for the involvement of carnitine-dependent long-chain acyltransferases in neuronal triglyceride and phospholipid fatty acid turnover. J Neurochem 1994; 62: 1530–8.
38. Goldstein S, Meyerstein D, Czapski G. The Fenton reagents. Free Rad Biol Med 1993; 15: 435–45.
39. Sevanian A, Muakkassah-Kelly F, Montestruque S. The influence of phospholipase $A_2$ and glutathione peroxidase on the elimination of membrane lipid peroxides. Arch Biochem Biophys 1983; 223: 441–52.
40. Ursini F, Maiorino M, Gregolin C. The selenoenzyme phospholipid hydroperoxide glutathione peroxidase. Biochim Biophys Acta 1985; 839: 62–71.
41. Porter WL. Recent trends in food applications of antioxidants. In: Simic MG, Karel M, editors. Autoxidation in food and biological systems. New York: Plenum Press, 1980: 295–365.
42. Burton GW, Ingold KU. Autoxidation of biological molecules. 1. The antioxidant activity of vitamin E and related chain-breaking phenolic antioxidants in vitro. J Am Chem Soc 1981; 103: 6472–7.
43. Arduini A, Fernandez E, Pallini R et al. Effect of propionyl-L-carnitine on rat spinal cord ischaemia and post-ischaemic reperfusion injury. Free Rad Res Commun 1990; 10: 325–32.
44. Ferrari R, Ciampalini G, Agnoletti G, Cargnoni A, Ceconi C, Visioli O. Effect of L-carnitine derivatives on heart mitochondrial damage induced by lipid peroxidation. Pharmacol Res Commun 1988; 20: 125–32.

45. Suzuki YJ, Packer L, Ford GD. Relationships between the effects of superoxide anion and palmitoyl-L-carnitine on the $Ca^{2+}$-ATPase of vascular smooth muscle sarcoplasmic reticulum. J Mol Cell Cardiol 1993; 25: 823–7.
46. Rice-Evans CA, Diplock AT. Current status of antioxidant therapy. Free Rad Biol Med 1993; 15: 77–96.
47. Yasuda M, Fujita T. Effect of lipid peroxidation on phospholipase $A_2$ activity of rat liver mitochondria. Jpn J Pharmacol 1977; 27: 429–35.
48. Kagan VE, Shvedova AA, Novikov KN. Participation of phospholipases in the "repair" of photoreceptor membranes exposed to peroxide oxidation. Biofizika 1978; 2: 279–84 (in Russian).
49. Fujimoto Y, Mino T, Fujita T. Effect of phospholipase C on lipid peroxidation of rat liver mitochondria. Res Commun Chem Pathol Pharmacol 1982; 35: 173–6.
50. Sevanian A, Kim E. Phospholipase $A_2$ dependent release of fatty acids from peroxidized membranes. Free Rad Biol Med 1985; 1: 263–71.
51. Beckman JK, Borowitz SM, Burr IM. The role of phospholipase A activity in rat liver microsomal lipid peroxidation. J Biol Chem 1987; 262: 1479–81.
52. Sevanian A, Wratten ML, McLeod LL, Kim E. Lipid peroxidation and phospholipase $A_2$ activity in liposomes composed of unsatured phospholipids: a structural basis for enzyme activation. Biochim Biophys Acta 1988; 961: 316–27.
53. Chakraborti S, Gurtner GH, Michael JR. Oxidant-mediated activation of phospholipase $A_2$ in pulmonary endothelium. Am J Physiol 1989; 257: L430–7.
54. Borowitz SM, Montgomery C. The role of phospholipase $A_2$ in microsomal lipid peroxidation induced with t-butyl hydroperoxide. Biochem Biophys Res Commun 1989; 158: 1021–8.
55. McLean LR, Hagaman KA, Davidson WS. Role of lipid structure in the activation of phospholipase $A_2$ by peroxidized phospholipids. Lipids 1993; 28: 505–9.
56. Natarajan V, Scribner WM, Taher MM. 4-Hydroxynonenal, a metabolite of lipid peroxidation, activates phospholipase D in vascular endothelial cells. Free Rad Biol Med 1993; 15: 365–75.
57. MacDonald JI, Sprecher H. Phospholipid fatty acid remodeling in mammalian cells. Biochim Biophys Acta 1991; 1084: 105–21.
58. Dise CA, Goodman DBP. t-Butyl hydroperoxide alters fatty acid incorporation into erythrocyte membrane phospholipid. Biochim Biophys Acta 1986; 859: 69–78.
59. Arduini A, Mancinelli G, Radatti GL, Dottori G, Ramsay RR. Potential role of carnitine palmitoyl transferase in membrane phospholipid repair mechanism. In: Ursini F, Cadenas E, editors. Biological free radical oxidations and antioxidants. Padova: CLEUP University Publisher, 1991: 159–71.
60. Allen DW, Newman LM, Okazaki IJ. Inhibition of arachidonic acid incorporation into erythrocyte phospholipids by peracetic acid and other peroxides. Role of arachidonoyl-CoA: 1-palmitoyl-sn-glycero-3-phosphocholine acyl transferase. Biochim Biophys Acta 1991; 1081: 267–73.
61. Sporn PHS, Marshall TM, Peters-Golden M. Hydrogen peroxide increases the availability of arachidonic acid for oxidative metabolism by inhibiting acylation into phospholipids in the alveolar macrophage. Am J Respir Cell Mol Biol 1992; 7: 307–16.
62. Zaleska MM, Wilson DF. Lipid hydroperoxides inhibit reacylation of phospholipids in neuronal membranes. J Neurochem 1989: 52: 255–60.
63. Idell-Wenger JA, Grotyohann LW, Neely JR. Coenzyme A and carnitine distribution in normal and ischemic hearts. J Biol Chem 1978; 253: 4310–8.
64. Mok AY, McMurray WC. Biosynthesis of phosphatidic acid by glycerophosphate acyltransferases in rat liver mitochondria and microsomes. Biochem Cell Biol 1990; 68: 1380–92.
65. Van der Vusse GJ, Glatz JF, Stam HCG, Reneman RS. Fatty acid homeostasis in the normoxic and ischemic heart. Physiol Rev 1992; 72: 881–940.
66. Bieber LL. Carnitine. Annu Rev Biochem 1988; 57: 261–83.

67. Arduini A, Tyurin V, Tyuruna Y et al. Acyl-trafficking in membrane phospholipid fatty acid turnover: the transfer of fatty acid from the acyl-L-carnitine pool to membrane phospholipids in intact human erythrocytes. Biochem Biophys Res Commun 1992; 187: 353–8.

*Corresponding Author*: Professor Arduino Arduini, Istituto di Scienze Biochimiche, Facoltà di Medicinà, Università degli Studi Gabriele D'Annunzio, Via dei Vestini, I-66100 Chieti, Italy

# 13. Experimental evidence of the anti-ischemic effect of L-carnitine

DENNIS J. PAULSON and AUSTIN L. SHUG

"It seems likely that L-carnitine will exhibit the greatest beneficial effect in situations where it is possible to pretreat the heart with this compound for sufficient duration such that intracellular L-carnitine levels are elevated."

## Introduction

There is considerable evidence that L-carnitine and some of its short-chain acyl derivatives (acetyl-L-carnitine, propionyl-L-carnitine and propionylcarnitine taurine amide) are capable of protecting the heart against ischemic/reperfusion injury. However, despite these findings, the efficacy of L-carnitine and these derivatives is still controversial. The purpose of this chapter is to review the experimental evidence concerning the anti-ischemic effects of L-carnitine, since other chapters in this book will deal with the actions of acetyl-L-carnitine and propionyl-L-carnitine. Possible mechanisms that may account for these beneficial effects of L-carnitine will be summarized. In addition, new data and explanations will be provided which may help clarify some of the experimental discrepancies reported among various studies.

## Anti-ischemic effect of L-carnitine

The first study to investigate the anti-ischemic effects of L-carnitine was performed by Folts et al. [1] in dogs. This group found that the epicardial S-T segment deviations produced by regional ischemia were diminished by infusion of L-carnitine into the ischemic tissue. The incidence of ventricular fibrillation was also dramatically decreased by L-carnitine. This anti-arrhythmic effect of L-carnitine has been subsequently confirmed by a number of studies using a variety of experimental protocols [2–5]. However, these results were not supported in a study by Gilmour et al. [6]. Using a protocol similar to that of Folts et al. [1], this study was unable to demon-

*J.W. de Jong and R. Ferrari (eds): The carnitine system*, 183–197.

strate a positive effect of L-carnitine in attenuating ischemia-induced ECG alterations. The explanation for this discrepancy remains uncertain, although it may be due to subtle, yet important, differences in the experimental protocols. For example, the Gilmour et al. [6] study used paced hearts; whereas in the Folts et al. [1] study, heart rate was not controlled. Therefore, the incidence of arrhythmias may have been different between these two studies.

Further evidence for the anti-ischemic effect of L-carnitine has been obtained in the ischemic dog heart. L-Carnitine has been reported to enhance the recovery of peak developed left ventricular pressure and prevent the postischemic increase in left ventricular diastolic pressure in dogs subjected to regional ischemia and reperfusion [7]. This study along with others [8] has shown the beneficial effects of L-carnitine during ischemia and reperfusion to be associated with an increased level of myocardial ATP. L-Carnitine has also been shown to preserve mitochondrial function in dogs and rabbits exposed to elevated plasma free fatty acids [9, 10]. The preservation of mitochondrial function may be related to a lowering of intracellular long-chain acyl-coenzyme A (CoA) and carnitine esters [8], as well as a preservation of myocardial phospholipid content [11].

In a series of studies using the swine model of ischemia, Liedtke et al. [12–15] provided additional data supporting the anti-ischemic effects of carnitine. In these experiments, plasma fatty acids were elevated with Intralipid infusion. This condition caused a greater impairment of cardiac contractile function when these hearts were exposed to a period of low-flow ischemia. D,L-Carnitine had no apparent effect in improving the overall hemodynamic performance in hearts with normal levels of plasma free fatty acids. However, in hearts with elevated free fatty acids, several hemodynamic parameters were improved during ischemia: left ventricular pressure, +dP/dt, regional wall shortening, and global mechanical efficiency. This beneficial effect of D,L-carnitine was associated with a decrease in fatty acid uptake, fatty acid oxidation and tissue stores of long-chain acyl-CoA, while high energy stores were increased.

Studies using the isolated perfused rat heart have produced inconsistent findings concerning the anti-ischemic effect of L-carnitine [16–21]. Paulson et al. [16] showed that adding L-carnitine to the perfusion medium had no effect on the recovery of mechanical function of isolated hearts subjected to 15 min of working heart perfusion, 90 min of low-flow ischemia, and 15 min of reperfusion. While L-carnitine was not beneficial in protecting the ischemic rat heart, acetyl-L-carnitine and propionyl-L-carnitine were able to significantly enhance the recovery of contractile function. These two compounds were also more effective in inhibiting the uptake of L-carnitine by isolated cardiac myocytes, suggesting that acetyl-L-carnitine and propionyl-L-carnitine may be taken up with greater affinity than L-carnitine. It was suggested that the slow rate of L-carnitine uptake by the cardiac myocyte may explain the lack of an anti-ischemic action in this model.

This hypothesis was supported by other studies. For example, Duan & Karmazyn [19] demonstrated a beneficial effect of D,L-carnitine using the isolated rat heart model. These hearts were exposed to 30 min constant flow perfusion, 30 min of low-flow ischemia, and followed by 30 min of reperfusion. D,L-Carnitine treatment produced no effect on either the contractile depression or elevation in resting tension during ischemia, but significantly decreased the incidence of arrhythmias at the termination of ischemia. The recovery of contractile function was also significantly increased. This beneficial effect of D,L-carnitine was associated with a partial preservation of mitochondrial function. The longer duration of pre-ischemic perfusion may explain why this study [19] was able to demonstrate an anti-ischemic effect of D,L-carnitine. Paulson et al. [16] may not have been able to demonstrate a significant anti-ischemic effect because the hearts were perfused for only 15 min prior to ischemia. Because of the slow rate of carnitine uptake by the perfused rat heart [22], it may take 30 min or longer before sufficient carnitine enters the myocyte to exert a measurable effect. This explanation is further supported recently by Broderick et al. [21]. In this study, hearts were initially perfused for a period of 60 min in the presence of 10 mM L-carnitine in order to increase intracellular carnitine content. The resultant elevation in myocardial carnitine content enhanced the recovery of mechanical function when these hearts were exposed to 35 min of zero-flow ischemia followed by 40 min of reperfusion. A stimulation of glucose oxidation without an increase in the rate of glycolysis was associated with the beneficial effect of L-carnitine.

A number of human studies have also provided evidence for the anti-ischemic effect of L-carnitine. Early studies evaluated the effect of D,L-carnitine in patients with coronary artery disease who were subjected to sequential rapid coronary sinus pacing protocols, 15 min apart [23]. The infusion of D,L-carnitine before the second pacing protocol resulted in significant increases in mean heart rate, pressure-rate product, and pacing duration. The treated group also had improvement in myocardial lactate extraction and left ventricular end-diastolic pressure, as well as less S-T segment depression. This anti-ischemic effect of L-carnitine was again confirmed in a study where patients with coronary artery disease were subjected to an exercise stress test [24, 25]. Subsequent studies using the L-isomer found similar beneficial results in patients given an exercise tolerance test [26] or pacing-induced ischemia [27–29]. In a double blind study on fifty six patients suffering from acute myocardial infarction, it was shown that L-carnitine administration reduced the number of premature ventricular beats after two days of treatment [30]. Another study showed that the administration of carnitine improved ECG alterations suggesting that the infarct size may have been limited [31]. In elderly patients with ischemia or hypertension-induced heart failure, L-carnitine therapy resulted in reduced heart rate, edema and dyspnea, increased diuresis and reduced digitalis consumption compared with the group not treated with carnitine [32].

In contrast to these positive studies, other clinical studies have suggested that L-carnitine's anti-ischemic effect may be small or absent. Martina et al. [33] studied the effects of intravenous L-carnitine treatment on patients with acute myocardial infarction. These treatments were initiated 4–12 h after onset of pain and were given over a 7-day period. Holter-ECG analysis indicated no differences in the incidence of ventricular premature beats over most of this period. However, on the second day after the acute myocardial infarction, there appeared to be some decrease in the number of arrhythmias compared to the placebo group. Demeyere et al. [34] investigated the cardioprotective effects of L-carnitine in patients undergoing multiple aortocoronary bypass grafting. Transmural left ventricular biopsy specimens were taken at the beginning and end of cardiopulmonary bypass, and assayed for ATP and creatine phosphate. Intravenous infusion of L-carnitine had no effect on either hemodynamic parameters or tissue levels of ATP and creatine phosphate. It was concluded that treatment with carnitine neither facilitates weaning from cardiopulmonary bypass in patients undergoing aortocoronay bypass surgery nor favorably affects hemodynamic function during the next 24 h. The lack of an anti-ischemic effect of L-carnitine in these studies may be due to the dose of carnitine used and whether tissue levels of L-carnitine were increased. As indicated by the animal studies, the slow rate of carnitine uptake may be a limiting factor in the beneficial effect of L-carnitine against ischemic/reperfusion injury. The rapid clearance of carnitine from the body is another limiting factor for the anti-ischemic effect of L-carnitine. In order to exert such an effect it may be necessary to pretreat the patient with sufficient L-carnitine at multiple treatment intervals prior to an ischemic episode to increase myocardial L-carnitine levels. Thus, L-carnitine pretreatment may be most beneficial in patients who will be undergoing bypass surgery or patients who are at a greater risk of developing ischemic heart disease such as diabetics or patients with a history of coronary artery disease.

## Mechanisms for the anti-ischemic effects of L-carnitine

There have been several mechanisms suggested by different investigators for the anti-ischemic effects of L-carnitine. We have broadly classified and summarized the evidence for the various proposed mechanisms below:

### *Loss of carnitine*

Carnitine and its esters are mainly located in the cytosolic compartment of the myocytes and are present in concentrations much higher than those found in plasma. Myocardial carnitine levels are lowered by a number of situations such as aging, pressure-overload hypertrophy, diabetes, and some forms of cardiomyopathy [35]. The cardiac depression associated with these conditions may be related to the lowered levels of intracellular carnitine. Carnitine is

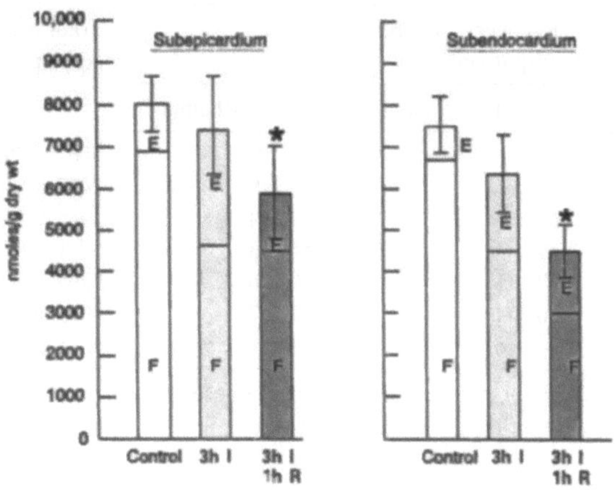


*Figure 1.* Changes in myocardial carnitine in subepicardium and subendocardium of open-chest dog heart subjected to 3 h of ischemia (I) with and without 1 h reperfusion (R). The lower portion of the bar represents free carnitine (F), and the upper portion represents the esterified form (E). Values represent the mean ± SE for = to 13. * indicates significant difference from control, $p < 0.05$.

also reportedly lost from the ischemic myocardium. However, there is some controversy in the literature as to whether this occurs in all animal models. A number of studies have shown that myocardial ischemia will induce a loss of carnitine in the dog [1, 36], pig [12] and human heart [37, 38]; but not all studies in the ischemic rat heart [39, 40] have been able to demonstrate a loss of carnitine with ischemia. There are also other issues which have not been resolved. It is uncertain whether carnitine is lost from both irreversibly and reversibly damaged tissues. Ischemia causes damage to the sarcolemmal membrane resulting in the diffusion of intracellular carnitine down its concentration gradient into the extracellular space. Presumably, reperfusion is required to washout this carnitine.

To help resolve these issues, we have examined the effects of 3 h of regional ischemia and 1 h of reperfusion on total myocardial L-carnitine content in the dog heart. Staining with p-nitro blue tetrazolium (pNBT) was used to determine whether the tissue was irreversibly or reversibly damaged. This duration of ischemia and reperfusion produced a 25% decrease in myocardial total carnitine content in reversibly damaged subepicardium tissue (Figure 1). In irreversibly damaged subendocardial tissue, the loss of carnitine was greater at 40%. Three hours of ischemia without reperfusion produced no change in total carnitine content, although free carnitine was significantly reduced with a commensurate increase in esterified carnitine.

*Table 1.* Effects of working heart perfusion, global ischemia, and anoxia on myocardial carnitine content in rats.

| Group (n) | Recovery of cardiac output (ml/min) | | Total myocardial carnitine content (nmol/g dry wt) |
|---|---|---|---|
| | Prior to ischemia | After ischemia | |
| Unperfused (12) | – | – | 5507 ± 293 |
| 85′WH (4) | 54 ± 6 | 52 ± 9 | 5437 ± 819 |
| 30′I, 15′R (6) | 67 ± 3 | 54 ± 4 | 4965 ± 502 |
| 45′I, 15′R (13) | 61 ± 3 | 54 ± 3 | 5011 ± 233 |
| 60′I, 15′R (13) | 59 ± 3 | 40 ± 6 | 5265 ± 309 |
| 90′I, 15′R (8) | 60 ± 4 | 37 ± 7 | 4815 ± 249 |
| 90′ Anoxia (8) | 64 ± 6 | 6 ± 1 | 4841 ± 464 |

All values are mean ± SE for number shown in parentheses.
WH = working heart; I = ischemia; R = reperfusion.

As indicated above, conflicting reports have appeared concerning carnitine levels in the ischemic rat heart [39, 40]. To resolve this controversy, the effects of global ischemia and reperfusion on myocardial carnitine content of isolated perfused rat, rabbit and guinea-pig hearts were determined. The effects of working rat-heart perfusion, global ischemia, and anoxia on myocardial carnitine content are shown in Table 1. Ischemic hearts were perfused with 1.2 mM palmitate and 5.5 mM glucose as substrates. Anoxic hearts were perfused in a nonrecirculating manner with 11 mM glucose as the only exogenous substrate. Cardiac output did not change with 85 min of normal working heart perfusion, nor did working heart perfusion affect tissue carnitine content. Increasing periods of global ischemia decreased the recovery of cardiac output, but had no statistically significant effect on total myocardial carnitine content despite the appearance of carnitine in the perfusion medium. On average, 170 μmol of carnitine were detected in the perfusion medium per heart. Nonrecirculating anoxic perfusion for 90 min caused a complete loss of cardiac contractile function, but did not change cardiac carnitine levels.

Myocardial carnitine content in vivo is higher in the dog, guinea- pig and rabbit than that found in the rat (8000, 6595, 7340, 5507 nmol/g dry wt, respectively). In the guinea-pig working heart perfusion for 60 min had no effect on cardiac output or carnitine content (Table 2). However, both 30 min of reversible and 90 min of irreversible ischemia significantly decreased myocardial carnitine content. Similarly, 60 min of anoxia in both guinea-pig and rabbit decreased myocardial carnitine content.

These findings are in line with previous reports of carnitine loss from the ischemic canine heart [1, 36]. In addition, it shows significant amounts of carnitine are lost from reversibly injured tissue as well as necrotic areas. This finding is of particular importance, because loss of carnitine from the reversibly damaged heart may render this tissue more vulnerable to a subsequent

*Table 2.* Effects of working heart perfusion, ischemia, and anoxia on myocardial carnitine content in guinea pigs and rabbits.

| Group (n) | Recovery of cardiac output (ml/min) | | Total myocardial carnitine content (nmol/g dry wt) |
|---|---|---|---|
| | Prior to ischemia | After ischemia | |
| *Guinea pig* | | | |
| Unperfused (5) | – | – | 6595 ± 394 |
| 60′WH (5) | 84 ± 3 | 81 ± 3 | 6284 ± 504 |
| 30′I, 15′R (5) | 71 ± 4 | 53 ± 10 | 5058 ± 634* |
| 90′ I, 15′R (6) | 77 ± 7 | 18 ± 3 | 4953 ± 448* |
| 60′ Anoxia | 64 ± 10 | 14 ± 1 | 5183 ± 388* |
| *Rabbit* | | | |
| Unperfused (4) | – | – | 6272 ± 645 |
| 60′ Anoxia (4) | – | – | 3777 ± 493* |

All values are mean ± SE for number shown in parentheses.
WH = working heart; I = ischemia; R = reperfusion.
* Significant difference from unperfused value, $p \leq 0.05$.

episode of ischemia. Although reflow is required for recovery of heart function, it may also exert a harmful effect by washing out important metabolites. Carnitine from ischemic tissue is a low-molecular weight, nonmetabolizable, water-soluble compound. Consequently it is most likely lost from oxygen-deficient tissues by leakage down its concentration gradient.

The isolated perfused rabbit and guinea-pig hearts both exhibited a loss of carnitine in response to oxygen deficiency. In contrast, the isolated perfused rat heart showed no significant loss of carnitine after ischemia/reperfusion or anoxia. Nevertheless, a small amount of carnitine was detected in the perfusion medium. This finding may explain why mean tissue carnitine levels were found to be lower in this, as well as previous studies [39] after ischemia and reperfusion. However, the loss of carnitine was not large enough to affect tissue levels significantly. In addition, previous studies, that suggested carnitine was lost from the ischemic/reperfused rat heart, expressed the data on a per gram wet weight. Since the ischemic heart may accumulate water due to capillary damage, this method of expressing the result may be inaccurate and may have led to an artifactual decrease in total carnitine content when expressed on a per gram wet weight basis.

These results indicate that L-carnitine is lost in most animal models of ischemia/reperfusion. Since carnitine participates in several metabolic pathways, the loss of carnitine may lead to a critical disruption in overall metabolic functions of the heart. This assertion is supported by the study by Patel et al. [41]. This study evaluated 40 patients with various types of acquired heart disease to assess the impact of myocardial carnitine concentration on prognosis. Myocardial carnitine concentration was normal in 22 patients and low in 18. Three patients in the normal myocardial carnitine group died; one of these was a cardiac death. In the low myocardial carnitine concentration group seven patients died; six from cardiac causes. These findings provide evidence that myocardial carnitine levels may have a significant prognostic implication in determining the long-term survival of patients with ischemia-induced depletion of myocardial carnitine content. L-Carnitine treatment may, therefore, be beneficial in preventing the ischemia-induced loss of carnitine.

*Lipid intermediates*

A number of investigators have shown that low-flow ischemia results in the accumulation of long-chain acylcarnitine and CoA esters [39, 40]. In contrast, zero-flow ischemia causes no change in the levels of long-chain acylcarnitine and only a small transient increase in long-chain acyl-CoA [42]. Since in vitro experiments (with isolated enzymes and subcellular fractions) have shown that these compounds can inhibit a number of enzyme activities and mitochondrial energy production, it has been suggested that the accumulation of these lipid intermediates may contribute to the genesis of irreversible ischemic damage. For example, long-chain acylcarnitine alters voltage-dependent calcium channels, inhibits sodium-pump activity, and interferes with gap-junction conductance [43, 44]. The increase of this compound during ischemia has been directly implicated in the ventricular arrhythmias induced by ischemia [45, 46]. Most of the total tissue CoA is located in the mitochondrial compartment [40]. Therefore the accumulation of long-chain acyl-CoA is thought to occur primarily within the mitochondria, although there may be a small but important accumulation between the inner and outer mitochondrial membrane. It has been shown that long-chain acyl-CoA can interfere with mitochondrial energy metabolism by inhibiting the adenine nucleotide translocator [47]. This inhibition occurs from both sides of the mitochondrial membrane. There is also evidence that accumulation of long-chain acyl-CoA may enhance free-radical production by damaged ischemic mitochondria when reperfusion is instituted [48].

L-Carnitine may protect the ischemic heart by preventing the accumulation of these lipid intermediates. L-Carnitine can react with long-chain acyl-CoA through the enzyme carnitine palmitoyltransferase I and can be converted into long-chain acylcarnitine. L-Carnitine may lower the accumulation of long-chain acyl-CoA within the mitochondrial compartment via the carnitine palmitoyltransferase II. The decrease in long-chain acylcarnitine may be mediated generalized improvement in fatty acid oxidation. An alternative explanation is that carnitine treatment may cause the efflux of excess acylcarnitine esters from the myocardium via an exchange transport mechanism. Studies by Rizzon et al. [30] have shown that carnitine treatment of ischemic patients results in an increased urinary excretion of long- and short-chain carnitine esters. Another explanation for the effects of L-carnitine on the accumulation of lipid intermediates may be through the enzyme carnitine acetyltransferase. By catalyzing the reaction of carnitine and acetyl-CoA to

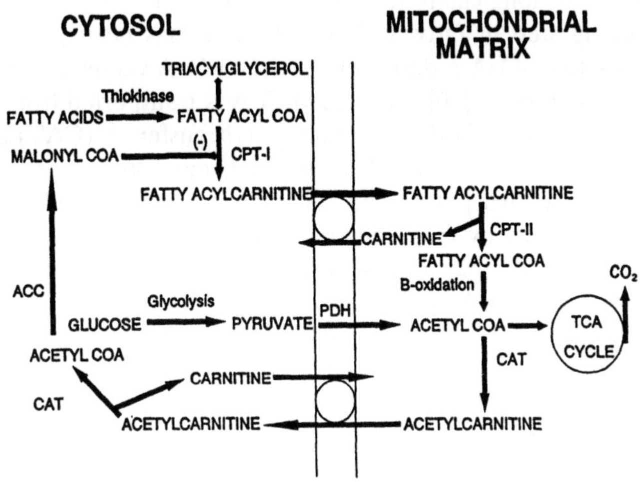


*Figure 2.* Malonyl-CoA regulation of fatty-acid oxidation in the heart. CPT-I and CPT-II: carnitine palmitoyltransferase-I and -II, respectively, CAT: carnitine acetyltransferase, PDH: pyruvate dehydrogenase, ACC: acetyl-CoA carboxylase, TCA: tricarboxylic acid.

produce acetylcarnitine, this reaction may lead to a stimulation of both glucose oxidation and glycolysis [21]. High levels of acetyl-CoA will feed back and inhibit these pathways. This mechanism is discussed in detail in the following section.

### *Substrate metabolism*

L-Carnitine is an essential component in the transport of long-chain fatty acids into mitochondria, where they undergo $\beta$-oxidation [49]. Although it is assumed that ischemic-induced alterations in myocardial carnitine content will impair the ability of the heart to oxidize fatty acids, there is little direct evidence to support this hypothesis. L-Carnitine will stimulate fatty-acid oxidation of tissue homogenates or isolated mitochondria, but when it is added to the perfusion medium of isolated perfused rat hearts, a depression in palmitate uptake and oxidation occurs [50, 51]. Glucose oxidation is increased in these carnitine-perfused hearts. As suggested by Broderick et al. [21], this stimulation of glucose oxidation by L-carnitine may be an important mechanism for the anti-ischemic effect of L-carnitine.

The mechanism by which carnitine can stimulate glucose oxidation has only recently been revealed. Carnitine, by modulating tissue malonyl-CoA synthesis, may reciprocally alter glucose and fatty-acid oxidation [52–55] (Figure 2). Malonyl-CoA has been shown to be a potent inhibitor of the enzyme carnitine palmitoyltransferase-I (CPT-I). This enzyme converts long-chain acyl-CoA to long-chain acylcarnitine which has been suggested to be the rate-limiting step in fatty acid oxidation. It has been postulated that cytoplasmic malonyl-CoA is derived from acetyl-CoA via the enzyme, acetyl-CoA carboxylase (ACC) [56]. The acetyl-CoA is transported from the mitochondria to the cytosol via the carnitine acetyltransferase (CAT) and carnitine acetyltranslocase pathway. The link between glucose and fatty acid oxidation occurs at this point. The acetyl-CoA used to produce malonyl-CoA is derived primarily from glucose oxidation via the enzyme pyruvate dehydrogenase (PDH) [52, 57, 58]. Carnitine stimulates glucose oxidation by converting acetyl-CoA to acetylcarnitine which removes the feedback inhibition on PDH. This, in turn, leads to increased production of cytoplasmic acetylcarnitine, acetyl-CoA, and malonyl-CoA. Thus, carnitine increases levels of cytoplasmic malonyl-CoA which causes an inhibition of fatty acid oxidation [52] and a stimulation of glucose utilization. In addition, the inhibition of CPT-I may be the mechanism by which carnitine prevents the accumulation of long-chain acylcarnitine and long-chain acyl-CoA esters within the ischemic heart.

*Free radical production*

Reperfusion of ischemic tissue is believed to cause cellular damage through the production of free radicals. A number of sources of free radicals have been suggested. Of these, mitochondria are probably a major source. Recent studies have indicated that the accumulation of long-chain acyl-CoA in ischemic tissue may facilitate the production of free radicals by damaged mitochondria [48, 59]. L-Carnitine, by preventing the accumulation of these esters, may be beneficial for this reason. It has also been suggested that L-carnitine may be involved in a repair mechanism for oxidative-induced damage to membrane phospholipids [60]. Free radical damage to membrane phospholipids is thought to cause the accumulation of lysophospholipids which may result in electrophysiologic derangement and malignant arrhythmias. Carnitine may facilitate the formation of long-chain acylcarnitine which can be used to reacylate these lysophospholipids. Several studies have suggested that the cardioprotective effect of propionyl-L-carnitine may also be mediated via mechanisms preventing free radical damage. Both propionyl-L-carnitine and propionyl-D-carnitine have been shown to inhibit hydroxyl radical production by chelating iron required for the generation of hydroxyl radicals [61].

*Membrane interaction*

It has been proposed that L-carnitine and its short-chain acyl derivatives may provide protection from ischemic reperfusion injury by rendering cell membranes more resistant to the harmful effects of free radicals or other toxic agents [44]. The basis for this hypothesis comes from a number of studies indicating that L-carnitine possesses a stabilizing effect on cell membranes. For example, Batelli et al. [62] provides indirect evidence that the positively charged molecule of carnitine may interact with the negatively charged molecule of cardiolipin in liver mitochondrial membranes. This binding may induce some changes in the phase organization of the cardiolipin molecule, within the membrane structure, which has a stabilizing effect. Arduini et al. [63] showed that both L-carnitine and acetylcarnitine enhances membrane stability of mature erythrocytes. This effect most likely occurs via a specific interaction with one or more cytoskeletal protein(s).

Further support for this hypothesis comes from a more recent study in cultured hepatocytes by Pastorino et al. [64]. In this study, L-carnitine and cyclosporin A protect cultured hepatocytes from the death produced by anoxia and rotenone. There is strong evidence that this protection is a consequence of the ability of these compounds to prevent mitochondrial permeability transition. The accumulation of long-chain acyl-CoA in the hepatocyte (as a result of the inhibition of fatty acid oxidation following mitochondrial de-energization) is an essential cofactor in the induction of the permeability transition. Thus, the ability of L-carnitine and its esters to stabilize cell membranes may also account for their anti-ischemic actions.

## Conclusions

Studies in both animals and humans provide evidence that L-carnitine does possess an anti-ischemic action. However, the cardioprotective effect appears to be relatively modest. There remains some inconsistencies in the literature which need to be examined further. First, the anti-ischemic effect of L-carnitine may be limited by the slow rate of uptake of carnitine by the heart. It seems likely that L-carnitine will exhibit the greatest beneficial effect in situations where it is possible to pretreat the heart with this compound for sufficient duration such that intracellular L-carnitine levels are first elevated [21]. During an acute emergency situation after an ischemic episode, carnitine may have limited efficacy due to the slow rate of transport. Since, short-chain carnitine esters may be taken up with greater affinity by the myocardium, they may have greater clinical efficacy. This assertion is strongly supported by a growing body of basic and clinical evidence on the anti-ischemic effects of propionyl-L-carnitine [65–70]. There are several mechanisms that may account for the anti-ischemic effect of L-carnitine, all of which may ultimately benefit the ischemic heart. The actions of L-carnitine on myocardial substrate metabolism are strongly supported. In addition, the effect of L-carnitine upon the sarcolemmal membrane is intriguing and certainly merits further investigation.

## References


1. Folts JD, Shug AL, Koke JR, Bittar N. Protection of the ischemic dog myocardium with carnitine. Am J Cardiol 1978; 41: 1209–14.
2. DiPalma JR, Ritchie DM, McMichael RF. Cardiovascular and antiarrhythmic effects of carnitine. Arch Int Pharmacodyn Ther 1975; 217: 246–50.
3. Kotaka K, Miyazaki Y, Ogawa K, Satake T, Sugiyama S, Ozawa T. Protection by carnitine against free fatty acid-induced arrhythmia in canine heart. J Appl Biochem 1981; 3: 292–300.
4. Suzuki Y, Kamikawa T, Yamazaki N. Effects of L-carnitine on ventricular arrhythmias in dogs with acute myocardial ischemia and a supplement of excess free fatty acids. Jpn Circ J 1981; 45: 552–9.
5. Duan J, Moffat MP. Protective effects of D,L-carnitine against arrhythmias induced by lysophosphatidylcholine or reperfusion. Eur J Pharmacol 1991; 192: 355–63.
6. Gilmour Jr RF, Williams ES, Farmer BB, Zipes DP. Effects of carnitine and atractyloside on canine cardiac electrical activity. Am J Physiol 1981; 241: H505–12.
7. Silverman NA, Schmitt G, Vishwanath M, Feinberg H, Levitsky S. Effect of carnitine on myocardial function and metabolism following global ischemia. Ann Thorac Surg 1985; 40: 20–4.
8. Suzuki Y, Kamikawa T, Kobayashi A, Masumura Y, Yamazaki N. Effects of L-carnitine on tissue levels of acyl carnitine, acyl coenzyme A and high energy phosphate in ischemic dog hearts. Jpn Circ J 1981; 45: 687–94.
9. Kotaka K, Miyazaki Y, Ogawa K et al. Protective effects of carnitine and its derivatives against free fatty acid-induced mitochondrial dysfunction. J Appl Biochem 1981; 3: 328–36.
10. Ferrari R, Ciampalini G, Agnoletti G, Cargnoni A, Ceconi C, Visioli O. Effect of L-carnitine derivatives on heart mitochondrial damage induced by lipid peroxidation. Pharmacol Res Commun 1988; 20: 125–32.
11. Nagao B, Kobayashi A, Yamazaki N. Effects of L-carnitine on phospholipids in the ischemic myocardium. Jpn Heart J 1987; 28: 243–51.
12. Liedtke AJ, Nellis SH, Whitesell LF. Effects of carnitine isomers on fatty acid metabolism in ischemic swine hearts. Circ Res 1981; 48: 859–66.
13. Liedtke AJ, Nellis SH, Copenhaver G. Effects of carnitine in ischemic and fatty acid supplemented swine hearts. J Clin Invest 1979; 64: 440–7.
14. Liedtke AJ, Vary TC, Nellis SH, Fultz CW. Properties of carnitine incorporation in working swine hearts. Effects of coronary flow, ischemia, and excess fatty acids. Circ Res 1982; 50: 767–74.
15. Liedtke AJ, Nellis SH, Whitesell LF, Mahar CQ. Metabolic and mechanical effects using L- and D-carnitine in working swine hearts. Am J Physiol 1982; 243: H691–7.
16. Paulson DJ, Traxler J, Schmidt M, Noonan J, Shug AL. Protection of the ischaemic myocardium by L-propionylcarnitine: effects on the recovery of cardiac output after ischaemia and reperfusion, carnitine transport, and fatty acid oxidation. Cardiovasc Res 1986; 20: 536–41.
17. Paulson DJ, Shug AL. Effects of carnitine on the ischemic arrested heart. Basic Res Cardiol 1982; 77: 460–3.
18. Hearse DJ, Shattock MJ, Manning AS, Braimbridge MV. Protection of the myocardium during ischemic arrest: Possible toxicity of carnitine in cardioplegic solutions. Thorac Cardiovasc Surg 1980; 28: 253–8.
19. Duan J, Karmazyn M. Effect of D,L-carnitine on the response of the isolated heart of the rat to ischaemia and reperfusion: Relation to mitochondrial function. Br J Pharmacol 1989; 98: 1319–27.
20. Paulson DJ, Schmidt MJ, Traxler JS, Ramacci MT, Shug AL. Improvement of myocardial function in diabetic rats after treatment with L-carnitine. Metabolism 1984; 33: 358–63.

21. Broderick TL, Quinney HA, Barker CC, Lopaschuk GD. Beneficial effect of carnitine on mechanical recovery of rat hearts reperfused after a transient period of global ischemia is accompanied by a stimulation of glucose oxidation. Circulation 1993; 87: 972–81.
22. Vary TC, Neely JR. Sodium dependence of carnitine transport in isolated perfused adult rat hearts. Am J Physiol 1983; 244: H247–52.
23. Thomsen JH, Shug AL, Yap VU, Patel AK, Karras TJ, DeFelice SL. Improved pacing tolerance of the ischemic human myocardium after administration of carnitine. Am J Cardiol 1979; 43: 300–6.
24. Kosolcharoen P, Nappi J, Peduzzi P et al. Improved exercise tolerance after administration of carnitine. Curr Ther Res 1981; 30: 753–64.
25. Canale C, Terrachini V, Biagini A et al. Bicycle ergometer and echocardiographic study in healthy subjects and patients with angina pectoris after administration of L-carnitine: semiautomatic computerized analysis of M-mode tracings. Int J Clin Pharmacol Ther Toxicol 1988; 26: 221–4.
26. Kamikawa T, Suzuki Y, Kobayashi A et al. Effects of L-carnitine on exercise tolerance in patients with stable angina pectoris. Jpn Heart J 1984; 25: 587–97.
27. Ferrari R, Cucchini F, DiLisa F, Raddino R, Bolognesi R, Visioli O. The effect of L-carnitine on myocardial metabolism of patients with coronary artery disease. Clin Trials J 1984; 21: 40–59.
28. Ferrari R, Cucchini F, Visioli O. The metabolical effects of L-carnitine in angina pectoris. Int J Cardiol 1984; 5: 213–6.
29. Reforzo G, De Andreis Bessone PL, Rebaudo F, Tibaldi M. Effects of high doses of L-carnitine on myocardial lactate balance during pacing-induced ischemia in aging subjects. Curr Ther Res 1986; 40: 374–83.
30. Rizzon R, Biasco G, DiBiase M et al. High doses of L-carnitine in acute myocardial infarction: metabolic and antiarrhythmic effects. Eur Heart J 1990; 10: 502–8.
31. Chiariello M, Brevetti G, Policicchio A, Nevola E, Condorelli M. L-carnitine in acute myocardial infarction, A multicenter randomized trial. In: Borum PR, editor. Clinical aspects of human carnitine deficiency. New York: Pergamon Press, 1986: 242–3.
32. Ghidini O, Azzurro M, Vita G, Sartori G. Evaluation of the therapeutic efficacy of L-carnitine in congestive heart failure. Int J Clin Pharmacol Ther Toxicol 1988; 26: 217–20.
33. Martina B, Zuber M, Weiss P, Burkart F, Ritz R. Antiarrhythmische Behandlung mit L-Carnitin beim akuten Myokardinfarkt. Schweiz Med Wochenschr 1992; 122: 1352–5.
34. Demeyere R, Lormans P, Weidler B, Minten J, Van Aken H, Flameng W. Cardioprotective effects of carnitine in extensive aortocoronary bypass grafting: A double-blind, randomized, placebo-controlled clinical trial. Anesth Analg 1990; 71: 520–8.
35. Rebouche CJ, Paulson DJ. Carnitine metabolism and functions in humans. Annu Rev Nutr 1986; 6: 41–66.
36. Suzuki Y, Kamikawa T, Yamazaki N. Carnitine distribution in subepicardial and subendocardial regions in normal and ischemic dog hearts. Jpn Heart J 1981; 22: 377–85.
37. Suzuki Y, Masumura Y, Kobayashi A, Yamazaki N, Harada Y, Osawa M. Myocardial carnitine deficiency in chronic heart failure. Lancet 1982; 1: 116 (Lett to the Ed).
38. Thomsen J, Holden J, Kosolcharoen P et al. Reduced myocardial carnitine levels in patients undergoing coronary artery bypass surgery. J Mol Cell Cardiol 1993; 25(Suppl III): S.63 (Abstr).
39. Shug AL, Thomsen JH, Folts JD, et al. Changes in tissue levels of carnitine and other metabolites during myocardial ischemia and anoxia. Arch Biochem Biophys 1978; 187: 25–33.
40. Idell-Wenger JA, Grotyohann LW, Neely JR. Coenzyme A and carnitine distribution in normal and ischemic hearts. J Biol Chem 1978; 253: 4310–8.
41. Patel AK, Thomsen JH, Kosolcharoen PK, Shug AL. Myocardial carnitine status: clinical prognostic and therapeutic significance. In: Ferrari R, DiMauro S, Sherwood G, editors. L-Carnitine and its role in medicine: From function to therapy. London: Academic Press, 1991: 325–35.

42. Paulson DJ, Schmidt MJ, Romens J, Shug AL. Metabolic and physiological differences between zero-flow and low-flow myocardial ischemia: effects of L-acetylcarnitine. Basic Res Cardiol 1984; 79: 551–61.
43. McMillin Wood J, Bush B, Pitts BJR, Schwartz A. Inhibition of bovine heart $Na^+$, $K^+$-ATPase by palmitylcarnitine and palmityl-CoA. Biochem Biophys Res Commun 1977; 74: 677–84.
44. Fritz IB, Arrigoni-Martelli E. Sites of action of carnitine and its derivatives on the cardiovascular system: Interactions with membranes. Trends Pharmacol Sci 1993; 14: 355–60.
45. Heathers GP, Su C-M, Adames VR, Higgins AJ. Reperfusion-induced accumulation of long-chain acylcarnitines in previously ischemic myocardium. J Cardiovasc Pharmacol 1993; 22: 857–62.
46. DaTorre SD, Creer MH, Pogwizd SM, Corr PB. Amphipathic lipid metabolites and their relation to arrhythmogenesis in the ischemic heart. J Mol Cell Cardiol 1991; 23(Suppl 1): 11–22.
47. Paulson DJ, Shug AL. Inhibition of the adenine nucleotide translocator by matrix-localized palmityl-CoA in rat heart mitochondria. Biochim Biophys Acta 1984; 766: 70–6.
48. Subramanian R, Plehn S, Noonan J, Schmidt M, Shug AL. Free radical-mediated damage during myocardial ischemia and reperfusion and protection by carnitine esters. Z Kardiol 1987; 76(Suppl 5): 41–5.
49. Bremer J. Carnitine-metabolism and functions. Physiol Rev 1983; 63: 1421–80.
50. Rodis SL, D'Amato PH, Koch E, Vahouny GV. Effect of carnitine on uptake, oxidation and esterification of palmitate by the perfused rat heart. Proc Soc Exp Biol Med 1970; 133: 973–7.
51. Broderick TL, Quinney HA, Lopaschuk GD. Carnitine stimulation of glucose oxidation in the fatty acid perfused isolated working rat heart. J Biol Chem 1992; 267: 3758–63.
52. Saddik M, Gamble J, Witters LA, Lopaschuk GD. Acetyl-CoA carboxylase regulation of fatty acid oxidation in the heart. J Biol Chem 1993; 268: 25836–45.
53. Cook GA, Lappi MD. Carnitine palmitoyltransferase in the heart is controlled by a different mechanism than hepatic enzyme. Mol Cell Biochem 1992; 116: 39–45.
54. McGarry JD, Mills SE, Long CS, Foster DW. Observations on the affinity for carnitine and malonyl-CoA sensitivity, of carnitine palmitoyltransterase I in animal and human tissues. Biochem J 1983; 214: 21–8.
55. Cook GA, Gamble MS. Regulation of carnitine palmitoyltransferase by insulin results in decreased activity and decreased apparent Ki values for malonyl-CoA. J Biol Chem 1987; 262: 2050–5.
56. Iverson AJ, Bianchi A, Nordlund AC, Witters LA. Immunological analysis of acetyl-CoA carboxylase mass, tissue distribution and subunit composition. Biochem J 1990; 269: 365–71.
57. Lysiak W, Lilly K, DiLisa F, Toth PP, Bieber LL. Quantitation of the effect of L-carnitine on the levels of acid-soluble short-chain acyl-CoA and CoASH in rat heart and liver mitochondria. J Biol Chem 1988; 263: 1151–6.
58. Lysiak W, Toth PP, Suelter CH, Bieber LL. Quantitation of the efflux of acylcarnitines from rat heart, brain and liver mitochondria. J Biol Chem 1986; 261: 13698–703.
59. Paulson DJ, Shug AL, Jurak R, Schmidt M. The role of altered lipid metabolism in the cardiac dysfunction associated with diabetes. In: Nagano M, Dhalla NS, editors. The diabetic heart. New York: Raven Press, 1991: 395–407.
60. Arduini A. Carnitine and its acyl esters as secondary antioxidants? Am Heart J 1992; 123: 1726–7 (Lett to the Ed).
61. Reznick AZ, Kagan VE, Ramsey R et al. Antiradical effects in L-propionyl carnitine protection of the heart against ischemia-reperfusion injury: the possible role of iron chelation. Arch Biochem Biophys 1992; 296: 394–401.
62. Battelli D, Bellei M, Arrigoni-Martelli E, Muscatello U, Bobyleva V. Interaction of carnitine with mitochondrial cardiolipin. Biochim Biophys Acta 1992; 1117: 33–6.

63. Arduini A, Rossi M, Mancinelli G et al. Effect of L-carnitine and acetyl-L-carnitine on the human erythrocyte membrane stability and deformability. Life Sci 1990; 47: 2395–400.
64. Pastorino JG, Snyder JW, Serroni A, Hoek JB, Farber JL. Cyclosporin and carnitine prevent the anoxic death of cultured hepatocytes by inhibiting the mitochondrial permeability transition. J Biol Chem 1993; 268: 13791–8.
65. Lagioia R, Scrutinio D, Mangini SG, et al. Propionyl-L-carnitine: A new compound in the metabolic approach to the treatment of effort angina. Int J Cardiol 1992; 34: 167–72.
66. Leasure JE, Kordenat K. Effect of propionyl-L-carnitine on experimental myocardial infarction in dogs. Cardiovasc Drugs Ther 1991; 5(Suppl 1): 85–96.
67. Leipälä JA, Bhatnagar R, Pineda E, Najibi S, Massoumi K, Packer L. Protection of the reperfused heart by L-propionylcarnitine. J Appl Physiol 1991; 71: 1518–22.
68. Liedtke AJ, DeMaison L, Nellis SH. Effects of L-propionylcarnitine on mechanical recovery during reflow in intact hearts. Am J Physiol 1988; 255: H169–76.
69. Micheletti R, Di Paola ED, Schiavone A et al. Propionyl-L-carnitine limits chronic ventricular dilation after myocardial infarction in rats. Am J Physiol Heart 1993; 264: H1111–7.
70. Di Biase M, Tritto M, Pitzalis MV, Favale S, Rizzon P. Electrophysiologic evaluation of intravenous L-propionylcarnitine in man. Int J Cardiol 1991; 30: 329–33.

*Corresponding Author*: Dr Dennis J. Paulson, Department of Physiology, Midwestern University, 555 31st Street, Downers Grove, IL 60515, USA

# 14. Carnitine metabolism during diabetes and hyperthyroidy

DANIELLE FEUVRAY

"L-Carnitine treatment of diabetic rats significantly reduced plasma glucose and lipid levels."

## Introduction

Interest in the function of carnitine has increased greatly since the demonstration in 1955 that carnitine could be reversibly acetylated by acetyl coenzyme A (acetyl-CoA) [1] and that it caused a marked stimulation of long-chain fatty acid oxidation in liver homogenates [2]. Since these initial studies, it has been clearly established that L-carnitine is a required participant in the transport of long-chain acyl-residues across the inner mitochondrial membrane, and the literature has been extensively covered in several reviews [3–5]. Because most tissues, which do not synthesize L-carnitine (the naturally occurring form of carnitine) such as cardiac muscle [7], have an intracellular concentration of free carnitine that is about 40–fold higher than the plasma concentration, active uptake must take place. This carnitine uptake in muscle has been studied in heart myocytes [6], and characterized in isolated perfused adult rat hearts [7]. The study by Vary and Neely [7] demonstrated that carnitine uptake across the sarcolemma occurs by both diffusion and carrier-mediated transport systems. Indeed, at physiological concentrations of extracellular carnitine (44 μM) about 80% of uptake occurs by the carrier-mediated system. Also, at this extracellular carnitine concentration, carnitine transport into the cell is slow and the turnover of total tissue carnitine would require about 60 h. Thus changes in the myocardial levels would not be expected to occur rapidly. This is consistent with the observation that the carnitine concentration in normal cardiac muscle is fairly constant over a wide range of physiological conditions [7]. However, carnitine deficiency has been recognized in several muscle tissues including cardiac muscle.

*J.W. de Jong and R. Ferrari (eds): The carnitine system*, 199–208.

*Table 1.* Effects of mild and acute diabetes on tissue levels of carnitine and CoA.

| Condition | Days | Tissue levels (nmol/g dry wt) | |
|---|---|---|---|
| | | Carnitine | CoA |
| Control | 0 (11) | 5785 ± 78 | 498 ± 17 |
| Mild diabetes | 7 (3) | 5476 ± 100 | 743 ± 30 |
| | 19 (4) | 4958 ± 239 | 589 ± 26 |
| | 34 (4) | 4246 ± 109 | 607 ± 27 |
| Acute diabetes | 2 (22) | 5114 ± 96 | 570 ± 19 |

Mild or acute diabetes was induced by injection of alloxan (37.5 and 60 mg/kg body weight, respectively). Mildly diabetic rats were maintained without insulin treatment for the times indicated before they were killed. The hearts were removed from the rats, trimmed and blotted, and the ventricles were quickly frozen in liquid nitrogen. Hearts from acutely diabetic rats were removed 48 h after alloxan injection. They were perfused for 10, 30 and 60 min with buffer containing 11 mM glucose, then quick-frozen in liquid nitrogen. Total tissue carnitine and CoA were determined on the homogenized tissue, after alkaline hydrolysis of the acyl esters. In acute diabetic perfused hearts, there were no noticeable differences in the levels at various perfusion times and the data have been combined. The number of hearts in each group is shown in parentheses. (Reprinted with permission from Circ Res [8].)

## Carnitine metabolism and diabetes

In cardiac muscle, such a deficiency is associated with diabetes and this has been demonstrated in mild or acute chemically-induced diabetic rats [8]. The decrease was observed after 48 h of severe diabetes and after several weeks of mild diabetes (Table 1). In this study [8], we found that total carnitine levels in rat hearts were decreased as a function of time after induction of diabetes (from 7 through 34 days). The level of total coenzyme A (CoA) on the other hand, another essential cofactor for fatty acid metabolism, had increased by 7 days and remained high throughout 34 days. When those diabetic rat hearts were perfused under aerobic conditions, they still had lower tissue levels of total carnitine and higher levels of total CoA (Table 1). In addition, this study first showed that this was associated with higher levels of both long-chain acyl-carnitine and acyl-CoA esters. Furthermore, this increase in long-chain acyl-carnitine and acyl-CoA levels was accentuated after a period of ischemia as compared to that of non-diabetic hearts (Table 2). Changes in the total tissue pool of CoA or carnitine in diabetic hearts may be expected to alter cellular metabolism. About 95% of the CoA in normal cardiac muscle is located in the mitochondrial matrix whereas 95% of the carnitine is cytosolic [9]. These distributions in normal hearts are likely to cause activated fatty acids to be funnelled toward oxidation rather than toward lipid synthesis. But, when carnitine is substantially decreased, fatty acids could be partially diverted from oxidation to esterification, contributing to the accumulation of triglycerides in the diabetic cardiac cells. As will be stated below, carnitine deficiency in diabetic hearts may have important

*Table 2.* Effects of short-term and long-term ischemia in vitro on long-chain acyl esters of carnitine and CoA in isolated hearts.

| Condition | | Tissue levels (nmol/g dry wt) | |
|---|---|---|---|
| | | Acyl-carnitine | Acyl-CoA |
| Normal hearts | Control (6) | 130 ± 13 | 92 ± 2 |
| | Short-term ischemia (9) | 240 ± 32 | 99 ± 8 |
| | Long-term ischemia (6) | 532 ± 58 | 104 ± 9 |
| Diabetic hearts | Control (8) | 277 ± 79 | 167 ± 10 |
| | Short-term ischemia (7) | 1087 ± 272 | 244 ± 15 |
| | Long-term ischemia (8) | 830 ± 109 | 243 ± 33 |

Control hearts from both groups of rats were perfused for 10 min. During short-term ischemia induced by reducing coronary flow by about 50% hearts were electrically paced, and the average time of ischemic perfusion was 3 min in the diabetic and 7 min in the normal hearts (corresponding to the beginning of ventricular failure). Long-term ischemia was induced by the same reduction in coronary flow, but without electrical pacing, and continued for 60 min. The number of hearts in each group is shown in parentheses. (Reprinted with permission from Circ Res [8].)

*Table 3.* Effects of diabetes on plasma and myocardial levels of carnitine.

| Condition | Duration | Serum carnitine ($nmol \cdot ml^{-1}$) | |
|---|---|---|---|
| | | Total | Free |
| Control | | 62 ± 2 | 44 ± 1 |
| Mild diabetes | 1 day | 69 ± 5 | 43 ± 3 |
| | 7 days | 59 ± 4 | 38 ± 2 |
| | 21 days | 47 ± 2* | 27 ± 2* |
| Acute diabetes | 6 h | 55 ± 4 | 26 ± 2 |
| | 12 h | 41 ± 4* | 17 ± 2* |
| | 48 h | 31 ± 2* | 12 ± 2* |

Mild or acute diabetes was induced by injection of alloxan (37.5 and 60 mg/kg body weight, respectively). Values are means ± SE of 12–22 animals at each time point. * $P < 0.005$ compared with control. (After Vary & Neely, with permission from Am J Physiol [10].)

functional consequences, especially if those hearts are submitted to ischemia or ischemia and reperfusion.

The demonstration of a reduction in myocardial carnitine levels during diabetes naturally led to examining possible mechanisms responsible for it. It was found that in each case, i.e. either after 48 h of severe diabetes induced by alloxan (60 mg alloxan/kg) as well as after several weeks of mild diabetes (37.5 mg alloxan/kg), low tissue levels of carnitine were associated with reduced plasma carnitine (both total and free carnitine) (Table 3) [10]. Similar changes in the plasma and myocardial levels of carnitine have been reported in rats rendered diabetic for 14 days by injection of streptozotocin [11]. Thus, a decreased rate of uptake due to lower plasma carnitine may contribute to the reduced levels of carnitine seen in diabetic hearts. However,

*Table 4.* Intracellular sodium activity ($a^iNa$) in quiescent papillary muscles from normal and diabetic rat hearts.

| Condition | Blood glucose (mM) | $a^iNa$ (mM) |
|---|---|---|
| Control | 9.52 ± 0.72 | 11.31 ± 0.40 |
| Diabetic | 35.74 ± 2.12* | 17.63 ± 0.54* |

Rats were made diabetic by injection of 40 mg/kg streptozotocin and hearts excised three weeks later. Samples for blood glucose were collected from control (n = 20) and diabetic (n = 20) animals at time of heart excision. Measurements of $a^iNa$ were obtained using microelectrodes filled with $Na^+$-selective ligand. * P < vs. control. (Reprinted with permission from Exp Physiol [13].)

defects or changes other than decreased plasma carnitine concentration may associate in order to account for myocardial carnitine deficiency associated with diabetes. Indeed, carnitine transport by cardiac cells has been demonstrated to be dependent on the $Na^+$ electrochemical gradient [12]. Using $Na^+$-selective microelectrodes in papillary muscles from hearts of rats with streptozotocin-induced diabetes, we have clearly shown that resting intracellular sodium activity was about 56% greater than in muscles from normal rats (Table 4) [13]. The decrease in transmembrane $Na^+$ gradient associated with the higher intracellular sodium activity may then be another contributing factor to decreased carnitine uptake by cardiac cells and consequently to carnitine deficiency.

## Carnitine metabolism and hyperthyroidy

The heart is a major target for the thyroid hormones and myocardial metabolism is influenced by the hyperthyroid state. Long-chain fatty acid metabolism is elevated in hyperthyroid guinea-pig [14] and rat [15] heart, and the activity of carnitine acyl transferase I, the rate-limiting enzyme in fatty acid metabolism, is enhanced [14]. Studies performed on two different species (i.e. mouse [16] and rat [15]) demonstrated a decrease in free carnitine levels associated with the hyperthyroid state. On the other hand, it has been reported that free carnitine was increased in hyperthyroid guinea-pig heart [14]. This discrepancy in free carnitine data may represent a species variation in thyroid-hormone mediated alterations in carnitine metabolism. Alternatively, and more likely, it may be related to differences in the thyroid hormone treatment (i.e. differences in doses and duration) and/or in the nutrition state at the time of measurement. This is illustrated by the results obtained in short-chain acyl-carnitine levels. Short-chain acyl-carnitine levels were found to be decreased in hyperthyroid mouse heart [16] whereas they were either increased or decreased in the rat heart, depending on the conditions of measurement, i.e. in the fed state or after starvation (48 h), respectively [17]. Such a discrepancy was also encountered for total myocardial carnitine content which was found to be either decreased [18] or unchanged [17]. On the other hand, myocardial long-chain acyl-carnitine levels were consistently found to be increased in the hyperthyroid guinea-pig [14] and rat heart [17]. In the latter study [17] that used thriiodothyronine ($T_3$)-treated rats, the significant increase in long-chain acyl-carnitine was associated, in the fed state, with an increase in short-chain acyl-carnitine and with a corresponding decrease in the concentration of free carnitine. As a consequence of changes in carnitine distribution, the ratio of free to acylated carnitine dramatically declined, whereas that of short-chain acyl-carnitine to free carnitine increased. In the heart, short-chain acyl-carnitine is predominantly acetylcarnitine [19]. Since the activity of carnitine acetyltransferase is high in these conditions of hyperthyroidism [20], it may be inferred that the acetyl-CoA/CoASH concentration ratio is increased by hyperthyroidism in the fed state [17]. The observed pattern of change in the concentration of free and acylated carnitine, and the implied change in the acetyl-CoA/CoASH concentration ratio, indicate an increased rate of cardiac fatty acid oxidation in the fed state in hyperthyroidism. This is, in addition, associated with an increased plasma fatty acid concentration in the fed state [17]. In contrast to the diabetic situation, the results of this study also suggested that rates of cardiac glycolysis remain at least as high in the hyperthyroid as in the euthyroid state [17]. Concomitant degradation of both fatty acids and glucose at high rates is consistent with the known effects of thyroid hormone to increase energy demand as a consequence of increased heart work.

Since lipid metabolic intermediates were shown to influence sarcoplasmic reticulum transport in the diabetic myocardium, (see below), the levels of carnitine derivatives were also determined in sarcoplasmic reticulum fractions isolated from $T_3$-treated rats [21]. The results showed a strong negative correlation between the level of long-chain acyl-carnitine associated with the sarcoplasmic reticulum and the calcium transport rate of the organelle, such that the $T_3$-treatment led to a reduction in the long-chain acyl-carnitine level in the sarcoplasmic reticulum concomitant with an increase in calcium transport activity. However, the temporal association of these alterations as well as the biochemical mechanism by which a reduction in long-chain acyl-carnitine could contribute to the augmented sarcoplasmic reticulum calcium transport rate in the $T_3$-treated rats remain to be determined.

## Effect of L-carnitine treatment on metabolism and cardiac performance

It has been suggested that carnitine therapy may be beneficial to the diabetic heart. Indeed, depressed contractile function associated with diabetes has been clearly shown. The defects in contractile performance include, in particular, a depressed velocity of shortening and a delayed onset of relaxation observed in the isolated papillary muscle [22], as well as a decreased ability to respond to increased filling pressures and increasing afterload in isolated working hearts from diabetic rats [23–25]. Adequate carnitine levels are required for normal fatty acid and energy metabolism in heart muscle, and changes in its level may affect energy production and thus mechanical performance. Rodrigues et al. [26] have studied the effects of L-carnitine administration ($3g \cdot kg^{-1}$ i.p., daily for 6 wk) in isolated perfused working hearts from streptozotocin-induced diabetic rats. They found that exogenous L-carnitine treatment of the diabetic rats increased myocardial free-carnitine levels, which were comparable with those of control rats. Moreover, after six weeks of diabetes in hearts from treated rats, there was no depression of mechanical performance whereas untreated diabetic rats exhibited depressed left ventricular developed pressure, cardiac contractility, and ventricular relaxation rates. In addition, L-carnitine treatment of diabetic rats significantly reduced plasma glucose and lipid levels but had no effect on control rats. The effects of L-carnitine on blood glucose were consistent with a previous observation that acute L-carnitine treatment of diabetic rats for 2 wk significantly reduced plasma glucose [27]. The ability of L-carnitine to lower blood lipids may participate in the improvement of cardiac function since the increased reliance of the diabetic heart on fatty acid metabolism is partly due to an increase in circulating fatty acids. Nevertheless, other possibilities to explain the beneficial effects of L-carnitine have been considered, such as its effects in reducing myocardial levels of long-chain acyl-CoA secondary to a stimulation in fatty acid oxidation. In fact, diabetes, which accelerates the rate of lipid metabolism in the heart, results in an elevation of the levels of intermediates of fatty acid metabolism as mentioned above, including long-chain acyl-CoA and acyl-carnitine [8]. The amphiphilic properties of these long-chain acyl-derivatives, especially long-chain acyl-carnitine, may facilitate their incorporation into membranes [28], with consequent perturbations in membrane proteins of sarcolemma and subcellular membranes, including those involved in transmembrane transport processes. For example, it has been reported that levels of long-chain acyl-carnitine are elevated in the microsomal sarcoplasmic reticulum preparations derived from chronically diabetic rats [29]. The increase in long-chain acyl-carnitine associated with the sarcoplasmic reticulum paralleled the increase in total tissue levels of long-chain acyl-carnitine observed previously in diabetic rat hearts [8]. Also, this study by Lopaschuk et al. [29] showed that cardiac sarcoplasmic reticulum microsomes isolated from chronically diabetic rats had a depressed ATP-dependent calcium transport. A further study by the same group [30] which determined various parameters of heart function, demonstrated that if carnitine treatment (i.e. a high oral dose of D,L-carnitine administered for a 42-day period) did not prevent the onset of heart dysfunction in diabetic rats, it did, however, prevent the depression in cardiac sarcoplasmic reticulum calcium transport from occurring. Furthermore, the accumulation of long-chain acyl-carnitine in the sarcoplasmic reticulum membrane of diabetic rats was prevented by carnitine treatment [30]. The depression in sarcoplasmic reticulum function probably explains the prolonged duration of the systolic calcium-transient that Lagadic-Gossmann et al. (unpublished) recently demonstrated in ventricular myocytes isolated from diabetic rats. This is also consistent with the decreased sarcoplasmic reticulum content reported in diabetes [31].

When intracellular concentration of long-chain acyl-carnitine is high, not only may acyl-carnitine redistribute in subcellular membranes such as the sarcoplasmic reticulum membrane but it may also, and even preferentially, concentrate in myocytes sarcolemma [28]. The resulting alterations in composition of the cardiac sarcolemmal membrane may then affect the function of membrane systems as already mentioned. In this respect, the diabetes-induced membrane defects that have been shown include a marked depression in the activity of the amiloride-sensitive $Na^+$-$H^+$ exchange in papillary muscles [32] and a decrease in the calcium-independent potassium permeability in isolated ventricular myocytes [33]. However, the mechanism of these defects is as yet unknown and long-chain acyl-carnitine accumulation may only be one contributing factor among others. In addition, neither of these defects has yet proved to be corrected by carnitine treatment.

One possibility that has been proposed for the beneficial effects of L-carnitine is related to its actions in reducing myocardial levels of long-chain acyl-CoA and consequently the long-chain acyl-CoA-induced inhibition of the mitochondrial adenine nucleotide translocase [34]. Indeed, studies by Pieper et al. [35] have shown that addition of L-carnitine to isolated working hearts from diabetic rats perfused with fatty acids, attenuated both the increase in long-chain acyl-CoA and the loss of ATP observed in these particular conditions. Paulson et al. [27] have also suggested that the beneficial effects of carnitine were due to its lowering of the myocardial acyl-CoA level.

More recently, the attractive explanation that the beneficial effect of L-carnitine may be mediated by a secondary stimulation of glucose utilization has been proposed by Broderick at al. [36]. As mentioned before, the increased reliance on fatty acids as an energy substrate is purported to be an important contributing factor to the development of biochemical changes that occur in the diabetic myocardium [37, 38]. Increasing evidence suggests that these detrimental effects of fatty acids are correlated with their ability to inhibit overall myocardial glucose utilization. In support of this, interventions aimed at overcoming fatty acid inhibition of glucose oxidation are beneficial to diabetic rat heart function [39–41]. It is known that the oxidation of glucose as a source of ATP production is essentially abolished in uncontrolled diabetes [42]. This is due to a marked inhibition of the pyruvate dehydrogenase complex [43]. L-Carnitine supplementation in non-diabetic rat hearts has been shown to decrease fatty acid oxidation in parallel with a stimulation of glucose oxidation [44]. The effects of L-carnitine on myocardial glucose metabolism probably occur secondary to a decrease in the intramitochondrial acetyl CoA/CoASH ratio [45], resulting in a stimulation of pyruvate dehydrogenase complex activity [46] and, consequently, in a stimulation of glucose oxidation. In working hearts from diabetic rats, an acute loading perfusion with L-carnitine significantly increases free and total myocardial carnitine content (by approximately twofold) [36]. Moreover, glucose oxidation rate during aerobic perfusion of these hearts loaded with carnitine and perfused with a high concentration of fatty acids (i.e. a concentration that can be seen in uncontrolled diabetic animals) was dramatically increased, whereas it was essentially abolished in untreated diabetic rat hearts. Treatment with L-carnitine also significantly increased glycolytic rates [36] which are depressed in diabetic rat hearts compared to control hearts [8].

In conclusion, on the basis of the data summarized here, carnitine treatment should be beneficial to cardiac muscle in which carnitine deficiency has been clearly demonstrated, especially as a consequence of diabetes. Among possible mechanisms that may be responsible for the beneficial effects of carnitine supplementation, its ability to overcome fatty acid inhibition of glucose metabolism is most likely.

## Acknowledgement

We wish to thank the valuable assistance of Mrs Françoise James in manuscript preparation.

## References

1. Friedman S, Fraenkel G. Reversible enzymatic acetylation of carnitine. Arch Biochem Biophys 1955; 59: 491–501.
2. Fritz IB. The effect of muscle extracts on the oxidation of palmitic acid by liver slices and homogenates. Acta Physiol Scand 1955; 34: 367–85.
3. Bremer J. Carnitine—metabolism and functions. Physiol Rev 1983; 63: 1420–80.
4. Fritz IB. Carnitine and its role in fatty acid metabolism. Adv Lipid Res 1963; 1: 285–334.
5. Frenkel RA, McGarry JD. Carnitine biosynthesis, metabolism and functions. New York: Academic Press, 1980; 1–356.
6. Bahl J, Navin T, Manian AA, Bressler R. Carnitine transport in isolated adult rat heart myocytes and the effect of 7,8-diOH chlorpromazine. Circ Res 1981; 48: 378–85.
7. Vary TC, Neely JR. Characterization of carnitine transport in isolated perfused adult rat hearts. Am J Physiol 1982; 242: H585–92.
8. Feuvray D, Idell-Wenger JA, Neely JR. Effects of ischemia on rat myocardial function and metabolism in diabetes. Circ Res 1979; 44: 322–9.
9. Idell-Wenger JA, Grotyohann LW, Neely JR. Coenzyme A and carnitine distribution in normal and ischemic hearts. J Biol Chem 1978; 253: 4310–8.
10. Vary TC, Neely JR. A mechanism for reduced myocardial carnitine levels in diabetic animals. Am J Physiol 1982; 243: H154–8.
11. Fogle PJ, Bieber LL. Effect of streptozotocin on carnitine and carnitine acyl transferases in rat heart, liver, and kidney. Biochem Med 1976; 22: 119–26.
12. Vary TC, Neely JR. Sodium dependence of carnitine transport in isolated perfused adult rat hearts. Am J Physiol 1983; 244: H247–52.
13. Lagadic-Gossmann D, Feuvray D. Intracellular sodium activity in papillary muscle from diabetic rat hearts. Exp Physiol 1991; 76: 147–9.

14. Bressler R, Wittels B. The effect of thyroxine on lipid and carbohydrate metabolism in the heart. J Clin Invest 1966; 45: 1326–33.
15. Fintel M, Burns AH. Effect of thyroxine treatment on exogenous myocardial lactate oxidation. Am J Physiol 1982; 243: H722–8.
16. Cederblad G, Engstrom G. Effect of thyroxine treatment on carnitine levels in mice. Acta Pharmacol Toxicol 1978; 43: 1–5.
17. Sugden MC, Holness MJ, Liu YL, Smith DM, Fryer LG, Kruszynska YT. Mechanisms regulating cardiac fuel selection in hyperthyroidism. Biochem J 1992; 286: 513–7.
18. Suzuki M, Tokuyama K, Yamane A. Carnitine metabolism in thyroid hormone treated rats and mice. J Nutr Sci Vitaminol 1983; 29: 413–28.
19. Pearson DJ, Tubbs PK. Carnitine and derivatives in rat tissues. Biochem J 1967; 105: 953–63.
20. Tanaka T, Morita H, Koide H, Kawamura K, Takatsu T. Biochemical and morphological study of cardiac hypertrophy. Effects of thyroxine on enzyme activities in the rat myocardium. Basic Res Cardiol 1985; 80: 165–74.
21. Black SC, McNeill JH, Katz S. Sarcoplasmic reticulum $Ca^{2+}$ transport and long chain acylcarnitines in hyperthyroidism. Can J Physiol Pharmacol 1988; 66: 159–65.
22. Fein FS, Kornstein LB, Strobeck JE, Capasso JM, Sonnenblick EH. Altered myocardial mechanics in diabetic rats. Circ Res 1980; 47: 922–33.
23. Penpargkul S, Schaible T, Yipintsoi T, Scheuer J. The effect of diabetes on performance and metabolism of rat hearts. Circ Res 1980; 47: 911–21.
24. Vadlamudi RVSV, Rodgers RL, McNeill JH. The effect of chronic alloxan- and streptozotocin-induced diabetes on isolated rat heart performance. Can J Physiol Pharmacol 1982; 60: 902–11.
25. Garber DW, Neely JR. Decreased myocardial function and myosin ATPase in hearts from diabetic rats. Am J Physiol 1983; 244: H586–91.
26. Rodrigues B, Xiang H, McNeill JH. Effect of L-carnitine treatment on lipid metabolism and cardiac performance in chronically diabetic rats. Diabetes 1988; 37: 1358–64.
27. Paulson DJ, Schmidt MJ, Traxler JS, Ramacci MT, Shug AL. Improvement of myocardial function in diabetic rats after treatment with L-carnitine. Metabolism 1984; 33: 358–63.
28. Knabb MT, Saffitz JE, Corr PB, Sobel BE. The dependence of electrophysiological derangements on accumulation of endogenous long-chain acyl carnitine in hypoxic neonatal rat myocytes. Circ Res 1986; 58: 230–40.
29. Lopaschuk GD, Katz S, McNeill JH. The effect of alloxan- and streptozotocin-induced diabetes on calcium transport in rat cardiac sarcoplasmic reticulum. The possible involvement of long chain acylcarnitines. Can J Physiol Pharmacol 1983; 61: 439–48.
30. Lopaschuk GD, Tahiliani AG, Vadlamudi RVSV, Katz S, McNeill JH. Cardiac sarcoplasmic reticulum function in insulin- or carnitine-treated diabetic rats. Am J Physiol 1983; 245: H969–76.
31. Bouchard RA, Bose D. Influence of experimental diabetes on sarcoplasmic reticulum function in rat ventricular muscle. Am J Physiol 1991; 260: H341–54.
32. Lagadic-Gossmann D, Chesnais JM, Feuvray D. Intracellular pH regulation in papillary muscle cells from streptozotocin diabetic rats: an ion-sensitive microelectrode study. Pflügers Arch 1988; 412: 613–7.
33. Jourdon P, Feuvray D. Calcium and potassium currents in ventricular myocytes isolated from diabetic rats. J Physiol (Lond) 1993; 470: 411–29.
34. Shug AL, Shrago E, Bittar N, Folts JD, Koke JR. Acyl-CoA inhibition of adenine nucleotide translocation in ischemic myocardium. Am J Physiol 1975; 228: 689–92.
35. Pieper GM, Murray WJ, Salhany JM, Wu ST, Eliot RS. Salient effects of L-carnitine on adenine-nucleotide loss and coenzyme A acetylation in the diabetic heart perfused with excess palmitic acid. A phosphorus-31 NMR and chemical extraction study. Biochim Biophys Acta 1984; 803: 241–9.
36. Broderick TL, Quinney HA, Lopaschuk GD. Protection of the ischemic diabetic myocar-

dium by L-carnitine: effects on glycolysis, glucose oxidation, and functional recovery. Diabetes 1994. In press.
37. Tahiliani AG, McNeill JH. Diabetes-induced abnormalities in the myocardium. Life Sci 1986; 38: 959–74.
38. Lopaschuk GD. Alterations in myocardial fatty acid metabolism contribute to ischemic injury in the diabetic. Can J Cardiol 1989; 5: 315–20.
39. Hékimian G, Feuvray D. Reduction of ischemia-induced acyl carnitine accumulation by TDGA and its influence on lactate dehydrogenase release in diabetic rat hearts. Diabetes 1986; 35: 906–10.
40. Lopaschuk GD, Spafford M. Response of isolated working hearts to fatty acids and carnitine palmitoyl transferase I inhibition during reduction of coronary flow in acutely and chronically diabetic rats. Circ Res 1989; 65: 378–87.
41. Nicholl TA, Lopaschuk GD, McNeill JH. Effects of free fatty acids and dichloroacetate on isolated working diabetic rat heart. Am J Physiol 1991; 261: H1053–9.
42. Wall SR, Lopaschuk GD. Glucose oxidation rates in fatty acid-perfused isolated working hearts from diabetic rats. Biochim Biophys Acta 1989; 1006: 97–103.
43. Kerbey AL, Vary TC, Randle PJ. Molecular mechanisms regulating glucose oxidation. Basic Res Cardiol 1985; 80(Suppl 2): 93–6.
44. Broderick TL, Quinney HA, Lopaschuk GD. Carnitine stimulation of glucose oxidation in the fatty acid perfused isolated working heart. J Biol Chem 1992; 267: 3758–63.
45. Lysiak W, Lilly K, DiLisa F, Toth PP, Bieber LL. Quantitation of the effect of L-carnitine on the levels of acid-soluble short-chain acyl-CoA and CoASH in rat heart and liver mitochondria. J Biol Chem 1988; 263: 1151–6.
46. Uziel G, Garavaglia B, Di Donato S. Carnitine stimulation of pyruvate dehydrogenase complex (PDHC) in isolated human skeletal muscle mitochondria. Muscle Nerve 1988; 11: 720–4.

*Corresponding Author:* Professor Danielle Feuvray, Laboratoire de Physiologie Cellulaire, Université Paris XI, Bât. 443, F-91405 Orsay Cédex, France

# 15. Carnitine and lactate metabolism

ROBERTO FERRARI and ODOARDO VISIOLI

"In all studies L-carnitine converted lactate production at peak pacing stress into extraction, thus preventing anaerobiosis. In addition, there was an increase in the uptake of free fatty acids as well as in that of glucose, confirming the role of L-carnitine as a metabolic modulator, improving not only FFA, but also carbohydrate metabolism."

## Introduction

The pharmacological activity of L-carnitine presents various aspects. The best known function of L-carnitine is the transport of long-chain fatty acids from cytosol across the inner mitochondrial membrane into the mitochondrial matrix, the site of $\beta$-oxidation [1]. Long-chain acyl-CoA cannot pass through the inner mitochondrial membrane, but their metabolic product, acylcarnitine, formed by the action of carnitine palmitoyltransferase I, an enzyme located on the outer surface of the inner mitochondrial membrane, can. Another enzyme, carnitine-acylcarnitine translocase exchanges carnitine (out) with acyl-carnitine (in) in a stoichiometric ratio of 1:1 and ensures the constancy of the intramitochondrial carnitine pool. The incoming acylcarnitine reacts with CoA. This reaction is catalyzed by the enzyme carnitine palmitoyltransferase II, which is attached to the inside of the inner membrane. Thus, acyl-CoA is reformed in the mitochondrial matrix, and carnitine is made available for exchange by the translocase [2, 3].

Another enzyme, carnitine-CoA acetyltransferase, is located at the inner surface of the inner mitochondrial membrane. This enzyme might be involved in: 1) "buffering" the mitochondrial pool of acetyl-moieties; 2) restoring the metabolic flux in the tricarboxylic acid cycle by releasing CoA from its thioesters; 3) stimulating the activity of pyruvate dehydrogenase which is also controlled by the acetyl-CoA/CoA ratio.

Thus, carnitine has the following functions [4]:

*J.W. de Jong and R. Ferrari (eds): The carnitine system*, 209–224.

- Facilitation of $\beta$-oxidation by transporting activated long-chain fatty acids into mitochondria.
- Enhancement of the metabolic flux in the tricarboxylic acid cycle by sparing free CoA.
- Activation of the transport of adenine nucleotides across the inner mitochondrial membrane by preventing adenylate translocase inhibition by long-chain acyl-CoA.
- Stimulation of the activity of pyruvate dehydrogenase by decreasing the acetyl-CoA/CoA ratio, thus enhancing the oxidative utilization of glucose.

It follows that besides lipid oxidation L-carnitine also influences glucose metabolism.

We have just mentioned its capacity to (re)activate pyruvate-dehydrogenase with a consequent increase in the flow of pyruvate in the citric acid cycle. Accordingly, administration of L-carnitine by various routes gives rise to a reduction in blood glucose levels induced by glucose loading [4].

L-Carnitine does not modify glycaemia in normal animals, though it reduces this parameter in animals with streptozotocin-induced diabetes, having an effect similar to that of insulin [5]. It is important to note that rats with streptozotocin- or alloxane-induced diabetes also have reduced levels of carnitine in heart, serum and muscle, and high levels in the liver [6].

Correctly, in the 1990s, carnitine was recognized to act as "metabolic modulator". This aspect is particularly important within the heart and skeletal muscles, which are respiring tissues second only to the brain in their obligate aerobic needs. The most important substrates for these muscles are fatty acids and carbohydrates although amino acids and ketone bodies also contribute under certain circumstances. A series of untoward metabolic and functional consequences of restricted oxygen delivery to both the heart and skeletal muscles has been reported, the production of lactate being one of typical metabolic markers of anaerobiosis [7, 8]. Administration of L-carnitine has been shown to reduce the production of lactate in several experimental and clinical conditions such as: sports medicine, peripheral arterial disease, acid-base disorders, myocardial ischaemia and heart failure.

These concepts form the basis of the present chapter. The existing data on the role of L-carnitine in reducing lactate production are critically examined to assess its therapeutic potential and to define its mechanism of action. For clarity the effects on skeletal and cardiac muscles are kept separate as, although similar, the metabolism and function of these tissues show some differences.

### *Effect of L-carnitine on lactate production from skeletal muscle*

In aerobic metabolism, the rate-limiting step for energy turnover by the muscle system is the maximum metabolic flux in the tricarboxylic acid cycle. This, in turn depends on three basic factors: 1) concentration of substrates;

2) mitochondrial mass and consequently concentration of rate-limiting enzymes; and 3) availability of oxygen which is related to maximal cardiac output and/or maximal local blood flow.

The flow of substrates or that of acetyl-CoA, the product of their degradation in general, is in excess compared to the potential of the tricarboxylic acid cycle. Acetyl-CoA supply varies with changes in the overall metabolic rate of the muscles. Acetyl-CoA production in the muscle is, in fact, almost exclusively lipid-dependent at rest and at moderate work loads. It is, however, up to 85% dependent on carbohydrates at maximal aerobic work rates [7].

When the muscle exceeds the so-called "anaerobic threshold", it starts utilizing energy obtained anaerobically. In this condition reoxidation of NADH by transfer of reducing equivalents through the respiratory chain to oxygen is prevented and pyruvate is reduced by lactate dehydrogenase to lactate, which is then released in the venous system. The reoxidation of NADH via lactate formation allows glycolysis to proceed in the absence of oxygen by regenerating sufficient $NAD^+$, for another cycle of the reaction catalyzed by glyceraldehyde 3–phosphate dehydrogenase. Thus the muscle tends to produce lactate, of which the rate of accumulation and release is an index of the rate of ATP generation by way of anaerobic glycolysis. In practice, 2 mol of ATP is generated per mol of glucose transformed into lactate.

In the course of maximal work load, not only glycolysis is stimulated to generate lactate but also lipolysis, as suggested by the increased glycerol and glycerol-3-phosphate levels found in human skeletal muscle [9]. Acetyl-CoA so generated tends to accumulate in the cytosol and within the mitochondrion, increasing the acetyl-CoA/CoA ratio and thus inhibiting the oxidative utilization of glucose which cannot operate at the rate required by the metabolic demand.

Carnitine, by functioning as an acetyl-group buffer [10–12], offers a number of potential advantages to cells functioning at or above their anaerobic threshold [9]:

- It allows the transport of long-chain fatty acids into the mitochondrion preventing "flooding" of the mitochondrial matrix by acetyl-CoA esters and further depletion of the free CoA pool.
- It maintains a viable pool of CoA even when the rate of acetyl-CoA formation exceeds that of condensation of the above metabolite with oxaloacetate. This is essential for assuring the oxidation of $\alpha$-ketoglutarate to succinate within the tricarboxylic acid cycle. Lack of CoA due to excessive accumulation of acyl-CoA would lead to a rapid fall in oxaloacetate and further accumulation of acetyl-CoA and, thus, to a vicious circle.
- It constitutes an additional sink for pyruvate. In fact acetylcarnitine is an accumulation product of anaerobic metabolism alternative to lactate and represents a pathway for increasing the oxygen debt of the muscles. In

practice, part of the pyruvate excess, instead of being transformed into lactate, can be decarboxylated and stored as acetylcarnitine.
- It improves the transport of adenine nucleotides across the inner mitochondrial membrane in conditions of "flooding" of long-chain acyl-CoA at the inner mitochondrial membrane level.

Thus, at least theoretically, L-carnitine in exercise metabolism provides a more efficient regulation of the energy flow from the different oxidative sources. During maximal work load it is expected to:

- increase the subject's maximal aerobic power ($VO_2$ max);
- spare glycogen degradation and utilization by enhancing $\beta$-oxidation of fatty acids;
- reduce the size of "lactic acid $O_2$ debt" contracted by the subject;
- improve an "acetylcarnitine $O_2$ debt" with the advantage of providing the subject with a greater working potential during short bursts of supramaximal exercise.

Equally, carnitine could optimize the oxidative pathway during muscular exercise carried out in acute or chronic hypoxic conditions.

*Effect of L-carnitine on lactate production in sports medicine*

The effects of L-carnitine on physical performance have been investigated on subjects in varying conditions of physical efficiency, at different dosages and for different periods of time [13–34]. For this reason it is difficult to draw inequivocal conclusions from the data obtained.

In general L-carnitine supplementation raises maximal aerobic power, and in well-trained individuals there is also evidence of increased $VO_2$ max [35, 36]. This most likely is due to a removal of part of the short-chain acyl-CoA by L-carnitine in the muscles heavily involved in exercise with a concurrent release of free CoA. The consequent stimulation of pyruvate dehydrogenase would enhance the flux in the Krebs cycle. Thus L-carnitine supplementation plays a positive role in augmenting the capacity of buffering pyruvate and reducing the accumulation of muscle lactate with a limitation of the adverse effects of an increased lactate accumulation in the muscle. The effects of L-carnitine on lactate production in athletes, however, are variable, although in general a reduction of lactate release has been found [7]. This probably is due to the existence of muscles or muscle regions undergoing loads exceeding their maximum aerobic potential, while most of the residual muscle mass is still below this threshold. Thus the lactate produced by anaerobic muscle is diluted and not easily detected in the venous circulation. The crucial experiment would be one in which the balance of the various energy sources of a muscle stimulated supramaximally could be established.

The results of studies carried out the following particular protocols or in special situations are also very interesting; Ferretti [23] examined the effects of administration of a single dose of L-carnitine in normal respiratory conditions and in acute or chronic hypoxia: the test consisted in carrying out physical exercise immediately after transfer of volunteers to an altitude of 5,050 meters and subsequently after a month's acclimatisation. Administration of L-carnitine (2 g per os) induced a reduction in venous lactate and an increase in the anaerobic threshold, implying the possibility of improving athletic performance during prolonged anaerobic activity; similar conclusions were drawn by Wyss et al. [34] in volunteers where high altitude physical exercise was simulated by a reduction in $O_2$ from 20.9 to 13.0% obtained by the gradual introduction of nitrogen in a closed circuit spirometer. These results can be connected to the data of Angelini et al. [13] who demonstrated that physical exercise carried out at high altitudes results in a significant increase in long- and short-chain acylcarnitine in plasma and urine, in relation to intense lipolysis and production of ketone bodies.

*Effect of L-carnitine on lactate production in peripheral arterial disease*

The impact of L-carnitine supplementation on skeletal muscle metabolism in patients with obstructive vascular disease is more pronounced than in evidently healthy trained or untrained individuals. This could be for two reasons: 1) as a consequence of that obstructed peripheral arterial circulation, patients with peripheral vascular disease develop ischaemia in their legs; and 2) patients with advanced peripheral vascular disease are in a condition of relative carnitine insufficiency or even of muscle carnitine deficiency [37]. Interestingly, even in patients with Fontaine's stage II, plasma short-chain acylcarnitine levels at maximally tolerated walking distance are increased, compared with resting values [38]. In these patients there is a statistically significant correlation between the concentration of short-chain acylcarnitine plasma level at rest and subsequent exercise performance. Therefore, patients who have the lowest walking capacity also have the greatest resting concentration of short-chain acylcarnitine. This suggests that the more severe the ischaemic heart disease, the greater the amount of carnitine required to remove the accumulation of acyl-CoA esters produced by chronic ischaemia. Equally, the more severe the disease is, the higher the production of lactate [37]. Thus, altered carnitine homeostasis may play a prime role in the pathophysiology of intermittent claudication and of the associated excess lactate production [37, 38]. This is strongly supported by the demonstration [39] that in claudicant patients, training-induced improvement in walking ability does not increase blood flow. The improvement is accompanied by a reduction in resting plasma concentrations of short-chain acylcarnitines and amelioration of lactate metabolism. Usually, the positive effects of training are attributed to increased blood flow. Hiatt and coworkers [39], however, have shown that training may improve walking capacity simply by ameliorating carnitine, and, consequently, lactate metabolism. They have also observed that, compared to normal subjects, patients with unilateral arterial disease of the lower limbs have normal muscle levels of carnitine (gastrocnemic muscle biopsy) and lactate in the healthy limb, whilst in the diseased leg long-chain acylcarnitine and lactate are significantly higher with a further increase during exercise [39].

Oral administration of L-carnitine (4 g/die for three weeks in a double-blind cross-over vs placebo study) in 20 patients with peripheral arterial disease with intermittent claudication increases the maximum distance walked by 75% compared to placebo and reduces the frequency and the severity of subjective symptoms such as paresthesia, feelings of cold extremities, asthenia and pain on walking. General and regional haemodynamic parameters show no change and the effects must therefore be attributed to metabolic changes. A reduction in the increase in lactate production during maximum exercise is always reported (from $107 \pm 16$ to $54 \pm 32\%$) [40, 41].

### *Effect of L-carnitine on lactate production in acid-base disorders*

Acid-base derangements are encountered frequently in clinical practice and many have life-threatening implications. Treatment is dependent on correctly identifying the acid-base disorder, and, whenever possible, repairing the underlying causal process.

The carbonic acid-bicarbonate buffer system plays a central role in acid-base balance because of its prevalence and its relation to physiological regulatory mechanisms. The hydration of dissolved $CO_2$ forms carbonic acid which then dissociates to yield bicarbonate and hydrogen ions. These chemical reactions rapidly achieve equilibrium conditions, allowing a simple expression of their relationship [42]: $[H^+] = 24\ (paCO_2/[HCO_3^-])$.

Thus, the hydrogen ion concentration is a function of the ratio of the arterial carbon dioxide tension ($paCO_2$) to the bicarbonate concentration. It follows that changes in hydrogen ion concentration, and, consequently, all acid-base disorders, result from changes in one or other of these two variables.

Acid-base disturbances are classified into four primary disorders: metabolic acidosis, metabolic alkalosis, respiratory acidosis and respiratory alkalosis – and various combinations of these disorders categorised as mixed disturbances.

Three distinct pathological processes work alone or in combination to produce metabolic acidosis: 1) loss of bicarbonate; 2) addition of acid (lactate); and 3) reduced capacity of the kidney to excrete acid [43]. Lactic acidosis in general is due to low cardiac-output states, septic shock and acute volume deficits.

Prompt treatment of low cardiac-output states, repletion of volume deficits, correction of shock and administration of antibiotics (in the septic patient) should be given the highest priority. The use of vasoconstrictive substances should be minimised to avoid aggravation of ischaemia of peripheral tissues. Invasive haemodynamic monitoring in an intensive care unit should be employed to guide the use of volume replacement, cardiotropic agents and pressors [44].

Controversy surrounds the use of alkali therapy in the treatment of lactic acidosis. Experimental studies in dogs suggest that administering sodium bicarbonate may augment the accumulation of lactic acid by stimulating its production or interfering with its metabolism by the liver [45].

Carnitine has been shown to be effective in normalizing blood levels of lactate in dogs with acidosis and ketosis induced by lipid perfusions during starvation [46]. In addition, the evidence in humans of an adverse effect of alkali therapy in lactic acidosis is not convincing. Sodium bicarbonate should be administered for severe, life-threatening lactic acidosis, that is, when the pH falls below 7.20 (which usually corresponds to a plasma bicarbonate below 10 mM). Certainly, at a pH below this value, the negative inotropic and arrhythmogenic effects of acidaemia are substantial, and alkali therapy gains time to address the principal disorder. No specific guidelines can be given regarding the amount of alkali to be administered because the rate of lactate production varies tremendously. As a general rule, however, it is desirable to give sufficient alkali to maintain plasma bicarbonate concentration at 15 to 18 mM. Acid-base parameters should be assessed frequently and therapy modified accordingly.

Adjunctive or alternative therapy to sodium bicarbonate in the treatment of lactic acidosis with dichloroacetate also has been proposed. While dichloracetate augments the oxidation of lactate to acetyl-CoA, and evidence from human studies is promising [47], toxicity may be a major drawback to its use. Thus, there is the need for other, safer compounds to be used to treat lactic acidosis.

Interestingly Corbucci et al. [48] administered L-carnitine to 80 patients with cardiogenic shock in an open study randomized against bicarbonate. L-Carnitine significantly improved survival over bicarbonate. Whilst these results are clearly of interest, particularly considering that mortality in cardiogenic shock is highly correlated to acidosis and to plasma lactate concentration [49], further evidence is needed before the role of L-carnitine as an adjuvant to the treatment of cardiogenic shock is defined.

### *Effect of L-carnitine on lactate production from the myocardium*

A crude estimate of the type of substrate used by the heart can be achieved by the respiratory quotient which is calculated by comparing the rate of oxygen uptake with the rate of carbon dioxide production. A respiratory quotient near one implies oxidation of glucose and/or lactate, whereas a lower value implies fatty acid oxidation. Because the myocardial respiratory quotient is frequently low, it is believed that lipids are the major myocardial fuel, although there is still some controversy concerning the preferred myocardial substrate. In this chapter we will briefly review myocardial glucose metabolism under aerobic and ischaemic conditions as it plays an essential role in lactate production.

Energy from glucose utilization is derived from anaerobic or the Embden-Meyerhof portion of glycolysis, the citric acid cycle, and the respiratory chain. The pentose phosphate shunt pathway is of little consequence in heart muscle. Glucose utilization is controlled at key steps outside the mitochondria, including glucose transport and reactions regulated by hexokinase, glycogen synthase, glycogen phosphorylase, fructokinase, and pyruvate kinase and dehydrogenase [50–53]. Glycolysis is limited by the rates of disposal of glycolytically produced NADH in the cytosol (true for both glucose and lactate as substrate) and, at high flux rates, the activity of glyceraldehyde-3-phosphate dehydrogenase is the first major regulatory restraint [54]. Glucose as a sole substrate in experiments in rat hearts fixes the availability of acetyl-CoA, and the citric acid cycle behaves in a "run-down" manner, as reflected by reduced levels of acetyl-CoA, citrate, and isocitrate; decreased mitochondrial NADH/$NAD^+$ ratios; and increased oxaloacetate levels [54]. Oxidative phosphorylation is limited by availability of NADH, and anaerobic glycolysis provides for only trivial amounts of ATP for myocardial energy needs.

Increasing plasma glucose concentrations, the presence of NADH in anoxia, high cardiac work, growth hormone, adrenaline, insuline and uncouplers of oxidative phosphorylation all stimulate transport. Conversely, availability of competing substrate and oxidation of fatty acids, ketones, and pyruvate all inhibit transport. Glucose-6-phosphate at concentrations that inhibit the hexokinase reaction does not inhibit transport.

The second major step for glucose transport in general is hexokinase-mediated phosphorylation, a rate-limiting step at high transport rates [50–53, 55]. Although it is accelerated by rising levels of intracellular glucose, its rate is strongly inhibited by glucose-6-phosphate as well as ATP, ADP, AMP, inorganic phosphate (Pi), oxidation of fatty acids and ketones, starvation, and diabetes. The reaction is essentially irreversible in heart muscle [51] and governed predominantly by two hexokinase isoenzymes bound to mitochondria [52].

Neely and Morgan [51] considered glucose metabolism (utilization) in five steps: glucose uptake (transport and the hexokinase reaction), glycogen metabolism (reactions between the formation of glucose-6-phosphate and glycogen); glycolysis (glucose-6-phosphate to pyruvate); pyruvate metabolism (pyruvate to either lactate, acetyl-CoA, or alanine); and the citric acid cycle. In brain, in the absence of glycogen formation, flux rates of glucose uptake are tightly linked to those of glycolysis. This is not the case in myocardium, in which glycogen formation exists in competition with glycolysis. Indeed, most of the glucose extracted by heart muscle is converted to glycogen both in fasting and fed states.

Glucose uptake may be influenced by free fatty acid, lactate and pyruvate. Lactate is normally extracted from the aerobic myocardium, and sufficiently high arterial concentrations (4.5 mM) of lactate have been shown in normal dog hearts to be preferentially utilized even with respect to both glucose and fatty acids [56]. It is well-known that lactate in plasma increases appreciably during exercise. Gertz et al. [57] reported in normal subjects during moderately intense exercise that myocardial lactate uptake rose more than threefold and that all of it underwent oxidative decarboxylation. Glucose oxidation increased twofold.

Pyruvate is a higly competitive substrate that, if utilized preferentially as an oxidizable substrate, could also allosterically influence glucose uptake and glycolysis. Pyruvate dehydrogenase is the key regulatory enzyme in its metabolism and is stimulated by increasing concentrations of pyruvate and inhibited by the presence of fatty acylcarnitine derivatives and ketones [58, 59].

Under ischaemic conditions not only is oxygen supply threatened in this aerobic organ, but critical washout of inhibitory intermediate products is curtailed, impairing overall flux within metabolic pathways. Key steps within glucose utilization and glycolytic pathway respond in a disparate fashion, with increases occurring in glucose transport and the hexokinase, fructokinase, and glycogen phosphorylase reactions and decreases in the glyceraldehyde-3-phosphate dehydrogenase and pyruvate dehydrogenase. Overall rates of glycolysis are determined by the absolute restriction of coronary flow. Glyceraldehyde-3-phosphate dehydrogenase, because of its inhibition by rising intracellular concentrations of NADH, hydrogen ion, and lactate with ischaemia, is considered the chief rate-limiting enzyme in glycolysis. Its continued activity is dependent on the washout of these metabolites.

Thus, at the beginning of ischaemia or under conditions of mild ischaemia (ex moderate reduction of coronary flow), lactate production and release from the heart into the coronary sinus is abundant. This phenomenon should be viewed as a positive, protective mechanism, lactate production contributing to a reduction of intracellular acidosis and providing two molecules of ATP produced in the absence of oxygen expenditure. When ischaemia persists or becomes more severe, intracellular acidosis inhibits glyceraldehyde-3-phosphate dehydrogenase and the flux of lactate production is reduced. This coincides with a deterioration of the ischaemic condition [60].

In the clinic, lactate production can be estimated by measuring the arterial-coronary sinus difference of this metabolite before, during and after an acute ischaemic insult. Usually ischaemia is induced by an atrial pacing. Exercise-induced ischaemia is usually carefully avoided as exercise induces production of lactate from skeletal muscle. Consequently plasma lactate rises and affects lactate uptake from the aerobic zone of the myocardium, thus making the interpretation of the overall results difficult.

Data on lactate production in coronary artery diseased (CAD) patients subjected to an atrial pacing are also difficult to interpret as several complications exist with the technique, such as changes in coronary flow, timing of sampling, positioning of the catheter in the coronary sinus, dilution of lactate and balance between lactate uptake from aerobic regions and production from the ischaemic one.

In general, however, it is the behaviour of lactate metabolism with a positive uptake before ischaemia converted into a net release during pacing and then gradually returning to a positive uptake during recovery which allows diagnosis of development of anaerobic metabolism [61].

A reduction of net uptake release in the same patients after drug treatment is considered a beneficial metabolic effect of the treatment. Usually, but not necessarily, this is associated with an increase in pacing time to angina and with a smaller increase in diastolic left ventricular pressure [62].

*Effects of L-carnitine on lactate metabolism at rest*

Although L-carnitine has little effect on heart function at rest, it exerts important changes on its metabolism. This information, however, is rare, mainly because measurements of heart metabolism in man are strictly dependent upon invasive methodology, i.e. catheterization of the coronary sinus.

We had the opportunity to investigate the effects of a central venous infusion of L-carnitine (40 mg $kg^{-1}$) on resting heart metabolism in 25 selected CAD patients without previous myocardial infarction and with normal myocardial function [63, 64]. Myocardial metabolism was measured in terms of arterial-coronary sinus difference, myocardial percentage of extraction and myocardial uptake of free fatty acids (FFA), lactate and glucose [65].

In our patients there was a linear relationship between myocardial arterial-coronary sinus difference for the different substrates and their availability in the arterial blood [63, 64]. As expected, the two major substrates metabolized from the heart before administration of L-carnitine were glucose and FFA. The oxygen extraction ratio for carbohydrates in the form of glucose and lactate accounted for up to 11% of oxygen consumption and that of FFA for about 39%.

Administration of L-carnitine had several important effects:

- a significant reduction in arterial concentration of FFA;
- a significant increase in myocardial uptake of FFA;
- a reduction in myocardial uptake of glucose;
- no major changes in myocardial uptake of lactate;
- no significant increase in the overall oxygen consumption of the heart.

The reduction of circulating FFA occurred in the absence of major systemic haemodynamic changes that could account for an increased utilization of these substrates. Thus, this effect appears to be primarily dependent on the metabolic and systemic action of L-carnitine.

The finding that acute administration of L-carnitine increases myocardial utilization of FFA at the expense of glucose is also a very important effect with clinical relevance, particularly in relation to chronic treatment with L-carnitine. Carbohydrates, especially those stored in the form of glycogen, represent a pool of emergency substrates utilized anaerobically by the myocardium under conditions of acute and severe energy need, like stress, exercise or, more relevant to this discussion, during acute ischaemia. The relatively high myocardial glucose utilization in CAD patients due to L-carnitine deficiency might lead to a depletion of glycogen stores, thus reducing the myocardial metabolic defence against attacks of acute ischaemia. Conversely, administration of L-carnitine for prolonged periods of time improves FFA consumption and reduces myocardial glucose utilization, thus restoring glycogen stores. A clear relationship between carnitine concentration and enzyme activities representative of different metabolic pathways has been demonstrated in skeletal muscle of healthy volunteers [66]. In particular, a highly significant correlation was found between carnitine concentration and muscle glycogen content as well as with the overall anaerobic glycolytic activity. We are at present investigating whether this correlation is also valid for heart muscle. If this turns out to be the case, chronic administration of L-carnitine to CAD patients will exert a favourable metabolic preconditioning, restoring the natural metabolic defence against ischaemia. It is also relevant to recall here that it has been suggested recently that shifting myocardial metabolism from lipids to carbohydrates induces cardiac hypertrophy and diastolic dysfunction [67]. L-Carnitine acting as a metabolic modulator restores a balanced myocadial metabolism and should avoid these negatives events.

### *Effects of L-carnitine on myocardial metabolism during induced tachycardia*

Myocardial metabolism under stress conditions has been investigated in controlled studies in CAD patients subject to atrial pacing [63, 64, 68, 69].

Single intravenous doses of L-carnitine (40 or 140 mg $kg^{-1}$) or DL-carnitine (20 or 40 mg $kg^{-1}$) decreased production of lactate from the myocardium or maintained a positive extraction of the substrate at peak sinus pacing relative to that seen in untreated or placebo-treated control groups. Moreover, the mean myocardial FFA extraction ratio increased significantly [63, 64].

In all studies L-carnitine converted lactate production at peak pacing stress into extraction, thus preventing anaerobiosis. In addition, there was an increase in the uptake of FFA as well as in that of glucose, confirming the role of L-carnitine as a metabolic modulator, improving not only FFA, but also carbohydrate metabolism. The finding that glucose and lactate uptake were maintained confirms the data of Lopaschuk [70] and supports the concept that L-carnitine is important in the critical condition of energy need. All these alterations are due to the metabolic properties of L-carnitine, since the increase in coronary flow induced by atrial pacing was not affected.

There is no simple explanation for these effects of L-carnitine. During ischaemia, the reduced oxygen availability leads to a decrease in mitochondrial electron transport, which, in turn, causes an accumulation of NADH and long-chain acyl-CoA. Increased NADH/NAD and acyl-CoA/CoA ratios inhibit the activity of pyruvate dehydrogenase, the enzyme which regulates the entry of pyruvate into the citric acid cycle. Under these conditions pyruvate is preferentially converted to lactate with consequent lactate production.

L-Carnitine can reduce lactate production under ischaemic conditions either by activating pyruvate dehydrogenase directly or by provoking a decrease in the acetyl-CoA/CoA ratio, as a consequence of acetyl removal from CoA, mediated by carnitine-CoA acetyltransferase. Therefore, the utilization of pyruvate and consequently of lactate in the oxidative pathway, induced by L-carnitine, may be attributed to enhanced pyruvate dehydrogenase activity, rather than to an increase in oxygen availability induced by L-carnitine.

Another possible explanation for the effects of L-carnitine on lactate metabolism is an indirect effect on fructokinase activity, the enzyme regulating the rate of anaerobic glycolysis. The activity of this enzyme is inhibited by a high cytosolic ATP concentration, whilst it is stimulated by low cytosolic ATP concentrations. Under ischaemic conditions the ATP concentration in the cytosol decreases either as a result of reduced oxidative metabolism or because the activity of adenine-nucleotide translocase is inhibited by long-chain acyl-CoA. This causes sequestration of ATP into the mitochondrial matrix with a decrease of cytosolic ATP and a consequent stimulation of fructokinase. By removing long-chain acyl-CoA, carnitine prevents its inhibitory action of adenine-nucleotide translocase, thus improving the ATP transfer to the cytosolic compartment. The increased cytosolic ATP levels might decrease fructokinase activity and lactate production.

### *Effects of L-carnitine in patients subjected to heart surgery and with cardiogenic shock*

Pre-operative administration of L-carnitine has been attempted to improve the metabolic alterations caused by total and global ischaemia imposed on CAD patients during aorto-coronary bypass surgery. The rationale of this use of carnitine is provided by the favourable results obtained in animal models with ischaemia [71–75] and in CAD patients exposed to tests aimed to induce ischaemia. Bohles et al. [76] studied 40 patients undergoing aorto-coronary bypass surgery who received either oral L-carnitine (1 g daily in three doses, for two days prior to surgery) plus 0.5 g intravenously immediately before surgery, or no treatment. The concentration of myocardial free carnitine significantly increased, and that of long-chain acylcarnitine decreased in patients receiving carnitine therapy. Treated patients showed a reduced concentration of lactate in the myocardium correlated with significantly elevated concentrations of ATP, suggesting that metabolism of long-chain FFA was improved.

Administration of L-carnitine reduces the increase in plasma lactate concentration caused by heart surgery [77].

The results indicate clearly that L-carnitine reduces the generalized cellular acidosis consequent to heart surgery, probably by reducing the inhibition of pyruvate dehydrogenase caused by accumulation of long-chain CoA.

## References

1. Pande SV, Parvin R. Carnitine-acylcarnitine translocase catalyzes an equilibrating undirectional transport as well. J Biol Chem 1980; 255: 2994–3001.
2. Bremer J. Biosynthesis of carnitine in vivo. Biochim Biophys Acta 1961; 48: 622–4.
3. Hülsmann WC, Siliprandi D, Ciman M, Siliprandi N. Effects of carnitine on the oxidation of $\alpha$-oxoglutarate to succinate in the presence of acetoacetate or pyruvate. Biochim Biophys Acta 1964; 93: 166–8.
4. Hoppel C. The physiological role of carnitine. In: Ferrari R, DiMauro S, Sherwood G, editors. L-Carnitine and its role in medicine: From function to therapy. San Diego: Academic Press, 1992: 5–21.
5. Pasini E, Cargnoni A, Condorelli E, Marzo A, Lisciani R, Ferrari R. Effect of prolonged treatment with propionyl-L-carnitine on erucic acid-induced myocardial dysfunction in rats. Mol Cell Biochem 1992; 112: 117–23.
6. Dhalla NS, Dixon IMC, Shah KR, Ferrari R. Beneficial effects of L-carnitine and derivatives on heart membranes in experimental diabetes. In: Ferrari R, DiMauro S, Sherwood G, editors. L-Carnitine and its role in medicine: From function to therapy. San Diego: Academic Press, 1992: 411–26.
7. Cerretelli P, Marconi C. L-Carnitine supplementation in humans. The effects on physical performance. Int J Sports Med 1990; 11: 1–14.
8. Ferrari R, Ceconi C, Curello S et al. Metabolic changes during post-ischaemic reperfusion. J Mol Cell Cardiol 1988; 20(Suppl 2): 119–33.
9. Harris RC, Foster CVL, Hultman E. Acetylcarnitine formation during intense muscular contraction in humans. J Appl Physiol 1987; 63: 440–2.
10. Alkonyi I, Kerner J, Sandor A. The possible role of carnitine and carnitine acetyl-transferase in the contracting frog skeletal muscle. FEBS Lett 1975; 52: 265–8.
11. Childress CC, Sacktor B, Traynor DR. Function of carnitine in the fatty acid oxidase-deficient insect flight muscle. J Biol Chem 1966; 242: 754–60.
12. Pearson DJ, Tubbs PK. Carnitine and derivatives in rat tissues. Biochem J 1967; 105: 953–63.
13. Angelini C, Vergani L, Costa L et al. Use of carnitine in exercise physiology. Adv Clin Enzymol 1986; 4: 103–10.
14. Caldarone G, Giampietro M, Berlutti G, Ciampini M. Sull'impiego di L-carnitina e complesso vitaminico in un gruppo di giovani canottieri di elevata performance. Rif Med 1987; 102: 149–55.
15. Canale C, Terrachini V, Biagini A et al. Bicycle ergometer and echocardiographic study in healthy subjects and patients with angina pectoris after administration of L-carnitine: semiautomatic computerized analysis of M-mode tracings. Int J Clin Pharmacol Ther Toxicol 1988; 26: 221–4.
16. Cooper MB, Jones DA, Edwards RHT, Corbucci GC, Montanari G, Trevisani C. The effect of marathon running on carnitine metabolism and on some aspects of muscle mitochondrial activities and antioxidant mechanisms. J Sport Sci 1986; 4: 79–87.
17. Corbucci GG, Montanari G, Cooper MB: Effetto dell'esercizio fisico prolungato sul metabolismo della L-carnitina nell'atleta. Atleticastudi 1984; 6: 507–16.
18. Dal Negro R, Pomari G, Zoccatelli G, Turco P. Changes in physical performance of untrained volunteers: effects of L-carnitine. Clin Trials J 1986; 23: 242–7.
19. Dragan AM, Vasiliu D, Eremia NMD, Georgescu E. Studies concerning some acute biologi-

cal changes after endovenous administration of 1 g L-carnitine in elite athletes. Physiologie 1987; 24: 231–4.
20. Dragan IG, Vasiliu A, Georgescu E, Eremia N. Studies concerning chronic and acute effects of L-carnitine in elite athletes. Physiologie 1989; 26: 111–29.
21. Dragan GI, Wagner W, Ploesteanu E. Studies concerning the ergogenic value of protein supply and L-carnitine in elite junior cyclists. Physiologie 1988; 25: 129–32.
22. Eclache JP, Quard S, Carrier E. Effets d'une adjonction de carnitine au régime alimentaire sur l'exercise intense et prolongé. Comptes Rendus du Colloque de Saint-Etienne. July 2–3, 1989.
23. Ferretti G. Effects of a single dose L-carnitine administration on maximal aerobic power and anaerobic threshold in normoxic and hypoxic (acute and chronic) humans. Proc Congress of Physiology Exercise, Switzerland. In press.
24. Fuentes E, Padilla C: Efecyos de la L-carnitina sobre metabolismo lipidico del deportista de alta competicion. Repercusion sobre su capacidad aerobica. Aplicaciones practicas. Proc 14th Int Congr Sport Medicine. Madrid, June 1–4, 1987.
25. Greig C, Finch KM, Jones DA, Cooper M, Sargeant AJ, Forte CA. The effect of oral supplementation with L-carnitine on maximum and submaximum exercise capacity. Eur J Appl Physiol 1987; 56: 457–60.
26. Marconi C, Sassi G, Carpinelli A, Cerretelli P. Effects of L-carnitine loading on the aerobic and anaerobic performance of endurance athletes. Eur J Appl Physiol 1985; 54: 131–5.
27. Narvaez Perez GE, Alvarez Casado JJ, Mollerach M et al. Action of L-carnitine on the submaximal work time and lipid metabolism in trained subjects. In: Borum PR, editor. Clinical aspects of human carnitine deficiency. New York: Pergamon Press, 1986: 44–5.
28. Otto RM, Shores KVM, Wygand JW, Perez HR. The effect of L-carnitine supplementation on endurance exercise. Med Sci Sports Exercise 1987; 19: S68 (Abstr).
29. Oyono-Enguelle S, Freund H, Ott C et al. Prolonged submaximal exercise and L-carnitine in humans. Eur J Appl Physiol 1988; 58: 53–61.
30. Poleszynski DV, Bohmer T. Improved oxygen uptake, blood pressure and triglyceride reduction with oral L-carnitine in healthy women. Int J Biospacial Med Res 1991; 13: 180–91.
31. Siliprandi N, Di Lisa F, Pieralisi G et al. Metabolic changes induced by maximal exercise in human subjects following L-carnitine administration. Biochim Biophys Acta 1990; 1034: 17–21.
32. Soop M, Bjorkman O, Cederblad G, Hagenfeldt L, Wahren J. Influence of carnitine supplementation on muscle substrate and carnitine metabolism during exercise. J Appl Physiol 1988; 64: 2394–9.
33. Vecchiet L, Di Lisa F, Pieralisi G et al. Influence of L-carnitine administration on maximal physical exercise. Eur J Appl Physiol 1990; 61: 486–90.
34. Wyss V, Ganzit GP, Rienzi A. Effects of L-carnitine administration on $VO_2$max and the aerobic-anaerobic threshold in normoxia and acute hypoxia. Eur J Appl Physiol 1990; 60: 1–6.
35. Oyono-Enguelle S, Freund H, Ott C et al. Prolonged submaximal exercise and L-carnitine in humans. Eur J Appl Physiol 1988; 58: 53–61.
36. Suzuki M, Kanaya M, Muramatsu S, Takahashi T. Effects of carnitine administration, fasting and exercise on urinary carnitine excretion in man. J Nutr Sci Vitaminol 1976; 22: 167–74.
37. Brevetti G, Perna S. Metabolic and clinical effects of L-carnitine in peripheral vascular disease. In: Ferrari R, DiMauro S, Sherwood G, editors. L-Carnitine and its role in medicine: From function to therapy. San Diego: Academic Press, 1992: 359–78.
38. Hiatt WR, Nawaz D, Brass EP. Carnitine metabolism during exercise in patients with peripheral vascular disease. J Appl Physiol 1987; 62: 2383–7.
39. Hiatt WR, Regensteiner JG, Hargarten ME, Wolfel EE, Brass EP. Benefit of exercise conditioning for patients with peripheral arterial disease. Circulation 1990; 81: 602–9.
40. Brevetti G, Chiariello M, Ferulano G et al. Increases in walking distance in patients with

peripheral vascular disease treated with L-carnitine: A double-blind, cross-over study. Circulation 1988; 77: 767–73.
41. Brevetti G, Attisano T, Perna S, Rossini A, Policicchio A, Corsi M. Effect of L-carnitine on the reactive hyperemia in patients affected by peripheral vascular disease: A double-blind, crossover study. Angiology 1989; 40: 857–62.
42. Henderson LJ. The theory of neutrality regulation in the animal organism. Am J Physiol 1908; 21: 427.
43. McLaughlin ML, Kassirer JP. Rational treatment of acid-base disorders. Drugs 1990; 39: 841–55.
44. Madias NE. Lactic acidosis. Kidney Int 1986; 29: 752–74.
45. Graf H, Leach W, Arieff AI. Evidence for a detrimental effect of bicarbonate therapy in hypoxic lactic acidosis. Science 1985; 227: 754–6.
46. Stacpoole PW, Harman EM, Curry SH, Baumgartner TG, Misbin RI. Treatment of lactic acidosis with dichloracetate. N Engl J Med 1983; 309: 390–6.
47. Broekhuysen J, Baudine A, Deltour G. Effect of carnitine on acidosis and ketosis induced by lipid perfusions in dogs during starvation. Biochim Biophys Acta 1965; 106: 207–10.
48. Corbucci GG, Gasparetto A, Antonelli M et al. Effects of L-carnitine administration on mitochondrial electron transport activity present in human muscle during circulatory shock. Int J Clin Pharm Res 1985; 5: 237–41.
49. Kette F, Weil MH, Von Planta M, Gazmuri RJ, Rackow EC. Buffer agents do not reverse intramyocardial acidosis during cardiac resuscitation. Circulation 1990; 81: 1660–6.
50. Neely JR, Rovetto MJ, Oram JF. Myocardial utilization of carbohydrate and lipids. Prog Cardiovasc Dis 1972; 15: 289–329.
51. Neely JR, Morgan HE. Relationship between carbohydrate and lipid metabolism and the energy balance of heart muscle. Annu Rev Physiol 1974; 36: 413–69.
52. Randle PJ, Tubbs PK. Carbohydrate and fatty acid metabolism. In: Steiner DF, Freinkel N, editors. Handbook of physiology. The cardiovascular system I. Washington, D.C.: American Physiological Society, 1979: 805–44.
53. Taegtmeyer H. Myocardial metabolism. In: Phelps M, Mazziotta J, Shelbert H, editors. Positron emission tomography and autoradiography: Principles and applications for the brain and heart. New York: Raven Press, 1986: 149–95.
54. Kobayashi K, Neely JR. Control of maximum rates of glycolysis in rat cardiac muscle. Circ Res 1979; 44: 166–75.
55. Liedtke AJ. Alterations of carbohydrate and lipid metabolism in the acutely ischemic heart. Progr Cardiovasc Dis 1981; 23: 321–36.
56. Drake AJ, Haines JR, Noble MIM. Preferential uptake of lactate by the normal myocardium in dogs. Cardiovasc Res 1980; 14: 65–72.
57. Gertz EW, Wisneski JA, Stanley WC, Neese RA. Myocardial substrate utilization during exercise in humans: Dual carbon-labeled carbohydrate isotope experiments. J Clin Invest 1988; 82: 2017–25.
58. Olson MS, Dennis SC, Routh CA, DeBuysere MS. The regulation of pyruvate dehydrogenase by fatty acids in isolated rabbit heart mitochondria. Arch Biochem Biophys 1978; 187: 121–31.
59. Dennis SC, Padasa A, DeBuysere MS, Olson MS. Studies on the regulation of pyruvate dehydrogenase in the isolated perfused rat heart. J Biol Chem 1979; 254: 1252–8.
60. Ferrari R, Cargnoni A, Gargano M et al. Molecular events during myocardial ischaemia and reperfusion. Adv Myochem 1987; 1: 354–6.
61. Visioli O, Bongrani S, Cucchini F, Di Donato M, Ferrari R. Myocardial lactic acid balance after left ventriculography. Eur J Cardiol 1980; 11: 357–65.
62. Ferrari R, Bolognesi R, Raddino R. Reproducibility of metabolical parameters during successive coronary sinus pacing in patients known to have coronary artery disease. Int J Cardiol 1985; 9: 231–4.
63. Ferrari R, Cucchini F, Di Lisa F, Raddino R, Bolognesi R, Visioli O. The effects of L-

carnitine on myocardial metabolism of patients with coronary artery disease. Clin Trials J 1984; 21: 41–58.
64. Ferrari R, Cucchini F, Visioli O. The metabolical effects of L-carnitine in angina pectoris. Int J Cardiol 1984; 5: 213–6.
65. Ferrari R. Carnitine and cardiac energy supply. In: De Jong JW, editor. Myocardial energy metabolism. Dordrecht-Boston-Lancaster: Martinus Nijhoff Publishers, 1988: 35–43.
66. Cederblad G, Lundholm K, Scherstén T. Carnitine concentration in skeletal muscle tissue from patients with diabetes mellitus. Acta Med Scand 1977; 202: 305–6.
67. Litwin SE, Raya TE, Gay RG et al. Chronic inhibition of fatty acid oxidation: new model of diastolic dysfunction. Am J Physiol 1990; 258: H51–6.
68. Thomsen JH, Shug AL, Yap VU, Patel AK, Karras TJ, De Felice SL. Improved pacing tolerance of the ischemic human myocardium after administration of carnitine. Am J Cardiol 1979; 43: 300–6.
69. Reforzo G, De Andreis-Bessone PL, Rebaudo F, Tibaldi M. Effects of high doses of L-carnitine on myocardial lactate balance during pacing-induced ischemia in aging subjects. Curr Ther Res 1986; 40: 1–11.
70. Lopaschuk GD. Effects of carnitine and carnitine acyltransferase inhibition on energy substrate utilization in the intact heart. In: Ferrari R, DiMauro S, Sherwood G, editors. L-Carnitine and its role in medicine: From function to therapy. San Diego: Academic Press, 1992: 403–10.
71. Folts JD, Shug AL, Koke JR, Bittar N. Protection of the ischemic dog myocardium with Carnitine. Am J Cardiol 1978; 41: 1209–14.
72. Shug AL, Thomsen JH, Folts JD, Bittar N, Klein MI, Koke JR, Huth PJ. Changes in tissue levels of carnitine and other metabolites during myocardial ischemia and anoxia. Arch Biochem Biophys 1978; 187: 25–33.
73. Suzuki Y, Masumura Y, Kobayashi A, Yamazaki N, Harada Y, Osawa M. Myocardial carnitine deficiency in chronic heart failure. Lancet 1982; i: 116 (Lett to the Ed).
74. Ferrari R, Di Lisa F, Raddino R et al. Factors influencing the metabolic and functional alterations induced by ischaemia and reperfusion. In: Ferrari R, Katz A, Shug A, Visioli O, editors. Myocardial ischaemia and lipid metabolism. New York: Plenum Press, 1984: 135–57.
75. Liedtke AJ. Lipid burden in ischemic myocardium. J Mol Cell Cardiol 1988; 20(Suppl 2): 65–74.
76. Bohles H, Noppency T, Akcetin Z et al. The effect of preoperative L-carnitine supplementation on myocardial metabolism during aorto-coronary-bypass surgery. Curr Ther Res 1986; 39: 429–35.
77. Ferrari R, Visioli O. Effects of L-carnitine in coronary artery disease patients. In: Ferrari R, DiMauro S, Sherwood G, editors. L-Carnitine and its role in medicine: From function to therapy. San Diego: Academic Press, 1992: 265–82.

*Corresponding Author*: Professor Roberto Ferrari, Cattedra di Cardiologia, Università degli Studi di Brescia, c/o Spedali Civili, P.le Spedali Civili, 1, I-25123 Brescia, Italy

# 16. Therapeutic potential of L-carnitine in patients with angina pectoris

CARL J. PEPINE and MICHAEL A. WELSCH

"There seems to be strong evidence to suggest that carnitine supplementation (immediate and short-term [<30 days]) may be beneficial in patients with ischemic heart disease. However, the mechanisms involved in the improved clinical status are presently not clear. There is evidence to suggest that long-term (up to 12 months) L-carnitine supplementation in patients with ischemic heart disease may have beneficial effects on clinical outcomes. Improvements in symptoms, exercise tolerance as well as reductions in mortality have been suggested".

## Introduction

Coronary artery disease (CAD) is responsible for a large proportion of the deaths and disabilities observed in western countries and its costs to the health care budgets of these societies are enormous. Typically, CAD exerts its deleterious effects through myocardial ischemia which may be symptomatic (e.g. angina or its equivalents) or asymptomatic. Usually myocardial ischemia results due to an increase in myocardial oxygen demand (e.g. secondary to increases in physical or emotional activity) with a limited myocardial oxygen delivery. Oxygen delivery is limited by an inability to raise coronary flow beyond the ceiling imposed by coronary artery atherosclerotic obstruction or dynamic constriction. This results in tissue hypoxia and inadequate energy production (e.g. ↓ [ATP]) in the myocardial regions supplied by the narrowed coronary artery. Myocardial ischemia leads to a regional accumulation of metabolites which are also potentially damaging [1]. The lack of ATP and the accumulation of metabolites result in mechanical dysfunction manifest by abnormalities in systolic and diastolic function. Thus, in addition to the chest pressure or tightness referred to as angina, patients may also complain of dyspnea or shortness of breath [2].

Inadequate production of ATP impairs regional myocardial function through a variety of mechanisms including loss of sodium and potassium gradients, calcium accumulation, and activation of phospholipases and proteases [1]. Accumulation of metabolites may also impair myocardial function

*J.W. de Jong and R. Ferrari (eds): The carnitine system*, 225–243.

as a result of inhibition of enzyme systems which block oxidative metabolism. Of particular importance is the acyl-CoA carnitine transferase (ACCT) system, which is pivotal in facilitating the transportation of free fatty acids (FFA) into mitochondria. A decrease in free carnitine and an increase in acyl groups occur in myocardial tissue of animals with experimentally induced ischemia [3]. and in the necrotic zone of infarcted myocardium in humans at autopsy [4, 5]. These findings suggest that lipid metabolism is altered in myocardial ischemia. Since the preferred pathway for energy production in the normally perfused heart is oxidation of FFA, an impaired carnitine transport system may play a key role in the loss of myocardial function.

The purpose of this chapter is to review the potential role of L-carnitine supplementation on lipid metabolism and the ACCT system in coronary artery disease. The specific objectives are to examine both the acute and long-term effects of L-carnitine supplementation in patients with angina pectoris.

**Lipid metabolism and the myocardium**

Cardiac muscle metabolism is reviewed in detail in several other chapters. In brief, cardiac muscle can use a wide variety of substrates for energy production; however, under normal coronary flow conditions, oxidation of FFA accounts for 60% to 90% of ATP production [6]. Uptake of FFA from plasma is concentration dependent [7] and utilization of FFA is dependent on many factors. Among these are the relationship between coronary flow and myocardial oxygen demand, the action of specific plasma hormones, the status of oxidative respiration, and the activity of several enzyme systems, including the ACCT system [8].

In addition to its role in energy production, FFA are essential components of membrane phospholipids, which play an important role in the structural integrity of the heart and blood vessels. FFA are involved in the maintenance of membrane stability, calcium storage, regulation of enzyme activity, signal transduction and propagation, storage and anchoring enzymes and proteins in the lipid bilayer of the membrane [8].

During episodes of myocardial ischemia striking changes in FFA metabolism occur. Beta-oxidation is inhibited when oxygen supply is reduced. During prolonged ischemia (>30 min) oxidation of FFA is further inhibited due to a loss in free carnitine [9, 10]. which results in a loss of activity in the ACCT system [6]. Moreover, there appears to be a "lipid paradox" in the myocardium, in the sense that low concentrations of FFA are essential for proper functioning of the heart whereas excessive amounts seem to be deleterious [8]. Inhibition of fat metabolism, in ischemic myocardium, results in accumulation of oxidative intermediates, in particular long-chain acyl-CoA esters which impair myocardial metabolic and mechanical function [11, 12]. The accumulation of acyl-CoA reduces the availability of intramitochondrial

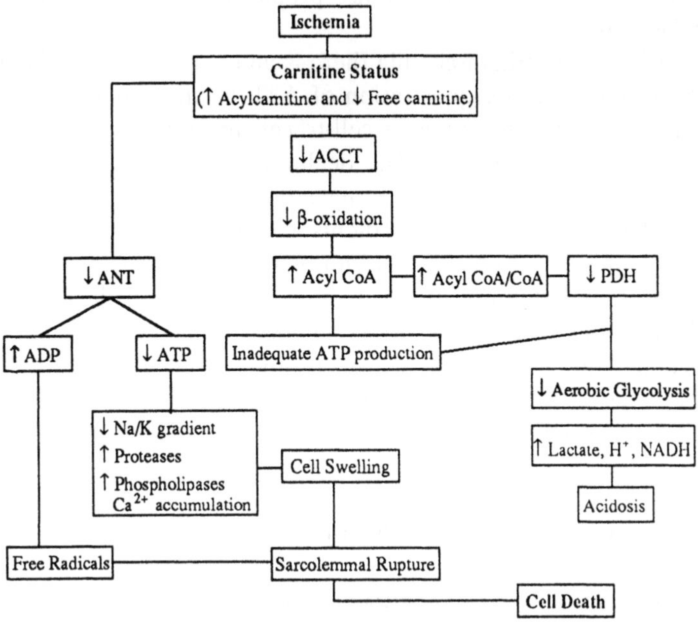


*Figure 1.* Proposed pathways through which carnitine deficiency may effect myocardial function in patients with ischemic heart disease. ACCT = Acyl-CoA carnitine transferase system; ANT = Adenine nucleotide translocase; PDH = Pyruvate dehydrogenase complex; ADP = Adenosine diphosphate; ATP = Adenosine triphosphate.

free CoA which is a regulatory step in the conversion of $\alpha$-ketoglutarate to succinyl-CoA [13]. The consequence of a reduction of free CoA is a slowing of oxidative metabolism. Under these conditions the intramitochondrial acetyl-CoA/CoA ratio also increases, which inhibits the activity of the pyruvate dehydrogenase (PDH) complex (an enzyme that facilitates oxidative utilization of pyruvate), and leads to the conversion of pyruvate to lactate (Figure 1). Long-chain acyl-CoA esters have been proposed to be a precursor to sarcolemmal damage and irreversible myocardial cell injury [3, 13–15]. Further accumulation of long-chain acyl-CoA esters impairs the activity of the adenine nucleotide translocase which mediates the exchange of ATP for ADP across the mitochondrial membrane [9, 13, 14]. As a result ATP is sequestered in the mitochondria, thereby slowing metabolism, and cytosolic [ATP] falls [14, 15].

During myocardial ischemia long-chain acylcarnitines also accumulate and contribute to a decrease in conduction velocity and cell-to-cell uncoupling [16]. The accumulation of long-chain acylcarnitines that occurs in response to ischemia has a detergent action, inhibits cell membrane function, and is associated with disturbances in impulse propagation and may lead to lethal arrhythmias [16].

Thus, during myocardial ischemia there is an overall slowing of mitochondrial ATP production, a sequestration of ATP in mitochondria, a decrease in conduction velocity resulting in cell-to-cell uncoupling, and subsequent decline in myocardial function. The declining [ATP] and accumulation of long-chain acyl-CoA esters have many adverse influences on myocellular function including loss of Na/K gradients, calcium accumulation, activation of phospholipases and proteases, and ultimately deterioration of myocardial function, structural stability of the membranes and cell death (Figure 1) [1]. Although, the precise link between impaired myocardial metabolism, mechanical function and angina pectoris is not fully understood, the additive effects of tissue hypoxia, reduced oxidative metabolism, and amphophilic metabolite (acyl esters) accumulation are thought to play a role.

## Carnitine status in ischemic heart disease

Carnitine is a naturally occurring compound ($\beta$-hydroxy-$\gamma$-trimethyl-amino butyric acid) found in high concentrations in myocardial and skeletal muscle [13]. Carnitine serves several important physiological functions in myocardial muscle [17]: 1) facilitation of $\beta$-oxidation by transporting activated long-chain acyl-CoA esters into the mitochondria; 2) enhancement of the metabolic flux in the citric acid cycle by sparing free CoA, and decreasing the acetyl-CoA/CoA ratio, thereby stimulating the activity of PDH and oxidation of glucose; and 3) by preventing inhibition of adenylate translocase by long-chain acyl-CoA, which transports adenine nucleotides (ATP, ADP) across the inner mitochondrial membrane.

Plasma carnitine in patients with ischemic heart disease is often within normal ranges or slightly elevated, although esterified carnitine (specifically long-chain acylcarnitine) is significantly elevated. The exact relevance of plasma (carnitine) and its derivatives to intracellular concentrations and function is not known except for patients with congenital defects. Elevated levels of plasma carnitine may reflect an impaired uptake or increased efflux of carnitine from the ischemic tissue [18].

Reduced levels of free carnitine have been observed in myocardial tissue of laboratory animals with experimentally induced ischemia [3] and patients with heart disease [4, 5, 18–21]. Generally, the reduced (carnitine) reflects free carnitine, whereas total myocardial carnitine is often normal due to an increase in esterified (bound) carnitine. After 10 to 30 min of ischemia, myocardial free carnitine decreases 30 to 50%, whereas acetyl- and long-chain acylcarnitine concentration (esterified) rises [3]. However, during prolonged ischemia (>30 min) the (total carnitine) decreases as well [3]. Bohles et al. [5] found that myocardial (free carnitine) in patients with serious heart disease was only 60% of that of patients who had died from causes other than heart disease. Spagnoli et al. [4] also found significantly reduced levels of free carnitine in the myocardium of humans who had died following acute myocardial infarction. The greatest reduction in myocardial (free carnitine) was seen in the necrotic area of the myocardium, with significant reductions in the peri-infarctual zone as compared to levels seen in the area of healthy tissue [4]. Reduced levels of carnitine have also been observed in the myocardium of patients with chronic heart failure [19, 21], dilated cardiomyopathy [19] or CAD [20], myocardial ischemia [5] and mitral valve disease [18].

In theory a decrease in [carnitine] can only have a role in the loss of myocardial function, if reductions in tissue carnitine are related to the Km of the ACCT system. In other words are the reduced levels of carnitine observed in ischemic myocardium low enough to inhibit the ACCT system? Pande [22] showed that maximal oxidation of acyl-CoA in heart mitochondria of the rat was achieved at a (carnitine) of 1.5 mM which is similar to normal tissue levels (1.0 μmol/g wet weight). Therefore, a 50% reduction in tissue carnitine during ischemia would be expected to significantly reduce $\beta$-oxidation [22]. In patients with heart failure, myocardial free carnitine was significantly lower than for controls, 4.5 vs. 9.7 nmol/g noncollagenous protein, respectively [20]. A significant relationship existed between myocardial free carnitine and myocardial function, with a threshold at 6.5 nmol/g per noncollagenous protein, below which ejection fractions were markedly depressed.

Despite evidence supporting a reduction in tissue (carnitine) in patients with heart disease, more recent research shows that even in the presence of normal levels of carnitine a significant elevation of long-chain acylcarnitine occurs during ischemic episodes [23, 24]. As previously stated long-chain acylcarnitine accumulation contributes to the loss in myocardial function due to a loss in various cell membrane functions, and a decrease in conduction velocity with subsequent cell-to-cell uncoupling [23]. Prevention of tissue accumulation of long-chain acylcarnitines markedly delays the onset and rate of electrical cell-to-cell uncoupling [23, 24].

The significant relationship between the concentration of carnitine and its derivatives and myocardial function and a decrease in (free carnitine) during ischemia, forms the rationale to hypothesize that exogenous carnitine supplementation may improve myocardial function in patients with ischemic heart disease. In addition, therapeutic interventions aimed at minimizing long-chain acylcarnitine accumulation may prevent or delay the functional (electrical and mechanical) derangements that occur during ischemia. This improved function may subsequently result in greater external activity before being limited by ischemia-related symptoms (e.g. angina) and could even impact on long-term outcome.

## Proposed mechanism of action of carnitine in ischemic heart disease

The precise mechanism by which exogenous carnitine exerts its action on myocardial metabolism and function in patients with ischemic heart disease is currently not well understood. However, based on the above it is postulated that exogenous carnitine restores free carnitine levels in the myocardium, acts as a modulator of metabolism and is a scavenger of acyl-esters that accumulate during ischemia. A recent study reported that cytosolic carnitine accounted for approximately 93% of the total cellular carnitine in the mitochondria of heart muscle [10]. Following 60 min of ischemia, the [free carnitine] in the cytosol decreased, whereas [long-chain acylcarnitine] was elevated in both the mitochondria and cytosol. Treatment with L-carnitine (30 and 100 mg/kg IV) significantly inhibited the accumulation of mitochondrial long-chain acylcarnitine levels with a concomitant improvement in metabolism [10].

Therefore, the therapeutic potential of carnitine supplementation may be to reverse the various pathways, described in Figure 1, that result in myocardial dysfunction and cell injury or death. The availability of free carnitine will stimulate the ACCT system and $\beta$-oxidation, thereby, reducing intramitochondrial levels of acyl-CoA and increasing ATP production. The reduction of intramitochondrial acyl-esters will also unblock the adenine nucleotide translocase and activate the transport of ADP and ATP across the mitochondrial membrane [13, 14, 25]. Furthermore, metabolic flux through the citric acid cycle will be enhanced and the oxidative utilization of pyruvate increased due to an increase in the availability of intramitochondrial free CoA, and a reduction in the acetyl-CoA/CoA ratio [13, 14, 25].

An increase in the oxidative utilization of pyruvate may be of particular importance in maintaining [ATP] during periods of ischemia. Recent data indicate that the choice of substrate may influence cardiac muscle efficiency during low coronary blood flow [23]. In experimental models myocardial function and high-energy phosphates fall to a greater extent during low-flow states when the heart is infused with a solution containing only FFA. These findings suggest that the myocardium is more efficient in its utilization of $O_2$ under low-flow conditions when glucose is metabolized compared with FFA. Fatty acid metabolism yields less ATP per mole of $O_2$ consumed than glucose metabolism (Table 1), so with equivalent $O_2$ delivery and consumption more ATP will be generated. Although, in vivo lipid and glucose metabolism cannot be separated (they are interrelated), glucose must provide the framework by which acetyl-CoA, derived from $\beta$-oxidation, is metabolized in the citric acid cycle. Thus, treatments which facilitate both lipid and glucose metabolism will yield more energy for cardiac work during periods of low flow (Table 2). Since carnitine indirectly preserves or stimulates PDH activity (by maintaining a low acetyl-CoA/CoA ratio) more glucose is available for metabolism. In support of this hypothesis other compounds (e.g. dichloroacetate) that activate PDH have shown to be beneficial in ischemia [26, 27]. Clinically this could be translated to more external activity before ischemia-related, activity-limiting symptoms (angina, shortness of breath) occur.

*Table 1.* Fatty acid metabolism yields less ATP per oxygen consumed than glucose metabolism.

| Hexanoic acid ($C_6H_{12}O_2$) | Glucose ($C_6H_{12}O_6$) |
|---|---|
| $H_3C—CH_2—CH_2—CH_2—CH_2—COOH$ | 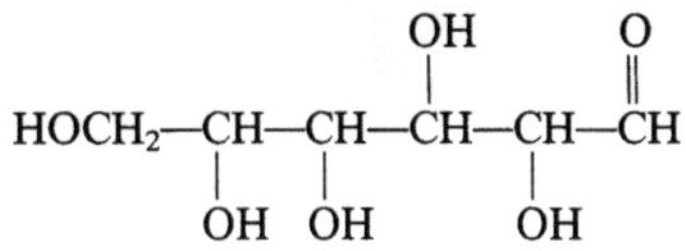 |
| Complete metabolism<br>$C_6H_{12}O_2 + 8O_2 \rightarrow 6CO_2 + 6H_2O$<br>Yield = 44 ATP | $C_6H_{12}O_6 + 6O_2 \rightarrow 6CO_2 + 6H_2O$<br>Yield = 38 ATP |
| Fuel efficiency<br>$ATP/O_2$ = 44 ATP/$8O_2$ = 5.5 ATP/$O_2$ | $ATP/O_2$ = 38 ATP/$6O_2$ = 6.3 ATP/$O_2$ |
| Oxygen demand to generate equivalent energy<br>44 ATP/5.5 ATP/$O_2$ = $8O_2$ | 44 ATP/6.3 ATP/$O_2$ = $7O_2$ |

*Table 2.* Carnitine, a novel mechanism of action in ischemia.

Preservation of lipid metabolism and lack of inhibition of glucose oxidation

- Improved energy production at equivalent oxygen delivery
- Potential generation of equivalent energy at lower oxygen demand without depression of blood pressure, heart rate or cardiac function
- Prevention or reduction of ischemia in cardiac or skeletal muscle
- Prolongation of exercise times in patients with angina or intermittent claudication

## Immediate and short-term (<30 days) effects of carnitine supplementation in patients with ischemic heart disease

Several studies have evaluated the immediate effects of carnitine on cardiodynamics in patients with ischemic heart disease (Table 3). In single-dose intravenous (IV) studies both L-carnitine and propionyl-L-carnitine are consistently associated with a favorable effect on cardiac function. However the magnitude of the response depends on such factors as: Type (L-carnitine, DL-carnitine or propionyl-L-carnitine), form (IV or oral) and dose of carnitine administered, infusion time, type and severity of disease, and methods used to stress and evaluate the heart.

Studies using pacing-induced tachycardia stress in patients with angina report a greater tolerance of tachycardia as well as significant increases in maximum heart rate (HR max) and rate-pressure product (RPP) after IV administration of either 20 or 40 mg/kg of DL-carnitine or L-carnitine [28, 29], or bolus infusion of 140 mg/kg of L-carnitine [30]. Left ventricular end diastolic pressure and magnitude of ST-segment depression during tachycardia were also reduced in carnitine-treated patients. Further evidence suggests

*Table 3.* Immediate and short-term (<30 days) effect of carnitine in patients with ischemic heart disease.

| Author | Ref. | Number of patients | Form, dose, duration | Cardiac function/metabolism | Exercise or pacing capacity/tolerance | ECG findings | Clinical findings |
|---|---|---|---|---|---|---|---|
| Thomsen 1979 | [28] | 11 | DLC, IV<br>20 or 30 mg/kg | ⇑HRmax, RPPmax<br>⇑FFA uptake, lactate extraction | ⇑Tolerance to tachycardia | | |
| Ferrari 1984 | [29] | 11 | LC, IV<br>40 mg/kg | ⇓LVEDP, ⇑FFA uptake lactate extraction | ⇑Tolerance to tachycardia | | |
| Reforzo 1986 | [30] | 19 | LC, bolus<br>140 mg/kg | ⇑Lactate extraction or ⇓production | ⇑Tolerance to tachycardia | ⇓ST depression | |
| Kosolcharoen 1981 | [31] | 18 | DLC, IV<br>40 mg/kg | Slower rise in HR, RPP | ⇑Time to angina | ⇓ST depression | Improved clinical status |
| Sciveres 1984 | [32] | 40 | LC, oral<br>2000 mg/d<br>2 weeks | | ⇑Increased time to angina | ⇓ST depression | ⇓NTG consumption |
| Cherchi 1985 | [33] | 44 | LC, oral<br>2000 mg/d<br>4 weeks | | ⇑Workload<br>⇑Time and workload to angina | ⇓ST depression at max and common workload | ⇓Anginal episodes |
| Canale 1988 | [34] | 16 | LC, oral<br>3000 mg/d<br>4 weeks | Improved LV function (echo) | ⇑Workload | ⇓ST depression | ⇓Anginal episodes |
| Cherchi 1990 | [35] | 18 | PLC, oral<br>1500 mg/d<br>4 weeks | | ⇑Exercise time | ⇓ST depression at max<br>⇑Time to 1 mm ST | |

*Table 3. Continued.*

| Author | Ref. | Number of patients | Form, dose, duration | Cardiac function/metabolism | Exercise or pacing capacity/tolerance | ECG findings | Clinical findings |
|---|---|---|---|---|---|---|---|
| Fujiwara 1991 | [36] | 30 | LC, IV<br>60 mg/kg<br>over 30 min | ⇑Coronary blood flow with exercise | | | |
| Lagioia 1992 | [37] | 12 | PLC<br>500 mg tid<br>15 days | | ⇑Total work, time and time to ischemia | ⇓ST segment depression at max common workload | |
| Bartels 1992 | [38] | PLC, IV<br>15 mg/kg<br>5 min<br>infusion | ⇑Contractility relaxation and pump function<br>Lactate uptake,<br>⇑AcylCoA/Free carnitine | | | | |
| Palazzuoli 1993 | [39] | LC, oral<br>6000 mg/d<br>2 weeks | | | ⇓PVCs | Improved clinical status | |

DLC = DL-carnitine, FFA = free fatty acids, HR = heart rate, IV = intraveneous, LC = L-carnitine, LV = left ventricular, LVEDP = left ventricular end-diastolic pressure, mg/d = milligram per day, NTC = nitroglycerin, PLC = propionyl-L-carntine, PVC = premature ventricular contraction, RPP = rate-pressure product, ST = ECG ST-segment, tid = 3× a day.

improvement in myocardial metabolism as indicated by decreased production or increased extraction of lactate, and increased uptake of FFA from arterial blood [28, 29].

Carnitine infusion prior to exercise stress is associated with improvements in exercise tolerance in patients with angina. Kosolcharoen et al. [31] used a 40 mg/kg infusate of DL-carnitine 30 min prior to a maximal exercise bout. Carnitine administration resulted in lower submaximal exercise heart rates and RPP, significantly less ST segment depression during exercise and recovery, and increased exercise time prior to the onset of angina [31].

More recent studies also report the beneficial effects of carnitine infusion in patients with CAD [36–38]. During resting conditions, IV infusions of propionyl-L-carnitine (15 mg/kg over 5 min) did not appear to affect coronary hemodynamics, but lactate uptake increased and myocardial contractility, relaxation and cardiac output improved significantly [38]. Fujiwara et al. [36] further assessed the protective effect of L-carnitine infusion in 30 patients with angina and ischemic changes as determined by electrocardiography (ECG) during exercise. Prior to exercise, 60 mg/kg of L-carnitine or placebo was infused over a period of 30 min. The effect of carnitine on resting cardiodynamics was minimal. During exercise subjects receiving carnitine had significantly higher coronary blood flow, arterio-venous oxygen difference and myocardial oxygen consumption ($MVO_2$) in comparison to untreated patients [36]. In addition, whereas determinants of myocardial function ($MVO_2$/left ventricular work index) decreased with exercise in the controls, they were unaffected in subjects receiving carnitine.

When carnitine is administered to patients with ischemic heart disease, for up to 30 days (1,500 to 6,000 mg/day, oral), exercise capacity, time to exhaustion, maximal workload, and time to onset of ischemia and angina all increase [32–35, 39], while the degree of ST segment depression (during submaximal exercise), premature ventricular contractions and episodes of angina pectoris decrease [32–35, 39, 40]. Typically, the magnitude of change for exercise time and workload ranges from 5 to 10% increase. Results from a short-term controlled study in patients with stable angina reports that 23% of patients were free of angina following 4 weeks of treatment with L-carnitine (2,000 mg/day) [33].

Few studies have evaluated myocardial mechanical function following short-term carnitine supplementation. Canale and co-workers [34] used M-mode echocardiography to evaluate myocardial function and noted improvements in a number of ventricular function parameters in patients with exercise-induced angina after only 30 days of treatment with 2,000 mg/day of L-carnitine.

Thus, there seems to be strong evidence to suggest that carnitine supplementation (immediate and short-term [<30 days]) may be beneficial in patients with ischemic heart disease. However, the mechanisms involved in the improved clinical status are presently not clear. Future studies designed to further identify the possible mechanisms involved as well as determine the optimal dosing strategies are needed.

## Long-term effect of carnitine therapy in patients with ischemic heart disease

Eight clinical trials have evaluated the effect of prolonged carnitine therapy (>30 days) in patients with CAD (Table 4) [40–47]. Results from such trials are consistent with the findings from short-term studies. Preliminary controlled studies, using small sample sizes, indicate a reduction in electrocardiographic manifestations of ischemia, an increase in maximal workload and total exercise time, an increase in the time to the onset of ischemia, and a decrease in anginal episodes [40, 41, 43]. L-carnitine (900 to 2,000 mg/day, oral) administered in combination with 40 mg/day of isosorbide dinitrate significantly decreased the frequency of angina attacks (95%) and consumption of nitroglycerin (96%) after 8 weeks of treatment [40]. Kawikawa et al. [41] administered L-carnitine (900 mg/day, oral) for 12 weeks in patients with stable effort angina and reported significant increases in maximal exercise time (from 11.4 to 12.8 min) and exercise time to 1 mm ST segment depression (from 6.4 to 8.8 min). In both studies, the greatest improvement was seen within the first 2 to 4 weeks of therapy.

In a large multicenter study conducted over a period of 12 months L-carnitine supplementation (2,000 mg/day, oral) resulted in significant improvements in clinical status in more than 1000 patients with stable angina pectoris [42]. The frequency of angina episodes decreased by 60%, approximately 70% of patients discontinued use of nitroglycerin, and about one-third reduced consumption of other cardioactive drugs (nifedipine, $\beta$-blockers) [42]. Again, the greatest improvement in symptoms was noted in the first 6 months of the study, with only minor additional improvements at 12 months.

Cacciatore et al. [44] reported improvements similar to Fernandez et al. [45], following 6 months of L-carnitine supplementation (2,000 mg/day, oral) in 200 patients with exercise-induced angina. In addition, they noted a decrease in premature ventricular contractions, a decrease in consumption of cardioactive drugs and an improved lipid profile. Recent clinical trials involving different dosages, patient populations and durations of therapy, continue to support the above-mentioned findings [45–47]. Perhaps the most encouraging result concerning the pharmacologic effect of L-carnitine comes from Davini et al. [47]. In this study, 160 patients who were post-myocardial infarction were randomized to standard care or standard care plus L-carnitine (4,000 mg/day, oral). Following 12 months of treatment, mortality was sig-

*Table 4.* Long-term (<30 days) effect of carnitine in patients with ischemic heart disease.

| Author | Ref. | Number of patients | Form, dose, duration | Cardiac function/metabolism | Exercise or pacing capacity/tolerance | ECG findings | Clinical findings |
|---|---|---|---|---|---|---|---|
| Garzya 1980 | [40] | 20 | LC, oral<br>50 mg/kg<br>8 weeks | | | | ⇓Anginal episodes<br>⇓NTG consumption |
| Kawikawa 1984 | [41] | 12 | LC, oral<br>900 mg/d<br>12 weeks | | ⇑Exercise time and time to onset ST depression | ⇓ST depression at same workload | |
| Fernandez 1985 | [42] | 1000 | LC, oral<br>2000 mg/d<br>52 weeks | | | | ⇓Anginal episodes<br>⇓NTG consumption<br>⇓CV drugs |
| Ghidini 1988 | [43] | 21 | LC, oral<br>2000 mg/d<br>45 days | | | ⇓Ischemia<br>⇓PVCs | |
| Cacciatore 1991 | [44] | 200 | LC, oral<br>2000 mg/d<br>26 weeks | ⇑HRmax, SPBmax<br>RPPmax | ⇑Time | ⇓ST depression at max<br>⇓PVCs | Improved NYHA class<br>⇓CV drugs<br>Improved lipids |

*Table 4. Continued.*

| Author | Ref. | Number of patients | Form, dose, duration | Cardiac function/metabolism | Exercise or pacing capacity/tolerance | ECG findings | Clinical findings |
|---|---|---|---|---|---|---|---|
| Fernandez 1992 | [45] | LC, oral<br>2000 mg/d<br>52 weeks | | ⇑Physical work capacity | ⇓PVCs | ⇓Anginal episodes<br>⇓NTG consumption<br>⇑QOL | |
| Kobayashi 1992 | [46] | 9 | LC, oral<br>12 weeks | | ⇑Tolerance | | Improved NYHA class<br>⇓NTG consumption |
| Davini 1992 | [47] | 160 | LC, oral<br>4000 mg/d<br>52 weeks | ⇓HR, BP | | ⇓Dysrhythmias | ⇓Anginal episodes<br>Improved clinical status<br>Improved lipids<br>⇓Mortality |

CV = cardiovascular, NYHA class = New York Heart Association classification, QOL = quality of life, SBP = systolic blood pressure. For other abbreviations, see Table 3.

nificantly lower in the carnitine treated patients (1.2%) versus those receiving standard care (12.5%).

In conclusion, there is evidence to suggest that long-term (up to 12 months) L-carnitine supplementation in patients with ischemic heart disease may have beneficial effects on clinical outcomes. Improvements in symptoms, exercise tolerance as well as reductions in mortality have been suggested. Clearly larger and longer controlled clinical trials are needed before definitive conclusions can be made. Since the greatest gains in clinical status are typically noted during the first few months of L-carnitine therapy, future trials should evaluate the need to continue to prescribe L-carnitine at high doses (>2,000 mg/day), or if supplementation using a lower dose is sufficient in maintaining the improved clinical status. In addition, long-term clinical trials (>12 months) should further assess the role of L-carnitine supplementation on cardiac function and adverse outcomes (e.g. death, myocardial infarction and need for hospitalization) in patients with ischemic heart disease.

## Effect of carnitine on skeletal muscle function in patients ischemic heart disease

There is increasing evidence that skeletal muscle undergoes considerable remodeling in patients with chronic and severe cardiovascular disease [48–54]. Particularly, in patients with chronic heart failure due to CAD, a reduction in oxidative capacity is due to ultrastructural abnormalities, a marked decline in mitochondrial volume and density [53, 54], and a decrease in skeletal muscle enzyme concentration and activity ([succinate dehydrogenase], [citrate synthetase], [cytochrome oxidase] and [$\beta$-hydroxyacyl-CoA dehydrogenase]) [49–51, 53, 54]. Histological changes include significant skeletal muscle atrophy with a pronounced decline in Type I muscle fibers [50, 51, 53, 54], whereas an accumulation of intracellular lipids has also been reported. Furthermore, reduced [free carnitine] have been observed in skeletal muscle of patients with heart disease [48–50], and peripheral vascular disease [52].

The mechanisms responsible for alterations in skeletal muscle metabolism in these patients are presently not completely understood. However, in patients with chronic heart failure, a number of mechanisms is involved in limiting blood flow to exercising muscle and contributes to the reduced exercise tolerance and capacity. These factors include: neurohumoral compensation (increased sympathetic nervous activity and activation of renin-angiotensin-aldosterone), alterations in vascular vasodilatory capacity (vascular stiffness and structural vascular changes), and chronic deconditioning [55]. Chronic deconditioning may be a key factor in the alterations in skeletal muscle metabolism of patients with heart disease, because many of the observed changes mimic those of detraining, and exercise training can improve exercise capacity and tolerance [56–58].

*Table 5.* How carnitine may modulate cardiac metabolism in patients with angina.

Summary:

- Prevents inhibition of glucose metabolism
- Preserves fatty acid metabolism
- Removes toxic long-chain fatty acid metabolites

However, a decrease in carnitine status must now also be considered as a possible factor in the impaired skeletal muscle oxidative metabolism. If a deficiency in skeletal muscle carnitine exists, it would follow that this could have a similar detrimental impact on oxidative metabolism and ATP production as observed in the myocardium. Furthermore, during submaximal exercise (55% of $VO_{2\,max}$) there may be a progressive increase in short- and long-chain acylcarnitines in skeletal muscle that could subsequently lead to a secondary depletion of free muscle carnitine and further impair oxidative metabolism [59].

Preliminary data [60] suggests that skeletal muscle carnitine deficiency in patients with chronic heart failure may, indeed, be significant and exogenous carnitine (propionyl-L-carnitine) may improve skeletal muscle metabolism by increasing the flux of pyruvate into the Krebs cycle and decreasing the production of lactate.

Thus, it may be hypothesized that the action of L-carnitine in patients with ischemic heart disease may both have a central (myocardial function) and peripheral (skeletal muscle) component. Perhaps an improvement in oxidative metabolism of skeletal muscle may indirectly improve cardiac function as a result of a lower myocardial oxygen demand at submaximal exercise workloads. If so, such a "dual" mechanism of action may well contribute to the improved functional status of the patient. Since exercise training increases skeletal muscle oxidative capacity in patients with heart disease [56–58], future studies should address the question whether exercise training alone can improve the carnitine status in patients with heart disease or if L-carnitine supplementation can further augment the exercise training response.

## Conclusions

A reduction in myocardial free carnitine could partially explain the rise in long-chain acyl CoA esters found in the ischemic heart. Reductions in free carnitine have been observed in myocardial tissue of laboratory animals with experimentally induced ischemia and patients with heart disease. Thus, exogenous carnitine supplementation could be a rational approach to improve cardiovascular function in patients with ischemic heart disease (Table 5).

There is increasing evidence that immediate, short- and long-term

L-carnitine supplementation in patients with ischemic heart disease may be beneficial and has the potential to result in significant improvement in exercise tolerance and capacity, as well as reductions in angina, use of cardioactive drugs, electrocardiographic manifestations of ischemia, dysrhythmias and perhaps even mortality.

L-carnitine may also have therapeutic potential in skeletal muscle, since there is evidence for muscle carnitine deficiency in patients with chronic heart failure. This potential "dual" effect on cardiac as well as skeletal muscle energetics could make carnitine an attractive alternative in the management of patient with ischemic heart disease.

Thus, further clinical evaluation of the therapeutic potential of L-carnitine in patients with CAD is warranted. Future studies should aim to identify how myocardial and skeletal muscle function is affected by carnitine supplementation and determine the optimal dosing strategies and assess effects on adverse outcomes.

## References

1. Reimer KA, Jennings RB. Myocardial ischemia, hypoxia and infarction. In: Fozzard HA, Jennings RB, Haber E, Katz AM, Morgan HE, editors. The heart and cardiovascular system. New York: Raven Press, 1986: 1163–99.
2. Pepine CJ, Wiener L. Relationship of anginal symptoms to lung mechanics during myocardial ischemia. Circulation 1972; 46: 863–9.
3. Shug AL, Thomsen JH, Folts JD et al. Changes in tissue levels of carnitine and other metabolites during ischemia and anoxia. Arch Biochem Biophys 1978; 187: 25–33.
4. Spagnoli LG, Corsi M, Villaschi S, Palmieri G, Maccari F. Myocardial carnitine deficiency in acute myocardial infarction. Lancet 1982; 1: 1419–20.
5. Bohles H, Noppeney T, Akcetin Z et al. The effect of pre-operative L-carnitine supplementation on myocardial metabolism during aorto-coronary-bypass surgery. Z Kardiol 1987; 76(Suppl 5): 14–8.
6. Camici P, Marraccini P, Lorenzoni R et al. Metabolic markers of stress-induced myocardial ischemia. Circulation 1991; 83(Suppl III): III8–13.
7. Neely JR, Bowman RH, Morgan HE. Effects of ventricular pressure development and palmitate on glucose transport. Am J Physiol 1969; 216: 804–11.
8. Dhalla NS, Elimban V, Rupp H. Paradoxical role of lipid metabolism in heart function and dysfunction. Mol Cell Biochem 1992; 116: 3–9.
9. Shrago E, Shug A, Sul H, Bittar N, Folts JD. Control of energy production in myocardial ischemia. Circ Res 1976; 38(Suppl 1): I75–9.
10. Fujisawa S, Kobayashi A, Hironaka Y, Yamazaki N. Effect of L-carnitine on the cellular distribution of carnitine and its acyl derivatives in the ischemic heart. Jpn Heart J 1992; 33: 693–705.
11. Katz AM, Messineo FC. Lipid-membrane interactions and the pathogenesis of ischemic damage in the myocardium. Circ Res 1981; 48: 1–16.
12. Vyska K, Machulla HJ, Stremmel W et al. Regional myocardial free fatty acid extraction in normal and ischemic myocardium. Circulation 1988; 78: 1218–33.
13. Opie LE. Role of carnitine in fatty acid metabolism of normal and ischemic metabolism. Am Heart J 1979; 97: 375–88.
14. Siliprandi N, Di Lisa F, Toninello A. Biochemical derangements in ischemic myocardium: the role of carnitine. G Ital Cardiol 1984; 14: 804–8.

15. Pepine CJ. The therapeutic potential of carnitine in cardiovascular disorders. Clin Ther 1991; 13: 2–21.
16. Yamada KA, McHowat J, Yan GX et al. Cellular uncoupling induced by accumulation of long-chain acylcarnitine during ischemia. Circ Res 1994; 74: 83–95.
17. Cerretelli P, Marconi C. L-Carnitine supplementation in humans. The effects on physical performance. Int J Sports Med 1990; 11: 1–14.
18. Bressler R, Gay R, Copeland J, Bahl J, Bedotto J, Goldman S. Chronic inhibition of fatty acid oxidation: new model of diastolic dysfunction. Life Sci 1989; 44: 1897–906.
19. Pierpont ME, Judd D, Goldenberg JF, Ring WS, Olivari MT, Pierpont GL. Myocardial carnitine in end-stage congestive heart failure. Am J Cardiol 1989; 64: 56–60.
20. Regitz V, Shug AL, Fleck E. Defective myocardial carnitine metabolism in congestive heart failure secondary to dilated cardiomyopathy and to coronary, hypertensive and valvular heart diseases. Am J Cardiol 1990; 65: 755–60.
21. Suzuki Y, Masumura Y, Kobayashi A, Yamazaki N, Harada Y, Osawa M. Myocardial carnitine deficiency in chronic heart failure. Lancet 1982; 1: 116 (Lett to the Ed).
22. Pande SV. On rate-controlling factors of long chain fatty acid oxidation. J Biol Chem 1971; 246: 5384–90.
23. Burkhoff D, Weiss RG, Schulman SP, Kalil Filho R, Wannenburg T, Gerstenblith G. Influence of metabolic substrate on rat heart, function and metabolism at different coronary flows. Am J Physiol 1991; 261: H741–50.
24. Corr PB, Creer MH, Yamada KA, Saffitz JE, Sobel BE. Prophylaxis of early ventricular fibrillation by inhibition of acylcarnitine accumulation. J Clin Invest 1989; 83: 927–36.
25. Goa KL, Brogden RN. L-Carnitine. A preliminary review of its pharmacokinetics, and its therapeutic use in ischaemic cardiac disease and primary and secondary carnitine deficiencies in relationship to its role in fatty acid metabolism. Drugs 1987; 34: 1–24.
26. Wargovich TJ, MacDonald RG, Hill JA, Feldman RL, Stacpoole PW, Pepine CJ. Myocardial metabolic and hemodynamic effects of dichloroacetate in coronary artery disease. Am J Cardiol 1988; 61: 65–70.
27. Pepine CJ. A new therapeutic approach to myocardial ischemia. J Myocard Isch 1993; 5: 9–10 (Editorial).
28. Thomsen JH, Shug AL, Yap VU et al. Improved pacing tolerance of the ischemic human myocardium after administration of carnitine. Am J Cardiol 1979; 43: 300–6.
29. Ferrari R, Cucchini F, Visioli O. The metabolic effect of L-carnitine in angina pectoris. Int J Cardiol 1984; 5: 213–6.
30. Reforzo G, De Andreis Bessone PL, Rebaudo F, Tibaldi M. Effects of high doses of L-carnitine on myocardial lactate balance during pacing-induced ischemia in aging subjects. Curr Ther Res 1986; 40: 374–83.
31. Kosolcharoen P, Nappi J, Peduzzi P et al. Improved exercise tolerance after administration of carnitine. Curr Ther Res 1981; 30: 753–64.
32. Sciveres G. Esperienza con L-carnitina nel trattamento della angina stabile da sforzo. Arch Med Interna 1984; 36: 61–6.
33. Cherchi A, Lai C, Angelino F et al. Effects of L-carnitine in exercise tolerance in chronic stable angina: a multicenter, double-blind, randomized, placebo controlled crossover study. Int J Clin Pharmacol Ther Toxicol 1985; 23: 569–72.
34. Canale C, Terrachini V, Biagini A et al. Bicycle ergometer and echocardiographic study in healthy subjects and patients with angina pectoris after administration of L-carnitine: Semiautomatic computerized analysis of M-mode tracing. Int J Clin Pharmacol Ther Toxicol 1988; 26: 221–4.
35. Cherchi A, Lai C, Onnis E et al. Propionyl carnitine in stable effort angina. Cardiovasc Drugs Ther 1990; 4: 481–6.
36. Fujiwara M, Nakano T, Tamoto S et al. Effect of L-carnitine in patients with ischemic heart disease. J Cardiol 1991; 21: 493–504 (in Japanese).
37. Lagioia R, Scrutinio D, Mangini S et al. Propionyl carnitine in stable effort angina. Cardiovasc Drugs Ther 1990; 4: 481–6.

38. Bartels GL, Remme WJ, Pillay M et al. Acute improvement of cardiac function with intravenous L-propionylcarnitine in humans. J Cardiovasc Pharmacol 1992; 20: 157–64.
39. Palazzuoli V, Mondillo S, Faglia S, D'Aprile N, Comporeale A, Gennari C. Valutazione dell'attività antiaritmica della L-carnitina e del propafenone nella cardiopatia ischemica. Clin Ter 1993; 142: 155–9.
40. Garzya G, Amico RM. Comparitive study on the activity of racemic and laevorotatory carnitine in stable angina pectoris. Int J Tissue React 1980; 2: 175–80.
41. Kawikawa T, Suzuki Y, Kobayashi A et al. Effects of L-carnitine on exercise tolerance in patients with stable angina pectoris. Jpn Heart J 1984; 25: 587–97.
42. Fernandez C, La Menza B, Pola P. Trials clinici difase IV e di farmacovigilanza-nuove ipotesi metodologiche-il terzo trial Ance. J Am Med Assoc (Italian edition) 1985; 2: 9–14.
43. Ghidini O, Azzurro M, Vita G, Sartori G. Evaluation of the therapeutic efficacy of L-carnitine in congestive heart failure. Int J Clin Pharmacol Ther Toxicol 1988; 26: 217–20.
44. Cacciatore L, Cerio R, Ciarimboli M et al. The therapeutic effect of L-carnitine in patients with exercise-induced stable angina: a controlled study. Drugs Exp Clin Res 1991; 17: 225–35.
45. Fernandez C, Proto C. La L-carnitina nel trattamento dell'ischemia miocardica cronico. Analisi dei risultati di tre studi multicentrici e rassegna bibliografica. Clin Ter 1992; 140: 353–77.
46. Kobayashi A, Masumura Y, Yamazaki N. L-Carnitine treatment for congestive heart failure – experimental and clinical study. Jpn Circ J 1992; 56: 86–94.
47. Davini P, Bigalli A, Lamanna F, Boem A. Controlled study on L-carnitine therapeutic efficacy in post-infarction. Drugs Exp Clin Rev 1992; 18: 355–65.
48. Dunnigan A, Pierpont ME, Smith SA, Biemingstall G, Benditt DG, Benson DW. Cardiac and skeletal myopathy associated with cardiac dysrhythmias. Am J Cardiol 1984; 53: 731–7.
49. Dunnigan A, Staley NA, Smith SA et al. Cardiac and skeletal muscle abnormalities in cardiomyopathy: comparison of patients with ventricular tachycardia or congestive heart failure. J Am Coll Cardiol 1987; 10: 608–18.
50. Caforio ALP, Rossi B, Risaliti R et al. Type I fiber abnormalities in skeletal muscle of patients with hypertrophic and dilated cardiomyopathy: evidence of subclinical myogenic myopathy. J Am Coll Cardiol 1989; 14: 1464–73.
51. Sullivan MJ, Green HJ, Cobb FR. Skeletal muscle biochemistry and histology in ambulatory patients with long-term heart failure. Circulation 1990; 81: 518–27.
52. Brevetti G, Angelini C, Rosa M et al. Muscle carnitine deficiency in patients with severe peripheral vascular disease. Circulation 1991; 84: 1490–5.
53. Mancini DM, Coyle P, Coggan A, et al. Contribution of intrinsic skeletal muscle changes to $^{31}$P NMR skeletal muscle metabolic abnormalities in patients with chronic heart failure. Circulation 1989; 80: 1338–46.
54. Drexler H, Riede U, Munzel T, Konig H, Funke E, Just H. Alterations of skeletal muscle in chronic heart failure. Circulation 1992; 85: 1751–9.
55. Drexler H. Reduced exercise tolerance in chronic heart failure and its relationship to neurohumoral factors. Eur Heart J 1991; 12(Suppl C): 21–8.
56. Clausen JP. Circulatory adjustments to dynamic exercise and effect of physical training in normal subjects and in patients with coronary artery disease. Progr Cardiovasc Dis 1976; 18: 459–95.
57. Hagberg JM. Central and peripheral adaptations to training in patients with coronary artery disease. In: Saltin B, editor. Biochemistry of exercise, VI. Champaign, IL: Human Kinetics Publishers, 1986: 267–77.
58. Sullivan MJ, Higginbotham MB, Cobb FR. Exercise training in patients with severe left ventricular dysfunction. Hemodynamic and metabolic effects. Circulation 1988; 78: 506–15.
59. Lennon DL, Stratman FW, Shrago E et al. Effects of acute moderate-intensity exercise on carnitine metabolism in men and women. J Appl Physiol 1983; 55: 489–95.

60. Ferrari R, Cargnoni A, deGiuli F, Pasini E, Anand I, Visioli O. Propionyl-L-carnitine improves skeletal muscle metabolism and exercise capacity of patients with congestive heart failure. Circulation 1993; 88(4, Part 2): I414 (Abstr).

*Corresponding Author*: Dr Carl J. Pepine, Division of Cardiology and Center for Exercise Science, University of Florida, Box 100277, Gainesville, FL 32610, USA

# 17. Carnitine and myocardial infarction

PAOLO RIZZON, MATTEO DI BIASE, GIUSEPPINA BIASCO
and MARIA VITTORIA PITZALIS

"These data suggest that L-carnitine reduces the extent of the necrotic area during acute myocardial infarction; this effect seems linked to an improvement of mitochondrial function in ischemic cells not yet irreversibly affected."

## Introduction

Experimental studies have demonstrated that some drugs can interact with the myocardial metabolism deranged by prolonged ischemia, particularly at the level of the stunned and hibernating cells. On the basis of these observations the possibility of performing a "metabolic intervention" for the treatment of acute and chronic myocardial infarction was suggested.

A drug which interacts with the cardiac metabolism is L-carnitine, a water soluble amino acid which plays an important role in fatty acid metabolism. This compound is extracted from the blood by the myocardium [1].

In this review the data available on a therapeutic action of L-carnitine administered to patients during the acute phase of myocardial infarction and in the post-infarction period are reported.

## Loss of carnitine during acute ischemia

A loss of free carnitine from the myocardium during acute ischemia, which creates a secondary deficiency of this compound, has been observed in experimental studies and in humans. Shug et al. [2] demonstrated a loss of free carnitine of between 20% and 60% in perfused anoxic rat hearts while Suzuki et al. [3] observed a loss of 40% of tissue carnitine in dog hearts after 25 min ligation of the descending coronary artery. This secondary deficit of carnitine has also been observed in humans. Spagnoli et al. [4] observed a carnitine deficiency in infarcted areas at necroscopy. The same phenomenon has been observed by Bartels & Remme [5] by evaluating carnitine concentra-

*J.W. de Jong and R. Ferrari (eds): The carnitine system*, 245–249.

tions in the coronary sinus blood of ischemic patients submitted to atrial pacing. Rizzon et al. [6] showed a progressive increase in serum free carnitine concentrations during acute myocardial infarction in a group of 28 patients, reaching a statistically significant difference when compared to a control group, after 16 h from the onset of symptoms with a peak concentration of 55 ± 17 μM (control 38 ± 13 μM) after 26 h. In the same group of patients the loss of carnitine was also confirmed by the abnormal increase in urinary excretion of free carnitine to 547 ± 631 μM (control group 126 ± 51 μM). From these data it becomes evident that during acute ischemia there is a loss of carnitine creating a secondary deficit of this compound.

### *The role of carnitine supplement during myocardial ischemia and infarction*

#### *Metabolic effects*

During acute myocardial ischemia the rate of ß-oxidation is reduced, leading to a decrease in the levels of acetyl-CoA and an increase in acyl-CoA. Free carnitine reacts with acyl-CoA resulting in an increase in long-chain acylcarnitine. The role of free carnitine, consisting primarily in the transport of long-chain fatty acids from the cytoplasm to the mitochondrion, is strongly depressed due to reduced availability of the compound, firstly because of the formation of acylcarnitine esters and secondly because of the loss of free carnitine from the cells.

In experimental studies the supplement of free carnitine during ischemia and anoxia gives rise to different changes: increased carnitine levels in infarcted and non-infarcted areas [3, 7]; decreased levels of long-chain acyl-CoA and long-chain acylcarnitine in ischemic and infarcted areas [2, 3, 7]; increase in ATP levels in infarcted areas [7]. It seems that the supplement of free carnitine enters the cells and exerts two main activities: 1) it improves the production of ATP by restoring fatty acid transport into the mitochondrion [3–7]; 2) it reduces the concentration of acyl-esters by transporting them outside the cells [3–7].

Based on these hypotheses, in 1988 Rizzon et al. [6] carried out a double-blind parallel and placebo controlled study in 56 patients with acute myocardial infarction admitted to the coronary care unit between 3 and 12 h from the onset of pain. Allocation of treatment was random after stratification based on the time from pain onset and infarction site. The treated group received intravenous L-carnitine at a dose of 100 mg/kg body weight every 12 h for 36 h.

In this group of patients a significant increase ($p < 0.01$) in serum and urine levels of short- and long-chain acylcarnitine esters was observed compared to the control group. This increase resulted from esterification of supplemented carnitine with acyl-CoA to form acylcarnitine and allowed the release of acyl compound into the blood and consequently into the urine. By this mechanism carnitine prevents negative effects of excessive acyl compounds and exerts a detoxifying effect.

*Antiarrhythmic effects*

Experimental studies have shown that an increased concentration of free fatty acids evokes arrhythmias, in particular a reduction of the ventricular fibrillation threshold, ventricular arrhythmias and conduction disturbances. Many mechanisms have been suggested to explain this phenomenon. One is a decreased mitochondrial $Ca^{2+}$ binding activity which increases the intracellular $Ca^{2+}$ concentrations [8]. Another is that fatty acids shorten the action potential duration during ischemia and interfere with glycolytically-derived ATP in the cytoplasm [9]. A further proposed mechanism is a biotoxic detergent effect on membranes enhanced by a high molar ratio of acid/albumin which leads to a cation loss [10].

Oliver et al. [11] reported that, in humans, high plasma levels of free fatty acids could be a predictive index of the vulnerability of patients with myocardial infarction to lethal arrhythmias. In the aforementioned study Rizzon et al. demonstrated some antiarrhythmic effects of L-carnitine during the acute phase of myocardial infarction [6]. In the treated group, during the second day of treatment there was a statistically significant reduction in:

- the mean number of premature ventricular beats per hour;
- hours with multiform premature ventricular beats;
- hours with couplets;
- hours with non-sustained ventricular tachycardias; and
- the total number of non-sustained ventricular tachycardias.

A statistically significant increase in patients with less than 10 premature ventricular beats per hour was observed.

Similar results, which confirm an antiarrhythmic effect of L-carnitine in man, were obtained by Martina et al. [12] in 12 patients with acute myocardial infarction treated with intravenous L-carnitine.

These effects could be explained by the detoxification effect of exogenous carnitine in ischemic myocardium exerted by removing excess fatty acids from the cells and transporting them into the blood and the urine.

*Effects on necrotic area size*

Experimental studies have shown that supplements of carnitine can reduce ST-segment elevation during ischemia. Two clinical studies, by Rebuzzi et al. [13] and Chiariello et al. [14], have been performed to evaluate the effects of L-carnitine in reducing the extent of the necrotic area.

Rebuzzi et al. [13], using creatine kinase (CK) and creatine kinase MB (MB-CK) plasma level measurements, showed a significantly reduced release in MB-CK and a minor CK maximum value, compared to placebo, in 12 patients with acute myocardial infarction treated with 40 mg/kg/die L-carnitine.

Chiariello et al. [14], using an electrocardiographic method to evaluate the extension of myocardial infarction, evaluated 177 patients with acute myocardial infarction treated with L-carnitine 9 g/die for 2 days. Compared to placebo, the treated group showed a statistically significant reduction in the decline of R-wave which was 26% less if the measurement was performed only in patients with acute anterior myocardial infarction.

These data suggest that L-carnitine reduces the extent of the necrotic area during acute myocardial infarction; this effect seems linked to an improvement of mitochondrial function in ischemic cells not yet irreversibly affected.

*Effects on myocardial contractility*

In experimental studies Neely et al. [15] and Liedtke & Nellis [16] observed that the administration of carnitine to ischemic hearts with excess free fatty acids induced several hemodynamic improvements compared to the untreated groups. In particular, left ventricular pressure (+25%), mean aortic pressure (+44%), left ventricular max dp/dt (+45%), regional left ventricular shortening (+24%) and left ventricular work (+72%) were increased at comparable levels of myocardial oxygen consumption.

On the basis of these data it has been suggested that carnitine could exert a positive effect on myocardial contractility in patients with acute myocardial infarction.

In an attempt to address this question the CEDIM study was undertaken [17, 18]. This is a double-blind, controlled, multicenter clinical trial, performed by 36 Italian universities and community hospital centers to assess the effect of L-carnitine on left ventricular function (left ventricular volumes and ejection fraction) after acute myocardial infarction. This study, begun in 1991, enrolled approximately five hundred patients with acute myocardial infarction of the anterior wall, with a follow-up period of one year. Patient enrollment criteria consisted in technically adequate two-dimensional echocardiographic imaging with randomized treatment with L-carnitine (9 g iv. for 5 days and 6 g o.s. during the following year) or placebo. Echocardiographic assessment was performed at admission ($3 \pm 1$ h after symptom onset) at discharge (<10 days from admission) and at 3, 6, 9 and 12 months after myocardial infarction. The results of this study are available in 1995.

## References


1. Sartorelli L, Ciman M, Rizzoli V. Siliprandi N. On the transport mechanisms of carnitine and its derivative in rat heart slices. Ital J. Biochem 1982; 31: 261–8.
2. Shug AL, Thomsen JH, Folts JD et al. Changes in tissue levels of carnitine and other metabolites during myocardial ischemia and anoxia. Arch Biochem Biophys 1978; 187: 25–33.
3. Suzuki Y, Kamikawa T, Kobayashi A, Masumura Y, Yamazaki N. Effects of L-carnitine on tissue levels of acyl carnitine, acyl coenzyme A and high energy phosphate in ischemic dog hearts. Jpn Circ J 1981; 45: 687–94.
4. Spagnoli LG, Corsi M, Villaschi S, Palmieri G, Maccari F. Myocardial carnitine deficiency in acute myocardial infarction. Lancet 1982; i: 1419–20.
5. Bartels GL, Remme WJ. Cardiac carnitine loss following myocardial ischaemia in patients

with coronary artery disease. Functional implications? Eur Heart J 1990; 11(Abstr Suppl): 41.
6. Rizzon P, Biasco G, Di Biase M et al. High doses of L-carnitine in acute myocardial infarction: metabolic and antiarrhythmic effects. Eur Heart J 1989; 10: 502–8.
7. Suzuki Y, Kamikawa T, Yamazaki N. Protective effects of L-carnitine on ischemic heart. In: Frenkel RA, McGarry JD, editors. Carnitine biosynthesis, metabolism, and functions. New York: Academic Press, 1980: 341–52.
8. Sugiyama S, Miyazaki Y, Kotaka K, Kato T, Suzuki S, Ozawa T. Mechanism of free fatty acid-induced arrhythmias. J Electrocardiol 1982; 15: 227–32.
9. Cowan JC, Williams EM, The effects of various fatty acids on action potential shortening during sequential periods of ischemia and reperfusion. J Mol Cell Cardiol 1980; 12: 347–69.
10. Kurien VA, Oliver MF. A metabolic cause for arrhythmias during acute myocardial hypoxia. Lancet 1970; i: 813–5.
11. Oliver MF, Kurien VA, Greenwood TW. Relation between serum-free-fatty acids and arrhythmias and death after acute myocardial infarction. Lancet 1988; i: 710–4.
12. Martina B, Zuber M, Burkart F, Weiss P, Ritz R. Antiarrhythmische Wirkung von L-Carnitin beim akuten Myocardinfarkt. Schweiz Med Wochenschr 1990; 32(Suppl II): 8 (Abstr).
13. Rebuzzi AG, Schiavoni G, Amico CM, Montenero AS, Meo F, Manzoli U. Beneficial effects of L-carnitine in the reduction of the necrotic area in acute myocardial infarction. Drugs Exptl Clin Res 1984; 10: 219–23.
14. Chiariello M, Brevetti G, Policicchio A, Nevola E, Condorelli M. L-Carnitine in acute myocardial infarction. A multicenter randomized trial. In: Borum PR, editor. Clinical aspects of human carnitine deficiency. New York: Pergamon Press, 1986: 242–3.
15. Neely JR, Garber D, McDonough K, Idell-Wenger J. Relationship between ventricular function and intermediates of fatty acid metabolism during myocardial ischemia: effects of carnitine. In: Winbury MM, Abiko J, editors. Perspectives in cardiovascular researches. New York: Raven Press, 1979: 225–34.
16. Liedtke AJ, Nellis SH. Effects of carnitine in ischemic and fatty acid supplemented swine hearts. J Clin Invest 1979; 64: 440–7.
17. Iliceto S, D'Ambrosio G, Scrutinio D, Marangelli V, Boni L, Rizzon P. A digital network for long-distance echocardiographic image and data transmission in clinical trials: The CEDIM study experience. J Am Soc Echocardiogr 1993; 6: 583–92.
18. Marangelli V, Sublimi Saponetti L et al. L'ecocardiografia nell'unità di terapia intensiva cardiologica: nuove tecnologie nella valutazione della cardiopatia ischemica. Cardiologia 1992; 37: 441–50.

*Corresponding Author*: Professor Paolo Rizzon, Department of Cardiology, Policlinico, University of Bari, Piazza G. Cesare, I-70124 Bari, Italy

PART THREE

# Therapeutic efficacy of propionyl-L-carnitine

# 18. Cardiac electrophysiology of propionyl-L-carnitine

ALESSANDRO MUGELLI, ELISABETTA CERBAI
and MARIO BARBIERI

"It could be hypothesized that action potential duration prolongation, which we have demonstrated to occur in papillary muscles isolated from the heart of rabbits treated for ten days with propionyl-L-carnitine (and which might be due to a reduction in transient outward potassium current), is somehow related to the observed improvement in the contractile properties of the intact myocardium."

## Introduction

Propionyl-L-carnitine (PLC) is a carnitine derivative which has been reported to exert relevant cardiovascular effects. Results from several animal studies seem to indicate that PLC may directly influence the myocardium, thus improving its performance [1, 2]. The aim of this article is to review the cellular electrophysiological effects provoked by PLC.

There are basically two kinds of studies in the literature: in the first kind, cardiac preparations, from which intracellular action potentials are recorded, are exposed to increasing concentrations of PLC. We will call these "acute" experiments. In a different kind of approach, experimental animals are treated for several days with PLC; after sacrifice, the papillary muscles are isolated and an electrophysiological study is performed. We will refer to these experiments as "chronic".

## Acute experiments

Acute experiments have been carried out in guinea-pig papillary muscles [3–5], dog [6] and sheep [7] Purkinje fibers. In these studies normal physiological conditions as well as conditions aimed at mimicking pathological situations (such as hypoxia, high $[K^+]_o$, low pH) or abnormal electrophysiological mechanisms (slow responses, delayed afterdepolarizations) have been used. This was done in an effort to explain the protective effect of PLC during

*J.W. de Jong and R. Ferrari (eds): The carnitine system*, 253–260.

ischemia-reperfusion which results in better recovery of cardiac output [8] and in an antiarrhythmic effect [9].

In ischemia there is an increase in extracellular $K^+$ concentrations [10] and the decrease in pH is unavoidable. PLC has practically no electrophysiologic effects under any of these different experimental conditions (low pH, high $[K^+]_o$, hypoxia). An increase in action potential duration at 90% of full repolarization is observed in both normal and pathological conditions at a 10 mM PLC. This high concentration is also able to lengthen the action potential duration recorded from dog Purkinje fibers exposed to a $pO_2$ of less than 40 mmHg as well as the duration of the slow response action potentials induced by high $K^+$ [6]. However, this concentration does not seem to be pharmacologically relevant. Thirty mM PLC causes, in fact, a clear-cut toxic effect: the action potential progressively deteriorates and the preparations become unexcitable.

The antiarrhythmic effect of PLC has been described in those arrhythmias associated with reoxygenation of the hypoxic myocardium [7, 9]. An antiarrhythmic effect is observed for concentrations as low as 1–10 μM. Since reoxygenation and reperfusion arrhythmias have often been associated with the occurrence of delayed afterdepolarizations, it was reasoned that even if PLC does not affect action potentials either in normal or in hypoxic conditions, it could affect an arrhythmogenic mechanism, such as the delayed afterdepolarizations (DADs) and result in an antiarrhythmic effect.

DADs can be induced by several interventions [11–13]; digitalis intoxication is certainly one of the most studied [11, 14, 15]. PLC at the concentration which exerts an antiarrhythmic effect on reoxygenation arrhythmias does not affect the action potential profile (as expected) nor the amplitude of DADs. The left panel of Figure 1 (upper traces) shows the last driven action potentials recorded from a Purkinje fiber intoxicated with strophanthidin. At the interruption of the stimulation a clear-cut delayed afterdepolarization superimposed on the diastolic depolarization and associated with an aftercontraction (bottom trace) develops.

PLC (10 μM) (middle and right panel) does not affect either DAD amplitude (which actually becomes bigger due to the progressive intoxication of the fiber) or contractile activity (bottom traces). It is concluded that appreciable and direct electrophysiological effects are not involved in the antiarrhythmic effect of PLC. However, it must be noted that the antiarrhythmic effect of PLC was apparent only on the late arrhythmias, i.e. those occurring during the 10th min of reperfusion. This behavior is completely different from that of other antiarrhythmic drugs, such as mexiletine, which promptly abolish reoxygenation or reperfusion arrhythmias [11, 14]. The antiarrhythmic effect of PLC is associated (and possibly due) to protective action on the hypoxic and reoxygenated myocardium: the release of creatine kinase (CK) and lactic dehydrogenase (LDH) is significantly reduced and the rise of diastolic left ventricular pressure prevented. The effect appears to be

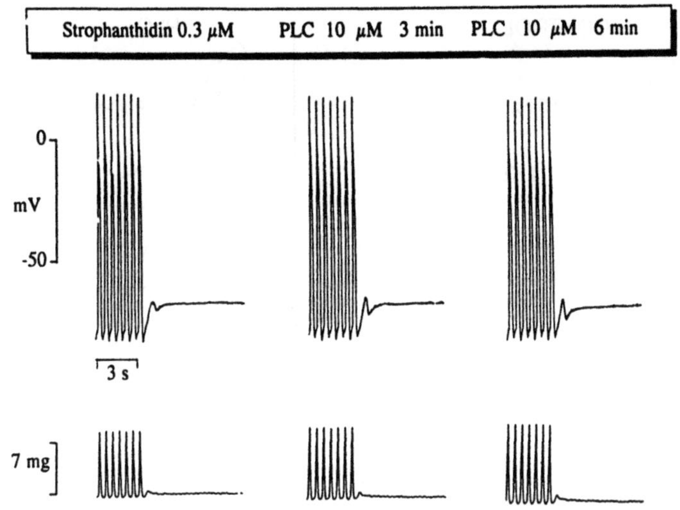


*Figure 1.* Effect of 10 μM propionyl-L-carnitine (PLC) on strophanthidin-induced delayed after-depolarizations. Each panel shows the electrical (upper traces) and mechanical (lower traces) activity recorded from a sheep Purkinje fiber during and after the interruption of stimulation, in control conditions (left) and 3 (middle) and 6 min (right panel) after PLC addition. Driving rate: 2 Hz.

specific since neither L-carnitine, propionyl-D-carnitine nor propionic acid has any significant effects under the same experimental conditions.

## Chronic experiments

In "chronic" experiments, rabbits were treated for 10 days with PLC (1 mmol/kg, i.p.), a schedule which provides plasma levels comparable to those obtained clinically [2]. L-carnitine- and saline-treated animals were used as controls. At the end of the treatment, animals were sacrificed and papillary muscles are isolated and mounted in a tissue bath for electrophysiological characterization. Basic electrophysiological characteristics were assessed by measuring intracellular action potentials from papillary muscles driven at 1 Hz and superfused with control (1.8 mM $Ca^{2+}$) Tyrode's solution. Action potentials recorded from preparations obtained from the propionyl-L-carnitine, control and L-carnitine groups were compared. A marked, statistically significant slowing of the repolarization phase between +20 and −40 mV, was observed in the PLC-treated group compared to the L-carnitine or

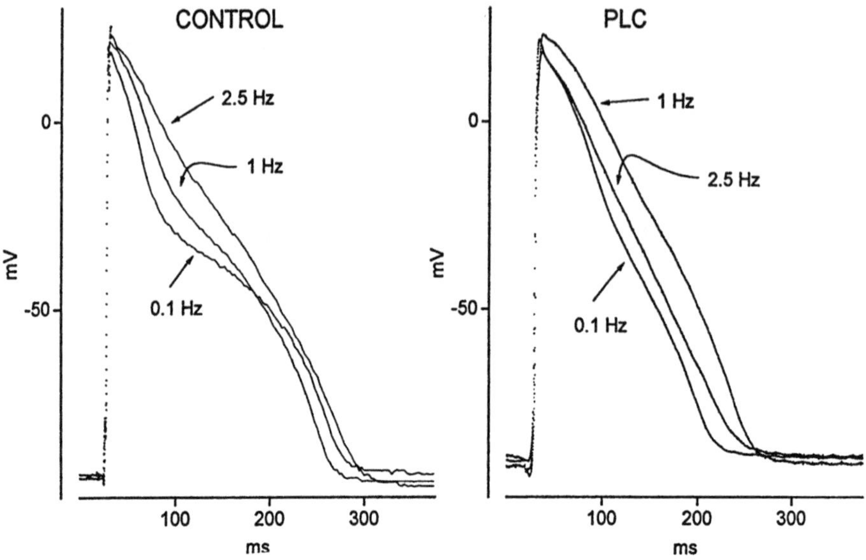


*Figure 2.* Effect of propionyl-L-carnitine (PLC) pre-treatment on action potential profile. Each panel shows superimposed action potentials recorded during pacing at 0.1, 1 and 2.5 Hz, from rabbit papillary muscles of the saline-treated group (control) and of the PLC-treated group (see [2]).

control groups (see Figure 3 of [2]). Consequently, the action potential duration measured at $-10$ mV ($APD_{-10}$) was significantly ($P < 0.001$) longer in the PLC-treated ($102 \pm 6$ ms) than in the saline- ($63 \pm 4$ ms) or the L-carnitine-treated groups ($56 \pm 7$ ms). This effect, which indicates an alteration in the plateau phase, was not accompanied by modifications in other electrophysiological parameters.

An interpretation of this finding in terms of changes in transmembrane ionic currents is not obvious, since the plateau phase of the action potential in rabbit ventricular myocytes is controlled by several different ionic fluxes [16]. No voltage clamp data are presently available. Suggestions may, however, come from the experiments shown in Figures 2 and 3, where the effects of changes in driving rate and extracellular calcium concentration on APD in control and "chronic" PLC preparations are reported. In Figure 2, action potentials recorded during stimulation at 0.1, 1 or 2.5 Hz in a control preparation (left panel) or in a PLC preparation (right panel) are superimposed. An increase in the stimulation rate from 0.1 to 1 Hz causes a lengthening of the action potential duration in both preparations. At 2.5 Hz, the duration of the plateau phase is further increased in the control papillary muscle (as

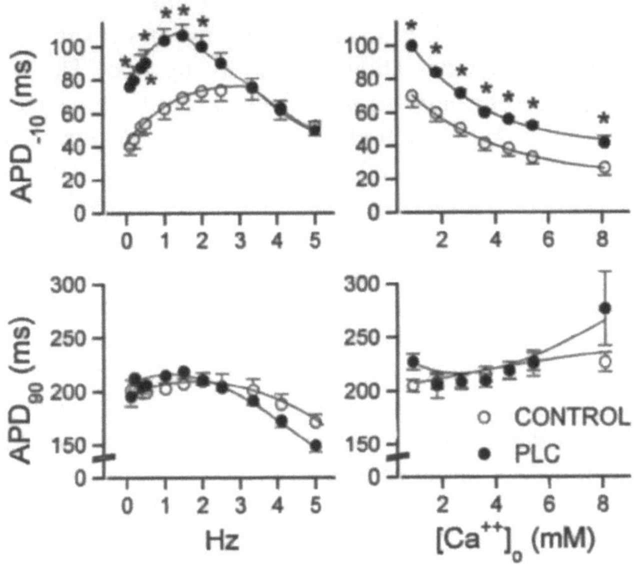


*Figure 3.* Effect of propionyl-L-carnitine (PLC) pre-treatment on action potential duration in rabbit papillary muscles: dependence on driving rate (left panels) and extracellular $Ca^{2+}$ concentration ($[Ca^{2+}]_o$) (right panels). Each point represents the mean±SEM value of the action potential duration measured at −10 mV ($APD_{-10}$) (upper panels) or at 90% repolarization ($APD_{90}$). * $P < 0.05$ PLC-treated group vs saline-treated group (Control). n = 8 to 9 (see [2]).

expected [16]); in the PLC-chronically treated papillary muscle it is, however, shortened. This finding is consistent: in 9 control preparations, $APD_{-10}$ was lengthened, compared to the value obtained at 0.1 Hz, by 66 ± 18% and 94 ± 29% at 1 and 2.5 Hz, respectively. In contrast, in 9 PLC preparations, $APD_{-10}$ was increased by 43 ± 7% and 36 ± 13% at 1 and 2.5 Hz, respectively, over the values measured at 0.1 Hz. The upper left panel of Figure 3 illustrates this frequency-dependent effect on $APD_{-10}$ measured over a wide range of driving rates. It is apparent that, at 0.1 Hz, the $APD_{-10}$ is significantly longer in chronic PLC preparations than in control preparations; the difference is maintained by increasing the driving rate up to 1.5 Hz, since in both groups a frequency-dependent increase in the duration of the plateau phase was observed. In the control group, the $APD_{-10}$ continued to increase with the stimulation frequency up to 3 Hz; then for higher driving rates it started to diminish. In the PLC group the shortening began earlier; the consequence is a progressive disappearance of the difference between the two groups and, at rates higher than 2 Hz, the effect of the treatment with PLC was completely lost.

A possible electrophysiological interpretation of this phenomenon is that the "chronic" treatment with PLC causes a change in density for a current which shows a marked modulation by the driving rate. The most likely candidate is the transient outward potassium current ($I_{to}$), which is responsible for the earlier repolarization phase of the action potential in the rabbit ventricle [17]. A decrease in $I_{to}$ density caused by the prolonged treatment of the rabbit with PLC would slow the early repolarization phase of the ventricular action potential, thus explaining the prolongation in APD in the plateau phase. Since the complete duration of the repolarization is not affected, one should not expect relevant electrocardiographic alterations. $I_{to}$ is markedly frequency-dependent [16, 18], being rapidly inactivated by increasing the driving rate. A smaller contribution of $I_{to}$, due to a decrease in its density, to the plateau phase duration is expected as the rate increases. This could also explain the smaller frequency-dependent increase in APD in the PLC group with respect to the control group.

APD control in the rabbit ventricle is under the influence of other ionic currents such as the L-type calcium current ($I_{Ca,L}$) [19], the calcium activated potassium current ($I_{K,Ca}$) and other potassium outward currents [17, 20, 21]. However, their contribution to the prolongation of APD in PLC-treated rabbits seems unlikely and can be excluded for at least three reasons.

First, the prolongation of the plateau phase of the action potential at short diastolic intervals seems to be attributable to the slower recovery from inactivation of $I_{to}$ rather than to an effect on $I_{Ca,L}$ [19]. Second, an increase in the extracellular calcium concentration (and consequently the Ca-dependent outward and inward currents) reduces in a similar fashion the $APD_{-10}$ in both control and PLC preparations (Figure 3, upper right panel); the difference between the two groups remains the same at both low (0.9 mM) and high (8.1 mM) $[Ca^{2+}]_o$. Since $I_{to}$ in the rabbit ventricle is a Ca-independent current [17], it should not be affected by changes in extracellular (and/or intracellular) calcium concentrations. Third, the last phase of action potential repolarization (90% repolarization, $APD_{90}$), which is also dependent on the balance of several currents (such as the delayed rectifier $I_K$ and the Na/Ca exchanger current) [22], is not modified by prolonged PLC treatment under the same experimental conditions (Figure 3, lower panels). These findings exclude their possible involvement in the slowing of the early phase of the repolarization process.

Finally, it is worth noting that interventions which are known to affect $I_{to}$ (such as 4-aminopyridine or rate) may affect contractility. It has been observed that superfusion with 4-aminopyridine [23], which selectively blocks the Ca-independent transient outward current, not only increases the action potential duration in rabbit papillary muscles, but also causes an increase in contractility. In the rabbit ventricle a decrease in the diastolic interval (in the range 10 to 0.2 s) causes an APD prolongation and an increase in developed force [20]; high rate, as previously discussed, inactivates $I_{to}$. Therefore, it could be hypothesized that APD prolongation, which we have demonstrated to occur in papillary muscles isolated from the heart of rabbits treated for 10 days with PLC (and which might be due to a reduction in $I_{to}$), is somehow related to the observed improvement in the contractile properties of the intact myocardium.

## Acknowledgement

Partly supported by a grant from MURST 60%, University of Ferrara.

## References

1. Ferrari R, Ceconi C, Curello S, Pasini E, Visioli O. Protective effect of propionyl-L-carnitine against ischaemia and reperfusion-damage. Mol Cell Biochem 1989; 88: 161–8.
2. Ferrari R, Di Lisa F, De Jong JW et al. Prolonged propionyl-L-carnitine pre-treatment of rabbit: biochemical, hemodynamic and electrophysiological effects on myocardium. J Mol Cell Cardiol 1992; 24: 219–32.
3. Aomine M, Arita M. Differential effects of L-propionylcarnitine on the electrical and mechanical properties of guinea pig ventricular muscle in normal and acidic conditions. J Electrocardiol 1987; 20: 287–96.
4. Aomine M, Arita M, Shimada T. Effects of L-propionylcarnitine on electrical and mechanical alterations induced by amphiphilic lipids in isolated guinea pig ventricular muscle. Heart Vessels 1988; 4: 197–206.
5. Carbonin PU, Ramacci MT, Pahor M et al. Antiarrhythmic effect of L-propionylcarnitine in isolated cardiac preparations. Cardioscience 1991; 2: 109–14.
6. Aomine M, Nobe S, Arita M. Electrophysiologic effects of a short-chain acyl carnitine, L-propionylcarnitine, on isolated canine Purkinje fibers. J Cardiovasc Pharmacol 1989; 13: 494–501.
7. Barbieri M, Carbonin PU, Cerbai E et al. Lack of correlation between the antiarrhythmic effect of L-propionylcarnitine on reoxygenation-induced arrhythmias and its electrophysiological properties. Br J Pharmacol 1991; 102: 73–8.
8. Paulson DJ, Traxler J, Schmidt M, Noonan J, Shug AL. Protection of the ischaemic myocardium by L-propionylcarnitine: effects on the recovery of cardiac output after ischaemia and reperfusion, carnitine transport, and fatty acid oxidation. Cardiovasc Res 1986; 20: 536–41.
9. Carbonin PU, Ramacci MT, Pahor M et al. Antiarrhythmic profile of propionyl-L-carnitine in isolated cardiac preparations. Abstr Symp Focus on Propionyl-L-Carnitine, Rome, Italy, 1988.
10. Kléber AG. Resting membrane potential, extracellular potassium activity, and intracellular sodium activity during acute global ischemia in isolated perfused guinea pig hearts. Circ Res 1983; 52: 442–50.
11. Amerini S, Carbonin P, Cerbai E, Giotti A, Mugelli A, Pahor M. Electrophysiological mechanisms for the antiarrhythmic action of mexiletine on digitalis-, reperfusion- and reoxygenation-induced arrhythmias. Br J Pharmacol 1985; 86: 805–15.
12. Ferrier GR, Moffat MP, Lukas A. Possible mechanisms of ventricular arrhythmias elicited by ischemia followed by reperfusion. Studies on isolated canine ventricular tissues. Circ Res 1985; 56: 184–94.

13. Mugelli A, Amerini S, Piazzesi G, Giotti A. Barium-induced spontaneous activity in sheep cardiac Purkinje fibers. J Mol Cell Cardiol 1983; 13: 697–711.
14. Amerini S, Bernabei R, Carbonin P, Cerbai E, Mugelli A, Pahor M. Electrophysiological mechanism for the antiarrhythmic action of propafenone: a comparison with mexiletine. Br J Pharmacol 1988; 95: 1039–46.
15. Ferrier GR. Digitalis arrhythmias: role of oscillatory afterpotentials. Progr Cardiovasc Dis 1977; 19: 459–74.
16. Kukushkin NI, Gainullin RZ, Sosunov EA. Transient outward current and rate dependence of action potential duration in rabbit cardiac ventricular muscle. Pflügers Arch 1983; 399: 87–92.
17. Giles WR, Imaizumi Y. Comparison of potassium currents in rabbit atrial and ventricular cells. J Physiol (Lond) 1988; 405: 123–45.
18. Boyett MR. Effect of rate-dependent changes in the transient outward current on the action potential in sheep Purkinje fibers. J Physiol (Lond) 1981; 319: 23–41.
19. Hiraoka M, Kawano S. Mechanism of increased amplitude and duration of the plateau with sudden shortening of diastolic intervals in rabbit ventricular cells. Circ Res 1987; 60: 14–26.
20. Wohlfart B. Relationships between peak force, action potential duration and stimulus interval in rabbit myocardium. Acta Physiol Scand 1979; 106: 395–409.
21. Cohen IS, Daytner NB, Gintant GA, Kline RP. Time-dependent outward currents in the heart. In: Fozzard HA, Haber E, Jennings RB, Katz AM, Morgan HE, editors. The heart and cardiovascular system. New York: Raven Press, 1986: 637–71.
22. Hilgemann DW, Noble D. Excitation-contraction coupling and extracellular calcium transients in rabbit atrium: reconstruction of basic cellular mechanisms. Proc R Soc Lond (Biol) 1987; 230: 163–205.
23. Wollmer P, Wohlfart B, Khan AR. Effects of 4-aminopyridine on contractile response and action potential of rabbit papillary muscle. Acta Physiol Scand 1981; 113: 183–7.


*Corresponding Author*: Professor Alessandro Mugelli, Dipartimento di Farmacologia, Università degli Studi di Firenze, Viale G.B. Morgagni, 65, I-50134 Firenze, Italy

# 19. Acute vs. chronic treatment with propionyl-L-carnitine: biochemical, hemodynamic and electrophysiological effects on rabbit heart

JAN WILLEM DE JONG, ALESSANDRO MUGELLI, FABIO DI LISA and ROBERTO FERRARI

"When added acutely to isolated and perfused heart, propionyl-L-carnitine does not improve mechanical function. When given chronically, for 10 days, to animals before the isolation of their heart, it results in a long-lasting increase of the Frank-Starling curve of the isolated myocardium."

## Introduction

In various animal models, propionyl-L-carnitine (PLC) attenuates myocardial ischemic damage, reduces the degree of stunning and improves contractile recovery during reperfusion [1–7]. In addition, PLC increases myocardial performance of animals with cardiac hypertrophy [8–10] or cardiomyopathies due to several experimental protocols [11–13].

When tested for its effect on cardiac function, PLC produced controversial results. After acute administration to isolated preparations or intact animals, no major changes of left ventricular performance are detected [10, 14, 15]. In contrast, when administered chronically, PLC increases the performance of the isolated aerobic myocardium [14].

In the present article we review the litterature and our own data [14, 16–18] on the effects of PLC on biochemistry, function and electrophysiology of the aerobic heart. The effects on the ischemic or failing heart are not discussed as they are extensively reviewed in other chapters of this volume.

## Acute short-term administration of propionyl-L-carnitine

The effect of acute administration of PLC has been tested in isolated and perfused rat or rabbit heart [14, 18], in intact animals [19], and in patients with coronary heart disease and normal left ventricular function [20, 21].

*J.W. de Jong and R. Ferrari (eds): The carnitine system*, 261–273.

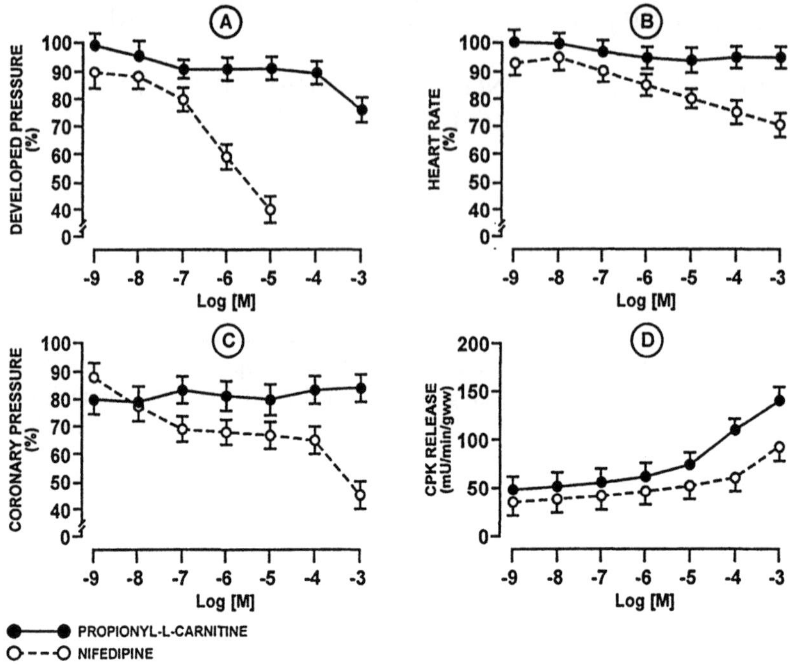


*Figure 1.* Millimolar concentrations of propionyl-L-carnitine, given to isolated and aerobically perfused rabbit hearts by the cumulative dose method, reduce left ventricular developed pressure to a minor extent, without an effect on heart rate and coronary pressure. The concomittant release of creatine kinase (CPK) suggests that this high dosage causes membrane damage. The calcium antagonist nifedipine, used as a pharmacological reference, affects function already in the micromolar range. Percentages refer to difference with untreated control hearts. Mean values ± SEM (n = 6). gww = gram wet weight. Data from [14].

## *Effects on in vitro preparations*

The effects of PLC administered to isolated and aerobically perfused rabbit hearts are shown in Figure 1. Propionyl-L-carnitine was administered directly to the perfusate, according to the cumulative dose method. Results are expressed as percent inhibition of left ventricular developed pressure (Figure 1A); heart rate (Figure 1B); and coronary pressure (Figure 1C). Also creatine kinase release is depicted (Figure 1D). In addition, the effects of nifedipine, a typical dihydropyridine calcium antagonist, are reported as pharmacological reference.

Clearly, PLC at doses of $10^{-9}$ to $10^{-4}$M does not modify the developed pressure. Only the highest dose of PLC ($10^{-3}$M) results in a minor reduction of developed pressure. This effect is independent of pH changes in the perfusion solution [14]. At this high dosage PLC causes a release of creatine kinase, suggesting a membrane-damaging effect. Furthermore, PLC does not alter heart rate or coronary perfusion pressure, even at the maximum dose tested. Similar effects are detected in rats [18].

*Effects on in vivo preparation*

The effects of intravenous administration of PLC were investigated in anesthetized dogs instrumented for the analysis of general hemodynamic and electrocardiographic data, peripheral and coronary blood flow, and oxygen consumption [19].

Propionyl-L-carnitine causes a dose-dependent, short-lasting enhancement of cardiac output, both in open- and closed-chest conditions [15, 19]. Arterial blood pressure, heart rate, and contractility vary slightly and unpredictably. The drug does not elicit electrocardiographic effects. These responses are not modified by $\alpha$- or $\beta$-adrenergic blockade, nor by administration of calcium antagonists. They are, however, abolished by the combination of all these interventions. Mesenteric and iliac blood flows are increased by both PLC and L-carnitine (LC); LC + propionic acid (P) increased these, and in addition increased renal blood flow. A strong diuresis obtained with PLC, LC, and LC + P was due to osmotic clearance following the administration of hyperosmotic solutions. Propionyl-L-carnitine elicited coronary vasodilation with hyperosmotic solutions. It elicited coronary vasodilation with reduced oxygen extraction; this effect lasted longer than the general hemodynamic effects, and was not seen with LC. All the cardiovascular actions of PLC can be attributed to its pharmacologic properties, rather than to its role as a metabolic intermediate.

*Effects in humans*

The hemodynamic effect of PLC has been evaluated in ten patients with coronary artery disease but having normal left ventricular function [20, 21]. The drug was administered intravenously at a dose of 15 mg/kg. Propionyl-L-carnitine improves the stroke volume and reduces the ejection impedence as a result of a decrease in systemic and pulmonary resistance and of an increase in arterial compliance. Total external heart power improves with a proportionally smaller increase in the energy requirement, suggesting that PLC has a positive inotropic property.

The same authors also investigated the effects of PLC on vasomotion of normal coronary arteries in 20 patients with atypical chest pain or angina but without previous myocardial infarction [20, 21].

Luminal area of the coronary arteries was measured using quantitative coronary arteriography in the proximal, middle, and distal thirds of a normal coronary vessel in basal conditions, and 2, 5, 10 min after 1.5 mg/kg intracoronary injection PLC or after intravenous administration of 30 mg/kg PLC. The data obtained were analyzed against a placebo group.

Propionyl-L-carnitine causes a significant vasodilation in the proximal segments already apparent 5 min after intracoronary injection and 15 min after intravenous administration. Placebo does not induce any changes. The drug does not modify heart rate, systolic blood pressure and coronary driving pressure.

## Chronic, prolonged administration of propionyl-L-carnitine

When administered chronically to animals several days before the isolation of the heart, PLC increases the performance of the aerobic myocardium, independently from changes of peripheral hemodynamics or coronary flow.

### *Effect on in vitro preparations*

We first noticed and reported that chronic administration of PLC improves the contractility of the isolated heart [18]. In further studies [14, 18], we investigated this effect in hearts which were isolated from rabbits or rats that had been treated for either two, five or ten days with saline or PLC (1 mmol/kg). During perfusion no further drug was administered to the isolated hearts. Two or five days of treatment with PLC has no effect on contractile function of the isolated hearts. On the contrary, the hearts harvested from rabbits that had been treated for ten days develop significantly more left ventricular pressure than controls. This effect is evident after 30, 60, or 90 min of aerobic perfusion [14]. The same prolonged treatment has no effect on heart rate or coronary perfusion pressure.

As these data can be strongly influenced by the positioning and degree of inflation of the intraventricular balloon, in a separate series of experiments we determined the rise in developed and diastolic pressure as a function of intraventricular volume, which was progressively increased by increments of 600 μl. The hearts utilized for these experiments were paced at the constant rate of 180 beats/min. The results obtained are shown in Figure 2. In control hearts volume loading from 0 to 1.8 ml results, as expected, in an incremental rise in developed pressure. Optimum developed pressure occurs with volumes of 1.2–1.8 ml. A further increase in volume results in a precipitous decline in developed pressure (Figure 2A) and in a steep rise in end-diastolic pressure (Figure 2D), suggesting that in our preparation the region of overstretch is between 1.8 and 3.6 ml.

Chronic treatment with PLC for two or five days fails to modify the volume-pressure curves of the left ventricle. However, when treatment is prolonged for as long as ten days, there is a significant increase of the optimum developed pressure (corresponding to volumes between 1.2 and 1.8 ml). Thereafter developed pressure remains constant, even after increasing the balloon volume from 2.4 to 3.6 ml, suggesting that overstretch is prevented. This suggestion is confirmed by the significant reduction in the

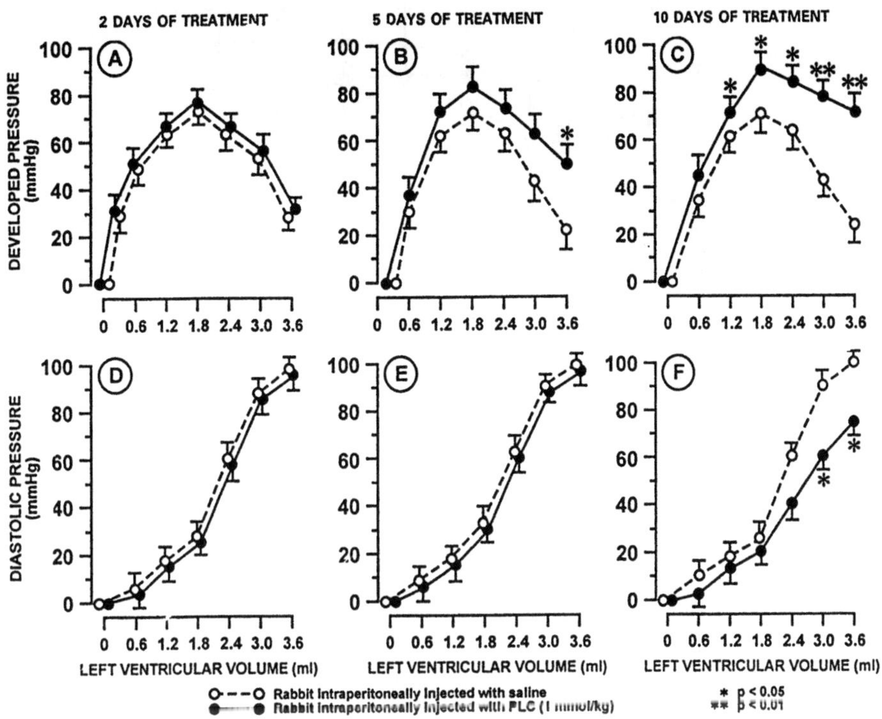


*Figure 2.* Pretreatment of rabbits for ten days with propionyl-L-carnitine (PLC) increases optimum developed pressure in their isolated, perfused hearts. The developed and end-diastolic pressures measured at higher filling pressures suggest that PLC prevents overstretch. Mean values ± SEM (n = 6). P-values refer to differences with saline-treated rabbits. Data from [14].

rise in end-diastolic pressure caused by prolonged treatment with PLC (Figure 2F). These results were later confirmed by us [16] and by other authors [9, 10].

### *Specificity of the chronic in vitro effect of propionyl-L-carnitine*

To investigate whether the chronic effect is specific for PLC or is due to L-carnitine or propionic acid, we have designed experiments in which rabbits were treated for ten days with either saline, L-carnitine (1 mmol/kg), propionic acid (1 mmol/kg) or PLC (1 mmol/kg). Propionic acid resulted in 98% mortality after ten days. No death could be detected after PLC or L-carnitine treatment. Figure 3 shows the volume-pressure curves which we obtained.

Treatment with L-carnitine fails to modify the volume-pressure curves of the isolated left ventricle (Figures 3A and B). Treatment with PLC prevents

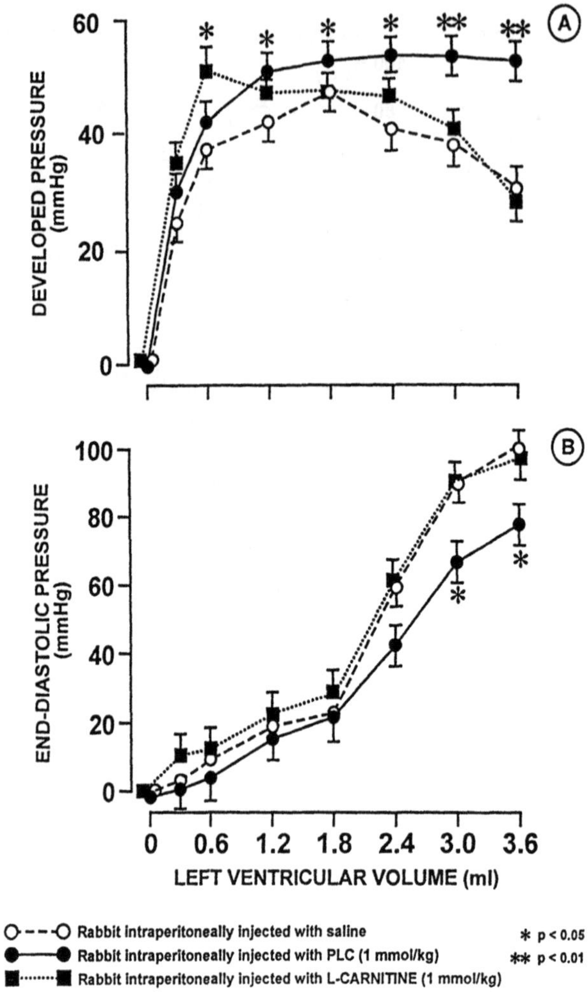


*Figure 3.* In isolated hearts L-carnitine is unable to modulate developed and end-diastolic pressures, as seen with propionyl-L-carnitine (PLC). Pretreatment of rabbits took place for ten days; propionic acid administration turned out to be deadly. Mean values ± SEM (n = 10). P-values refer to differences with saline- and carnitine-treated animals. From [16].

the decrease of the optimum developed pressure, which remains constant even after increasing volumes to as much as 3.6 ml (Figure 3A). In addition, it reduces the rise in end-diastolic pressure, again suggesting partial prevention of overstretch (Figure 3B).

To investigate whether the availability of calcium is a rate-limiting factor for the hemodynamic effects caused by prolonged treatment with PLC, the

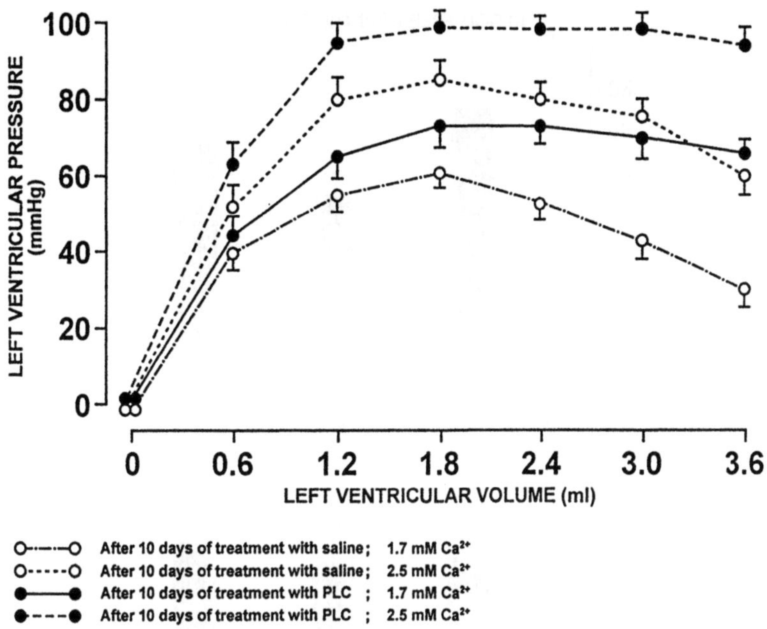


*Figure 4.* Calcium availability is unlikely to be a rate-limiting factor for the effect of propionyl-L-carnitine pretreatment (PLC, 1 mmol/kg rabbit i.p.): The increase in calcium concentration in the perfusion buffer results in a comparable positive inotropic effect in the hearts of control and PLC-treated animals. Mean ± SEM (n = 7–8). For details, see [16].

calcium concentration of the perfusion fluid was increased from 1.7 to 2.5 mM in another series of experiments. The data obtained are reported in Figure 4. Increasing the calcium content of the buffer results in a positive inotropic effect in the hearts of both control and PLC-treated animals. The shape of the volume-pressure curves is identical to that obtained with normal calcium, and the increase in contractility is of the same entity in both groups, independently from active treatment.

This suggests that calcium availability is not a rate-limiting factor for the effects of PLC.

### *Pharmacokinetics of propionyl-L-carnitine*

We measured the levels of (acyl)carnitine found in rabbit blood and tissue after one, five and ten days of treatments [16]. Blood and tissue content of free carnitine increases at least twofold after administration of PLC (Figures 5 and 6). Blood (Figure 5) and tissue [16] short-chain acylcarnitines also increase. Radiochromatographic analysis of the short-chain acylcarnitine

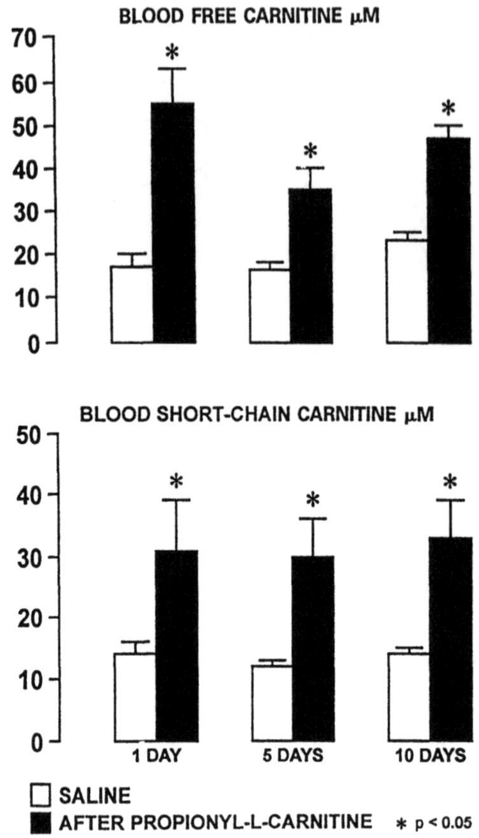


*Figure 5.* Intraperitoneal treatment of rabbits with propionyl-L-carnitine for one to ten days more than doubles the blood concentration of free carnitine and short-chain acylcarnitines. The latter fraction consists almost entirely of acetyl-L-carnitine, indicating that rapid conversion of PLC took place. Mean ± SEM (n = 7–9). * $P < 0.05$ vs. saline. Data from [16].

fraction, performed after the isotopic exchange with [$^3$H]carnitine, demonstrates that it consists almost completely of the acetyl-ester. Long-chain acylcarnitines (Figure 6) and acyl-CoAs [16] also increase in the heart after PLC administration. Several other CoA esters, as well as free CoA, remain unchanged; succinate also remains constant [16]. The increase in tissue carnitine demonstrates that PLC is taken up and utilized by the heart [1, 18]. Clearly this rise in myocardial carnitine content cannot be attributed to the increase in plasma carnitine in the same experimental group. If one considers an average water content of 0.4 ml/heart, of which 40% is present in the extracellular space, plasma carnitine values <100 µM cannot contribute more than 4% of the observed augmented tissue carnitine content.

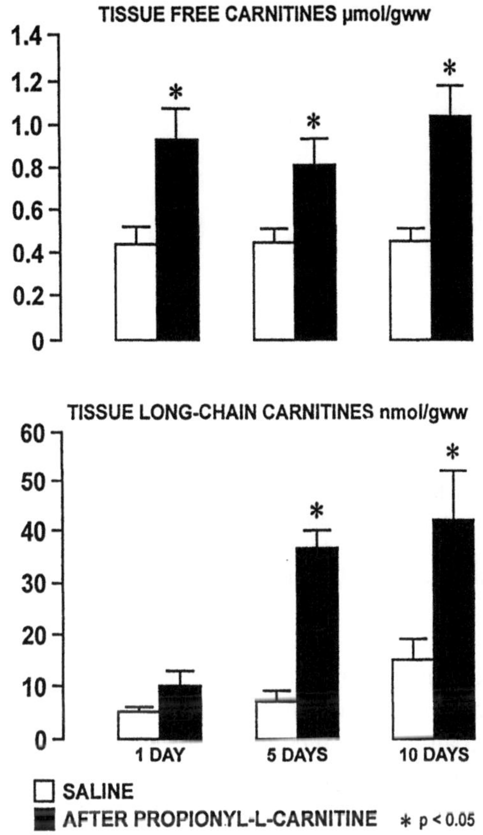


*Figure 6.* Treatment of rabbits with propionyl-L-carnitine increases the cardiac concentration of carnitine and long-chain acylcarnitines. The short-chain acylcarnitines increase as well, but to a smaller extent. gww = gram wet weight. For other details, see legend to Figure 5.

When measured 6 h after sacrifice of the animals, myocardial content of PLC is increased, but only after prolonged (ten days) treatment [14]. However, the increase in PLC content is no longer apparent when the assay is performed 24 h after sacrifice [16]. Due to the high carnitine acetyltransferase activity of cardiac muscle mitochondria, short-chain acylcarnitines and short-chain acyl-CoAs are rapidly metabolized.

## *Metabolical effect of propionyl-L-carnitine*

In our previous studies [14, 16], we postulated that PLC enhances the mechanical performance of the heart by improving its metabolism. It is known that pyruvate increases heart contractility allowing a more efficient energy

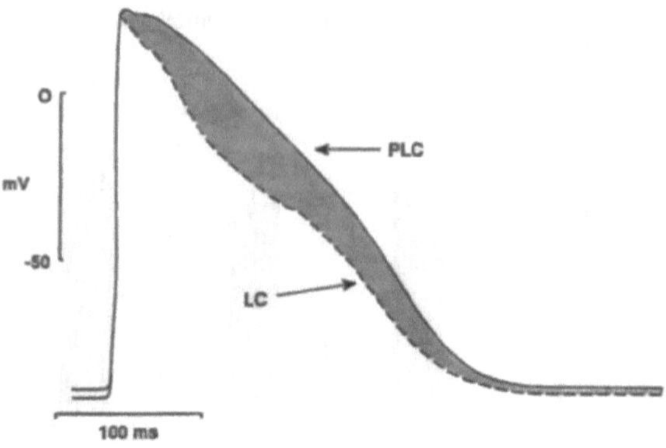


*Figure 7.* Treatment of rabbits with propionyl-L-carnitine (PLC) causes a statistically significant prolongation of the action potential duration in the plateau phase in comparison to treatment with L-carnitine (LC) or saline (data not shown). Average values are given for 8–9 individual records, obtained from isolated papillary muscles, driven at 1 Hz. Adapted from [16].

utilization [22]. Administration of pyruvate in fact leads to a higher cytosolic phosphorylation potential which in conjunction with a reduced $P_i$ concentration translates into an increased contraction.

We investigated whether a similar mechanism is at the basis of the PLC effects. Energy metabolism does not seem to be involved, since high-energy phosphates, $P_i$ and mitochondrial function remain unchanged after chronic PLC administration [16]. These findings, however, lead to some important implications. Usually, typical inotropic agents such as digitalis, calcium and adrenergic compounds stimulate contractility by increasing myofibrillar energy utilization at the expense of energy supply. Consequently these agents cause a decline in phosphocreatine/$P_i$ ratio, suggesting that they place the heart in a supply/demand imbalance [22]. This is not the case for PLC. Energy metabolism remains unchanged despite the increase in myocardial performance.

### *Electrophysiological effect of propionyl-L-carnitine*

Finally, we investigated whether the mechanical effects of prolonged treatment with propionyl-L-carnitine are related to electrophysiological changes. It is known that acute administration of PLC does not influence the action potential parameters [23]. In contrast, chronic treatment causes a prolongation of the action potential duration in the plateau phase; other action potential parameters are not modified (Figure 7). This action appears to be a specific effect for PLC, as prolonged treatment with L-carnitine does not cause any electrophysiological change [16]. During the repolarization phase, which is modified by PLC, important events occur which are able to influence contractility [24]. It is attractive to correlate the effect on papillary muscle action potential duration with that on cardiac mechanical performance, since PLC, but not L-carnitine, affects both of them. In addition, the electrophysiological and mechanical effects of propionyl-L-carnitine are not dependent on the extracellular calcium concentration. However, the improvement in mechanical performance is observed not only in the spontaneously beating heart, but also at driving rates (180 beats/min) at which the effect on action potential duration is completely lost. Consequently the relationship between the two events remains uncertain, and the intrinsic difference between the two experimental preparations does not allow to draw a definitive link.

## Conclusion

When added acutely to isolated and perfused heart, PLC does not improve mechanical function. When given chronically, for ten days, to animals before the isolation of their heart, it results in a long-lasting increase of the Frank-Starling curve of the isolated myocardium. This effect, although modulated by an increase of the calcium concentration, is unlikely to depend upon a modification to calcium sensitivity of the various $Ca^{2+}$-ATPases of the myocardium. The effect is also independent of changes in phosphorylation potential.

The effects of propionyl-L-carnitine are evident 8 h after the last injection to the animals and remain for at least 24 h. Overstretching is also completely prevented. It is therefore possible that prolonged treatment with PLC causes molecular conformational changes of the contractile component, and this hypothesis should be carefully investigated.

Prolonged administration of PLC also alters the action potential. However, a clear relationship between electrophysiological and mechanical effects of PLC, although suggestive, cannot be drawn at the moment.

## Acknowledgements

This work was supported by the Italian Research Council (C.N.R.) target projects "Prevention and Control Disease Factors" and "Biotechnology and Bioinstrumentation". We thank Ms Roberta Bonetti and Ms Carina Poleon-Weghorst for secretarial help.

## References

1. Paulson DJ, Traxler J, Schmidt M, Noonan J, Shug AL. Protection of the ischaemic myocardium by L-propionylcarnitine: effects on the recovery of cardiac output after ischaemia and reperfusion, carnitine transport, and fatty acid oxidation. Cardiovasc Res 1986; 20: 536–41.
2. Liedtke AJ, DeMaison L, Nellis SH. Effects of L-propionylcarnitine on mechanical recovery during reflow in intact hearts. Am J Physiol 1988; 255: H169–76.
3. Di Lisa F, Menabò R, Siliprandi N. L-Propionyl-carnitine of mitochondria in ischemic rat hearts. Mol Cell Biochem 1989; 88: 169–73.
4. Raddino R, Ceconi C, Curello S, Bigoli MC, Ferrari R, Visioli O. The effects of DL-carnitine, DL-acetylcarnitine and L-propionyl-carnitine on the haemodynamic function of isolated rabbit hearts. J Mol Cell Cardiol 1983; 15(Suppl 3): 17 (Abstr).
5. Packer L, Valenza M, Serbinova E, Starke-Reed P, Frost K, Kagan V. Free radical scavenging is involved in the protective effect of L-propionyl-carnitine against ischemia-reperfusion injury of the heart. Arch Biochem Biophys 1991; 288: 533–7.
6. Sassen LMA, Bezstarosti K, Koning MMG, Van der Giessen WJ, Lamers JMJ, Verdouw PD. Effects of administration of L-propionylcarnitine during ischemia on the recovery of myocardial function in the anesthetised pig. Cardioscience 1990; 1: 155–61.
7. Molaparast-Saless F, Nellis SH, Liedtke AJ. The effects of propionylcarnitine taurine on cardiac performance in aerobic and ischemic myocardium. J Mol Cell Cardiol 1988; 20: 63–74.
8. Yang XP, Samaja M, English E et al. Hemodynamic and metabolic activities of propionyl-L-carnitine in rats with pressure-overload cardiac hypertrophy. J Cardiovasc Pharmacol 1992; 20: 88–98.
9. El Alaoui-Talibi Z, Bouhaddioni N, Moravec J. Assessment of the cardiostimulant action of propionyl-L-carnitine on chronically volume-overloaded rat hearts. Cardiovasc Drugs Ther 1993; 7: 357–63.
10. Micheletti R, Giacalone G, Reggiani C, Canepari M, Bianchi G. Effect of propionyl-L-carnitine treatment on mechanical properties of papillary muscles from pressure-overload rat hearts. J Mol Cell Cardiol 1992; 24(Suppl 5): S.41 (Abstr).
11. Pasini E, Comini L, Ferrari R, de Giuli F, Menotti A, Dhalla NS. Effect of propionyl-L-carnitine on experimentally induced cardiomyopathy in rats. Am J Cardiovasc Pathol 1992; 4: 216–22.
12. Pasini E, Cargnoni A, Condorelli E, Marzo A, Lisciani L, Ferrari R. Effect of prolonged treatment with propionyl-L-carnitine on erucic acid-induced myocardial dysfunction in rats. Mol Cell Biochem 1992; 112: 117–23.
13. Micheletti R, Di Paola ED, Schiavone A et al. Propionyl-L-carnitine limits chronic ventricular dilation after myocardial infarction in rats. Am J Physiol 1993; 264: H1111–7.
14. Ferrari R, Pasini E, Condorelli E et al. Effect of propionyl-L-carnitine on mechanical function of isolated rabbit heart. Cardiovasc Drugs Ther 1991; 5(Suppl 1): 17–23.
15. Cevese A, Schena F, Cerutti G. Short-term hemodynamic effects of intravenous propionyl-L-carnitine in anesthetized dogs. Cardiologia 1989; 34: 95–101.
16. Ferrari R, Di Lisa F, De Jong JW et al. Prolonged propionyl-L-carnitine pretreatment of rabbit: biochemical, hemodynamic and electrophysiological effects on myocardium. J Mol Cell Cardiol 1992; 24: 219–32.
17. Siliprandi N, Di Lisa F, Pivetta A, Miotto G, Siliprandi D. Transport and function of L-carnitine and L-propionylcarnitine: relevance to some cardiomyopathies and cardiac ischemia. Z Kardiol 1987; 76(Suppl 5): 34–40.
18. Pasini E, Ceconi C, Curello S, Cargnoni A, Condorelli E, Ferrari R. Effect of acute and chronic treatment with propionyl-L-carnitine on mechanical performance of isolated heart. J Mol Cell Cardiol 1989; 21(Suppl 4): S.91 (Abstr).
19. Cevese A, Schena F, Cerutti G. Short-term hemodynamic effects of intravenous propionyl-L-carnitine in anesthetized dogs. Cardiovasc Drugs Ther 1991; 5(Suppl 1): 45–56.

20. Chiddo A, Musci S, Bortome A et al. Effetti emodinamici e sul circolo coronarico della propionil-l-carnitina. Cardiologia 1989; 54(Suppl 1): 111–7.
21. Chiddo A, Gaglione A, Musci S et al. Hemodynamic study of intravenous propionyl-L-carnitine in patients with ischemic heart disease and normal left ventricular function. Cardiovasc Drugs Ther 1991; 5(Suppl 1): 107–12.
22. Zweier JL, Jacobus E. Substrate-induced alteration of high energy phosphate metabolism and contractile function in the perfused heart. J Biol Chem 1987; 262: 8015–21.
23. Barbieri M, Carbonin PU, Cerbai E et al. Lack of correlation between the antiarrhythmic effect of L-propionylcarnitine on reoxygenation-induced arrhythmias and its electrophysiological properties. Br J Pharmacol 1991; 102: 73–8.
24. Noble D. The surprising heart: a review of recent progress in cardiac electrophysiology. J Physiol (Lond) 1984; 353: 1–50.

*Corresponding Author*: Dr Jan Willem de Jong, Cardiochemical Lab., Thorax Center Ee2371, Erasmus University Rotterdam, P.O. Box 1738, 3000 DR Rotterdam, The Netherlands

# 20. Mechanical recovery with propionyl-L-carnitine

A. JAMES LIEDTKE

"Some effects noted with propionyl-L-carnitine are associative epiphenomena with no clear insight into pharmacological mechanism(s), some are in direct conflict in terms of observational content, others are in agreement and at face value would seem to rule out possible mechanisms, while still others are suggestive of attractive leads to possible explanations of drug effect in enhancing myocardial contractility."

## Introduction

Carnitine has received attention in the past as a naturally occurring compound which can transiently improve or preserve cardiac performance during myocardial ischemia [1]. An example of this preservative effect on mechanical function is shown in Figure 1 from experimental observations in extracorporeally perfused pig hearts at conditions of moderate global ischemia and excess fatty acids (FFA) in coronary perfusion fluid [2]. The exact mechanism for this therapeutic influence has never been identified but hypotheses have centered on a sparing effect from a deleterious *lipid burden* known to occur in ischemic heart muscle which drains essential energy stores and disrupts membrane integrity and enzyme functions via accumulation of amphiphiles [3, 4]. Unfortunately, by definition, the benefits of carnitine (or any other therapy) in ischemic myocardium must be short-lived since any treatment strategy in the absence of expeditious repletion of oxygen delivery cannot be sustained for long. Irreversible cardiac injury with its accompanying mechanical dysfunction must occur in a matter of minutes-to-hours.

Perhaps a better experimental environment to study therapeutic agents with potential enhancing properties on mechanical function is myocardial reperfusion following reversible ischemic injury. Here the cell is capable of complete, eventual recovery and may be assisted by a variety of adjunctive metabolic schemes to hasten the repair and recuperative process. One interesting legacy of ischemic stress which carries over into the reperfusion condition in reversibly injured tissue is mechanical stunning. The mechanism(s) for this phenomenon, which may take hours-to-days to resolve, has/have

*J.W. de Jong and R. Ferrari (eds): The carnitine system*, 275–289.

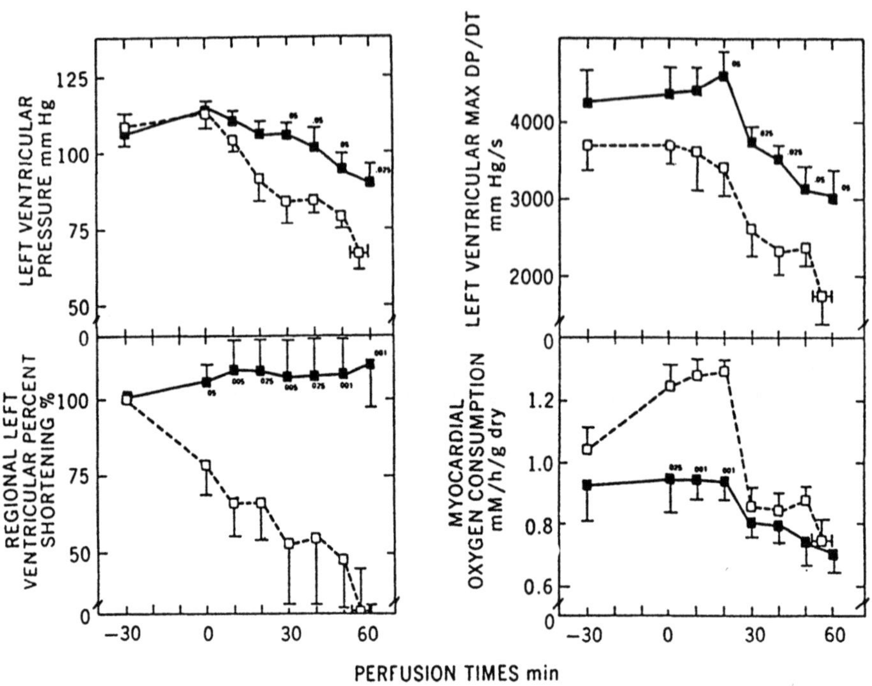


*Figure 1.* Mechanical and metabolic functions in FFA-supplemented pig hearts at control (−30 to +20 min perfusion time) and globally ischemic (+20 to +60 min perfusion time; average reduction in total coronary flow by 41%) conditions. Treatments with DL-carnitine (closed symbols) improved global and regional mechanical performance, particularly during ischemia, and decreased oxygen consumption during normal flows. Numbers above data points refer to statistical P values between treated and untreated hearts [2].

been the focus of wide debate, and arguments supporting altered bioenergetics, injury from free radical formation, and imbalance of calcium homeostasis have been proposed. Stunning, both regional and global in compass depending on the protocol and preparation, makes an excellent backdrop for testing agents designed to enhance mechanical performance since it is quickly reversed when exposed to standard inotropic drugs (see Figure 2) [5]. The theme of the text of this monograph deals with the transition from carnitine to propionyl-L-carnitine as treatment choices in impaired myocardium. The topic of this chapter will review the efficacy of this latter compound in restoring contractile function in reperfusion following myocardial ischemia.

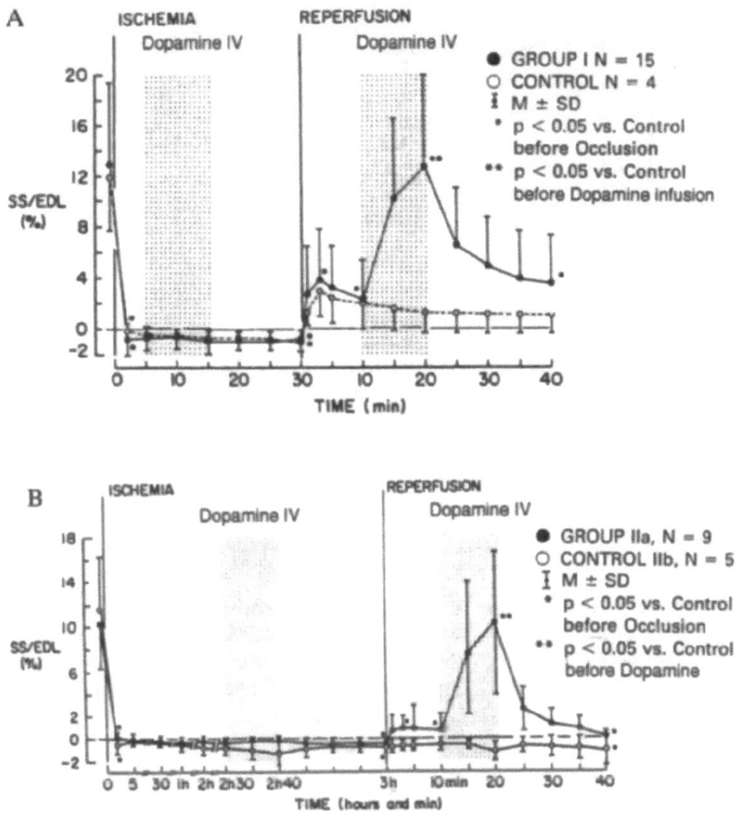


*Figure 2. Panel A.* Solid circles indicate segmental function in dogs during 30 min of LAD coronary occlusion, after reperfusion, and during infusions of dopamine. SS = systolic shortening; EDL = end-diastolic length. Open circles indicate segmental function during ischemia and after reperfusion in control dogs that did not receive dopamine. *Panel B.* Solid circles indicate segmental function during 3 h of LAD coronary occlusion, after reperfusion, and during infusions of dopamine in nine dogs. In these dogs, necrosis was limited to the subendocardial one-third of wall thickness. Open circles indicate segmental function in five dogs not receiving dopamine. In these dogs, necrosis in the area of the length gauge was virtually transmural [5].

## Mechanical benefits of propionyl-L-carnitine

The first report which clearly described a therapeutic advantage with propionyl-L-carnitine in myocardial reflow was by Paulson and co-workers [6] who demonstrated a cardioprotective effect in isolated perfused rat hearts.

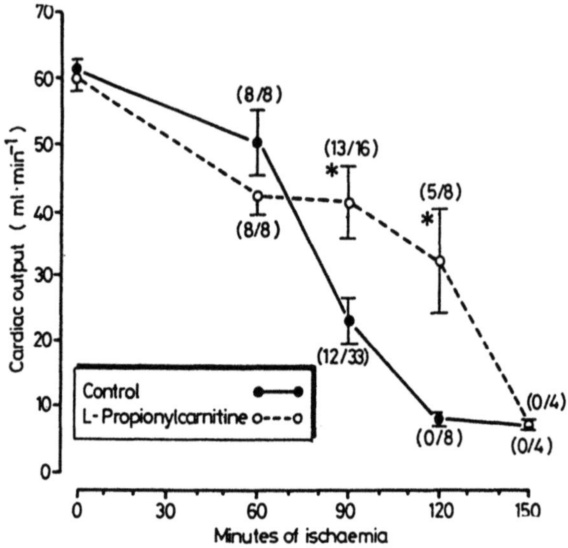


*Figure 3.* Effects of 5.5 mM propionyl-L-carnitine on the recovery of cardiac output in isolated perfused rat hearts subjected to various durations of ischemia followed by fifteen min of reperfusion. In parentheses is the proportion of hearts in each group that recovered the ability to work. Values are mean ± SEM. * $p < 0.05$ for difference between 0 and 5.5 mM propionyl-L-carnitine groups [6].

Propionyl-L-carnitine at two different dosing schedules (5.5 and 11 mmol $\times$ $l^{-1}$), when administered to coronary perfusion fluid beginning in the preischemic perfusion interval, improved mechanical function as estimated by increased cardiac output (see Figure 3), left ventricular pressure and left ventricular dp/dt during 15 min of reperfusion following several perfusion intervals of ischemia. This compound and two of its synthetic derivatives, propionylcarnitine taurine amide and butyrylcarnitine taurine amide, proved equally efficacious in restoring function and were superior in effect when compared with those of carnitine, taurine or taurine plus propionyl-L-carnitine [7].

Substantiating evidence supporting improved functional recovery in ischemia-reperfusion protocols using isolated, perfused heart preparations has since been reported from one other laboratory (see Figure 4) [8] and suggested in data from a second laboratory [9]. There is also information to suggest that propionyl-L-carnitine has an enhancing influence on cardiac performance (cardiac output and heart rate-pressure product) at conditions of aerobic perfusion in an isolated heart model [10].

Propionyl-L-carnitine has further been tested using a more physiological

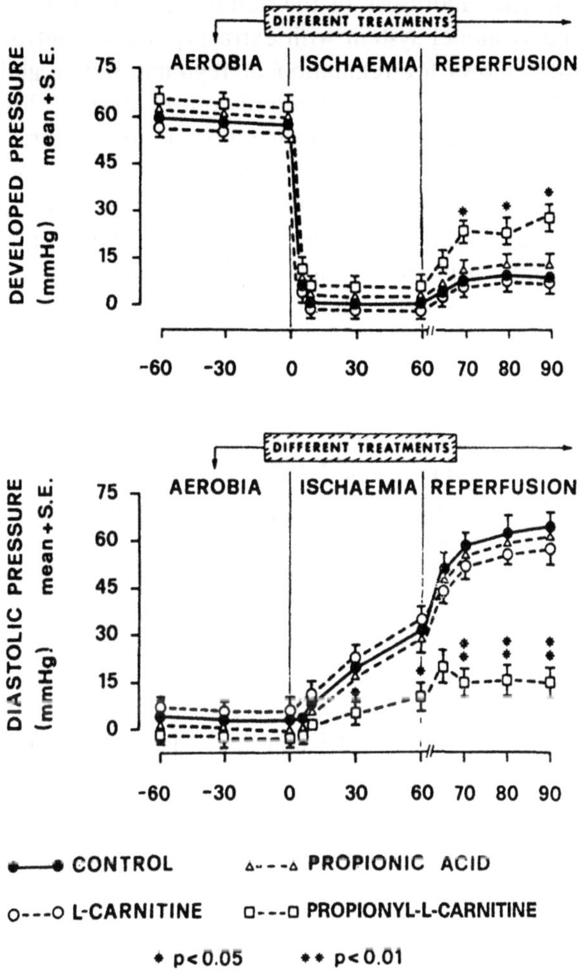


*Figure 4.* Effect of L-carnitine, propionic acid and propionyl-L-carnitine on the ischemia and reperfusion induced alterations of left ventricular pressure. Under aerobic and reperfusion conditions isolated rabbit hearts were perfused at a mean coronary flow of 25 ml/min. Ischemia was induced (at time 0) by reducing coronary flow to 1 ml/min. The administration of the different substances was started 30 min before the onset of ischemia. Each point is the mean of at least six separate experiments. P relates to the significance of the difference between the controls and the relative treated groups [8].

preparation with features of whole blood coronary perfusion fluid in intact, working hearts. The laboratory of the author has for some time employed an intact pig heart model system with extracorporeally-controlled coronary perfusion, which allows for regulation of regional and/or global coronary flows. In a perfusion protocol of 20 min of control, aerobic flows; 45 min of regional ischemia rendered by reducing flow in the anterior descending (LAD) perfusion bed by 60%; and 35 min reperfusion at aerobic preischemic levels, hearts underwent predictable mechanical stunning during reperfusion with accompanying declines in myocardial oxygen consumption. Pretreatment with propionylcarnitine (50 mg/kg at 0 min perfusion and 40 mg/kg at 40 min perfusion), infused IV, improved function (albeit not significantly) during ischemia but essentially restored function to near preischemic values during reperfusion (see Figure 5) [11]. $MVO_2$ was also higher in treated hearts (see Figure 6, panel A). A different yet complementary response in contractile function was reported by Sassen et al. [12] in another intact, pig heart preparation. In this protocol, LAD flow was more significantly curtailed (−80Δ% for 60 min) and then reperfused for 120 min. Hemodynamics in saline-treated hearts during reflow were further impaired below ischemic values (−20Δ% in mean arterial blood pressure; −25Δ% in left ventricular $dp/dt_{max}$; −25Δ% in cardiac output; −24Δ% in stroke volume; about +17Δ% in left ventricular end-diastolic pressure; and +30Δ% in systemic vascular resistance). Propionyl-L-carnitine pretreatment (50 mg/kg) attenuated these deteriorating trends with clear increases in cardiac output and left ventricular $dp/dt_{max}$ and decreases in left ventricular end-diastolic pressure and systemic vascular resistance (see Figure 7).

A final result suggesting a mechanical enhancing property of propionyl-L-carnitine, although brief, was reported in a canine infarct preparation which featured a thrombus-forming technique induced by a copper wire placed in the LAD artery [13]. At 30 and 180 min following coronary occlusion, treated animals evinced a more stabilized cardiac output and improved left ventricular pressures. The hemodynamic effects were associated with a reduction in infarct size.

**Possible mechanisms**

The above reports have also provided additional descriptions of other treatment effects noted with propionyl-L-carnitine. Some are associative epiphenomena with no clear insight into pharmacological mechanism(s), some are in direct conflict in terms of observational content, others are in agreement and at face value would seem to rule out possible mechanisms, while still others are suggestive of attractive leads to possible explanations of drug effect in enhancing myocardial contractility. The record is as yet incomplete and obviously subject to revision and recategorization. The following is the

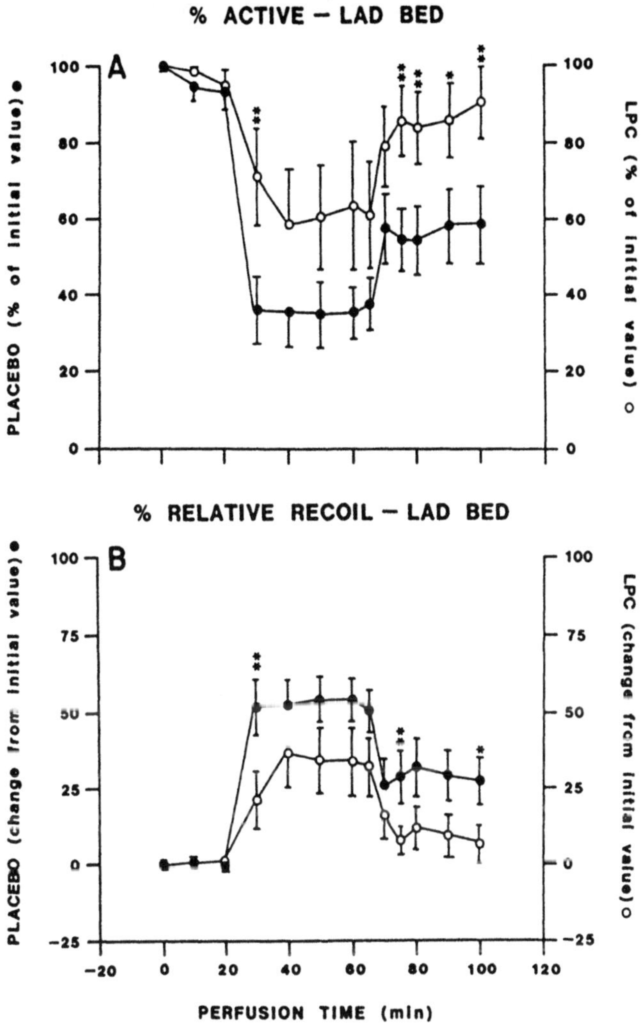


*Figure 5.* Regional motion changes showing% active contraction (Panel A) and relative recoil (Panel B) displayed as a function of perfusion times in the LAD extracorporeally perfused bed. Control aerobic flows were maintained from 0 to 20 min perfusion time; LAD flow thereafter was reduced by 60% from 20 to 65 min perfusion time; reperfusion at preischemic aerobic flows was restored again during the final 35 min reperfusion.% Active contraction is expressed as% of initial values at 0 min perfusion, whereas% relative recoil is expressed as the difference from initial values. Significant loss of motion and failure of mechanical recovery were noted in placebo porcine hearts. Treatment with propionyl-L-carnitine (LPC) spared motion loss during ischemia and reversed mechanical stunning during reflow. Asterisk symbols for statistical difference of intergroup comparisons are: * $p \leq 0.05$; ** $p \leq 0.025$ [11].

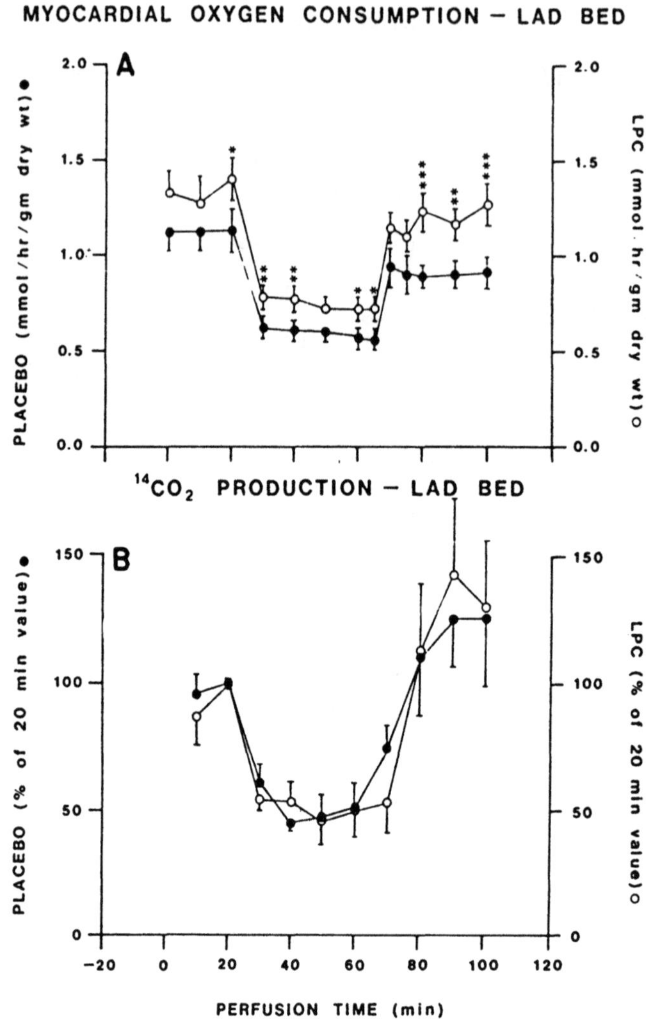


*Figure 6.* Indexes of aerobic metabolism including myocardial oxygen consumption ($M\dot{V}O_2$; Panel A) and $^{14}CO_2$ production rates from labeled palmitate (Panel B). Perfusion conditions and perfusion times are those described in Figure 5. Values of $M\dot{V}O_2$ closely paralleled those changes noted in regional motion in the LAD perfusion bed. Rates of fatty acid oxidation were almost identical between treatment groups during ischemia and the reperfusion portions of the perfusion trials. Relative uncoupling of substrate utilization from aerobic levels of performance in placebo porcine hearts was restored or partially reversed in treated hearts. LPC, L-propionyl-carnitine. Asterisk symbols for statistical difference of intergroup comparisons are: * $p \leq 0.05$; ** $p \leq 0.025$; *** $p \leq 0.005$ [11].

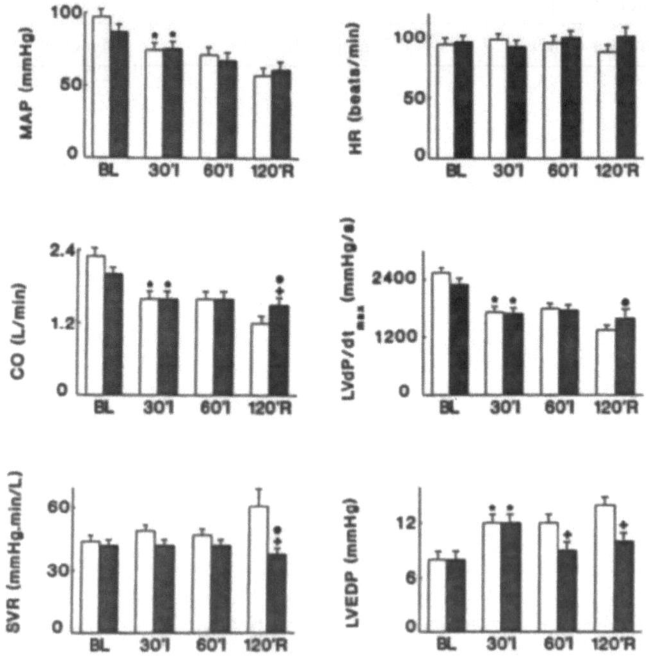


*Figure 7.* Systemic hemodynamic effects in saline-treated (open bar, n = 10) and propionyl-L-carnitine-treated (hatched bar, n = 9) pigs at baseline (BL) and after 30 and 60 min of ischemia (I) and 120 min of reperfusion (R). * $P \leq 0.05$ vs. BL, only presented for 30 and 60 min of I; + changes vs. BL in propionyl-L-carnitine-treated animals are significantly different from changes vs. BL in saline-treated animals; • changes vs. 60 min of I in propionyl-L-carnitine-treated animals are significantly different from changes vs. 60 min of I in saline treated animals. MAP, mean arterial blood pressure; HR, heart rate; CO, cardiac output; $LVdP/dt_{max}$, maximal rate of rise of left ventricular pressure; SVR, systemic vascular resistance; and LVEDP, left ventricular end-diastolic pressure [12].

author's attempt to survey this adjunctive material with personal interpretations where obvious.

Benefits with propionyl-L-carnitine do not seem related to shifts in myocardial amphiphile concentrations. Problems with lipid burden clearly occur in ischemic myocardium [4] with, among other consequences, accumulations of long-chain esters of both carnitine and CoA. The latter has particularly been incriminated as having nonspecific detergent-like properties which may accelerate ischemic injury by disrupting the integrity of cellular and organelle membranes. Prior studies with carnitine treatment in ischemic myocardium demonstrated reductions in tissue fatty acyl-CoA levels commensurate with a decrease in FFA cellular uptake [3]. However, with reperfusion following

*Table 1.* Effects of L-carnitine, acetyl-L-carnitine, and propionyl-L-carnitine on $^{14}C$-palmitate oxidation (in pmol · $mg^{-1}$ · $min^{-1}$) by cardiac homogenate and myocytes. Data are mean of three determinations [6].

| Drug concentration (mM) | Homogenate | Myocyte |
|---|---|---|
| Control | 202 | 72 |
| L-Carnitine: | | |
| 1 | 234 | 76 |
| 5 | 286 | 78 |
| 10 | 353 | 77 |
| Acetyl-L-carnitine: | | |
| 1 | 44 | 72 |
| 5 | 35 | 82 |
| 10 | 38 | 69 |
| Propionyl-L-carnitine: | | |
| 1 | 555 | 91 |
| 5 | 398 | 94 |
| 10 | 309 | 136 |

myocardial ischemia, these esters in the absence of therapeutic interventions are rapidly returned to aerobic values [14], presumably via the powerful action of washout, and thus on face value would not seem to be a principal mechanism to explain the mechanical benefits of propionyl-L-carnitine. Indeed, two studies in isolated and intact hearts confirmed that the protective effects of propionylcarnitine were not associated with alterations in long-chain acyl CoA levels [6, 11]. That is not to say, however, that propionyl-L-carnitine does not contain membrane-stabilizing properties, which may or may not involve local interactions with endogenous phospholipids in membranes. In isolated hearts, pretreatment with the carnitine derivative clearly spared release of creatine kinase in reperfusion following the ischemic period, suggesting that the compound maintained membrane integrity during the stress of ischemia and reperfusion [8, 15]. Aomine et al. [16] reported using a protocol of acidosis simulating possible ischemia in isolated guinea pig ventricular papillary muscles that pretreatment with propionyl-L-carnitine prevented most of the mechanical and electrical changes provoked by experimental exposure to the amphiphiles, lysophosphatidyl-L-choline and palmitoyl-L-carnitine. This electrophysiological benefit, however, could not be confirmed by Yamada and coworkers [17] studying hypoxia and reoxygenation in isolated adult canine myocytes and epicardial tissue.

Evidence is in conflict also regarding the role of propionyl-L-carnitine in influencing substrate utilization and calcium homeostasis in heart muscle. Paulson et al. [6] demonstrated an increase in $^{14}C$-palmitate oxidation in treated, aerobic, isolated cardiomyocytes and heart homeogenates (Table 1). The authors also reported a "rebound" in $CO_2$ production from [U-$^{14}C$]

palmitate above aerobic values during reperfusion but this increase was not confined to treated hearts and was equally obvious in untreated, placebo hearts (Figure 6, panel B) [11]. We have subsequently pursued this "rebound" of fatty acid oxidation in separate experiments independent of carnitine therapies and have determined that this trend may in part relate to distributions of radioactivity from exogenous labeling into and out of large free fatty acid (triacylglycerol) pools [18]. At conditions of severe ischemia followed by reperfusion, propionyl-L-carnitine also failed to alter by implication glucose utilization as reflected by comparable release of lactate into coronary effluent in treated and untreated hearts using an isolated rabbit heart preparation [8, 15]. However, this deduction conflicts with the direct measurements of Broderick et al. [19, 20] who reported in both aerobic and reperfused myocardium that L-carnitine (without the propionyl attachment) stimulated glycolysis and particularly, glucose oxidation, presumably by relieving the allosteric inhibition of fatty acids on a key glycolytic regulatory site, the pyruvate dehydrogenase complex. In reperfused hearts, this relief was associated with improved mechanical recovery.

Another area of controversy dealing with drug effect is calcium homeostasis. In ischemia the loss of function of calcium sequestration in sarcoplasmic reticulum (SR) as well as $Ca^{2+}$ ATPase activity in sarcolemma lead to calcium overload in cytosol and mitochondria. These abnormal levels conspire to activate lysosomal and membrane $Ca^{2+}$ dependent enzyme systems which further accelerate ischemic injury and to alter adversely the local environment of the actin-myosin contractile proteins [21, 22]. Studies with amiloride, an $H^+$ – $Na^+$ exchange inhibitor, suggest that this exchanger, which is stimulated in the presence of ischemic acidosis, also contributes to calcium overload in ischemia and early reflow [23, 24]. In reperfusion, calcium dynamics do not immediately revert to normal, in part because of damage to the SR and its calcium transport system, such that cytosolic calcium overload is not cleared by SR sequestration [25]. Taken in conjunction with an observed insensitivity of calcium responsiveness by myofilaments [26], an imbalance of calcium homeostasis has become one of the leading candidates to explain mechanical stunning in reperfusion. This was an obvious source of interest in evaluating the mechanical enhancing properties of propionyl-L-carnitine but as yet this interaction is uncertain. Sassen et al. [27] showed that SR was not selectively protected in terms of calcium uptake or $^{32}P$ incorporation into phospholamban as a function of carnitine treatment. On the other hand, Ferrari and coworkers [8, 15] showed evidence that propionyl-L-carnitine reduced the deterioration in mitochondrial respiration and improved in vitro ATP production which was in part affected by a reversal of mitochondrial calcium overload in ischemia and reperfusion.

A more attractive lead into possible therapeutic mechanisms was also provided in the studies of Sassen and coworkers [12, 27]. They demonstrated that the above described preservations in mechanical function were associated with an improvement in the "no-reflow" phenomenon with a 31%

increase in transmural myocardial blood flow during reperfusion as measured by radioactive microspheres. The drug prevented the increase in coronary vascular resistance which typically characterizes postischemic myocardium and which has been recently termed microvascular "stunning" [28]. The increase in coronary flow provided by propionylcarnitine may have improved mechanical recovery as has been demonstrated in experiments using vasodilator-enhanced coronary blood flow to enhance function in postischemic myocardium [29]. The vasodilator properties of propionyl-L-carnitine were also observed in the Sassen reports in other organ beds (skin and small intestine) and in systemic vascular resistance as compared with saline-treated animals.

Another etiology which has attracted great interest is the formation of reactive oxygen metabolites during myocardial ischemia and reperfusion. These oxidants have been strongly implicated as an alternate or additional mechanism to explain mechanical stunning [30]. Although these metabolites were initially difficult to measure, increasingly sophisticated techniques have provided convincing evidence that superoxide anion, hydrogen peroxide, and hydroxyl radicals all contribute to mechanical stunning, perhaps via adverse interactions on cellular proteins and lipids. Peroxidation of lipid constituents within cell membranes with loss of membrane integrity and resulting compromise of critical enzyme systems in membranes have been hypothesized. The causative role of free radical formation on mechanical stunning has been further strengthened by the improvement in function afforded by treatment with the antioxidants, superoxide dismutase and catalase [31]. With this as a lead, Packer et al. [9] have tested the possible antioxidant properties of propionyl-L-carnitine. Using an isolated, Langendorff perfused rat heart preparation, they observed improvements in mechanical function and high energy phosphate stores in treated hearts during a protocol of ischemia and reperfusion. More to the point, they documented in propionylcarnitine-treated hearts that oxidative injury to proteins as estimated by the accumulation of protein carbonyl groups and hydroxyl radicals, as measured by the spin trap, 5,5-dimethyl-1-pyrroline-1-oxide method (see Figure 8), were less. These findings were corroborated by Ferrari et al. [8] in independent studies examining the effects of propionyl-L-carnitine on the oxidative stress to myocardium subjected to ischemia and reperfusion. These investigators described this stress in terms of the "release of reduced and oxidized glutathione (GSH and GSSG, respectively) from the myocardium into the coronary effluent, the decrease in the tissue GSH/GSSG ratios, and the shift of the tissue redox-state toward oxidation". In hearts pretreated with propionyl-L-carnitine, oxidative stress was less as reflected by a decrease in GSH and GSSG release and a reduction in tissue GSSG. Taken together, these two reports provide a basis for pursuing further studies to validate this very attractive hypothesis regarding a possible drug mechanism of action.

A final hypothesis of drug efficacy centers on improved bioenergetics with propionyl-L-carnitine therapy. Despite the disparate trends in fatty acid oxidation (see above) observed with propionylcarnitine treatment and the

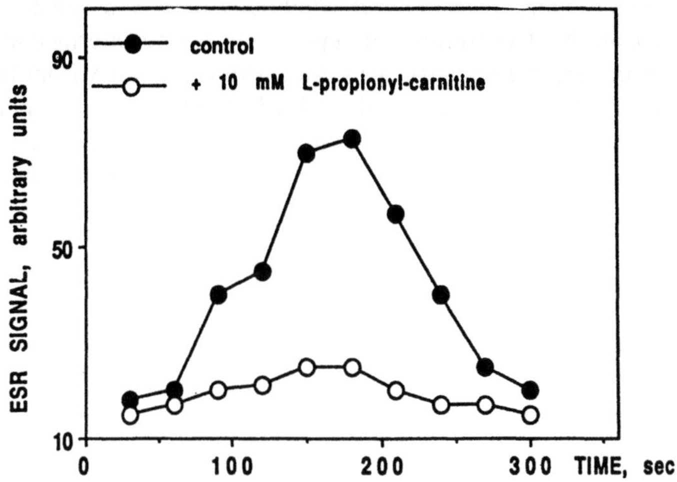


*Figure 8.* Effect of 10 mM propionyl-L-carnitine on the time course of ESR signal of 5, 5-dimethyl-1-pyrroline-1-oxide (DMPO) spin adducts in the rat-heart perfusate after 40 min of global ischemia [9].

reports of Sassen and coworkers [12, 27] which showed further deterioration in the adenosine nucleotide pool with such treatment, four other laboratories have reported clear improvements in energy metabolism and high energy phosphate production under a variety of experimental conditions, protocols, and perfusion models and preparations [6, 9, 10, 15]. Such trends may be explained by the added contributions of the propionyl group attached to the carnitine base of the propionylcarnitine compound. Sherry et al. [32] using high-resolution $^{13}C$ n.m.r. spectroscopy noted that enriched propionate initially enters the citric acid cycle at succinyl-CoA, proceeds to malate where it transiently exits the cycle via dispersal pathways, and then reenters the total pyruvate pool for eventual further processing in oxidative metabolism. Propionate may account for 27% of this pyruvate pool. A clue as to the benefits of the propionyl moiety in myocardial ischemia was provided by Di Lisa and colleagues [33] who described in hypoperfused rat hearts increases in measured propionyl CoA, relative retention of succinyl CoA, and an associated decrease in acetyl CoA, all compatible with an increase in flux of the tricarboxylic-acid cycle by contributions from the anaplerotic pathways. The participation of propionate in oxidative metabolism was confirmed in the laboratory of the author [34] who showed a clear dose-response relationship of $^{14}CO_2$ production from labeled propionate with a 12-fold increase in propionate oxidation between the doses of 0.1 and 10 mM propionate in coronary perfusate of isolated, working rat hearts. Propionate as a sole substrate was shown to be inadequate to support mechanical function alone. However, under more physiological conditions of mixed substrate availability, particularly at conditions of reperfusion where intermediate metabolism is again charging the citric acid cycle with acetyl CoA from fatty acid and glucose substrate utilizations, the added "bonus" of propionate as an anaplerotic contributor of Krebs cycle intermediates may increase cycle flux rates and in so doing, accelerate energy production and mechanical recovery.

## Acknowledgement

This work was supported in part by Sigma Tau Pharmaceuticals, the Oscar Mayer Cardiovascular Research Fund and U.S. Public Health Service Grants HL-41914 and HL32350.

## References

1. Liedtke AJ, Saless-Molaparast FMS. Secondary carnitine deficiency in cardiac disease. In: Borum PR, editor. Clinical aspects of human carnitine deficiency. Elmsford, New York: Pergamon Press, 1986: 204–15.
2. Liedtke AJ, Nellis SH. Effects of carnitine in ischemic and fatty acid supplemented swine hearts. J Clin Invest 1979; 64: 440–7.
3. Liedtke AJ, Nellis SH, Whitesell LF. Effects of carnitine isomers on fatty acid metabolism in ischemic swine hearts. Circ Res 1981; 48: 859–66.
4. Liedtke AJ. Lipid burden in ischemic myocardium. J Mol Cell Cardiol 1988; 20(Suppl 2): 65–74.
5. Mercier JC, Lando U, Kanmatsuse K et al. Divergent effects of inotropic stimulation on the ischemic and severely depressed reperfused myocardium. Circulation 1982; 66: 397–400.
6. Paulson DJ, Traxler J, Schmidt M, Noonan J, Shug AL. Protection of the ischaemic myocardium by L-propionylcarnitine: Effects on the recovery of cardiac output after ischaemia and reperfusion, carnitine transport, and fatty acid oxidation. Cardiovasc Res 1986; 20: 536–41.
7. Regitz V, Paulson DJ, Noonan J, Fleck E, Shug AJ. Protection of the ischemic myocardium by propionylcarnitine taurine amide. Comparison with other carnitine derivatives. Z Kardiol 1987; 76(Suppl 5): 53–8.
8. Ferrari R, Ceconi C, Curello S, Pasini E, Visioli O. Protective effect of propionyl-L-carnitine against ischaemia and reperfusion-damage. Mol Cell Biochem 1989; 88: 161–8.
9. Packer L, Valenza M, Serbinova E, Starke-Reed P, Frost K, Kagan V. Free radical scavenging is involved in the protective effect of L-propionyl-carnitine against ischemia-reperfusion injury of the heart. Arch Biochem Biophys 1991; 288: 533–7.
10. Bertelli A, Conte A, Ronca G, Zucchi R. Effect of propionyl carnitine on cardiac energy metabolism evaluated by the release of purine catabolites. Drugs Exp Clin Res 1991; 17: 115–8.
11. Liedtke AJ, DeMaison L, Nellis SH. Effects of L-propionylcarnitine on mechanical recovery during reflow in intact hearts. Am J Physiol 1988; 255: H169–76.
12. Sassen LMA, Bezstarosti K, Van der Giessen WJ, Lamers JMJ, Verdouw PD. L-propionyl-carnitine increases postischemic blood flow but does not affect recovery of energy charge. Am J Physiol 1991; 261: H172–80.

13. Leasure JE, Kordenat K. Effect of propionyl-L-carnitine on experimental myocardial infarction in dogs. Cardiovasc Drugs Ther 1991; 5(Suppl 1): 85–96.
14. Liedtke AJ, DeMaison L, Eggleston AM, Cohen LM, Nellis SH. Changes in substrate metabolism and effects of excess fatty acids in reperfused myocardium. Circ Res 1988; 62: 535–42.
15. Ferrari R, Ceconi C, Cargnoni A, Pasini E, Boffa GM, Curello S, Visioli O. The effect of propionyl-L-carnitine on the ischemic and reperfused intact myocardium and on their derived mitochondria. Cardiovasc Drugs Ther 1991; 5(Suppl 1): 57–66.
16. Aomine M, Arita M, Schimada T. Effects of L-propionylcarnitine on electrical and mechanical alterations induced by amphiphilic lipids in isolated guinea pig ventricular muscle. Heart Vessels 1988; 4: 197–206.
17. Yamada KA, Dobmeyer DJ, Kanter EM, Priori SG, Corr PB. Delineation of the influence of propionylcarnitine on the accumulation of long-chain acylcarnitines and electrophysiologic derangements evoked by hypoxia in canine myocardium. Cardiovasc Drugs Ther 1991; 5(Suppl 1): 67–76.
18. Nellis SH, Liedtke AJ, Renstrom B. Distribution of carbon flux within fatty acid utilization during myocardial ischemia and reperfusion. Circ Res 1991; 69: 779–90.
19. Broderick TL, Quinney HA, Lopaschuk GD. Carnitine stimulation of glucose oxidation in the fatty acid perfused isolated working rat heart. J Biol Chem 1992; 25: 3758–63.
20. Broderick TL, Quinney HA, Barker CC, Lopaschuk GD. Beneficial effect of carnitine on mechanical recovery of rat hearts reperfused after a transient period of global ischemia is accompanied by a stimulation of glucose oxidation. Circulation 1993; 87: 972–81.
21. Nayler WG, Daly MJ. Calcium and the injured cardiac myocyte. In: Sperelakis N, editor. Physiology and pathophysiology of the heart. Boston: Martinus Nijhoff Publishing, 1984: 477–92.
22. Braunwald E. Mechanism of action of calcium-channel-blocking agents. New Engl J Med 1982; 307: 1618–27.
23. Tani M, Neely JR. Role of intracellular $Na^+$ in $Ca^{2+}$ overload and depressed recovery of ventricular function of reperfused ischemic rat hearts: Possible involvement of $H^+$-$Na^+$ and $Na^+$-$Ca^{2+}$ exchange. Circ Res 1989; 65: 1045–56.
24. Murphy E, Perlman M, London RE, Steenbergen C. Amiloride delays the ischemia-induced rise in cytosolic free calcium. Circ Res 1991; 68: 1250–8.
25. Krause SM, Jacobus WE, Becker LC. Alterations in cardiac sarcoplasmic reticulum calcium transport in the postischemic "stunned" myocardium. Circ Res 1989; 65: 526–30.
26. Marban E. Myocardial stunning and hibernation: The physiology behind the colloquialisms. Circulation 1991; 83: 681–8.
27. Sassen LMA, Bezstarosti K, Koning MMG, Van der Giessen WJ, Lamers JMJ, Verdouw PD. Effects of administration of L-propionylcarnitine during ischemia on the recovery of myocardial function in the anesthetised pig. Cardioscience 1990; 1: 155–61.
28. Bolli R, Triana JF, Jeroudi MO. Prolonged impairment of coronary vasodilation after reversible ischemia: Evidence for microvascular "stunning". Circ Res 1990; 67: 332–43.
29. Stahl LD, Aversano TR, Becker LC. Selective enhancement of function of stunned myocardium by increased flow. Circulation 1986; 74: 843–51.
30. Bolli R. Mechanism of myocardial "stunning". Circulation 1990; 82: 723–38.
31. Jeroudi MO, Triana JF, Patel BS, Bolli R. Effect of superoxide dismutase and catalase, given separately, on myocardial "stunning". Am J Physiol 1990; 259: H889–901.
32. Sherry AD, Malloy CR, Roby RE, Rajagopal A, Jeffrey FMH. Propionate metabolism in the rat heart by $^{13}C$ n.m.r. spectroscopy. Biochem J 1988; 254: 593–8.
33. Di Lisa F, Menabo R, Siliprandi N. L-propionylcarnitine protection of mitochondria in ischemic rat hearts. Mol Cell Biochem 1989; 88: 169–73.
34. Bolukoglu H, Nellis SH, Liedtke AJ. Effects of propionate on mechanical and metabolic performance in aerobic rat hearts. Cardiovasc Drugs Ther 1991; 5(Suppl 1): 37–44.

*Corresponding Author*: Dr A. James Liedtke, Section of Cardiology, Hospital and Clinics, Highland Avenue, H6/339, University of Wisconsin, Madison, WI 53792–3248, USA

# 21. Dissociation of hemodynamic and metabolic effects of propionyl-L-carnitine in ischemic pig heart

PIETER D. VERDOUW, LOES M.A. SASSEN,
DIRK J. DUNCKER and JOS M.J. LAMERS

"Post-ischemic blood flow to the myocardium perfused by the left anterior descending coronary artery was higher in the propionyl-L-carnitine-treated than in the untreated animals. It has been well established that reperfusion after a prolonged period of myocardial ischemia does not always result in a complete return of blood flow since the vasculature of the ischemic myocardium can become obstructed by extravascular compression or by intravascular obstructions (the "no-reflow"-phenomenon). It could then be argued that propionyl-L-carnitine treatment attenuated this no-reflow phenomenon."

## Introduction

Propionyl-L-carnitine protects ischemic myocardium in the isolated working rat heart preparation [1] and enhances recovery of myocardial function following mild myocardial ischemia in an in vivo porcine model [2]. The enhanced recovery of post-ischemic myocardium has been ascribed to a positive inotropic action of the drug [2]. Stimulation with catecholamines will also improve function of depressed post-ischemic ("stunned") myocardium [3, 4], but this is accompanied by a shift in substrate utilization in favor of free fatty acids [4, 5]. Free fatty acid oxidation could, however, still be impaired because of a carnitine deficiency, depletion of mitochondrial pools of citric acid cycle intermediates and fatty acyl-CoA inhibition-induced adenine nucleotide translocation across the inner mitochondrial membrane. Propionyl-L-carnitine administration has the potential to relieve the deficiency of carnitine, to supply propionyl-CoA as a source of succinyl-CoA and to reduce fatty acyl- CoA accumulation (as we discussed in [6]). In this respect it is of interest that carnitine has increased the tolerance to pacing-induced myocardial ischemia in patients undergoing diagnostic cardiac catheterization [7]. Thus, propionyl-L-carnitine could, because of its metabolic actions, be useful during inotropic stimulation of post-ischemic myocardium with catecholamines.

In the present communication we review the results of our studies [6, 8,

*J.W. de Jong and R. Ferrari (eds): The carnitine system*, 291–306.

9] on the effects of propionyl-L-carnitine on recovery of function and metabolism (i) following severe myocardial ischemia in open-chest anesthetized pigs, and (ii) in view of the metabolic effects of catecholamine stimulation, during chronotropic (atrial pacing) and inotropic stimulation of post-ischemic myocardium with dobutamine.

## Materials and methods

Cross-bred Landrace × Yorkshire pigs (n = 47, 22–40 kg) were sedated, anesthetized and intubated for artificial ventilation, before the animals were catheterized for continuous administration of the anesthesia, the measurement of the arterial and left ventricular blood pressures and the withdrawal of blood samples [6, 8, 9]. After opening the thorax with a midsternal split, an electromagnetic flow probe (Skalar, Delft, The Netherlands) was placed around the ascending aorta for the measurement of aortic blood flow, while the left anterior descending coronary artery (LADCA) was dissected free distal from its first diagonal branch for placement of an inflatable balloon (R.E. Jones, Silver Spring, MD, USA) or an atraumatic clamp, depending on the study protocol. The vein accompanying the LADCA was cannulated for the withdrawal of blood samples in which coronary venous oxygen content was determined. To determine regional blood flows, the left atrial appendage was catheterized for injection of radioactively labelled microspheres.

In the first protocol regional myocardial function was assessed from myocardial wall thickness recordings, which were obtained using a 5 MHz ultrasonic transducer (Krautkamer-Branson, Lewistown, PA, USA) sutured onto a part of the epicardial surface in the distribution area of the LADCA. Systolic wall thickening (SWT,%) was calculated as 100 × (EST-EDT)/EDT, in which EST and EDT are wall thickness at end systole and end diastole, respectively. In two other protocols two pairs of ultrasonic crystals (Triton Technology, San Diego, CA, USA) were implanted in the distribution area of the LADCA and the LCXCA (left circumflex coronary artery) to record regional myocardial segment length changes. Systolic segment length shortening was calculated as: SLS (%) = 100 × (EDL-ESL)/EDL, in which EDL and ESL are the segment length at end diastole and end systole, respectively.

## Experimental protocols

*General.* In all protocols, values of systemic hemodynamics, regional myocardial function and regional blood flows were taken at baseline and during the course of the experiments, while biopsies for the measurement of high energy phosphates, (acyl)carnitine, short-chain fatty acylcarnitine and sarcoplasmic reticulum function were collected as described below.

*Protocol* 1. After administration of propionyl-L-carnitine (50 mg/kg over a period of 10 min, n = 9) or an equal volume of saline (20 ml, n = 10) the flow in the LADCA was gradually reduced by inflation of the balloon, until SWT had virtually been abolished. After 60 min, the balloon was deflated and the myocardium reperfused for two h. At the end of this reperfusion period, biopsies were taken from both the area perfused by the LADCA and from the control area for the determination of sarcoplasmic reticulum function (not discussed in this paper; for results see [8]) and tissue levels of free carnitine and short-chain fatty acylcarnitine.

*Protocol* 2. In this protocol, 7 animals were pretreated p.o. with propionyl-L-carnitine (50 mg/kg b.i.d. for 3 days). On the day of the experiments the animals received 50 mg/kg p.o. and 50 mg/kg iv. (administered over a period of 20 min), three h prior to the experiment. Seven other animals served as control. The myocardium perfused by the LADCA underwent two periods of 10 min of ischemia seperated by 30 min of myocardial reperfusion. Thirty min after the second reperfusion, heart rate was increased by 50 beats/min via left atrial pacing and 5 min later, a 10 min intravenous infusion of 2 μg/kg/min dobutamine was started, while the heart rate was kept constant.

*Protocol* 3. This protocol included 6 propionyl-L-carnitine-pretreated animals (for the dose see protocol 2) and 8 controls. These animals also underwent two sequences of 10 min coronary artery occlusion and 30 min of myocardial reperfusion, but at the end of the second reperfusion period, biopsies were taken from both the area perfused by the LADCA and from the control area for the determination of sarcoplasmic reticulum function (not discussed in this paper; for results see [6]) and tissue levels of free carnitine and short-chain fatty acylcarnitine.

*Drugs*. Propionyl L carnitine HCl and $NaHCO_3$, in an equimolar ratio, were dissolved in distilled water. pH of the final solution was between 6.8 and 7.3.

*Statistics*. All data have been presented as means ± SE. Because the animals in protocol 2 and 3 underwent the same procedure during the two sequences of 10 min LADCA occlusion and 30 min of reperfusion, the data obtained at baseline and after induction of stunning have been pooled. The significance of the propionyl-L-carnitine-induced changes was determined by comparing these changes with those observed in the saline-treated animals at corresponding time-points. Statistical significance was accepted for $P < 0.05$.

## Results

### Protocol 1

*Effects of propionyl-l-carnitine during 60 min of ischemia and 2 h of reperfusion*

Propionyl-L-carnitine administration before the flow reduction did not affect any of the systemic hemodynamic variables and regional blood flow data. The data determined after administration have therefore been presented as baseline before the flow reduction.

*Ventricular arrhythmias.* Seven of the 9 propionyl-L-carnitine-treated and 8 of the 10 saline-treated animals encountered an episode of ventricular fibrillation during the 60 min of flow reduction. All animals could be defibrillated within 30 s and resumed pre-fibrillation values and were therefore included in the analysis of the study. There was not a single period of ventricular fibrillation in either group during reperfusion.

*Systemic hemodynamics.* In the saline-treated animals mean arterial blood pressure (MAP), cardiac output (CO), stroke volume (SV) and $LVdP/dt_{max}$ decreased, while left ventricular end diastolic pressure (LVEDP) increased ($P < 0.05$) during the flow reduction. There was no recovery during the 2 h of reperfusion as MAP (by 20%), $LVdP/dt_{max}$ (by 25%), CO (by 25%) and SV (by 24%) further decreased, while LVEDP increased. Systemic vascular resistance (SVR) also increased (by 30%) and this prevented a larger fall in MAP (Figure 1).

Pretreatment with propionyl-L-carnitine had no effect on the ischemia-induced changes in MAP, CO, SV and $LVdP/dt_{max}$, while LVEDP increased from 8 ± 1 to 12 ± 1 mmHg during the first 30 min of flow reduction but, at variance with the saline-treated pigs, decreased to 9 ± 1 mmHg during the last 30 min of the 60 min flow reduction (Figure 1). CO did not deteriorate further during reperfusion, while SVR did not increase.

*Regional myocardial wall thickening.* During the coronary blood flow reduction, EDT decreased in both the saline-treated and the propionyl-L-carnitine-treated animals (8–12%, $P < 0.05$), while there was a complete loss of systolic thickening (baseline values of 29 ± 2% and 31 ± 4%, respectively). In both groups of animals, EDT had increased to values above baseline (10%) after two h of reperfusion, but there was no recovery of systolic wall thickening.

*Regional blood flows.* During the 60 min of ischemia the changes in flow to various organs were not the same in the saline-treated animals: kidneys (−12%), the iliopsoas muscle (−39%), the skin (−29%), the small intestine (−22%) and the brains (−18%). Propionyl-L-carnitine did not modify these

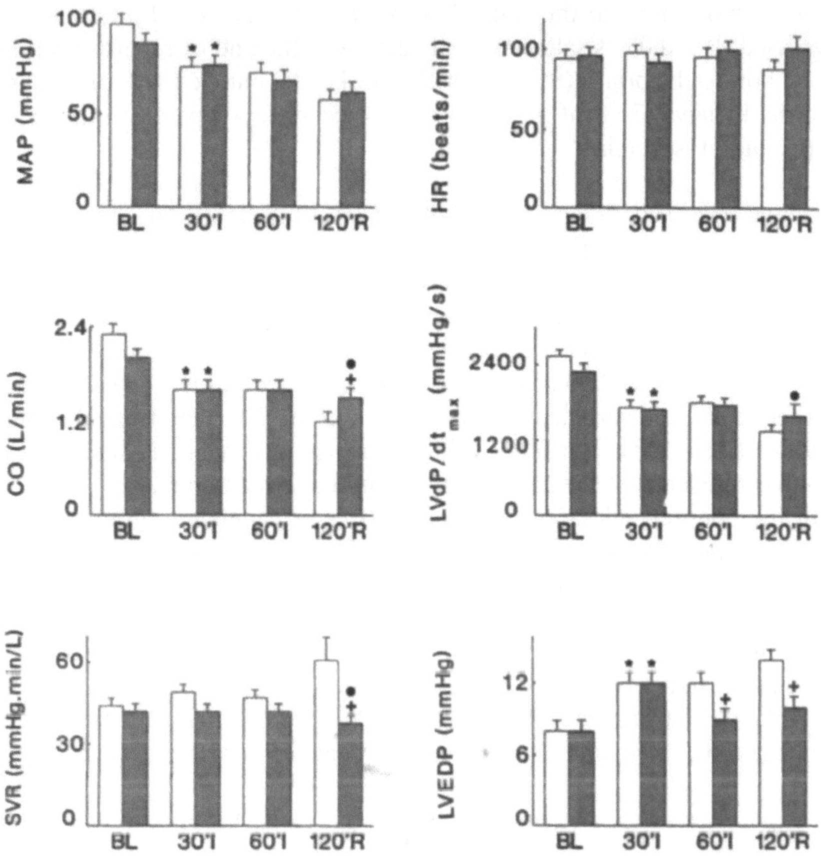


*Figure 1.* Systemic hemodynamic effects in saline-treated (open bars, n = 10) and propionyl-L-carnitine-treated (hatched bars, n = 9) pigs at baseline (BL), after 30 and 60 min of ischemia (I) and 120 min of reperfusion (R). * $P < 0.05$ vs baseline, only presented for 30 and 60 min of ischemia; + changes vs baseline in the propionyl-L-carnitine-treated animals are significantly different from the changes vs baseline in the saline-treated animals; • changes vs 60 min of ischemia in the propionyl-L-carnitine-treated animals are significantly different from the changes vs 60 min of ischemia in the saline-treated animals. MAP = mean arterial blood pressure, HR = heart rate, CO = cardiac output, $LVdPdt_{max}$ = maximal rate of rise of left ventricular pressure, SVR = systemic vascular resistance, LVEDP = left ventricular end diastolic pressure. (With permission from American Journal of Physiology [8].)

ischemia-induced changes. Reperfusion did not lead to recovery of regional blood flows in the saline-treated animals. In the propionyl-L-carnitine-treated animals, blood flow to the skin (79% of baseline) and small intestine were increased ($P < 0.05$ vs saline-treated animals at the end of reperfusion), but blood flow to the brains (96% of baseline), skeletal muscle (84% of baseline) and the kidneys (74% of baseline) did not change, compared to the values at the end of ischemia.

*Regional myocardial blood flow.* Inflation of the balloon caused similar decreases in transmural myocardial blood flow in the saline-treated and the propionyl-L-carnitine-treated animals during ischemia (Figure 2). After two h of reperfusion, transmural perfusion in the distribution area of the LADCA had returned to only 50% of baseline in the saline-treated animals, but to 80% of baseline ($P < 0.05$) in the propionyl-L-carnitine-treated animals. Consequently, the coronary vascular resistance of the saline-treated animals had increased (25%, $P < 0.05$), while there was no change in the coronary vascular resistance of the propionyl-L-carnitine-treated animals.

In the non-ischemic area of the saline-treated animals transmural blood flow decreased because of the fall in arterial blood pressure during ischemia. During reperfusion coronary vascular resistance did not change. In the propionyl-L-carnitine-treated animals blood flow at the end of reperfusion was not different from that at baseline (Figure 2). In these animals coronary vascular resistance decreased during the experiment from $0.9 \pm 0.1$ mmHg · min/mL/100 g at baseline, $0.8 \pm 0.1$ mmHg · min/mL/100 g after 60 min of ischemia to $0.7 \pm 0.1$ mmHg · min/mL/100 g at the end of reperfusion.

*Regional myocardial oxygen consumption.* Oxygen saturation of the coronary venous blood of the LADCA-perfused myocardium was $25 \pm 5$ and $24 \pm 2$% at baseline and $64 \pm 5$ and $61 \pm 8$% after 2 h of reperfusion in the saline-treated and the propionyl-L-carnitine-treated animals, respectively. Hence myocardial oxygen consumption remained directly related to coronary blood flow and was, because of the higher blood flow, higher in the propionyl-L-carnitine-treated animals at the end of reperfusion.

*High-energy phosphates.* Except for a 15% decrease ($P < 0.05$ vs baseline) in the adenine nucleotide pool of the propionyl-L-carnitine-treated animals at the end of reperfusion, there were no changes in the tissue high energy phosphate levels of the non-ischemic segment of either group (Figures 3 and 4). In the distribution area of the LADCA, there were similar decreases in ATP, creatine phosphate (CP), energy charge and the sum of the adenine nucleotides of both groups during the flow reduction. During this condition, the sum of CP and creatine remained unchanged, however. ATP levels did not recover during reperfusion in either group, but especially in the saline-treated animals there were improvements in CP and in the energy charge. The adenine nucleotide pool did not recover in the saline-treated animals,

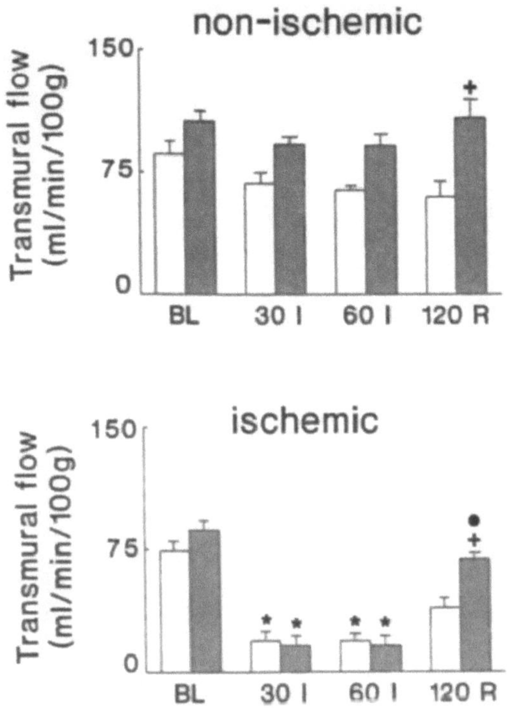


*Figure 2.* Transmural myocardial blood flow of non-ischemic and ischemic myocardium in saline-treated (open bars, n = 10) and propionyl-L-carnitine-treated (hatched bars, n = 9) pigs at baseline (BL), after 30 and 60 min of ischemia (I), and 120 min of reperfusion (R). * $P < 0.05$ vs baseline, only presented for 30 min and 60 min of ischemia; + changes vs baseline in the propionyl-L-carnitine-treated animals are significantly different from the changes vs baseline in the saline-treated animals; • changes vs 60 min of ischemia in the propionyl-L-carnitine-treated animals are significantly different from the changes vs 60 min of ischemia in the saline-treated animals. (With permission from American Journal of Physiology [8].)

while there was a further decrease in the propionyl-L-carnitine-treated animals (Figure 4). The sum of CP and creatine remained constant in the saline-treated pigs but decreased to 47% of baseline ($P < 0.05$ vs baseline) in the propionyl-L-carnitine-treated animals.

*Free carnitine and short-chain fatty acylcarnitine levels in plasma and myocardium.* The plasma levels short-chain fatty acylcarnitine increased from $1.20 \pm 0.49$ to $682 \pm 65$ μM after the bolus of 50 mg/kg of propionyl-L-carnitine. During ischemia and reperfusion the plasma level of short chain fatty acylcarnitine gradually decreased to $67 \pm 6$ μM at the end of reper-

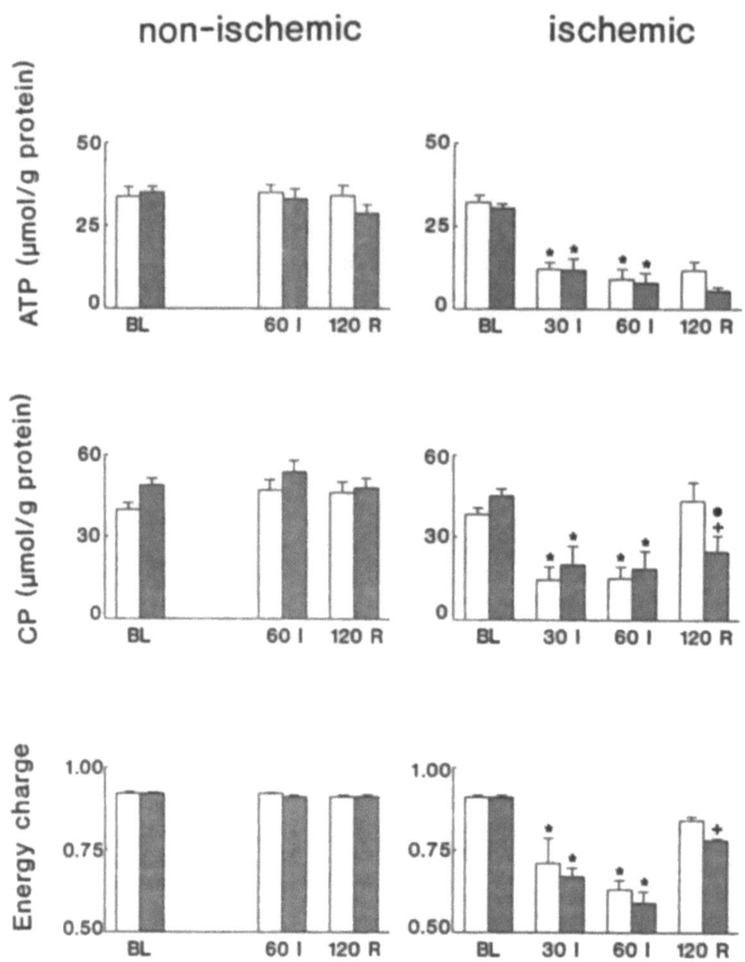


*Figure 3.* ATP and creatine phosphate (CP, µmol/g protein) and energy charge [(ATP + 0.5ADP)/(ATP + ADP + AMP)] of the ischemic and non-ischemic myocardium of saline-treated animals (open bars, n = 10) and propionyl-L-carnitine-treated (hatched bars, n = 9) at baseline (BL), at 30 (only for the ischemic segment) and 60 min of ischemia (I) and after 120 min of reperfusion (R). * $P < 0.05$ vs baseline, only presented for 30 and 60 min of ischemia; + changes vs baseline in the propionyl-L-carnitine-treated animals are significantly different from the changes vs baseline in the saline-treated animals; ● changes vs 60 min of ischemia in the propionyl-L-carnitine-treated animals are significantly different from the changes vs 60 min of ischemia in the saline-treated animals. (With permission from American Journal of Physiology [8].)

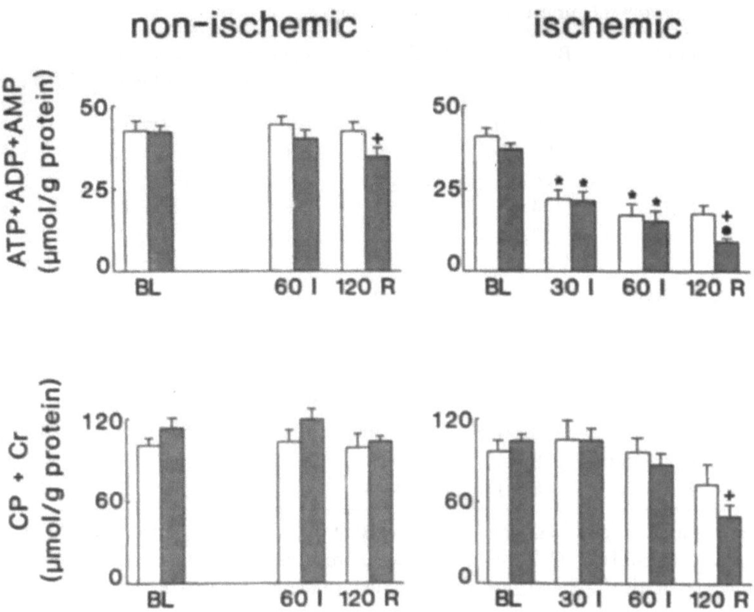


*Figure 4.* Total adenine nucleotide (ATP + ADP + AMP) and creatine (CP + creatine) pools of the non-ischemic and ischemic myocardium in saline-treated (open bars, n = 10) and propionyl-L-carnitine-treated (hatched bars, n = 9) pigs at baseline (BL), at 30 (only for the ischemic segment) and 60 min of ischemia (I) and after 120 min of reperfusion (R). * $P < 0.05$ vs baseline, only presented for 30 and 60 min of ischemia; + changes vs baseline in the propionyl-L-carnitine-treated animals are significantly different from the changes vs baseline in the saline-treated animals; • changes vs 60 min of ischemia in the propionyl-L-carnitine-treated animals are significantly different from the changes vs 60 min of ischemia in the saline-treated animals. CP = creatine phosphate. (With permission from American Journal of Physiology [6].)

fusion. The free carnitine levels increased rapidly by a factor of 10 after the bolus infusion of propionyl-L-carnitine and remained at this level until the end of reperfusion.

The myocardial free carnitine levels had decreased to about 60% of baseline value at the end of reperfusion. In both the saline- and the propionyl-L-carnitine treated animals, the levels of short-chain fatty acylcarnitine remained unchanged [8].

*Table 1.* The hemodynamic effects of two sequences of proximal left anterior descending coronary artery occlusion (10 min) and reperfusion (30 min) in untreated (n = 14) and with propionyl-L-carnitine (PLC, n = 13) pretreated anesthetized pigs.

| | Group | Baseline | Occlusion-1 (10 min) | Reperfusion-1 (30 min) | Occlusion-2 (10 min) | Reperfusion-2 (30 min) |
|---|---|---|---|---|---|---|
| CO | Untreated | 2.9 ± 0.2 | 2.5 ± 0.2[a] | 2.5 ± 0.2[a] | 2.2 ± 0.2[a] | 2.3 ± 0.2[a] |
| | PLC | 2.9 ± 0.2 | 2.4 ± 0.3[a] | 2.5 ± 0.2[a] | 2.2 ± 0.3[a] | 2.3 ± 0.3[a] |
| HR | Untreated | 106 ± 5 | 107 ± 5 | 106 ± 3 | 109 ± 4 | 109 ± 4 |
| | PLC | 114 ± 5 | 116 ± 6 | 115 ± 6 | 120 ± 6 | 119 ± 6 |
| SV | Untreated | 27 ± 1 | 23 ± 1[a] | 23 ± 1[a] | 20 ± 1[a] | 22 ± 1[a] |
| | PLC | 26 ± 2 | 21 ± 2[a] | 22 ± 2[a] | 19 ± 2[a] | 20 ± 2[a] |
| MAP | Untreated | 90 ± 4 | 85 ± 4[a] | 86 ± 3 | 79 ± 4[a] | 86 ± 3 |
| | PLC | 89 ± 3 | 79 ± 2[ab] | 80 ± 4[a] | 69 ± 3[ab] | 74 ± 4[ab] |
| SVR | Untreated | 33 ± 2 | 36 ± 3[a] | 37 ± 3[a] | 37 ± 2[a] | 39 ± 3[a] |
| | PLC | 32 ± 2 | 35 ± 2[a] | 33 ± 2 | 34 ± 2 | 34 ± 2[b] |
| $LVdP/dt_{max}$ | Untreated | 2150 ± 130 | 1720 ± 100[a] | 1680 ± 90[a] | 1570 ± 90[a] | 1640 ± 100[a] |
| | PLC | 2130 ± 105 | 1670 ± 100[a] | 1670 ± 110[a] | 1590 ± 110[a] | 1590 ± 120[a] |
| LVEDP | Untreated | 8 ± 1 | 14 ± 2[a] | 7 ± 1 | 13 ± 2[a] | 9 ± 1 |
| | PLC | 8 ± 1 | 13 ± 2[a] | 8 ± 1 | 11 ± 2[a] | 8 ± 1 |

CO, cardiac output (L/min); HR, heart rate (beats/min); SV, stroke volume (mL); MAP, mean arterial pressure (mmHg); SVR, systemic vascular resistance (mmHg · min/L); LV $dP/dt_{max}$, maximum rate of rise of left ventricular pressure (mmHg/s); LVEDP, left ventricular end-diastolic pressure (mmHg).
[a] $P < 0.05$ vs baseline; [b] PLC-induced changes vs baseline are significantly different from changes vs baseline in untreated pigs. (Adapted from Journal Cardiovascular Pharmacology [6].)

### *Protocols 2 and 3*

### *Effect of propionyl-l-carnitine during two sequences of 10 min coronary artery occlusion and 30 min of reperfusion*

*Systemic hemodynamics*. CO fell from 2.9 ± 0.2 to 2.5 ± 0.2 L/min ($P < 0.05$, Table 1) in the control group during the first coronary artery occlusion. MAP decreased only by about 5% because of an increase in SVR. Because HR did not change, a decrease in SV was therefore the cause of the reduction in CO. $LVdP/dt_{max}$ decreased by 20% ($P < 0.05$), while LVEDP doubled. Except for LVEDP, there was no recovery during the first reperfusion period. During the second coronary artery occlusion there were additional decreases in CO, MAP, SV, while LVEDP again doubled. During the second reperfusion, MAP recovered partially because of vasoconstriction.

In the propionyl-L-carnitine-treated animals, the large decrease in MAP at the end of the two coronary artery occlusion-reperfusion sequences was the major difference with the findings in the untreated animals. In concert with the series of experiments in protocol 1, SVR did not increase in the propionyl-L-carnitine-treated animals.

*Table 2.* Myocardial segment length shortening (SLS) during two sequences of proximal left anterior descending coronary artery (LADCA) occlusion and reperfusion in untreated and with propionyl-L-carnitine (PLC) pretreated anesthetized pigs.

| | Group | n | Baseline | Occlusion-1 (10 min) | Reperfusion-1 (30 min) | Occlusion-2 (10 min) | Reperfusion-2 (30 min) |
|---|---|---|---|---|---|---|---|
| Control Area | | | | | | | |
| SLS (%) | Untreated | 13 | 14.9 ± 1.2 | 16.9 ± 1.6 | 14.2 ± 1.3 | 16.4 ± 1.7 | 14.4 ± 1.3 |
| | PLC | 13 | 11.4 ± 1.0 | 12.0 ± 1.5 | 10.2 ± 1.1 | 11.3 ± 1.1 | 10.6 ± 1.1 |
| LADCA-Perfused Area | | | | | | | |
| SLS (%) | Untreated | 14 | 18.5 ± 1.5 | $-3.7 \pm 2.0^{a}$ | $6.2 \pm 1.6^{a}$ | $-3.8 \pm 0.6^{a}$ | $5.1 \pm 1.5^{a}$ |
| | PLC | 13 | 17.4 ± 1.2 | $-4.5 \pm 0.7^{a}$ | $7.6 \pm 1.1^{a}$ | $-5.0 \pm 0.8^{a}$ | $6.8 \pm 1.2^{ab}$ |

[a] $P < 0.05$ vs baseline; [b] $P = 0.056$ for the PLC-induced change from baseline vs change from baseline in untreated pigs. (Adapted from Journal Cardiovascular Pharmacology [6].).

*Regional myocardial segment length shortening.* Segment length shortening (SLS) in the distribution area of the LADCA was abolished in the untreated animals during the first coronary artery occlusion and recovered only partially during the first reperfusion (Table 2). A similar pattern was observed during the second occlusion-reperfusion sequence. In the propionyl-L-carnitine-treated animals, there was also a complete loss in SLS during the occlusions, but systolic segment shortening of the LADCA-perfused area was 39% of baseline at the end of reperfusion-2 in the treated animals, while this was 28% of baseline ($P = 0.056$) in the untreated animals.

*Regional myocardial blood flow.* Subendocardial blood flow in the not stunned segment (LCXCA) was reduced at the end of reperfusion-2, while subepicardial blood flow remained unchanged (see [6]). This resulted in a decrease in the subendocardial/ subepicardial blood flow ratio from 1.11 ± 0.09 at baseline to 0.92 ± 0.17 at the end of reperfusion 2 ($P < 0.05$). In the LADCA-perfused area, the two occlusion-reperfusion sequences reduced transmural myocardial blood flow from 164 ± 38 to 101 ± 23 mL/min/100 g ($P < 0.05$). In this segment, the subendocardial/subepicardial blood flow ratio did not change. Pretreatment with propionyl-L-carnitine did not affect baseline blood flow in any of the layers of the stunned and the not stunned myocardium. The changes in blood flow to the LADCA-perfused area in the propionyl-L-carnitine group were not different from those in the untreated animals (see [6]).

*Regional myocardial oxygen ($MVO_2$) and lactate consumption ($MV_{lactate}$).* In the untreated group, transmural $MVO_2$ of the myocardium perfused by the LADCA decreased by 39% ($P < 0.05$). This was accompanied by a marked reduction in $MV_{lactate}$ from 118 ± 40 μmol/min/100 g at baseline to 11 ± 14 μmol/min/100 g at the end of reperfusion-2 ($P < 0.05$). Pretreatment with propionyl-L-carnitine had no affect on $MVO_2$ and $MV_{lactate}$ at baseline. Also, ischemia- and reperfusion-induced changes in $MVO_2$ and $MV_{lactate}$ were similar to those in the control group.

*Protocol 2*

*Effect of propionyl-L-carnitine during chronotropic and inotropic stimulation of post-ischemic myocardium*

*Systemic hemodynamics.* In the untreated animals, atrial pacing at 50 beats/min above spontaneous sinus rhythm did not change any of the hemodynamic variables, with the exception of SV, which decreased by 39% (see [6]). The additional infusion of dobutamine improved $LVdP/dt_{max}$ and CO with 90% and 30% respectively ($P < 0.05$). As a result, CO returned to baseline level and $LVdP/dt_{max}$ was increased by 60% compared to baseline. SV increased slightly but remained depressed compared to baseline, while LVEDP decreased by 30% during dobutamine administration. SVR, which was elevated, returned to baseline values, leaving MAP unchanged.

In the animals pretreated with propionyl-L-carnitine, SVR was not affected by either chronotropic or inotropic stimulation.

*Regional myocardial segment length shortening.* Atrial pacing reduced SLS in the not stunned myocardium of the untreated animals from $13.4 \pm 4.2\%$ (at the end of reperfusion-2) to $9.3 \pm 3.0\%$ ($P < 0.05$), while there was no additional change after dobutamine. In the stunned area, SLS was not affected by atrial pacing, but increased to $14.7 \pm 6.0\%$ during dobutamine. Pretreatment with propionyl-L-carnitine had no effect on the response of SLS to atrial pacing and dobutamine.

*Myocardial blood flow.* Atrial pacing had no affect on transmural myocardial blood flow in the stunned and not stunned area of the untreated group. During infusion of dobutamine, however, perfusion of the not stunned and the stunned areas increased with 25 and 50%, respectively (see [6]). These increases were equally distributed over the subepicardial and subendocardial layers. In the stunned area of the propionyl-L-carnitine treated animals, the changes in flow in the LADCA-perfused area were similar to those in the untreated group, but in the not stunned segment they did not reach levels of statistical significance during either atrial pacing or dobutamine infusion.

*Regional myocardial oxygen and lactate consumption.* Atrial pacing had no effect on $MVO_2$ and $MV_{lactate}$ of the stunned myocardium of the untreated animals, but during the additional infusion of dobutamine, $MVO_2$ increased from $369 \pm 93$ μmol/min/100 g (during atrial pacing), to $518 \pm 107$ μmol/min/100 g ($P < 0.05$), while $MV_{lactate}$ almost doubled ($P < 0.05$). $MVO_2$ of the treated animals did not change significantly during chronotropic and inotropic stimulation, while changes in $MV_{lactate}$ were similar to those in the untreated animals.

*Protocol* 3

*Effect of propionyl-L-carnitine on myocardial depletion of high energy phosphates and carnitine produced by ischemia and reperfusion*

*Regional myocardial high energy phosphates.* After induction of stunning, ATP and ADP in the stunned area were reduced by 35% (see [6]). The energy charge, however, was unchanged, while CP increased by 50%, which suggests an intact oxidative phosphorylation potential. Pretreatment with propionyl-L-carnitine had no effect on these variables, before or after induction of stunning.

*Regional myocardial free carnitine and short chain acylcarnitine levels.* In the control group the occlusion-reperfusion sequences resulted in 25% lower levels of free carnitine in the stunned area than in the not stunned area at the end of reperfusion-2, while levels of short chain acylcarnitine in the ischemic area remained unchanged. Pretreatment with propionyl-L-carnitine increased the myocardial levels of free carnitine. The ischemia-reperfusion-induced decreases in the tissue levels of free carnitine in the LADCA perfused myocardium were similar for both groups. However, due to the initial increase, the tissue levels in the post-ischemic myocardium of the treated animals did not fall below the baseline values of the untreated animals.

## Discussion

The most important finding in the first protocol was that post-ischemic blood flow to the myocardium perfused by the LADCA was higher in the propionyl-L-carnitine-treated than in the untreated animals. It has been well established that reperfusion after a prolonged period of myocardial ischemia does not always result in a complete return of blood flow since the vasculature of the ischemic myocardium can become obstructed by extravascular compression or by intravascular obstructions (the "no-reflow"-phenomenon [10]). It could then be argued that propionyl-L-carnitine-treatment attenuated this no-reflow phenomenon. It has been suggested that propionyl-L-carnitine has a stabilizing action on plasma membrane, during ischemia and associated acidosis of the Langendorff-perfused rat heart [11]. Not only the sarcolemma of the cardiomyocytes, but also the vascular endothelium may be protected by the compound. Van Hinsberg and Scheffer [12] demonstrated in fura-2-loaded human endothelial cells in culture that propionyl-L-carnitine decreased the resting cytoplasmic free $Ca^{2+}$ concentration, which also indicates a direct effect on the plasma membrane of these cells. The mechanism by which propionyl-L-carnitine decreases resting levels of $Ca^{2+}$ is not known, but could be mediated by a direct effect on the plasma membrane $Na^{+}/Ca^{2+}$ exchanger and/or $Ca^{2+}$ pump. Propionyl-L-carnitine could also have produced a direct coronary vasodilatory effect. The data obtained immediately after administration of the drug do not support this idea but at the end of the experiment, the higher blood flows to the non-ischemic myocardium and some organs may have been due to a direct vasodilatory action of propionyl-L-carnitine.

The higher post-ischemic blood flow in the propionyl-L-carnitine-treated animals was not accompanied by a return of systolic contractile function during early reperfusion. The explanation for this can be threefold: (i) The myocardial tissue is irreversibly injured. In our laboratory we have shown that four weeks after a 60 min total coronary artery occlusion recovery of contractile function is absent [13, 14], but others found that 30–40% of the myocardium at risk was still viable [10]. In the present study we decreased coronary blood flow with 80% for 60 min. A significant fraction of the affected myocardium should therefore have still been viable. (ii) The low level of ATP of the post-ischemic myocardium prevented contractile function. However, it has repeatedly been demonstrated that enhanced recovery of function occurs while the low ATP levels are not affected [15, 16]. It is unlikely that the stunning of the myocardium was so severe that the myocardium became resistant to any stimulation, as even after two h of total occlusion, recruitment of regional contractile function of the stunned myocardium is still possible [15]. Previous studies on the effects of propionyl-L-carnitine on function of post-ischemic myocardium have yielded variable results. Thus, in isolated rat [11] and rabbit [17] hearts, subjected to low-flow global ischemia, propionyl-L-carnitine improved the recovery of developed left ventricular pressure during subsequent reperfusion. This effect has been ascribed to the increase in fatty acid oxidation [11] and the attenuation of oxidative stress as reflected by the oxidized/reduced glutathion ratio [17]. However, e.g. myocardial long chain acyl-CoA levels were not affected [11]. Observations in isolated perfused rat hearts subjected to low-flow ischemia, suggested that preservation of fatty acid oxidation may occur as propionyl-L-carnitine attenuated the ischemia-induced decrease in succinyl-CoA. Propionyl-L-carnitine stimulates the tricarboxylic-acid cycle via this anaplerotic action [18]. (iii) In extracorporeal blood-perfused pig hearts subjected to 45 min of low-flow ischemia (60% flow reduction) propionyl-L-carnitine improved myocardial contractile function during reflow [2]. This effect was ascribed to positive inotropic actions of propionyl-L-carnitine as long chain acylcarnitine and long-chain-acyl-CoA levels were unchanged. In contrast, in the first protocol propionyl-L-carnitine had no effect on post-ischemic metabolism or function in a model of 60 min of low-flow ischemia (80% flow reduction).

SLS recovered to 39% of baseline in the untreated animals after the two occlusion-reperfusion sequences (Protocols 2 and 3), but to 56% of baseline after pretreatment with propionyl-L-carnitine. This positive effect is not necessarily caused by the putative inotropic action as the prevention of peripheral systemic vasoconstriction by propionyl-L-carnitine, and the subsequent lower arterial blood pressure (afterload) could have been responsible for the slight improvement in SLS at the end of the second reperfusion period. During atrial pacing and dobutamine infusion, there were no apparent differences between propionyl-L-carnitine-pretreated and untreated animals in any of the metabolic or functional variables, which also points towards a mechanism not related to an inotropic action.

Although pretreatment with propionyl-L-carnitine did increase myocardial carnitine levels, it did not result in the expected shift from anaerobic to aerobic metabolism during stimulation with dobutamine [6]. These results indicate that myocardial depletion of carnitine or mitochondrial citric acid cycle intermediates does not contribute to myocardial hypofunction in the present model of repetitive ischemia and reperfusion.

## References


1. Paulson DJ, Traxler J, Schmidt M, Noonan J, Shug AL. Protection of the ischaemic myocardium by L-propionylcarnitine: effects on the recovery of cardiac output after ischaemia and reperfusion, carnitine transport, and fatty acid oxidation. Cardiovasc Res 1986; 20: 536–41.
2. Liedtke AJ, DeMaison L, Nellis SH. Effects of L-propionylcarnitine on mechanical recovery during reflow in intact hearts. Am J Physiol 1988; 255: H169–76.
3. Becker LC, Levine JH, DiPaula AF, Guarnieri T, Aversano T. Reversal of dysfunction in postischemic stunned myocardium by epinephrine and postextrasystolic potentiation. J Am Coll Cardiol 1986; 7: 580–9.
4. McFalls EO, Duncker DJ, Krams R, Sassen LMA, Hoogendoorn A, Verdouw PD. Recruitment of myocardial work and metabolism in regionally stunned porcine myocardium. Am J Physiol 1992; 263: H1724–31.
5. Van der Vusse GJ, Glatz JFC, Stam HCG, Reneman RS. Fatty acid homeostasis in the normoxic and ischemic heart. Physiol Rev 1992; 72: 881–940.
6. Duncker DJ, Sassen LMA, Bartels GL et al. L-Propionylcarnitine does not affect myocardial metabolic or functional response to chronotropic and inotropic stimulation after repetitive ischemia in anesthetized pigs. J Cardiovasc Pharmacol 1993; 22: 488–98.
7. Thomson JH, Shug AL, Yap VU, Patel AK, Karras TJ, DeFelice SL. Improved pacing tolerance of the ischemic human myocardium after administration of carnitine. Am J Cardiol 1979; 43: 300–6.
8. Sassen LMA, Bezstarosti K, Van der Giessen WJ, Lamers JMJ, Verdouw PD. L-propionylcarnitine increases post-ischemic blood flow but does not affect recovery of the energy charge. Am J Physiol 1991; 261: H172–80.
9. Sassen LMA, Duncker DJ, Hogendoorn A et al. L-propionylcarnitine and myocardial performance in stunned porcine myocardium. Mol Cell Biochem 1992; 116: 147–53.
10. Kloner RA, Ganote CE, Jennings RB. The "no-reflow" phenomenon after temporary coronary occlusion in the dog. J Clin Invest 1974; 54: 1496–508.
11. Hülsmann WC. Biochemical profile of propionyl-L-carnitine. Cardiovasc Drugs Ther 1991; 5(Suppl 1): 7–9.
12. Van Hinsberg VWM, Scheffer MA. Effect of propionyl-L-carnitine on human endothelial cells. Cardiovasc Drugs Ther 1991; 5(Suppl 1): 97–105.

13. Post JA, Lamers JMJ, Verdouw PD, Ten Cate FJ, Van der Giessen WJ, Verkleij AJ. Sarcolemmal destabilization and destruction after ischaemia and reperfusion and its relation with long-term recovery of regional left ventricular function in pigs. Eur Heart J 1987; 8: 423–30.
14. Van der Giessen WJ, Verdouw PD, Ten Cate FJ, Essed CE, Rijsterborgh H, Lamers JMJ. In vitro cyclic AMP induced phosphorylation of phospholamban: an early marker of long-term recovery of function following reperfusion of ischaemic myocardium? Basic Res Cardiol 1988; 10: 714–8.
15. Arnold JMO, Braunwald E, Sandor T, Kloner RA. Inotropic stimulation of reperfused myocardium with dopamine: Effects on infarct size and myocardial function. J Am Coll Cardiol 1985; 6: 1026–34.
16. Van der Giessen WJ, Schoutsen B, Tijssen JGP, Verdouw PD. Iloprost (ZK 36374) enhances recovery of regional myocardial function during reperfusion after coronary artery occlusion in the pig. Br J Pharmacol 1986; 87: 23–7.
17. Ferrari R, Ceconi C, Curello S, Pasini E, Visioli O. Protective effect of propionyl-L-carnitine against ischaemia and reperfusion-damage. Mol Cell Biochem 1989; 88: 161–8.
18. Di Lisa F, Menabò R, Siliprandi N. L-propionyl-carnitine protection of mitochondria in ischemic rat hearts. Mol Cell Biochem 1989; 88: 169–73.

*Corresponding Author*: Professor Pieter D. Verdouw, Experimental Cardiology, Thoraxcenter, Ee2351, Erasmus University Rotterdam, P.O. Box 1738, 3000 DR Rotterdam, The Netherlands

# 22. Effect of propionyl-L-carnitine on rats with experimentally induced cardiomyopathies

ROSELLA MICHELETTI, ANTONIO SCHIAVONE and GIUSEPPE BIANCHI

"Propionyl-L-carnitine stands out as possessing some new and interesting features that could be of value in the management of congestive heart failure. It is also evident that, because the biochemical alterations accompanying congestive heart failure may differ according to either the stage of the disease or its aetiology, propionyl-L-carnitine should be particularly useful in those conditions where its biochemical activity could compensate for an existing metabolic deficiency."

## Introduction

Congestive heart failure (CHF) is the functional definition of a clinical syndrome with a heterogeneous underlying aetiology, but a common manifestation characterised by the "exhaustion of the reserve force of the heart" [1]. Despite the different classes of drugs employed in its treatment, CHF maintains a poor prognosis [2] and is considered a malignant disease. It is noteworthy that some recent editorials published in leading medical journals emphasise the "failure to treat heart failure" [3, 4]. Thus, despite the progress that has unequivocally been achieved in recent years with the development of pharmacological classes of agents acting through novel mechanisms, the need to develop effective and safe drugs remains strong [5].

The available therapy comprises drugs that exert their primary effect directly on the myocardial cell, and drugs that act mainly on the heart load. The former group includes the inotropes. Digitalis glycosides, the longest running inotropes, are undoubtedly effective in improving symptoms in a subset of patients [6], but whether they are also effective in the prognosis of heart failure is still being investigated. Disappointing results have been yielded by the new inotropic agents, aimed at elevating cAMP levels: both the phosphodiesterase inhibitors, milrinone and enoximone, and the partial $\beta$-adrenoceptor agonist, xamoterol, shorten survival in CHF patients [7–9], limiting their usefulness to an acute setting.

Agents aimed at reducing preload and afterload, i.e. diuretics, vasodilators and angiotensin converting enzyme inhibitors (ACE), include the only drugs

*J.W. de Jong and R. Ferrari (eds): The carnitine system*, 307–322.

effective both on symptoms and on prognosis [10]. It is likely that the therapeutic efficacy of ACE inhibitors does not depend solely on the reduction of work load. In fact, in experimentally induced cardiomyopathies, they have been shown to directly inhibit the cardiac ACE system [11], affecting remodelling and the development of fibrosis [12] and to exert effects unrelated to inhibition of angiotensin II production [13].

Heart function may fail due to a primary impairment of myocardial contractility or following the imposition of an excessive load. Depending on the aetiology of heart failure, different cardiac adaptive mechanisms and different pathological evolutions ensue. These will be considered separately below. Irrespective of the nature of the overload, the fundamental mechanism elicited in the attempt to reduce the increase in wall stress is the development of hypertrophy. However, the initial compensation consequent to the unloading of cardiomyocytes does not ultimately prevent the transition from hypertrophy to failure. Katz postulated that the failing heart is in a state of energy starvation, due to an imbalance between energy production and use [14]. Possible causes, found independently of the aetiology of hypertrophy, include an inadequate capillary proliferation, with diminished oxygen diffusion and subendocardium hypoperfusion [15]; at the cellular level, an unfavourable ratio of mitochondria to myofibrils [15], that may result in the inability of the energy-producing system to fulfil the needs of the contractile apparatus. Several observations sustain the possibility that impairment of cardiac performance depends on an energetic imbalance. Firstly, mitochondria from end-stage cardiomyopathic patients have decreased cytochrome content and activity [16]. Secondly, the newly available technique of $^{31}P$ nuclear magnetic resonance has provided experimental evidence for the energy-starved state of the failing heart. A decrease in both creatine phosphate (CP) content and creatine kinase reaction velocity was shown to be coupled with a significant impairment of contractile reserve in the failing rat heart [17, 18]. Thirdly, a negative correlation between CP to ATP ratio and severity of heart failure could be demonstrated in patients; importantly, in these patients, clinical amelioration was associated with an increase in CP/ATP ratio [19].

Alterations in energy production may also result from changes in the metabolic substrate used. In this regard, fatty acid oxidation was found reduced in homogenates from failing hearts [20] and an altered ratio of oxidative vs glycolytic metabolism was demonstrated in hypertrophied hearts [21].

No therapeutic approach has yet been aimed at preventing or opposing the consequences of metabolic imbalance. Propionyl-L-carnitine (PLC) represents the prototype of a novel class of therapeutic agents capable of stimulating substrate oxidation with a consequent increase in energy production. PLC, a naturally occurring L-carnitine derivative, has some advantages over carnitine, namely, it is able to replenish mitochondria with intermediates of the citric acid cycle [22]; it stimulates palmitate oxidation to a greater extent than carnitine, both in heart homogenates [23] and in isolated hypertrophied myocytes [24]; in the latter preparation, only PLC, but not carnitine, does increase the ATP to ADP ratio; moreover, the hydrolysed propionate improves the flux of intermediates throughout the tricarboxylic acid cycle [25, 26]. These metabolic properties are mirrored by a greater stimulation of cardiac performance than with L-carnitine, as shown by mechanical recovery of isolated hearts after ischemia [23, 27]; in normal hearts, PLC, but not carnitine pretreatment, improves contractility at high left ventricular filling volumes [28].

This review will discuss some of the findings recently accumulated on the effect of PLC in different models of cardiomyopathy in the rat. Each model will be considered separately, because of its histological, mechanical and metabolic peculiarities [29].

**Pressure overload cardiomyopathy**

We developed a model of heart hypertrophy in adult rats by constricting the abdominal aorta (AC) to an extent that produced an average 40% increase in total heart weight, and a depletion in the myocardial content of carnitine and high energy phosphates. This model was employed to investigate: 1) the haemodynamics of conscious rats; 2) the haemodynamics of anaesthetised rats during increased preload and during increased afterload; 3) the performance and energetic state of Langendorff perfused hearts; 4) the mechanics of isolated papillary muscles; 5) the left ventricle content in myosin isoenzymes; 6) the metabolic and energetic state of isolated cardiomyocytes.

When pressure overloaded conscious rats were treated with PLC (50 mg/kg intraarterially) for 4 days, they showed an improvement of haemodynamics, that was more significant for the rats with the higher cardiac hypertrophy and greater carnitine deficiency (Figure 1). Treatment produced an increase in cardiac output, cardiac work, stroke volume and stroke work, accompanied by a decrease in total peripheral resistance (Figure 2); these haemodynamic changes subsided after treatment withdrawal (Figure 2) [30].

In urethane anaesthetised rats, AC did not affect basal haemodynamics. However, an acute preload stress induced by i.v. infusion of saline (40 ml/kg in 1 min) disclosed an overt impairment of cardiac function, evident as lower CO and SV (Donato di Paola, unpublished). Treatment with PLC (oral administration for 3 weeks, beginning on the third week after aortic banding) improved the performance of AC animals, while leaving unchanged the response in sham-operated rats. Peak cardiac output reached during infusion was significantly increased compared to untreated clip rats (Figure 3). The same was observed for cardiac work and stroke work calculated at peak cardiac output. In these experimental conditions, PLC displayed a dose-dependent effect in the range 30–180 mg/kg. These results compared favourably with those obtained with enalapril (3 mg/kg).

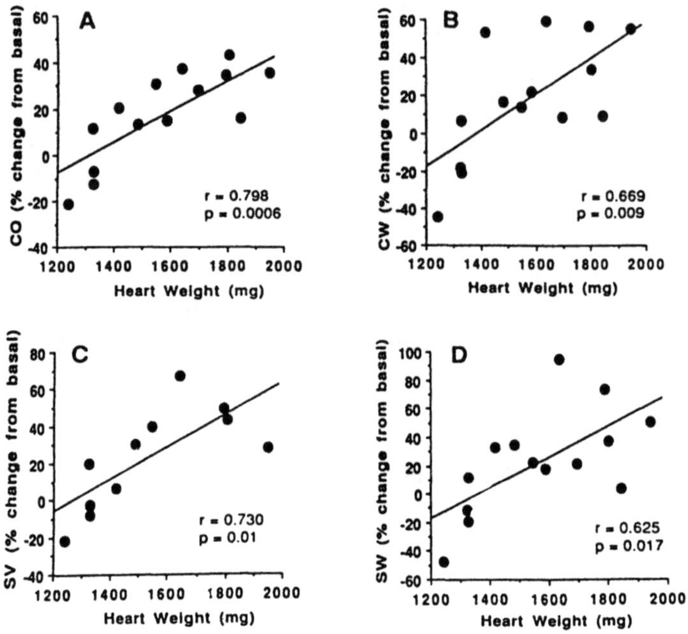


*Figure 1.* Scatter plot of relation between heart weight and haemodynamic variables in rats with abdominal aorta constriction treated with PLC 50 mg/kg i.a. for 4 days. Values are the maximum percent change recorded on any of the 4 days of treatment. Panel A: cardiac output (CO, $y = -90.6 + 0.069x$). Panel B: cardiac work (CW, $y = -130.4 + 0.095x$). Panel C: stroke volume (SV, $y = -84.9 + 0.068x$). Panel D: stroke work (SW, $y = 135.9 + 0.10x$). (From [30], with permission.)

Langendorff perfused hearts obtained from PLC-treated animals displayed a lower diastolic pressure and a faster relaxation rate than hearts from untreated animals. At the biochemical level, a significantly greater myocardial content in ATP (20.7 ± 1.3 μmol/g dry weight, in treated animals vs 15.5 ± 1.0 μmol/g in untreated, $p < 0.01$) total adenine nucleotides (30.5 ± 0.2 vs 23.7 ± 1.4 μmol/g, $p < 0.05$) and CP (27.2 ± 1.5 vs 20.5 ± 2.4 μmol/g, $p < 0.05$) was demonstrated [30]; the content in high energy phosphates was directly correlated to the relaxation rate and inversely correlated to the end-diastolic pressure [31]. In Langendorff perfused hypertrophied hearts we also demonstrated that PLC activity depended on the presence of fatty acids (palmitate) in the perfusion medium [31].

Papillary muscles were studied from rats that had been treated with 180 mg/kg PLC for 8 weeks, starting from weaning. AC was performed at 8 weeks of age, and lasted 4 weeks. AC muscles displayed a prolongation of timing parameters, i.e. of time-to-peak tension and time from peak tension

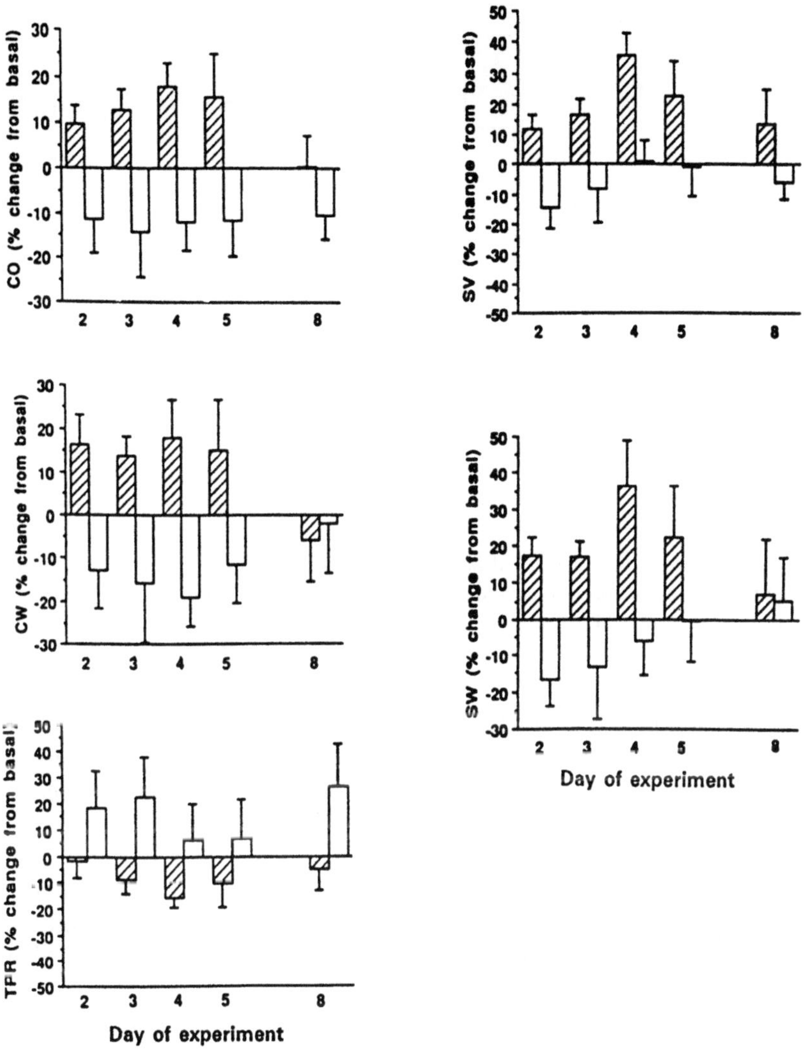


*Figure 2.* Histograms of time-course of haemodynamic changes observed during PLC or saline treatment (hatched and open bars, respectively), and on the 4th day after treatment withdrawal (day 8), in rats with abdominal aorta constriction. Data refer to rats with a heart weight >1400 mg. Values are expressed as a percentage of pretreatment baseline values and plotted as means ± SEM (n = 9–10). Upper left panel: cardiac output (CO); middle left: cardiac work (CW); lower left: total peripheral resistance (TPR); upper right panel: stroke volume (SV); lower right: stroke work (SW). (From [30], with permission.)

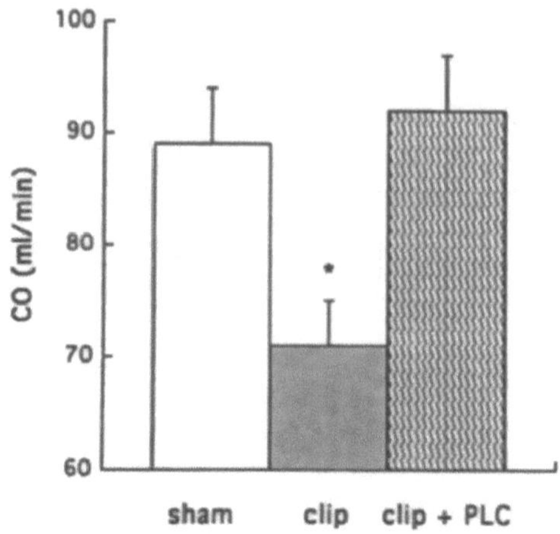


*Figure 3.* Peak cardiac output reached during acute volume expansion in anaesthetised sham-operated rats, untreated aortic-constricted rats (clip) and PLC (180 mg/kg for 3 wk) treated clip rats. Values are means ± SEM (n = 9–14). * $p < 0.05$ vs sham and treated rats.

to 30% relaxation, and a reduced peak rate of tension rise and decay (Figure 4, lower right) [32]. PLC treatment normalised the timing parameters and the rate of tension decay (Figure 4, lower left). The isotonic velocity of contraction was also maintained by PLC treatment as fast as in sham preparation (Figure 5). At the biochemical level, the slowing of muscle mechanics in AC preparations corresponded to a significant reduction in $\alpha$ myosin heavy chain and increase in the foetal $\beta$ isoform. PLC substantially prevented the reduction in fast myosin isoform seen in hypertrophied preparation (Figure 5) [32].

Isolated myocytes obtained from AC rats showed a 21% reduction in ATP content and a 30% reduction in ATP to ADP ratio [24]. Palmitate oxidation in hypertrophied cells was also depressed compared to normal cells, with a reduction in Vmax of 27%, while $K_M$ remained unchanged. PLC (25 μM) increased palmitate oxidation by 21%, increasing also the ATP to ADP ratio.

In conclusion, PLC improves the function of the AC preparations described above. Experiments in papillary muscle show that the compound is able to affect directly muscle function. Data in isolated myocytes suggest that PLC corrects the defective energy producing system of hypertrophied cells. The maintenance of a normal myosin isoenzyme proportion is in accordance with the higher energy availability afforded by PLC.

In agreement with the present view that pressure-overloaded hearts undergo a shift in substrate utilisation [20, 21], we found a depressed fatty acid oxidation in hypertrophied myocytes. PLC improvement of substrate

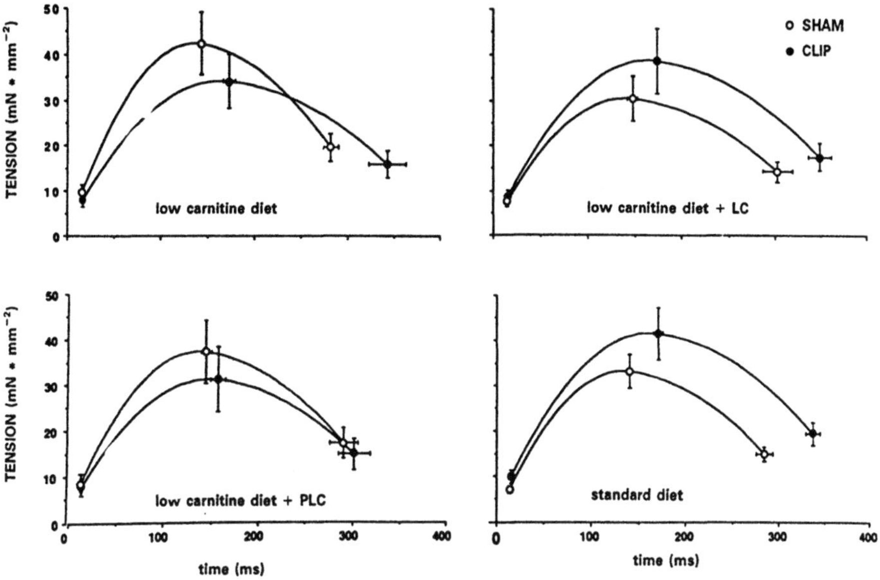


*Figure 4.* Average isometric contractions of left papillary muscles from sham and aortic-constricted rats maintained on a low-carnitine diet (upper left), low-carnitine diet +180 mg/kg PLC (lower left), low-carnitine diet +13.4 mg/kg L-carnitine (LC, upper right), and standard diet (lower right). Values are means ± SEM (n = 11–15). (From [32], with permission.)

use might explain the effect on myosin composition, since substrate utilisation has been suggested to influence expression of myosin isoforms [33]. The haemodynamic improvement may be explained by an amelioration of diastolic filling due to enhanced relaxation. This explanation is consistent with data from Langendorff perfused hearts and papillary muscles. Amelioration of diastolic relaxation is likely to favour the diastolic filling of the heart, resulting in greater stroke volume and stroke work. The parallel improvement in diastolic function and myocardial ATP levels supports the notion that the two findings may be causally related. In pressure-overload hypertrophy, diastolic function is often compromised earlier and more severely than systolic function, due to the peculiar characteristics of concentric hypertrophy, i.e. increase in wall thickness and development of fibrosis. Diastolic filling is impaired by the alteration in the passive visco-elastic properties caused by collagen network remodelling [34]. Moreover, cardiac relaxation is also influenced by the rate of $Ca^{2+}$ uptake by the sarcoplasmic reticulum, the rate of dissociation from troponin C and rate of uptake by the sarcolemmal $Ca^{2+}$ pump. Evidence that $Ca^{2+}$ handling is altered in hypertrophied myocytes, leading to diastolic cytosolic $Ca^{2+}$ overload has been provided [35]. A

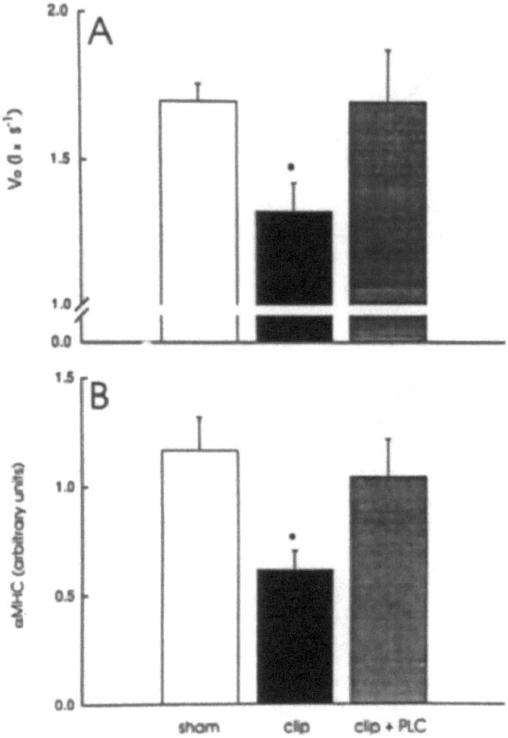


*Figure 5.* A. Unloaded shortening velocity (Vo) of skinned trabeculae from left ventricles of sham, untreated aortic-constricted and PLC-treated (180 mg/kg for 8 wk) aortic-constricted rats. B. Relative amounts of fast myosin heavy chain isoform ($\alpha$ MHC) determined in left ventricle free wall of the same animals as above. Values are means ± SEM (n = 6–7). * $p < 0.05$ from both other groups. (From [32], with permission.)

reduced $Ca^{2+}$ ATPase activity and uncoupling between ATP hydrolysis and $Ca^{2+}$ transport have been reported, as well as a lower ATPase density on the sarcoplasmic reticulum membrane. Whether PLC affects any of these biochemical functions remains to be demonstrated, although evidence for a greater $Ca^{2+}$ uptake and $Ca^{2+}$ stimulated ATPase activity have been provided in the diabetic heart [36].

## Volume overload

At variance with the findings obtained in pressure overloaded rats, when PLC was administered at the same dose (50 mg/kg intraarterially for 4 days) in conscious rats with a two-month aortocaval fistula, no appreciable effect on the haemodynamic parameters recorded could be found (Schiavone, unpublished). Heart rate, cardiac output, stroke volume and stroke work, as well as total peripheral resistance of animals with volume overload did not differ from those of untreated animals.

Papillary muscle function was also studied. Right and left muscles were obtained from rats with a 4 to 5 month aortocaval fistula (Micheletti, unpublished) that had been treated for 2 weeks with 180 mg/kg PLC in the drinking water. The fistula determined a considerable hypertrophy of both the left and right ventricles (64% and 77% increase in wet weight, respectively), but no alteration in left papillary muscle function. On the contrary, the right papillary muscle exhibited longer timing parameters. The time from peak tension to 30% relaxation was prolonged from 146 ± 16 to 195 ± 14 ms ($p < 0.05$). PLC prevented these changes (148 ± 13 ms), but did not affect left papillary muscle function.

Hypertrophy consequent to volume overload is characterised by a dramatic increase in chamber volume and heart weight due to the series addition of new sarcomeres, with a moderate increase in wall thickness. In spite of these remarkable changes, clear signs of left ventricular dysfunction are rare [37]. In the rat, aortocaval fistula was recently shown to induce a time-dependent cardiac hypertrophy that reaches its maximum (nearly doubling of heart wet weight) at 1 month without appreciable deterioration of the pumping function [38, 39]. The mechanical and biochemical consequences of volume overload are surprisingly negligible compared to the morphological ones; a recent study failed to show any change in either the energetic parameters or in the $Ca^{2+}$ ATPase activity of rabbit papillary muscles from animals with clinical signs of overt failure [40]. The lack of impairment seen in the left papillary muscle is consistent with these observations. Since these muscles were derived from animals with a volume overload of 2 months longer duration than animals for the in vivo study, it is reasonable to conclude that in the latter left ventricle function was fully maintained. If this is the case, the failure of PLC to exert any effect is not surprising.

The mechanical alterations affecting the right papillary muscles may be accounted for by a pressure overload on the right ventricle [39]. It is presumable that to evidence PLC efficacy, more extreme experimental conditions are needed. This could explain the efficacy of PLC in isolated volume-overloaded hearts with a clearly depressed function [41] (see also elsewhere in this book). It should also be emphasised that volume overload is not a single pathological entity, as profound differences in term of prognosis exist between the high-pressure volume overload (arterovenous anastomosis, aortic insufficiency) and the low-pressure volume overload (mitral regurgitation) [37].

## Myocardial infarction

In rats with moderate-size infarcts, comprising about 40% of the left ventricular free wall, we demonstrated that PLC (60 mg/kg per os given for 5 months) positively influenced ventricular remodelling, being equieffective to enalapril (1 mg/kg per os) in limiting the magnitude of the left ventricular dilatation estimated by passive pressure-volume curves [42]. Both drugs significantly shifted the pressure-volume curves towards that of sham rats. In untreated infarcted rats, chamber volume at 4 mmHg pressure increased by 117% as compared to sham (from $0.44 \pm 0.02$ to $0.95 \pm 0.11$ ml/kg, $p < 0.05$), while the changes seen in PLC and enalapril treated rats (+36 and +43%, respectively) were not statistically significant. Finally, PLC was able to limit the changes in ventricular chamber stiffness induced by infarction: both at low and at high filling pressures, the estimated stiffness constants in PLC-treated rats did not differ from those of sham rats.

Global cardiac performance under basal condition was not affected by either PLC or enalapril treatment. However, the beneficial effect of PLC treatment on pumping function of the heart was disclosed by acute volume expansion. During this manoeuvre we observed that a sensitive index of myocardial function, peak +dP/dt, was depressed in untreated infarcted rats; conversely, in infarcted rats treated with PLC, it reached values no different from those recorded in control animals (Figure 6).

In isolated rat myocytes obtained 1 month after coronary artery occlusion, 25 μM PLC increased peak shortening (20%), shortening velocity (23%) and peak systolic calcium (14%) [43].

The model of coronary ligation in the rat has been useful in detecting ACE inhibitors' ability to prolong survival after myocardial infarction [44, 45]. While ACE inhibitors reduce afterload via their vasodilating effect, PLC does not seem to affect ventricular loading. Although not directly demonstrated under these conditions, the positive effect of PLC on remodelling might be due to enhanced generation of ATP that could ameliorate the cellular mechanical behaviour. This would facilitate ventricular emptying, resulting in a reduction in systolic and diastolic loading. The results on isolated myocytes are compatible with an improved function of the energy producing system.

## Effect of PLC on myocardial carnitine stores

A reduction of myocardial carnitine levels has been demonstrated both in human and experimental cardiomyopathies, and has been suggested to play a causal role in the altered cardiac performance seen in these conditions [46]. In this context, we have been interested in investigating if PLC may act by replenishing myocardial carnitine stores. Data obtained with PLC in AC conscious rats, showing a haemodynamic activity in parallel with normalis-

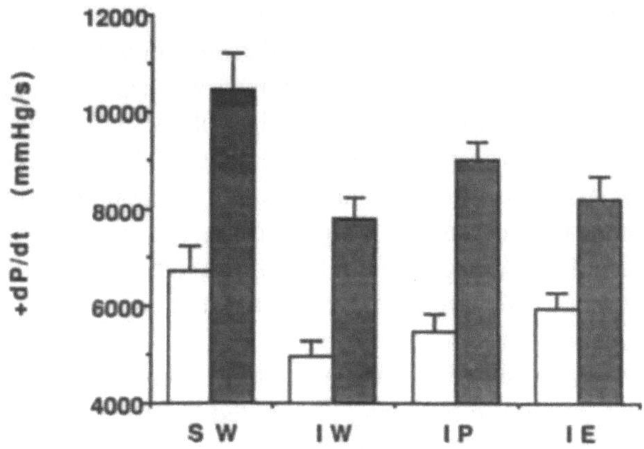


*Figure 6.* Maximum velocity of left ventricular pressure rise before (open bars) and during acute volume expansion (closed bars) in anaesthetised untreated sham-operated rats (SW), and infarcted rats that had been untreated (IW) or treated with 60 mg/kg PLC (IP) and 1 mg/kg enalapril (IE). Basal +dP/dt was significantly reduced by infarction in all groups. +dP/dt during volume expansion was reduced in all groups with the exception of PLC-treated rats. Values are means ± SEM (n = 8).

ation of myocardial carnitine levels, were consistent with a relevant role of myocardial carnitine stores [30] and led to conclude that PLC activity depended on stores replenishment. Subsequent data in other experimental conditions allowed less clearcut conclusions.

To investigate whether myocardial carnitine depletion is capable of directly affecting function, we studied papillary muscles obtained from animals fed a diet with a low total carnitine content, corresponding to 2 nmol/g pellet [32]. This condition, compared with a standard diet (56 nmol/g total carnitine), led to a significant depletion of myocardial carnitine levels, but comparable papillary muscle mechanics [32]. On the other hand, despite AC in the low carnitine group resulted in an even greater carnitine depletion than in animals fed the standard diet, function was not further altered. It is therefore likely that myocardial carnitine depletion has a different pathophysiological meaning whether due to reduced dietary intake or to AC. Moreover, it may not be excluded that carnitine is stored in different cytoplasmic compartments, possibly with different roles. The complexity of the cellular metabolic state is likely to need a wider and detailed biochemical characterisation, although the myocardial carnitine depletion per se obviously constitutes an index of metabolic derangement.

## Diabetes

Ferrari and coworkers investigated the effect of PLC (250 mg/kg, i.p. for 2 months) on the perfused heart obtained from animals with streptozotocin induced diabetes [47]. Cardiac performance was studied under basal conditions and during stepwise increases in the volume of a saline-filled balloon inserted into the left ventricle. Results demonstrated that treatment normalised diabetes associated tachycardia while not affecting developed pressure. PLC prevented the decrease in developed pressure and the increase in diastolic pressure due to progressive filling of the left ventricular balloon [47].

The same schedule of treatment prevented the depression in sarcoplasmic reticulum function observed in untreated rats: $Ca^{2+}$-stimulated ATPase activity, $Ca^{2+}$ uptake and $Mg^{2+}$-ATPase activity were similar to those of non-diabetic animals. On the contrary, PLC failed to reverse the diabetes associated changes in $Ca^{2+}$ ATPase found at the sarcolemmal level [36]. There is presently no explanation for this variable effect of PLC, nor for the mechanism underlying it. In this context it should be mentioned that treatment with a high dose of carnitine (3 g/kg daily in the drinking water for 6 weeks) suppressed accumulation of long-chain acylcarnitines, but did not improve global cardiac function, leading the authors to conclude that sarcoplasmic reticulum alterations were not the main determinants for impaired cardiac function [48].

Diabetes involves widespread metabolic alterations, including disorders of carbohydrate, lipid and protein metabolism; the diabetic heart has a near total dependency on fatty acids as energy substrate. The accelerated rate of lipid metabolism in the heart is responsible for elevated cytosolic levels of triglycerides, free fatty acids and CoA, and consequently for elevation in acylCoA and acylcarnitines [49]. PLC efficacy could depend on its ability to stimulate glucose oxidation. Carnitine has been shown to increase glucose oxidation in normal hearts, where fatty acids are the major energy source [50]. Similarly, the stimulation of glucose oxidation, induced by dichloroacetate, positively affects function of the diabetic working heart [51].

$Ca^{2+}$ handling in the diabetic cardiomyopathy is impaired. The ATP-dependent $Ca^{2+}$ transport of the sarcoplasmic reticulum was found significantly decreased in diabetes, and both $Ca^{2+}$ uptake and $Ca^{2+}$-stimulated ATPase activity were reduced [52].

## Conclusions

In conclusion, PLC displayed a therapeutic effect in a number of CHF models, namely, pressure overload, infarction and diabetic cardiomyopathies. Less clearcut results were observed in the volume-overload model.

The most relevant findings are: in *pressure overload*,

- improvement of cardiac work and stroke work in conscious and in anaesthetised animals;
- improvement of the diastolic function of isolated hearts from pretreated animals;
- normalisation of timing parameters and relaxation rate of isolated papillary muscles from pretreated rats;
- preservation of the ratio $\alpha/\beta$ myosin heavy chain;
- increase in high energy phosphates in perfused hearts and isolated cardiomyocytes.

In the *infarct model* of CHF, PLC treatment reduced ventricular dilatation and tended to restore the diastolic chamber stiffness towards normal values. Thus, PLC seems to act primarily on the diastolic function, both in its active component, due to sarcoplasmic reticulum function (increase in −dP/dt in isolated hearts and papillary muscles from pressure-overloaded rats; reduced relaxation time in right papillary muscles from volume-overloaded rats) and in its passive component, due to tissue remodelling (improvement of diastolic stiffness in myocardial infarcts). Importantly, in several instances (AC anaesthetised rats, isolated hearts from AC rats, diabetic cardiomyopathy), PLC effect was apparent under conditions of high energy demand, induced by increase in work load. It seems thus plausible that PLC is able to correct some metabolic steps of the process that leads to CHF. CHF is an evolving pathology in which impairment of biochemical activities occurs progressively: presumably, the relative importance of the diverse biochemical steps involved varies during the evolution of the syndrome and plays a different role according to the stage of the disease.

The metabolic derangement occurring in CHF is far from clear. The increased levels of high energy phosphates induced by PLC suggest that, in the experimental conditions employed, energy supply was inadequate. In normal conditions, PLC is ineffective, as evidenced from data in sham animals and in volume overload.

From the studies performed in different models of CHF so far discussed, PLC stands out as possessing some new and interesting features that could be of value in the management of CHF. It is also evident that, because the biochemical alterations accompanying CHF may differ according to either the stage of the disease or its aetiology, PLC should be particularly useful in those conditions where its biochemical activity could compensate for an existing metabolic deficiency. Future work should be aimed at clarifying the relationship between the PLC efficacy and the biochemical abnormalities in different settings and stages of heart failure.

## References


1. Mackenzie J. Disease of the heart. Oxford: Oxford Medical Publications, 1908.
2. Ho KKL, Pinsky JL, Kannel WB, Levy D. The epidemiology of heart failure: the Framingham Study. J Am Coll Cardiol 1993; 22(Suppl A): 6A–13A.
3. Curfman GD. Inotropic therapy for heart failure – An unfulfilled promise. N Engl J Med 1991; 325: 1509–10.
4. Anonymous. Failure to treat heart failure. Lancet 1992; 339: 278–9.
5. Armstrong PW, Moe GW. Medical advances in the treatment of congestive heart failure. Circulation 1994; 88: 2941–52.
6. DiBianco R, Shabetai R, Kostuk W, Moran J, Schlant RC, Wright R. A comparison of oral milrinone, digoxin, and their combination in the treatment of patients with chronic heart failure. N Engl J Med 1989; 320: 677–83.
7. Packer M, Carver JR, Rodheffer RJ et al. Effect of oral milrinone on mortality in severe chronic heart failure. The PROMISE Study Research Group. N Engl J Med 1991; 325: 1468–75.
8. Uretsky BF, Jessup M, Konstam MA et al. Multicenter trial of oral enoximone in patients with moderate to moderately severe congestive heart failure: lack of benefit compared with placebo. Enoximone Multicenter Trial Group. Circulation 1990; 82: 774–80.
9. The Xamoterol in Severe Heart Failure Study Group. Xamoterol in severe heart failure. Lancet 1990; 336: 1–6.
10. Cohn JN, Johnson G, Ziesche S et al. A comparison of enalapril with hydralazine-isosorbide dinitrate in the treatment of chronic congestive heart failure. N Engl J Med 1991; 325: 303–10.
11. Schunkert H, Dzau VJH, Tang SS, Hirsch AT, Apstein CA, Lorell BH. Increased rat cardiac angiotensin converting enzyme activity and mRNA expression in pressure overload left ventricular hypertrophy. Effects on coronary resistance, contractility, and relaxation. J Clin Invest 1990; 86: 1913–20.
12. Pfeffer JM. Progressive ventricular dilation in experimental myocardial infarction and its attenuation by angiotensin-converting enzyme inhibition. Am J Cardiol 1991; 68: 17D–25D.
13. Sunman W, Sever PS. Non-angiotensin effects of angiotensin-converting enzyme inhibitors. Clin Sci 1993; 85: 661–70.
14. Katz AM. Metabolism of the failing heart. Cardioscience 1993; 4: 199–203.
15. Anversa P, Ricci R, Olivetti G. Quantitative structural analysis of the myocardium during physiologic growth and induced cardiac hypertrophy: a review. J Am Coll Cardiol 1986; 7: 1140–9.
16. Buchwald A, Till H, Unterberg C et al. Alterations of the mitochondrial respiratory chain in human dilated cardiomyopathy. Eur Heart J 1990; 11: 509–16.
17. Bittl JA, Balshi JA, Ingwall JS. Contractile failure and high-energy phosphate turnover during hypoxia: $^{31}$P-NMR surface coil studies in living rat. Circ Res 1987; 60: 871–8.
18. Ingwall JS. Is cardiac failure a consequence of decreased energy reserve? Circulation 1993; 87(Suppl VII): 58–62.
19. Neubauer S, Krahe T, Schindler R et al. $^{31}$P magnetic resonance spectroscopy in dilated cardiomyopathy and coronary artery disease. Altered cardiac high-energy phosphate metabolism in heart failure. Circulation 1992; 86: 1810–8.
20. Wittels B, Spann Jr JF. Defective lipid metabolism in the failing heart. J Clin Invest 1968; 47: 1787–94.
21. Allard MF, Schönekess BO, Henning SL, English DR, Lopaschuk GD. Contribution of oxidative metabolism and glycolysis to ATP production in hypertrophied hearts. Am J Physiol 1994; 267: H742–50.
22. Davis EJ, Spydevold O, Bremer J. Pyruvate carboxylase and propionyl-CoA carboxylase as anaplerotic enzymes in skeletal muscle mitochondria. Eur J Biochem 1980; 110: 255–62.
23. Paulson DJ, Traxler J, Schmidt M, Noonan J, Shug AL. Protection of the ischaemic

myocardium by L-propionylcarnitine: Effects on the recovery of cardiac output after ischaemia and reperfusion, carnitine transport, and fatty acid oxidation. Cardiovasc Res 1986; 20: 536–41.
24. Torielli L, Conti F, Cinato E, Bianchi G, Ferrari P. Effect of propionyl-L-carnitine on oxidative and energetic metabolism in hypertrophic isolated rat cardiomyocytes. Eur Heart J 1993; 14(Suppl): 46 (Abstr).
25. Siliprandi N, Di Lisa F, Menabò R. Propionyl-L-carnitine: biochemical significance and possible role in cardiac metabolism. Cardiovasc Drugs Ther 1991; 5(Suppl 1): 11–5.
26. Latipaa PM, Peuhkurinen KJ, Hiltunen JK, Hassinen IE. Regulation of pyruvate dehydrogenase during infusion of fatty acids of varying chain lengths in the perfused rat heart. J Mol Cell Cardiol 1985; 17: 1161–71.
27. Ferrari R, Ceconi C, Curello S, Pasini E, Visioli O. Protective effect of propionyl-L-carnitine against ischaemia and reperfusion-damage. Mol Cell Biochem 1989; 88: 161–8.
28. Ferrari R, Di Lisa F, De Jong JW et al. Prolonged propionyl-L-carnitine pre-treatment of rabbit: biochemical, hemodynamic and electrophysiological effects on myocardium. J Mol Cell Cardiol 1992; 24: 219–32.
29. Morgan HE, Baker KM. Cardiac hypertrophy. Mechanical, neural, and endocrine dependence. Circulation 1991; 83: 13–25.
30. Yang X-P, Samaja M, English E et al. Hemodynamic and metabolic activities of propionyl-L-carnitine in rats with pressure-overload cardiac hypertrophy. J Cardiovasc Pharmacol 1992; 20: 88–98.
31. Motterlini R, Samaja M, Tarantola M, Micheletti R, Bianchi G. Functional and metabolic effects of propionyl-L-carnitine in the isolated perfused hypertrophied rat heart. Mol Cell Biochem 1992; 116: 139–45.
32. Micheletti R, Giacalone G, Canepari M, Salardi S, Bianchi G, Reggiani C. Propionyl-L-carnitine prevents myocardial mechanical alterations due to pressure overload in rats. Am J Physiol 1994; 266: H2190–7.
33. Sheer D, Morkin E. Myosin isoenzyme expression in the rat ventricle: effect of thyroid hormone analogs, catecholamines, glucocorticoids and high carbohydrate diet. J Pharmacol Exp Ther 1984; 229: 872–9.
34. Weber KT, Janicki JS, Shroff SG et al. Collagen remodeling of the pressure-overloaded, hypertrophied nonhuman primate myocardium. Circ Res 1988; 62: 757–65.
35. Morgan JP. Abnormal intracellular modulation of calcium as a major cause of cardiac contractile dysfunction. N Engl J Med 1991; 325: 625–32.
36. Ferrari R, Shah KR, Hata T, Beamish RE, Dhalla NS. Subcellular defects in diabetic myocardium: influence of propionyl-L-carnitine on $Ca^{2+}$ transport. In: Nagano M, Dhalla NS, editors. The diabetic heart. New York: Raven Press, 1991: 167–81.
37. Braunwald E. Valvular heart disease. In: Braunwald E, editor. Heart disease. Philadelphia: Saunders, 1992: 1007–77.
38. Liu Z, Hilbelink DR, Crockett WB, Gerdes AM. Regional changes in hemodynamics and cardiac myocyte size in rats with aortocaval fistulas. I. Developing and established hypertrophy. Circ Res 1991; 69: 52–8.
39. Liu Z, Hilbelink DR, Gerdes AM. Regional changes in hemodynamics and cardiac myocyte size in rats with aortocaval fistulas. II. Long-term effects. Circ Res 1991; 69: 59–65.
40. Gibbs CL, Wendt IR, Kotsanas G, Young IR. Mechanical, energetic, and biochemical changes in long-term volume overload of rabbit heart. Am J Physiol 1992; 262: H819–27.
41. El Alaoui-Talibi Z, Moravec J. Assessment of the cardiostimulant action of propionyl-L-carnitine on chronically volume-overloaded rat hearts. Cardiovasc Drugs Ther 1993; 7: 357–63.
42. Micheletti R, Donato di Paola E, Schiavone A et al. Propionyl-L-carnitine limits chronic ventricular dilation after myocardial infarction in rats. Am J Physiol 1993; 264: H1111–7.
43. Li P, Park C, Micheletti R, Li B, Cheng W, Sonnenblick EH, Anversa P, Bianchi G. Myocyte performance during the evolution of myocardial infarction in rats: effects of propionyl-L-carnitine. Am J Physiol, in press.

44. Pfeffer MA, Pfeffer JM, Steinberg C, Finn P. Survival after an experimental myocardial infarction: beneficial effects of long-term therapy with captopril. Circulation 1985; 72: 406–12.
45. Sweet CS, Emmert SE, Stabilito II, Ribeiro LGT. Increased survival in rats with congestive heart failure treated with enalapril. J Cardiovasc Pharmacol 1987; 10: 636–42.
46. Regitz V, Fleck E. Role of carnitine in heart failure. In: Ferrari R, Di Mauro S, Sherwood G, editors. L-Carnitine and its role in medicine: from function to therapy. New York: Academic Press, 1992: 295–323.
47. Pasini E, Comini L, Ferrari R, de Giuli F, Menotti A, Dhalla NS. Effect of propionyl-L-carnitine on experimentally induced cardiomyopathy in rats. Am J Cardiovasc Pathol 1992; 4: 216–22.
48. Lopaschuk GD, Tahiliani AG, Vadlamudi RVSV, Katz S, McNeill JH. Cardiac sarcoplasmic reticulum function in insulin- or carnitine-treated diabetic rats. Am J Physiol 1983; 245: H969–76.
49. Lopaschuk GD, Katz S, McNeill JH. The effect of alloxan- and streptozotocin-induced diabetes on calcium transport in rat sarcoplasmic reticulum. The possible involvement of long chain acylcarnitines. Can J Physiol Pharmacol 1983; 61: 439–48.
50. Broderick TL, Quinney HA, Lopaschuk GD. Carnitine stimulation of glucose oxidation in the fatty acid perfused isolated working rat heart. J Biol Chem 1992; 267: 3758–63.
51. Nicholl TA, Lopaschuk GD, McNeill JH. Effects of free fatty acids and dichloroacetate on isolated working diabetic rat heart. Am J Physiol 1991; 261: H1053–9.
52. Heyliger CE, Prakash A, McNeill JH. Alterations in cardiac sarcolemmal $Ca^{2+}$ pump activity during diabetes mellitus. Am J Physiol 1987; 252: H540–4.

*Corresponding Author*: Dr. Rosella Micheletti, Prassis Istituto di Ricerche Sigma Tau, Via Forlanini 3, I20019 Settimo Milanese, Italy

# 23. Utilization of propionyl-L-carnitine for the treatment of heart failure

ROBERTO FERRARI and INDER ANAND

---

"Studies in carefully selected homogenous small groups of patients show that oral administration of propionyl-L-carnitine improves exercise capacity and skeletal muscle metabolism of patients with heart failure (classes II and III of NYHA). This provides a logical basis for individualizing therapy in order to improve exercise tolerance and quality of life in patients with heart failure."

---

## Introduction

Heart failure is a common and disabling disease with a poor prognosis, but it can be easily diagnosed and recent data show that treatment of chronic heart failure reduces mortality, morbidity and improves quality of life [1, 2].

The usual treatment of chronic heart failure is with angiotensin-converting enzyme inhibitors, diuretics and digitalis. Other drugs are often used such as calcium antagonists, vasodilators, antiarrhythmics, positive inotropic agents, etc. Basically, all these compounds aim at reducing the progression of the disease, improving the hemodynamic profile of patients and/or reducing the generalized neuroendocrine response.

Several landmark trials in the treatment of heart failure have been published in the last decade [3–12]. From these studies a consolidated body of evidence has emerged: the addition of angiotensin-converting enzyme inhibitors to conventional therapy provides benefit in terms of symptomatic improvement, reduces hospital admissions, prevents the progression of heart failure, delays death and, most surprisingly, prevents coronary events. Though such large-scale studies provide a clear answer to simple global questions, many practical issues of interest for the physician are still unresolved. Patients often do not accept to be treated solely because benefit was shown in a study. They are interested in knowing whether that particular form of treatment is appropriate for them and for their individual condition. They also expect to have a short-term benefit from treatment such as an improvement in the standard of living which in chronic heart failure is severely limited.

*J.W. de Jong and R. Ferrari (eds): The carnitine system*, 323–335.

In general, a patient with heart failure complains of fatigue and shortness of breath. Usually these symptoms are alleviated by the administration of diuretic, digoxin and vasodilator therapy, and the majority of patients experiences resolution of symptoms at rest. Unhappily, however, many patients continue to experience exertional symptoms [13–15] and frequently report that they are unable to perform regular activities of daily life. During maximal exercise testing, patients with heart failure terminate exercise earlier than normal subjects of comparable weight and gender.

Traditionally, fatigue and shortness of breath have been related to a low cardiac output and increased end-diastolic pressure. Although there may be some validity to this claim, recent studies have demonstrated that the origin of symptoms in heart failure is much more complex [16, 17]. In chronic heart failure, unlike acute heart failure, shortness of breath is not related simply to end-diastolic pressure either at rest or at peak exercise [16, 17]. Many subtle changes do take place in the lungs such as an increased ventilation for a given carbon dioxide production and weakness of the diaphragm. Equally, exertional fatigue is not simply the result of skeletal muscle underperfusion [14]. Muscle deconditioning and altered peripheral muscle metabolism have recently been shown to substantially contribute to exertional fatigue in heart failure.

The understanding of the role of these multiple potential contributors to the major symptoms of heart failure has generated new forms of interventions which are at the present under extensive evaluation such as, for example, the use of exercise programs to avoid deconditioning and of propionyl-L-carnitine to specifically improve metabolism and function of cardiac and skeletal muscle in heart failure.

In this chapter, the alterations of peripheral skeletal muscle during heart failure, the advantages of propionyl-L-carnitine versus carnitine and the rationale for its use in chronic heart failure as well as the available clinical data are reviewed. Propionyl-L-carnitine, in addition to its effect on skeletal muscle, positively affects the metabolism and function of the failing cardiac muscle and, without doubt, these effects are an important component of its overall pharmacological effect. These aspects, however, are extensively addressed elsewhere in this volume (Chapter 24). Therefore, in the present chapter, attention is limited to the effects of propionyl-L-carnitine on skeletal muscle.

## Causes of exertional fatigue in chronic heart failure

Patients with chronic heart failure commonly experience limitation of activity secondary to fatigue and shortness of breath. The underlying fear that regular physical activity may result in further deterioration and symptoms have traditionally led patients to avoid exercise and other physical efforts. There are three major mechanisms of exertional fatigue:

### *1. Inadequate muscle flow and arterial dilatation*

Recent observations support the concept that inadequate muscle flow is an important contributor to exertional fatigue [13, 18, 19]. Using a femoral venous thermodilution catheter, we measured leg flow responses to a low work-load bicycle exercise mimicking the small amount of activity encountered in daily life in a group of patients with heart failure. We found a significant decrease in flow responses to exercise in the majority of patients, although in some the flow response was within the normal limits [20]. The failure of muscle blood flow to increase normally during exercise in patients with heart failure is primarily due to an abnormality of arterial vasodilation as evidenced by a failure of leg vascular resistances to decrease normally during exercise [13, 20, 21]. This, in turn, could be due to: a) compression of capillaries and intrinsic vascular changes due to fluid and sodium retention [22]; b) increase in sympathetic activation and angiotensin II, impairing dilatation [23]; c) abnormality of vascular endothelium and vascular remodelling [24].

### *2. Muscle deconditioning*

While abnormalities of the peripheral circulation in chronic heart failure patients have been frequently described as a major determinant of exercise limitation, Wiener et al. have found no relationship between plethysmographic measurements of exercising forearm blood flow and muscle metabolism measured by phosphorus nuclear magnetic resonance (NMR) spectroscopy [25]. These workers suggest that other mechanisms, such as alterations in mitochondrial population or substrate utilization may be responsible for the depressed exercise performance [25]. It is also of interest that in patients with peripheral disease and impaired arterial blood flow, the pathophysiology of muscle dysfunction in the limbs affected is not just due to arterial obstruction but also involves type II fiber atrophy [26].

Perhaps the best evidence that muscle deconditioning contributes to exertional fatigue comes from studies demonstrating that participation in a home exercise or in a formal rehabilitation program can improve the maximal exercise capacity of patients with heart failure by 15% to 30% [27, 28].

Patients with chronic heart failure are deconditioned and the resulting muscle atrophy can easily lead to tiredness because the work or exercise falls disproportionately upon the remaining viable muscle fibers [27]. However, Mancini et al. studied the contribution of skeletal muscle atrophy to exercise limitation in chronic heart failure and found that muscle atrophy contributes only modestly to the reduced exercise capacity [30]. Using NMR spectroscopy, it was found that abnormalities of intrinsic oxidation by exercising muscles correlated weakly with muscle volume used as an index of muscle atrophy [29].

### *3. Altered muscle metabolism and function*

Metabolic studies have been conducted using phosphorus-31 nuclear magnetic resonance, a technique that permits the non-invasive monitoring of phosphocreatine, inorganic phosphate, adenosine triphosphate (ATP) and pH in working muscle. Patients with heart failure have a more pronounced increase in the organic phosphate/phosphocreatine ratio and a more pronounced decrease in muscle pH than do normal subjects performing comparable work loads [25, 29–33].

In other studies, the arterial-venous femoral difference has been measured at rest and during exercise. By using this technique, we could demonstrate that in patients with heart failure, the uptake of FFA is impaired both at rest and during exercise [34]. In addition, we and others have shown that the skeletal muscle metabolism in heart failure is abnormally dependent on anaerobic glycolysis, resulting in excessive accumulation and release of lactic acid [34]. These biochemical changes could not be explained by impaired blood flow or reduced oxygen delivery alone. One metabolic change that would explain these observations is a decrease in glucose and FFA oxidation relative to glycolysis [35]. Such an imbalance would accelerate tissue acidosis during exercise.

Support for metabolic changes in chronic heart failure also comes from studies using a pyruvate dehydrogenase complex stimulator, dichloroacetate (DCA). Wilson et al. gave 35 mg/kg of DCA intravenously to 18 chronic heart failure patients before exercise in a double-blind cross-over trial. No adverse effects were observed. In this acute intravenous study, a significant improvement in serum lactate levels during exercise was noted but exercise time did not change [36]. Wargovich et al. studied the effects of DCA on cardiac haemodynamic and coronary blood flow during cardiac catheterization and found beneficial acute haemodynamic effects [37]. Bersin et al. compared the effects of intravenous DCA (50 mg/kg) with clinically optimized doses of dobutamine in 7 patients with severe heart failure and found that DCA significantly improved myocardial mechanical efficiency [38].

These findings suggest that oxidative phosphorylation of glucose is reduced in heart failure. The data from metabolic studies are confirmed by biochemical and histological examination of muscle biopsy specimens taken from patients with chronic heart failure [39, 40]. Compared to normal controls, Sullivan et al. found that during submaximal exercise, patients with heart failure have reduced skeletal muscle aerobic activity resulting in early onset of anaerobic metabolism and lactate accumulation which is inversely related to rest aerobic enzyme activity [39]. Drexler et al. further showed that reduced oxidative enzyme activity, associated with decreased aerobic enzyme activity, occurred in all skeletal muscle fibers in chronic heart failure patients [40]. This reduction in oxidative enzyme activity seemed to cause a shift to type IIB skeletal muscle fibers that are more adapted to anaerobic metabolism [40].

*Table 1.* **Advantages of propionyl-L-carnitine over L-carnitine for the improvement of cardiac muscle function and metabolism in heart failure and skeletal muscle abnormalities.**

Propionyl-L-carnitine

- is superior for increase to cellular content of L-carnitine
- has greater affinity for cardiac and skeletal muscle carnitine transferase
- allows delivery of the propionyl-moiety to the mitochondria, a group otherwise toxic to muscle metabolism
- allows the anaplerotic utilization of the propionyl-group
- has specific pharmacological effects

In addition to the fiber atrophy noted in muscle biopsy specimens, patients also exhibit generalized muscle atrophy, again consistent with inactivity and deconditioning. This was clearly demonstrated by means of anthropometric studies and measurement of muscle volume with magnetic resonance imaging [41].

In general, the available evidence suggests that altered skeletal muscle metabolism can substantially influence exertional fatigue in chronic heart failure.

## Advantages of propionyl-L-carnitine over L-carnitine for improving skeletal and cardiac muscle function and metabolism

The abnormal skeletal muscle metabolism in failing patients and particularly the excessive dependence on glycolysis, delayed utilization of free fatty acids (FFA), and excessive accumulation of lactate might potentially be reversed by L-carnitine or its derivatives through its effects on oxidative metabolism.

Propionyl-L-carnitine is a naturally occurring component of the carnitine pool which exists endogenously and is maintained in a homeostatic balance.

The empirical formula is $C_{10}H_{20}O_4NCl$, and the molecular weight is 253.7. Propionyl-L-carnitine is formed by means of carnitine acetyltransferase from propionyl-CoA, a product of methionine, threonine, valine and isoleucine as well as of odd-chain fatty acids.

Pharmacokinetic studies have demonstrated that, in humans, the plasma concentration of propionyl-L-carnitine increases following intravenous administration and then decreases to baseline values within 6 to 24 h [42]. This timespan varies with dosage. The plasma concentrations of L-carnitine follow a similar pattern, but in a more sustained fashion. Urinary excretion of all these compounds increases after the intravenous dose of propionyl-L-carnitine and the excretion rates reach their highest values during the first 24 h following administration.

Propionyl-L-carnitine has several advantages over L-carnitine for improving the abnormal skeletal muscle metabolism of patients with chronic heart failure (Table 1). First of all, as just mentioned, it increases plasma and

*Table 2.* Rationale for the use of propionyl-L-carnitine to improve cardiac and skeletal muscle function in heart failure.

1. *Data obtained from experiments under physiological conditions*

- Prolongation of plateau phase of cardiac action potential [49]
- Long-lasting positive inotropism after prolonged administration [47, 49]
- Dose-dependent reduction of arterial peripheral resistances [48]

2. *Data obtained from experiments under ischemic conditions*

- Reduction of degree of post-ischemic left ventricular dysfunction [44]
- Improvement of contractile recovery during post-ischemic reperfusion [52]
- Improvement of remodelling [53]

3. *Data obtained from experiments under failing conditions*

- Improvement of myocardial performance in several models of heart failure
- volume overloaded [56]
- erucic acid cardiomyopathy [54]
- pressure overload [55]
- cardiac hypertrophy [57]

cellular carnitine content thus enhancing FFA oxidation in carnitine-deficient states as well as increasing glucose oxidation rates [43].

Secondary, muscular carnitine transferases have higher affinity for propionyl-L-carnitine than for L-carnitine or its other derivatives. Therefore, propionyl-L-carnitine is highly specific for both skeletal and cardiac muscle [44]. Third, while having a similar mechanism of action as carnitine, it carries the propionyl group, and enhances the uptake of this agent by the myocardial cell [45]. This is particularly important as propionate can be used by mitochondria as an anaplerotic substrate, thus providing energy in the absence of oxygen consumption [46]. Interestingly, propionate alone cannot be administered due to its toxicity [47]. Finally, due to the particular structure of the molecule with a long lateral tail, propionyl-L-carnitine has a specific pharmacological action independently of its effect on muscle metabolism resulting in peripheral dilatation and positive inotropism [48, 49].

## Rationale for the use of propionyl-L-carnitine in chronic heart failure

Data obtained from experimental work provide the rationale for the use of propionyl-L-carnitine to improve cardiac and skeletal muscle function and metabolism in chronic heart failure (Table 2).

Our work on rabbits has demonstrated that propionyl-L-carnitine improves the mechanical function of the isolated aerobic heart. This improvement was noted to persist for at least 24 h after the final administration of this agent and to be present even when propionyl-L-carnitine was undetectable in either blood or tissue [49]. Research conducted with ischaemic myocardium by

Broderick et al. [50] on isolated rat hearts and global no-flow ischaemia showed that, during the reperfusion of previously ischaemic hearts, carnitine stimulated glucose oxidation and significantly improved the functional recovery, as measured by heart rate and peak systolic pressure. This work supports the theory that the beneficial effects of carnitine on ischaemic myocardium is a result of its ability to overcome the inhibition of glucose oxidation induced by increased levels of fatty acids. Other work also suggests an intracellular mechanism of action and implies that better protection is provided if the agent is administered prior to the ischaemic insult [51]. Paulson et al. [44] studied isolated rat hearts subjected to global, low-flow ischaemia. During reperfusion, the propionyl-L-carnitine group exhibited a significantly greater recovery of all haemodynamic variables: interestingly, a concentration of 1 mM of propionyl-L-carnitine had no significant protective effect, while concentrations of 5.5 mM and 11 mM significantly improved recovery of cardiac output. This beneficial effect was determined to be greater than that of L-acetylcarnitine or L-carnitine on a molar basis. Equally, propionyl-L-carnitine has been found to directly improve post-ischaemic stunning [44].

Furthermore, specific experimental studies have been conducted on the efficacy of this agent with respect to congestive heart failure [52–57]. They are reviewed in detail in Chapter 24. This study reports a clear improvement in cardiac function and metabolism.

Finally, in dogs, propionyl-L-carnitine causes a dose-dependent enhancement of cardiac output, reduction of arterial peripheral resistances and increase in mesenteric and iliac blood flow, confirming its pharmacological properties besides action on metabolism [49].

Interestingly, systemic carnitine deficiency produces a reversible form of cardiomyopathy [58, 59]. Low levels of carnitine have been found in adults with heart failure [60, 61] and in the explanted heart of patients with dilated [62] or ischaemic [63] cardiomyopathy. Carnitine therapy has also been found to be beneficial in the treatment of diphtheric [64] and anthracycline induced myocarditis [65].

## Effects of propionyl-L-carnitine in patients with chronic heart failure

There are several pilot studies in which administration of propionyl-L-carnitine to patients with chronic heart failure has been shown to exert several positive clinical effects. Furthermore, there are two current and ongoing large-scale national and international trials, the results of which should be available in a short time [66].

We have studied the effects of acute and chronic administration of propionyl-L-carnitine (1.5 g/day) on haemodynamics, hormonal levels, exercise capacity and oxygen consumption measurements in 15 patients with chronic heart failure (NYHA classes II and III and left ventricular ejection fraction

<40%). There were no changes in the haemodynamic or neurohormonal levels after either acute or chronic administration [67]. After one month of treatment, however, a significant increase in exercise capacity and peak $VO_2$ was observed, suggesting a possible improvement of peripheral muscle metabolism.

In a subsequent study, we examined the effects of propionyl-L-carnitine (15 g/day for one month) on limb metabolism both at rest and during exercise [68].

Skeletal muscle metabolism was assessed as femoral arterial-venous (A-V) difference for lactate, pyruvate and FFA. At rest, propionyl-L-carnitine caused a reduction of arterial and venous blood level of FFA but did not change overall muscle extraction of FFA, lactate or pyruvate. After maximal exercise propionyl-L-carnitine decreased the negative A-V difference for lactate, restored a positive A-V difference for pyruvate and did not change that for FFA. We concluded that propionyl-L-carnitine improves skeletal muscle metabolism in patients with idiopathic dilated cardiomyopathy by increasing pyruvate flux into the Krebs cycle and decreasing lactate production. This effect which occurs in the absence of major haemodynamic and neuroendocrine changes may underlie the ability of propionyl-L-carnitine to increase exercise performance in patients with heart failure.

Very recently, Caponetto et al. [69] have reported the effects of propionyl-L-carnitine on 50 patients with mild chronic heart failure in NYHA class II, symptomatic, despite therapy with digitalis and diuretics, with ejection fraction <45%. They were randomized to receive 1.5 g of propionyl-L-carnitine or placebo as oral treatment for six months. Maximal exercise time in the treated group was significantly increased (1 min more than placebo), whilst lactate production was significantly reduced. Left ventricular shortening fraction and left ventricular ejection fraction showed a significant increase in propionyl-L-carnitine group ($p < 0.0001$) while no difference was apparent in the placebo group. Stroke volume index and cardiac index had significant increments in the treated group ($p < 0.05$) and systemic vascular resistance was lowered ($p < 0.05$). No haemodynamic variations were observed in the placebo group. The clinical score showed a significant improvement in the propionyl-L-carnitine treated group.

The greatest changes occurred after the first month of treatment and persisted throughout the entire period of treatment.

These authors envisage two possible mechanisms of action: an improvement of skeletal muscle function and metabolism as well as a positive effect on cardiac muscle which explains the enhancement of the haemodynamic parameters.

Finally, it is interesting to note that propionyl-L-carnitine in patients with severe heart failure (NYHA IV) was able to reduce the increase of soluble receptors of TNF-$\alpha$ [70].

The specific role of TNF-$\alpha$ and of its soluble receptors in patients with heart failure is not clear and still under investigation. It is of interest for the present discussion, however, to recall that an increased tissue necrosis factor has been implicated in the skeletal muscle changes of heart failure [71].

## Conclusion

Skeletal muscle changes contribute to exertional fatigue in heart failure. Skeletal muscle metabolism is also altered in heart failure. Glucose and FFA oxidation is impaired, energy production is abnormally dependent on anaerobic glycolysis and lactate is produced in excess. Often there is a carnitine deficiency. Studies in carefully selected homogenous small groups of patients show that oral administration of propionyl-L-carnitine improves exercise capacity and skeletal muscle metabolism of patients with heart failure (classes II and III of NYHA).

This provides a logical basis for individualizing therapy in order to improve exercise tolerance and quality of life in patients with heart failure. The results of the ongoing larger trials will tell us whether this treatment should be recommended to all patients with heart failure.

## References

1. Garg R, Packer M, Pitt B, Yusuf S. Heart failure in the 1990s: Evolution of a major public health problem in cardiovascular medicine. J Am Coll Cardiol 1993; 22(Suppl A): 3A–5A.
2. Bourassa M, Gurne O, Bangdiwala SI et al. Natural history and patterns of current practice in heart failure. The Studies of Left Ventricular Dysfunction (SOLVD) Investigators. J Am Coll Cardiol 1993; 22(Suppl A): 14A–19A.
3. Cohn JN, Archibald DG, Ziesche S et al. Effect of vasodilator therapy on mortality in chronic congestive heart failure: results of a Veterans Administration Cooperative Study. N Engl J Med 1986; 314: 1547–52.
4. The CONSENSUS Trial Study Group. Effects of enalapril on mortality in severe congestive heart failure: results of the Cooperative North Scandinavian Enalapril Survival Study (CONSENSUS). N Engl J Med 1987; 316: 1429–35.
5. Cohn JN, Johnson G, Ziesche S et al. A comparison of enalapril with hydralazine-isosorbide dinitrate in the treatment of chronic congestive heart failure. N Engl J Med 1991; 325: 303–10.
6. The SOLVD Investigators. Effect of enalapril on survival in patients with reduced left ventricular ejection fractions and congestive heart failure. N Engl J Med 1991; 325: 293–302.
7. The SOLVD Investigators. Effect of enalapril on mortality and the development of heart failure in asymptomatic patients with reduced left ventricular ejection fractions. N Engl J Med 1992; 327: 685–91.
8. Pfeffer M, Braunwald E, Moye LA et al. Effect of captopril on mortality and morbidity in patients with left ventricular dysfunction after myocardial infarction: results of the survival and ventricular enlargement trial. The SAVE investigators. N Engl J Med 1992; 327: 669–77.
9. Swedberg K, Held P, Kjekshus J, Rasmussen K, Ryden L, Wedel H, for the CONSENSUS II Study Group. Effects of the early administration of enalapril on mortality in patients

with acute myocardial infarction: results of the Cooperative New Scandinavian Enalapril Survival Study II (CONSENSUS II). N Engl J Med 1992; 327: 678–84.

10. Yusuf S, Pepine CJ, Garces C et al. Effect of enalapril on myocardial infarction and unstable angina in patients with low ejection fractions. Lancet 1992; 340: 1173–8.
11. Kleber FX, Niemoller L, Doering W. Impact of converting enzyme inhibition on progression of chronic heart failure: results of the Munich Mild Heart Failure Trial. Br Heart J 1992; 67: 289–96.
12. Lindsay DC, Poole-Wilson PA. Angiotensin-converting enzyme inhibitors or vasodilators as therapy in chronic heart failure: a review of the trials. J Cardiovasc Pharmacol 1992; 19(Suppl 4): S45–55.
13. Wilson JR, Martin JL, Schwartz D, Ferraro N. Exercise intolerance in patients with chronic heart failure: role of impaired nutritive flow to skeletal muscle. Circulation 1984; 69: 1079–87.
14. Lipkin DP, Poole-Wilson PA. Symptoms limiting exercise in chronic heart failure. Br Med J Clin Res Ed 1986; 292: 1030–1.
15. Wilson JR, Mancini DM. Factors contributing to exercise limitation of heart failure. J Am Coll Cardiol 1993; 22(Suppl A): 93A–8A.
16. Lipkin DP, Canepa-Anson R, Stephens MR, Poole-Wilson PA. Factors determining symptoms in heart failure: comparison of fast and slow exercise tests. Br Heart J 1986; 55: 439–45.
17. Poole-Wilson PA. Relation of pathophysiologic mechanisms to outcome in heart failure. J Am Coll Cardiol 1993; 22(Suppl A): 22A–9A.
18. Drexler H, Banhardt U, Meinertz T, Wollschlager H, Lehmann M, Just H. Contrasting peripheral short-term and long-term effects of converting enzyme inhibition in patients with congestive heart failure. A double-blind, placebo-controlled trial. Circulation 1989; 79: 491–502.
19. Mancini DM, Davis L, Wexler JP, Chadwick B, LeJemtel TH. Dependence of enhanced maximal exercise performance on increased peak skeletal muscle perfusion during long-term captopril therapy in heart failure. J Am Coll Cardiol 1987; 10: 845–50.
20. Sullivan MJ, Knight JD, Higginbotham MB, Cobb FR. Relation between central and peripheral hemodynamics during exercise in patients with chronic heart failure. Muscle blood flow is reduced with maintenance of arterial perfusion pressure. Circulation 1989; 80: 769–81.
21. LeJemtel TH, Maskin CS, Lucido D, Chadwick BJ. Failure to augment maximal limb blood flow in response to one-leg versus two-leg exercise in patients with severe heart failure. Circulation 1986; 74: 245–51.
22. Anand IS, Ferrari R, Kalra GS, Wahi PL, Poole-Wilson PA, Harris P. Edema of cardiac origin. Studies of body water and sodium, renal function, hemodynamic indexes, and plasma hormones in untreated congestive cardiac failure. Circulation 1989; 80: 299–305.
23. Wilson JR, Ferraro N, Wiener DH. Effect of the sympathetic nervous system on limb circulation and metabolism during exercise in patients with heart failure. Circulation 1985; 72: 72–81.
24. Gibbons GH, Dzau VJ. The emerging concept of vascular remodeling. N Engl J Med 1994; 330: 1431–8.
25. Weiner DH, Fink LI, Maris J, Jones RA, Chance B, Wilson JR. Abnormal skeletal muscle bioenergetics during exercise in patients with heart failure: role of reduced muscle blood flow. Circulation 1986; 73: 1127–36.
26. Regensteiner JG, Wolfel EE, Brass E et al. Chronic changes in skeletal muscle histology and function in peripheral arterial disease. Circulation 1993; 87: 413–21.
27. Sullivan MJ, Higginbotham MB, Cobb FR. Exercise training in patients with severe left ventricular dysfunction: hemodynamic and metabolic effects. Circulation 1988; 78: 506–15.
28. Coats AJ, Adamopoulos S, Radaelli A et al. Controlled trial of physical training in chronic heart failure. Exercise performance, hemodynamics, ventilation, and autonomic function. Circulation 1992; 85: 2119–31.

29. Wilson JR, Fink L, Maris J et al. Evaluation of energy metabolism in skeletal muscle of patients with heart failure with gated phosphorus-31 nuclear magnetic resonance. Circulation 1985; 71: 57–62.
30. Mancini DM, Ferraro N, Tuchler M, Chance B, Wilson JR. Detection of abnormal calf muscle metabolism in patients with heart failure using phosphorus-31 nuclear magnetic resonance. Am J Cardiol 1988; 62: 1234–40.
31. Massie B, Conway M, Yonge R et al. Skeletal muscle metabolism in patients with congestive heart failure: relation to clinical severity and blood flow. Circulation 1987; 76: 1009–19.
32. Massie BM, Conway M, Rajagopalan B et al. Skeletal muscle metabolism during exercise under ischemic conditions in congestive heart failure. Evidence for abnormalities unrelated to blood flow. Circulation 1988; 78: 320–6.
33. Minotti JR, Johnson EC, Hudson TL et al. Skeletal muscle response to exercise training in congestive heart failure. J Clin Invest 1990; 86: 751–8.
34. Ferrari R, Pasini E, de Giuli F, Opasich C, Cobelli F, Tavazzi L. Limb uptake of substrate in patients with congestive heart failure (CHF). Can J Cardiol 1994: 10: 73.
35. Lopaschuk GD, Wambolt RB, Barr RL. An imbalance between glycolysis and glucose oxidation is a possible explanation for the detrimental effects of high levels of fatty acids during aerobic reperfusion of ischemic hearts. J Pharmacol Exp Ther 1993; 264: 135–44.
36. Wilson JR, Mancini DM, Ferraro N, Egler J. Effect of dichloroacetate on the exercise performance of patients with heart failure. J Am Coll Cardiol 1988; 12: 1464–9.
37. Wargovich TJ, MacDonald RG, Hill JA, Feldman RL, Stacpoole PW, Pepine CJ. Myocardial metabolic and hemodynamic effects of dichloroacetate in coronary artery disease. Am J Cardiol 1988; 61: 65–70.
38. Bersin R, Kwasman M, Wolfe C et al. Improved hemodynamic function in congestive heart failure with the metabolic agent sodium dichloroacetate (DCA). J Am Coll Cardiol 1990; 15(Suppl A): 157A (Abstr).
39. Sullivan MJ, Green HJ, Cobb FR. Altered skeletal muscle metabolic response to exercise in chronic heart failure: relation to skeletal muscle aerobic enzyme activity. Circulation 1991; 84: 1597–607.
40. Drexler H, Funke E, Riede U. The oxidative enzyme activity decreases in all fiber types in skeletal muscle of patients with chronic heart failure. Circulation 1991; 84(Suppl II): II-74 (Abstr)
41. Mancini DM, Nazzaro D, Georgopoulos L, Wagner N, Mullen JL, Wilson JR. Skeletal muscle atrophy contributes to exercise intolerance in heart failure. J Am Coll Cardiol 1991; 17(Suppl A): 88A (Abstr).
42. Marzo A, Cardace G, Corbelleta C, Bassani E, Morabito E, Arrigoni-Martelli E. Homeostatic equilibrium of L-carnitine family before and after i.v. administration of propionyl-L-carnitine in humans, dogs and rats. Eur J Drug Metab Pharmacokinet 1991; 3: 357–63.
43. Siliprandi N, Di Lisa F, Menabò R. Propionyl-L-carnitine: biochemical significance and possible role in cardiac metabolism. Cardiovasc Drugs Ther 1991; 5(Suppl 1): 11–5.
44. Paulson DJ, Traxler J, Schmidt M, Noonan J, Shug AL. Protection of the ischaemic myocardium by propionyl-L-carnitine: effects on the recovery of cardiac output after ischaemia and reperfusion, carnitine transport, and fatty acid oxidation. Cardiovasc Res 1986; 20: 536–41.
45. Hülsmann WC. Biochemical profile of propionyl-L-carnitine. Cardiovasc Drugs Ther 1991; 5(Suppl 1): 7–9.
46. Tassani V, Cattapan F, Magnanimi L, Peschechera A. Anaplerotic effect of propionyl carnitine in rat heart mitochondria. Biochem Biophys Res Commun 1994; 199: 949–53.
47. Ferrari R, Pasini E, Condorelli E et al. Effect of propionyl-L-carnitine on mechanical function of isolated rabbit heart. Cardiovasc Drugs Ther 1991; 5(Suppl 1): 17–23.
48. Cevese A, Schena F, Cerutti G. Short-term hemodynamic effects of intravenous propionyl-L-carnitine in anesthetized dogs. Cardiovasc Drugs Ther 1991; 5(Suppl 1): 45–56.
49. Ferrari R, Di Lisa F, De Jong JW et al. Prolonged propionyl-L-carnitine pretreatment of

rabbit: biochemical, hemodynamic and electrophysiological effects on myocardium. J Mol Cell Cardiol 1992; 24: 219–32.
50. Broderick TL, Quinney HA, Barker CC, Lopaschuk GD. Beneficial effect of carnitine on mechanical recovery of rat hearts reperfused after a transient period of global ischemia is accompanied by a stimulation of glucose oxidation. Circulation 1993; 87: 972–81.
51. Leipala JA, Bhatnagar R, Pineda E, Najibi S, Massoumi K, Packer L. Protection of the reperfused heart by L-propionylcarnitine. J Appl Physiol 1991; 71: 1518–22.
52. Moravec J, El Alaoui Talibi Z. Effect of propionyl-L-carnitine on the energy turnover and mechanical performance of chronically overloaded rat hearts. Cardiovasc Drugs Ther 1991; 5(Suppl 3): 102 (Abstr).
53. Micheletti R, Di Paola ED, Schiavone A et al. Propionyl-L-carnitine limits chronic ventricular dilation after myocardial infarction in rats. Am J Physiol 1993; 264: H1111–7.
54. Pasini E, Cargnoni A, Condorelli E, Marzo A, Lisciani L, Ferrari R. Effect of prolonged treatment with propionyl-L-carnitine on erucic acid-induced myocardial dysfunction in rats. Mol Cell Biochem 1992; 112: 117–23.
55. Micheletti R, Giacalone G, Reggiani C, Canepari M, Bianchi G. Effect of propionyl-L-carnitine treatment on mechanical properties of papillary muscles from pressure-overload rat hearts. J Mol Cell Cardiol 1992; 24(Suppl 5): S41 (Abstr).
56. El-Alaoui-Talibi Z, Bouhaddioni N, Moravec J. Assessment of the cardiostimulant action of propionyl-L-carnitine on chronically volume-overloaded rat hearts. Cardiovasc Drugs Ther 1993; 7: 357–63.
57. Yang XP, Samaja M, English E et al. Hemodynamic and metabolic activities of propionyl-L-carnitine in rats with pressure-overload cardiac hypertrophy. J Cardiovasc Pharmacol 1992; 20: 88–98.
58. Tripp ME, Katcher ML, Peters HA et al. Systemic carnitine deficiency presenting as familial endocardial fibroelastosis. A treatable cardiomyopathy. N Engl J Med 1981; 305: 385–90.
59. Waber LJ, Valle D, Neill C, DiMauro S, Shug A. Carnitine deficiency presenting as familial cardiomyopathy: a treatable defect in carnitine transport. J Pediatr 1982; 101: 700–5.
60. Suzuki Y, Masumuro Y, Kobayashi A, Yamazaki L, Harada Y, Osawa M. Myocardial carnitine deficiency in congestive heart failure. Lancet 1982; 1: 116 (Lett to the Ed).
61. Regitz V, Shug AL, Fleck E. Defective myocardial metabolism in congestive heart failure secondary to dilated cardiomyopathy and to coronary, hypertensive and valvular heart diseases. Am J Cardiol 1990; 65: 755–60.
62. Regitz V, Shug AL, Schuler S, Yankah AC, Hetzer R, Fleck E. Herzinsuffizienz bei dilatativer Kardiomyopathie and koronarer Herzkrankheit – Beitrag biochemischer Parameter zur Beurteilung der Prognose. Dtsch Med Wochenschr 1988; 113: 781–6.
63. Regitz V, Muller M, Schuler S et al. Carnitinstoffwechsel – Veranderungen im Eindstadium der dilatativen Kardiomyopathie und der ischämischen Herzmuskelerkrankung. Z Kardiol 1987; 76(Suppl 5): 1–8.
64. Figuelredo Ramos ACM, Elias PRP, Barrucand L, Da Silva JAF. The protective effect of carnitine in human diphtheric myocarditis. Pediatr Res 1984; 18: 815–9.
65. De Leonardis V, Neri B, Bacalli S, Cinelli P. Reduction of cardiac toxicity of anthracyclines by L-carnitine: preliminary overview of clinical data. Int J Clin Pharmacol Res 1985; 5: 137–42.
66. Garg R, Yusuf S. Current and ongoing randomized trials in heart failure and left ventricular dysfunction. J Am Coll Cardiol 1993; 22(Suppl A): 194A–7A.
67. Anand IS, Chandrashekhar Y, Sarma PR, Ferrari R, Corsi M. Effect of propionyl-L-carnitine on the hemodynamics, peak $VO_2$ and hormones in CHF. Chest 1991; 100(Suppl): 110S (Abstr).
68. Ferrari R, Cargnoni A, de Giuli F, Pasini E, Anand I, Visioli O. Propionyl-L-carnitine improves skeletal muscle metabolism and exercise capacity of patients with congestive heart failure. Circulation Atlanta, Georgia 1993; 88(2): 2223(Abstr).
69. Caponetto S, Canale C, Masperone MA, Terrachini V, Valentini G, Brunelli C. Efficacy

of L-propionylcarnitine treatment in patients with left ventricular dysfunction. Eur Heart J 1994; 15: 1267–73.
70. Bachetti T, Corti A, Cassani G, Confortini R, Mazzoletti A, Ferrari R. Cytokines in end stage congestive heart failure: effect of propionyl-L-carnitine. Can J Cardiol 1994; 10: 66.
71. Levine B, Kalman J, Mayer L, Fillit HM, Packer M. Elevated circulating level of tumor necrosis factor in severe chronic heart failure. N Engl J Med 1991; 323: 236–41.

*Corresponding Author*: Professor Roberto Ferrari, Cattedra di Cardiologia, Università degli Studi di Brescia, c/o Spedali Civili, P.le Spedali Civili, 1, I-25123 Brescia, Italy

# 24. Hemodynamic and metabolic effect of propionyl-L-carnitine in patients with heart failure

SALVATORE CAPONNETTO and CLAUDIO BRUNELLI

---

"Compared to the control group, the patients treated with propionyl-L-carnitine showed significant increases in the values of exercise capacity and ejection fraction, which became even more evident after 90 and 180 days."

---

## Introduction

Congestive heart failure (CHF) is an insidious disease process associated with profound symptoms and a poor long-term prognosis. Today the increasing geriatric population may in part explain the high prevalence, and mortality rate, of the syndrome, despite improved therapeutic strategies in the treatment of hypertension, coronary artery disease, severe arrhythmias and sudden death. According to the Framingham study, five-year mortality for CHF is 60% in men and 45% in women; if we consider patients in NYHA class IV, one year mortality reaches almost 50% [1].

Heart failure alters the myocardium in several animal models and in humans in a comparable way. Initial damage to the myocardium is followed by a period of myocardial hyperfunction that is associated with qualitative and quantitative alterations of cardiac metabolism and composition. This state can only be tolerated for a limited time, progressive deterioration of function leading ultimately to end-stage heart failure [2, 3].

The most attractive hypotheses to explain myocardial dysfunction in heart failure postulate biochemical defects in the myocardium other than the alteration in loading conditions or ischemic damage as a cause for the functional deficits. These biochemical causes of contractile dysfunction have not yet been conclusively described. The assessment of changes in the complex metabolism of the myocardium may improve the understanding of both the underlying biochemical defects and the role of exogenous factors in a multifactorial system.

A low cardiac output and a high cardiac filling pressure are the hemodynamic hallmarks of heart failure. They are thought to be the major circulat-

*J.W. de Jong and R. Ferrari (eds): The carnitine system*, 337–351.

ory determinants of the fatigue and breathlessness that characterize heart failure. Therefore it has been assumed that drug therapy that favorably influences the cardiac output and filling pressure should relieve symptoms in patients with CHF. A variety of oral drug regimes appear to have a favorable chronic effect on left ventricular hemodynamics. Nevertheless, not all of these agents appear to alleviate symptoms or improve quality of life.

Exertional fatigue is a major limiting symptom in patients with CHF. Usually, this fatigue has been attributed to skeletal muscle underperfusion. However, increased cardiac output during exercise exerted by vasodilators cannot be translated immediately into increased exercise capacity and peak oxygen consumption [4–6]. Even when oxygen delivery to skeletal muscle is improved by pharmacological intervention, the oxygen utilization is not augmented acutely.

Carnitine plays an important role in fatty acid oxidation as well as in other metabolic pathways. Myocardium and skeletal muscle depend on fatty acid oxidation and thus require carnitine to maintain energy metabolism; they are highly dependent on carnitine transport from its sites of synthesis [7]. Propionyl-L-carnitine (PLC) is an ester of propionic acid and L-carnitine (LC). PLC is more lipophilic and penetrates better into the myocytes; has enhanced affinity for carnitine acetyltransferase, the key enzyme for free fatty acid transport into myocardial matrix; and stimulates the Krebs cycle, increasing the efficiency of ischemic mitochondria [8].

This article outlines present concepts on carnitine and its derivatives in animal models of heart failure, the characteristic features of myocardial carnitine loss, and the chemical effects of PLC in heart failure. After a short note about carnitine and propionate metabolism in cardiac and skeletal muscle, we report a profile on long-term L-carnitine therapy in patients with mild chronic heart failure.

## Carnitine and its derivatives in animal models of heart failure

Several animal species models as well as functional and metabolic parameters of injury have been utilized to test the possible beneficial effects of LC and PLC on heart failure. Suzuki et al. [9] gave anesthetized closed-chest dogs an intravenous infusion of 80 mg/kg/min of LC for 8 min. They demonstrated a 17% decrease in heart rate, a 20% increase in aortic and left ventricular pressure and a 35% increase in peak positive left ventricular dP/dt. The positive inotropic activity of LC seen in these animals was accompanied by a 60% rise in coronary blood flow and a 25% reduction in coronary vascular resistence. Brooks et al. [10] found that the positive inotropic effect of LC, administered intravenously, became more pronounced with increasing doses. A 60 mg/kg/min infusion rate was associated with a doubling of stroke volume, a nearly threefold increase in left ventricular end-diastolic pressure, a rise in left ventricular dP/dt max, and a 38% increase in left ventricular contractile force. Failure of either propranolol or reserpine to reverse changes induced by LC infusion indicated that the hemodynamic effects seen in these animals were not mediated by catecholamines.

Paulson et al. [11] showed comparable positive inotropic properties in isolated perfused rat hearts subjected to 90 min global ischemia followed by 15 min of reflow. Specifically, cardiac output, left ventricular pressure, left ventricular dP/dt max, and tissue high-energy phosphates were all increased following treatment with PLC. Liedtke et al. [12] too saw improvements in mechanical function during reperfusion in the intact working pig heart.

In subcellular studies, Kotaka et al. [13] and Ferrari et al. [14] noted improvements in mitochondrial function and respiration in the presence of PLC via favorable shifts in mitochondrial fatty acyl CoA and protection from lipid peroxidation.

Ferrari et al. [15] studied the acute and chronic effects of PLC on mechanical function of isolated rabbit heart. When administered acutely PLC had no effect on inotropism, heart rate, or coronary perfusion pressure. When administered chronically it induced a positive inotropic effect, with no changes in heart rate or in coronary perfusion pressure, and it ameliorated the pressure-volume relationship.

To understand these results the same authors [16] pretreated 253 rabbits up to ten days with daily doses of 1 mmol/kg intraperitoneally of PLC or LC, using saline-treated animals as control. The studies carried out with perfused hearts and isolated mitochondria failed to show an effect of PLC pretreatment on high-energy phosphate metabolism or respiration. Perfused hearts displayed positive inotropy in prolonged treatment with PLC, but not with LC.

Leipala et al. [17] studied the effects of PLC on mechanical function, creatine phosphate and ATP content, and lactate dehydrogenase leakage in isolated perfused rat hearts exposed to global, no-flow ischemia for 30 min followed by reperfusion for 20 min. Five and 10 mM PLC resulted in a 100% recovery of left ventricular developed pressure, whereas the recovery was only 40% in the hearts perfused without this agent. PLC provided protection for the post-ischemic, reperfused heart in a dose-dependent manner. The optimal time for administration was prior to the ischemic insult. High doses of this compound may perturb all membrane integrity.

Cevese et al. [18] investigated the effects of iv. administration of PLC in anesthetized dogs instrumented for the analysis of general hemodynamic and electrocardiographic data, peripheral blood flow, coronary blood flow and oxygen consumption, urine flow and renal function. PLC was administered as a bolus or by infusion. In some cases LC and LC plus propionate were also administered in doses equimolar to those of PLC. PLC elicited dose-dependent, short-lasting enhancements of cardiac output, arterial blood pressure, heart rate; contractility varied slightly and impredictably. These responses were not changed by $\alpha$- or $\beta$-adrenergic blockade, nor by the administration of a calcium antagonist, but they were abolished or reversed by the combination of such blocking interventions. Mesenteric and iliac blood flows were increased by both PLC and LC; LC plus propionate increased these flows in addition to renal blood flow. PLC elicited coronary vasodilation with reduced oxygen extraction; this effect lasted longer than the general hemodynamic effects and was not seen with LC.

Therefore experimental animal data show beneficial effects of LC and PLC on cardiac mechanical function, coronary flow, peripheral vascular resistance, high-energy phosphates, mitochondrial and respiratory function. Usually, these effects correlate with an increase in myocardial content of carnitine and its metabolites.

## Characteristic features of myocardial carnitine loss

Whitmer [19] studied Syrian hamsters with inborn hypertrophic and dilated cardiomyopathy associated with carnitine deficiency. This investigator demonstrated that LC administration restored myocardial carnitine concentration to normal and improved myocardial function.

Keene et al. [20] reported the occurrence of spontaneous dilated myocardiopathy in dogs. Two of 6 animals affected by severe CHF had strongly diminished myocardial concentrations of carnitine. In these dogs high doses of LC produced an increased myocardial concentration of carnitine and improved their clinical status and myocardial function.

Regitz et al. [21] investigated myocardial carnitine content in explanted hearts of patients with end-stage heart failure. To assess whether myocardial carnitine deficiency was specific for dilated cardiomyopathy or if it also occurred in heart failure of other origins, myocardial carnitine levels were determined from biopsies in patients with coronary or valvular heart disease. The patients with heart failure due to dilated cardiomyopathy and those with coronary or valvular heart diseases showed a significant reduction of myocardial carnitine levels in all areas of the explanted hearts.

Decreased myocardial carnitine is not simply an answer to hemodynamic changes because different compensatory mechanisms may be effective in different patients and may lead to differing responses to metabolic stress. Measurements in endomyocardial biopsies enable changes in the stages of cardiomyopathies to be monitored, to investigate possible correlations between metabolic changes and function, and to obtain normal tissue for comparisons. In patients with dilated cardiomyopathy and mild to severe heart failure, myocardial carnitine was significantly reduced compared with normal controls and the extent of the reduction in patients with coronary heart disease was not different from the carnitine loss in patients with dilated cardiomyopathy [22].

Therefore, carnitine levels do not only reflect functional impairment but, as already stated, probably also represent the metabolic response of the myocardium to stress.This response can vary in different ways depending on duration of the injury, genetic disposition, nutritional status, availability of substrates, hormonal regulations and the availability of other compensatory mechanisms.

**Clinical effects of carnitine and propionyl-L-carnitine in heart failure and in peripheral vascular disease**

The positive inotropic activity of LC observed in laboratory animals has also been detected in man. This effect, however, appears to be much more pronounced in patients with evidence of ischemic heart disease than in normal individuals.

Schiavoni et al. [23] showed that an iv. dose of 40 mg/kg of LC administered to healthy volunteers over a period of 2 min produced only modest variations in heart rate, arterial pressure, preejection period (PEP), left ventricular ejection time (LVET), PEP/LVET rate and in echocardiographic indices of cardiac performance. In 10 patients with presumed coronary artery disease slight changes in arterial pressure and heart rate were observed during LC infusion; a maximal decrease in PEP/LVET ratio of approximately 15% occurred at 10 min of infusion and persisted 45 min after completion. Echocardiographic examination of these patients showed significant increases in left ventricular wall motion.

Giordano et al. [24] reported a similar beneficial effect on cardiac performance after a more prolonged period of LC administration to 18 patients with angina pectoris and mild heart failure symptoms.

Ghidini et al. [25] studied 38 elderly patients with CHF secondary to ischemia or hypertensive heart disease. All of them received traditional therapy with digitalis and diuretic and, when necessary, antiarrhythmic agents. Twenty-one patients were also treated with oral LC at a dose of 1 g twice daily for 15 days. The other 17 patients constituted the control group. Both groups demonstrated similar improvement in subjective and objective clinical parameters and NYHA functional class. Echocardiographic measures of left ventricular wall motion and size also indicated a favorable effect. The LC group experienced a reduction in the incidence of cardiac arrhythmias and a more marked decrease in digoxin requirements.

Hiatt et al. [26, 27] found in patients with peripheral vascular obstruction that skeletal-muscle oxidative metabolism was impaired as reflected by significant elevation in plasma acylcarnitines, detected after a relatively short duration of low-level exercise.

Brevetti et al. [28] reported that, in one double-blind crossover trial, 20 patients with peripheral vascular disease were treated alternately with either placebo or oral LC 2 g b.i.d. for three weeks. The absolute walking capacity achieved by these patients after LC therapy was 75% greater than that observed after placebo administration. Walking time increased by 67% in a separate series of eight patients given a short iv. course of LC.

Brevetti et al. [29] evaluated the effect of severe peripheral insufficiency on carnitine concentrations and carnitine acetyl- and palmitoyltransferase activity in the ischemic skeletal muscles of patients with severe peripheral vascular disease. In biopsies, ischemic muscles showed a significant reduction in total carnitine from the control value; a significantly lower free carnitine and acylcarnitine content contributed to this reduction.

Brevetti et al. [30], in an acute, iv., double-blind, crossover study, compared the effects of PLC vs LC on walking capacity in patients with peripheral vascular disease. Results indicated that neither drug affected the blood velocity nor the blood flow rate in the ischemic leg, suggesting that the beneficial effect on walking capacity was dependent on a metabolic action. On a molar basis, this beneficial effect was greater than with LC.

**Exercise capacity, hemodynamic and metabolic features in patients with congestive heart failure**

Mechanisms responsible for exercise intolerance in patients with CHF are not known, but do not seem directly related to a decrement of cardiac output or an increase in left ventricular filling pressure.

Franciosa et al. [31] reported that hemodynamic measurements at rest, including left ventricular filling pressure and cardiac output, do not correlate with exercise capacity in patients with CHF and that the relation between exercise capacity and measurements of cardiac performance at rest is hardly known. Repeated studies of treatment of heart failure failed to show correlations between changes in exercise capacity and changes in left ventricular performance at rest; thus, measures of left ventricular performance obtained at rest do not accurately reflect exercise tolerance and symptomatic status of patients with CHF.

Holloszy et al. [32] found that regularly performed endurance exercise induces major adaptations in skeletal muscle, including increase in the mitochondrial content and respiratory capacity in the muscle fibers. As a consequence of the increase in mitochondria, exercise of the same intensity results in a disturbance in homeostasis that is smaller in trained than in untrained muscles. The major metabolic consequences of the adaptations of muscles to endurance exercise are a slower utilization of muscle glycogen and blood glucose, a greater reliance on fat oxidation and less lactate production during exercise of a given intensity. Moreover, regional blood flow to exercising skeletal muscle is reduced in patients with CHF, but the histology, biochemistry, and contractile function are also abnormal.

The mechanisms for the intrinsic abnormality of the skeletal muscle are unknown. The interpretation of experimental data is complicated by the different etiology of heart failure, drug treatment, exercise protocols, limitations of methods for the measurements of blood flow and metabolism in intact humans and by selection of particular groups of muscles for study that can not reflect changes in other muscles in the body.

Many observations have demonstrated that intrinsic abnormalities of skeletal muscle emerge in CHF. Histological abnormalities of skeletal muscle in patients with hemodynamic evidence of cardiomyopathy have been recognized and an association between abnormal skeletal muscle histological findings and cardiac disturbances has been reported [33].

Furthermore, Dunnigan et al. [34] compared cardiac and skeletal muscle histology and biochemistry in patients whose initial manifestations of cardiac disease were due to symptoms resulting from CHF. Cardiac histological studies revealed a spectrum of abnormalities including fibrosis, dilated sarcoplasmic reticulum, increased numbers of intercalated disks and mitochondrial abnormalities. Histological abnormalities of skeletal muscle consisted of endomysial fibrosis and increased lipid deposits; slightly more than half of these patients also had a low concentration of skeletal muscle long-chain acylcarnitine. These data suggested that patients with CHF may have a generalized myopathy.

Massie et al. [35, 36] demonstrated by $^{31}P$ nuclear magnetic resonance (NMR) during fatigue-limited exercise that patients with CHF exhibit phosphocreatine depletion and increased glycolytic metabolism. Similar findings have been reported by Wilson et al. [37] during less strenuous steady-state exercise. The metabolic consequences of impaired nutritive blood flow during exercise may ultimately influence respiratory gas exchange to the extent that breathlessness and fatigue are experienced.

Douglas [38] described a ventilatory response during submaximal exercise, that was related to excess $CO_2$ production relative to oxygen utilization and to an increase in blood lactate. This response was termed the "anaerobic threshold" [39] based on the reasoning that excess $CO_2$ production resulted from buffering of lactate released from exercising muscles that had switched on anaerobic glycolysis. The anaerobic (or ventilatory) threshold has received attention as a nonmotivational submaximal measurement that correlates with aerobic activity in both normal subjects and patients with CHF [40]. In addition, since exercise in normal subjects can be sustained for a long period below, but not above the anaerobic threshold, this threshold may be a measure of functional capacity in patients with CHF that is pertinent to daily activity. Simonton et al. [41] studied optimal ventilatory criteria and exercise protocols for determining the anaerobic threshold, and the day-to-day reproducibility of the anaerobic threshold and its relation to peak oxygen uptake ($VO_2$) and blood lactate concentration in patients with CHF. Their results showed that the lactate increment at the anaerobic threshold occurred within a narrow range in patients with CHF (as well as in normal subjects), although the anaerobic threshold did not predict a precise lactate level for individual subjects.

### Carnitine and propionate metabolism in cardiac and skeletal muscles

A reasonable interpretation of these metabolic effects may be the following. Oxygen availability in skeletal muscle is critical for the conversion of pyruvate either to lactate or to acetylcoenzyme A (acetyl-CoA). The latter reaction is catalyzed by pyruvate dehydrogenase, the activity of which is controlled by the acetyl-CoA/CoA ratio [42, 43].

Bieber et al. [44] showed that, through the action of carnitine acetyltransferase, carnitine may decrease such a ratio and stimulate the activity of pyruvate dehydrogenase, consequently preventing the formation of lactate. In patients with peripheral vascular disease, pyruvate oxidation is presumably limited by two conditions: the inadequacy of oxygen supply and the accumulation of acetyl-CoA caused by a decreased flux into the Krebs cycle. Alkonyi et al. [45] and Bremer [46] demonstrated that carnitine does not apparently influence tissue oxygen supply but is able to decrease the acetyl-CoA concentration due to the presence of a very active carnitine acetyltransferase in the muscular tissue. This assumption is supported by the finding that administration of LC resulted in a significant increase of total carnitine in muscles of the affected leg.

Both the increase in free carnitine and short-chain acylcarnitine contributed to such an increase. These changes in the concentrations of muscular carnitine fractions indicate that part of the administered carnitine was taken up by muscles of the affected leg and that a consistent portion was transformed into short-chain acylcarnitine, presumably acetylcarnitine. This implies that a corresponding amount of short-chain acyl-CoA was removed along with a concurrent release of free CoA. The consequent decrease of the acetyl-CoA/CoA ratio would explain the above-mentioned stimulation of pyruvate dehydrogenase.

Bjorkman [47] demonstrated that in short-term exercise the preferentially utilized substrate was conceivably muscle glycogen or blood glucose. As a consequence, a large increase of pyruvate production should be expected. The stimulation of pyruvate dehydrogenase activity by the increased availability of carnitine might explain both the decreased production of lactate and the higher yield of energy, resulting from pyruvate oxidation. It is well known that 1 molecule glucose used for anaerobic glycolysis yields 2 ATP, whereas its utilization in the aerobic pathway produces 36 ATP. This enhancement of pyruvate oxidation, hence in energy production, may result in an improvement of walking capacity after treatment with LC.

Di Lisa et al. [48] also showed that treatment with LC might be beneficial because it removed the long-chain acyl-CoA. An accumulation of these metabolites in oxygen deficient conditions may be detrimental to the cellular membrane stability.

A few studies [23–25] are available documenting the efficiency of carnitine as a therapeutic agent with positive inotropic and vasodilator properties in aerobic and ischemic heart muscle. Carnitine has been shown to block the entry of fatty acids into cardiomyocytes, to reduce intracellular concentrations of neutral lipids and long-chain acyl-CoA, and to improve the action of at least one critical enzyme system, adenine nucleotide translocase. The role of propionate is less certain.

Fafournoux et al. [49] showed that propionate in liver, and by inference myocardium, is taken up by a carrier-mediated process that is responsive to external pH. An acid environment stimulates uptake, while an alkaline pH hinders entry. Following uptake, propionate moves into mitochondria. The citric acid cycle requires activation to the CoA derivative with conversion first to methylmalonyl-CoA and finally to succinyl-CoA. This transformation from propionyl-CoA to succinyl-CoA is in part catalyzed by propionyl-CoA carboxylase, which is energy dependent.

Brass et al. [50, 51] showed that propionate can inhibit enzyme functions and oxidative metabolism in noncardiac tissue; these effects are relieved with the addition of carnitine. Sundqvist et al. [52] and Latipaa et al. [53] found that propionate can stimulate enzyme activity in cardiac tissue, and in the case of pyruvate dehydrogenase contributes to pyruvate synthesis. The role of propionate as a substrate for oxidative phosphorylation in heart muscle is little understood.

Peuhkurinen [54] thought that propionate was a minor substrate, making only marginal contributions to cardiac ATP production. Sundqvist et al. [52] noted that propionate increased the citric acid cycle pool size fourfold but did little to influence myocardial oxygen consumption. Furthermore, as substrate, propionate is capable by feedback inhibitions to decrease the rates of glycolysis, as well as glucose, palmitate and pyruvate oxidation.

Bolukoglu et al. [55] reported that the profile of substrate oxidation with substrate dosage did suggest a rate-limiting step in propionyl-CoA carboxylase and that there was a decline in mechanical function at the maximum dose of propionate (with glucose present) or when used as sole substrate. Therefore, the optimal advantages of propionate seem best expressed at moderate doses and in an environment of mixed substrate availability.

### Profile of long-term propionyl-L-carnitine therapy in patients with congestive heart failure

Few studies exist about treatment with PLC in CHF. Mancini et al. [56] studied, in a double-blind study vs placebo, 60 patients with mild to moderate CHF, who had been undergoing chronic treatment with digitalis and diuretics for 3 or 6 months. Compared to the control group, the patients treated with PLC showed significant increases in the values of exercise capacity and ejection fraction (EF), which became even more evident after 90 and 180 days.

Pucciarelli et al. [57] studied the clinical and hemodynamic effects of the PLC in 50 patients with mild to moderate CHF. The study was carried out

*Table 1.* Long-term propionyl-L-carnitine therapy increases maximal exercise capacity of patients with congestive heart failure.

| Treatment | Duration of treatment (days) | | | | | |
|---|---|---|---|---|---|---|
| | Basal | 15 | 30 | 60 | 120 | 180 |
| Placebo | 5.35 ± 1.87 | 5.27 ± 2.07 | 5.32 ± 1.78 | 5.40 ± 1.98 | 5.52 ± 1.98 | 5.52 ± 1.94 |
| PLC | 5.50 ± 1.75 | 6.27 ± 1.71 | 6.72 ± 1.78 | 6.95 ± 1.99 | 6.67 ± 1.88 | 6.62 ± 2.00 |

Data (mean ± SEM) in minutes. PLC = propionyl-L-carnitine. Analysis of variance: treatment, $p < 0.01$; treatment × time interaction, $p < 0.01$.

in a double-blind, randomized vs placebo way, lasting for 180 days of treatment with 1.5 g of PLC. At the end of the experimental period, exercise duration and EF were significantly increased while peripheral vascular resistance was reduced.

Caponnetto et al. [58, 59] evaluated the effects of PLC vs placebo on exercise capacity, lactate and pyruvate metabolism before and after exercise test, and hemodynamics, in 80 patients who were symptomatic despite therapy with digitalis and diuretics. The study was of a parallel, randomized design, carried out single-blind during a basal period of 8 days and double-blind during a randomized controlled vs treatment period of 180 days. PLC and placebo were administered at a dose of 1.5 g/day orally, three times a day. At recruitment the patients who had satisfied all inclusion/exclusion criteria were assessed by an accurate analysis of physical activity. Then M-B mode and Doppler echocardiography and a symptom-limited incremental test were performed. Seven days after this last test the patients were given placebo; the first maximal steady-state test was performed and venous blood lactate and pyruvate concentration were determined before and after exercise. The randomized treatment period was begun and the patients presented themselves to the investigators at day 15, 30, 60, 90, 120 and 180 to carry out the tests required from the experimental protocol. The endpoints were: 1. maximal exercise capacity; 2. venous peripheral blood concentrations of lactate and pyruvate before and after exercise stress test; 3. hemodynamic values, i.e. left ventricular shortening fraction (SF), left ventricular ejection fraction (EF), stroke volume index (CWI), systolic wall stress (SWS), systemic vascular resistence (SVR).

The results indicated that maximal exercise capacity was significantly better after the 30th day of treatment with PLC (Table 1). In relation to exercise capacity, expressed in terms of work performed, the PLC group achieved the highest loads. Analysis of the percentage changes in comparison with basal values showed a definite increase in the PLC group. This increment was statistically significant in comparison to that obtained with placebo.

Figure 1 shows that the PLC group, at the end of the treatment period, had a reduction of lactate and pyruvate concentrations compared to placebo and to basal values ($p < 0.001$). Analysis of variance showed a significant

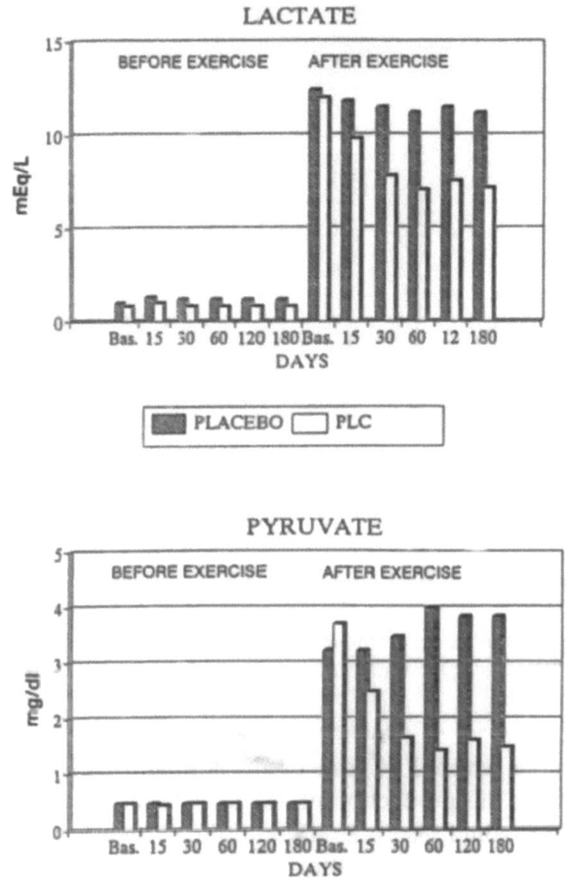


*Figure 1.* Propionyl-L-carnitine treatment reduced lactate and pyruvate concentrations in venous peripheral blood after submaximal exercise in patients with congestive heart failure. For details, see text. Bas., Basal values; PLC, propionyl-L-carnitine.

difference between the two groups, for both lactate ($p < 0.001$) and pyruvate ($p < 0.005$).

Left ventricular SF and EF showed an increase in the PLC group ($p < 0.001$), while no difference was apparent in the placebo group. SWS and SVR (Figure 2) lowered only in the PLC group; statistical analysis showed a significant difference between the two groups (for SWS $p < 0.001$, and for SVR $p < 0.005$). SVI, CI and CWI had significant increments in the PLC group in comparison with the placebo group ($p < 0.005$).

Our results can be explained by two possible mechanisms. The first is that

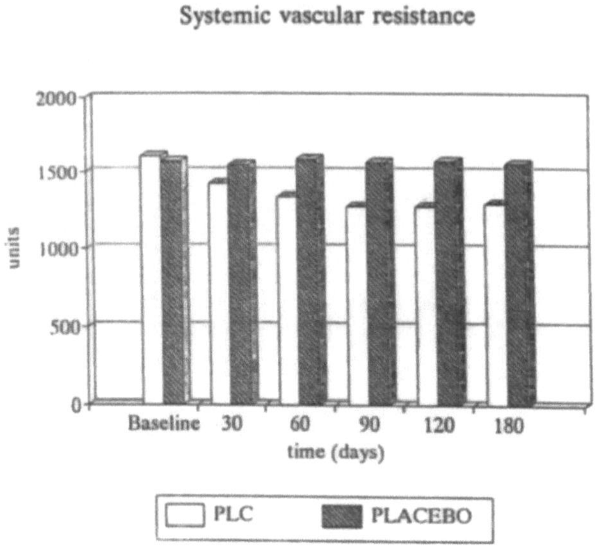


*Figure 2.* Propionyl-L-carnitine treatment lowered systemic vascular resistance of patients with congestive failure. For details, see text. PLC, propionyl-L-carnitine.

skeletal muscle, in which PLC can easily normalize the concentration of carnitine, increases the production of ATP for both aerobic and anaerobic pathways. So PLC could play a key role in the capacity of ischemic muscle to tolerate a larger workload, less affected by the conditions of hypoperfusion under which it was working. Our data support this hypothesis: they demonstrated lower levels of lactate and pyruvate in venous blood during the exercise stress test.

The second proposed mechanism considers the action of cardiac muscle. Hemodynamic parameters demonstrated an increase of SF, EF, SVI, CI and CWI, as well as a reduction of SWS and SVR. It remains to be explained whether these effects of PLC on the hemodynamics can be interpreted as a metabolic action on cardiac muscle or as a direct pharmacological action with a positive inotropic or vasodilator effect.

The therapeutic action of propionyl-L-carnitine described may occur through correction of aberrations in fatty acid metabolism underlying muscle dysfunction, unlike other mechanisms, as do most of the inotropic agents so far available.

## References


1. Kannel WB, Belanger AJ. Epidemiology of heart failure. Am Heart J 1991; 121: 951–7.
2. Brunelli C, Ghigliotti G, Martini U, Caponnetto S. New therapeutic strategies in the management of congestive heart failure. Eur Heart J 1991; 12(Suppl G): 53–7.
3. Poole-Wilson PA, Buller NP. Causes of symptoms in chronic congestive heart failure and implications for treatment. Am J Cardiol 1988; 62: 31A–4A.
4. Wilson JR, Ferraro N. Exercise intolerance in patients with chronic left heart failure: relation to oxygen transport and ventilatory abnormalities. Am J Cardiol 1983; 51: 1358–63.
5. Wilson JR, Martin JL, Ferraro N, Weber KT. Effect of hydralazine on perfusion and metabolism in the leg during upright bicycle exercise in patients with heart failure. Am J Cardiol 1983; 68: 425–32.
6. Wilson JR, Martin JL, Ferraro N. Impaired skeletal muscle nutritive flow during exercise in patients with congestive heart failure: role of cardiac pump dysfunction as determined by the effect of dobutamine. Am J Cardiol 1984; 53: 1308–15.
7. Siliprandi N, Sartorelli L, Ciman M, Di Lisa F. Carnitine: metabolism and clinical chemistry. Clin Chim Acta 1989; 183: 3–11.
8. Di Lisa F, Menabò R, Siliprandi N. L-Propionyl-carnitine protection of mitochondria in ischemic rat hearts. Mol Cell Biochem 1989; 88: 169–73.
9. Suzuki Y, Kamikawa T, Yamazaki N. Effect of L-carnitine on cardiac and peripheral hemodynamics. Jpn Heart J 1981; 22: 219–25.
10. Brooks H, Goldberg L, Holland R, Klein M, Sanzari N, DeFelice S. Carnitine-induced effects on cardiac and peripheral hemodynamics. J Clin Pharmacol 1977; 17: 561–8.
11. Paulson DJ, Traxler J, Schmidt M. Protection of the ischaemic myocardium by L-propionyl-carnitine: effects on the recovery of the cardiac output after ischaemia and reperfusion, carnitine transport, and fatty acid oxidation. Cardiovasc Res 1986; 20: 536–41.
12. Liedtke AJ, DeMaison L, Nellis SH. Effects of L-propionylcarnitine on mechanical recovery during reflow in intact heart. Am J Physiol 1988; 255: H169–76.
13. Kotaka K, Miyazaki Y, Ogawa K, Satake T, Sugiyama S, Ozawa T. Reversal of ischemia-induced mitochondrial dysfunction after coronary reperfusion. J Mol Cell Cardiol 1982; 14: 223–31.
14. Ferrari R, Ciampalini G, Agnoletti G, Cargnoni A, Ceconi C, Visioli O. Effect of L-carnitine derivatives on heart mitochondrial damage induced by lipid peroxidation. Pharmacol Res Commun 1988; 20: 125–31
15. Ferrari R, Pasini E, Condorelli E et al. Effect of propionyl-L-carnitine on mechanical function of isolated rabbit heart. Cardiovasc Drugs Ther 1991; 5(Suppl 1): 17–23.
16. Ferrari R, Di Lisa F, De Jong JW et al. Prolonged propionyl-L-carnitine pretreatment of rabbit: biochemical, hemodynamic and electrophysiological effects on myocardium. J Mol Cell Cardiol 1992; 24: 219–32.
17. Leipala JA, Bhatnagar R, Pineda E, Najibi S, Massommi K, Packer L. Protection of the reperfused heart by L-propionylcarnitine. J Appl Physiol 1991; 71: 1518–22.
18. Cevese A, Schena F, Cerrutti G. Short-term hemodynamic effects of intravenous propionyl-L-carnitine in anesthetized dogs. Cardiovasc Drugs Ther 1991; 5(Suppl 1): 45–56.
19. Whitmer JT. L-Carnitine treatment improves cardiac performance and restores high-energy phosphate pools in cardiomyopathic Syrian hamsters. Circ Res 1987; 61: 396–408.
20. Keene BW, Panciera DP, Atkins CE, Regitz V, Schmidt MJ, Shug AL. Myocardial L-carnitine deficiency in a family of dogs with dilated cardiomyopathy. J Am Vet Med Assoc 1991; 198: 647–50.
21. Regitz V, Muller M, Schuler S et al. Carnitinstoffwechsel-Veränderungen im Endstadium der dilatativen Kardiomyopathie und der ischämischen Herzmuskelerkrankung. Z Kardiol 1987; 76(Suppl 5): 1–8.
22. Regitz V, Shug AL, Fleck E. Defective myocardial carnitine metabolism in congestive heart

failure secondary to dilated cardiomyopathy and to coronary, hypertensive and valvular heart disease. Am J Cardiol 1990; 65: 755–60.
23. Schiavoni G, Pennestri F, Mongiardo R. Cardiodynamic effects of L-carnitine in ischaemic cardiopathy. Drugs Exptl Clin Res 1983; 9: 171–85.
24. Giordano MP, Corsi M, Roncarolo P, Falcone M, Gabasio C. Effect of L-carnitine on systolic time intervals in coronary artery disease. Curr Ther Res 1983; 33: 305–11.
25. Ghidini O, Azzurro M, Vita G, Sartori G. Evaluation of the therapeutic efficacy of L-carnitine in congestive heart failure. Int J Clin Pharmacol Ther Toxicol 1988; 26: 218–20.
26. Hiatt WR, Nawaz D, Brass EP. Carnitine metabolism during exercise in patients with peripheral vascular disease. J Appl Physiol 1987; 62: 2383–7.
27. Hiatt WR, Regensteiner JG, Hargarten ME, Wolfel EE, Brass EP. Benefit of exercise conditioning for patients with peripheral arterial disease. Circulation 1990; 81: 602–9.
28. Brevetti G, Chiarello M, Ferulano G et al. Increases in walking distance in patients with peripheral vascular disease treated with L-carnitine: A double-blind cross-over study. Circulation 1988; 77: 767–73.
29. Brevetti G, Angelini C, Rosa M et al. Muscle carnitine deficiency in patients with severe peripheral vascular disease. Circulation 1991; 84: 1490–5.
30. Brevetti G, Perna S, Sabb C et al. Superiority of L-propionylcarnitine vs L-carnitine in improving walking capacity in patients with peripheral vascular disease: an acute, intravenous, double-blind, cross-over study. Eur Heart J 1992; 13: 251–5.
31. Franciosa HA, Park M, Levine TB. Lack of correlation between exercise capacity and indexes of resting left ventricular performance in heart failure. Am J Cardiol 1981; 47: 33–9.
32. Holloszy JO, Coyle EF. Adaptations of skeletal muscle to endurance exercise and their metabolic consequences. J Appl Physiol 1984; 56: 831–8.
33. Dunnigan A, Pierpont ME, Smith SA, Breningstall G, Benditt DG, Benson Jr DW. Cardiac and skeletal myopathy associated with cardiac dysrhythmias. Am J Cardiol 1984; 53: 731–7.
34. Dunnigan A, Staley NA, Smith SA et al. Cardiac and skeletal muscle abnormalities in cardiomyopathy: comparison of patients with ventricular tachycardia or congestive heart failure. J Am Coll Cardiol 1987; 10: 608–18.
35. Massie BM, Conway M, Yonge R et al. Skeletal muscle metabolism in patients with congestive heart failure: relation to clinical severity and blood flow. Circulation 1987; 76: 1009–19.
36. Massie BM, Conway M, Yonge R et al. $^{31}$P Nuclear magnetic resonance evidence of abnormal skeletal muscle metabolism in patients with congestive heart failure. Am J Cardiol 1987; 60: 309–15.
37. Wilson JR, Fink L, Maris J et al. Evaluation of energy metabolism in skeletal muscle of patients with heart failure with gated phosphorus-31 nuclear magnetic resonance. Circulation 1985; 71: 57–62.
38. Douglas CG. Oliver-Sharpey lectures on the coordination of the respiration and circulation with variations in bodily activity. Lancet 1927; ii: 213–8.
39. Wasserman K. The anaerobic threshold measurement to evaluate exercise performance. Am Rev Respir Dis 1984; 129(Suppl): S35–40.
40. Wasserman K, Mcllroy MB. Detecting the threshold of anaerobic metabolism in cardiac patients during exercise. Am J Cardiol 1964; 14: 844–52.
41. Simonton CA, Higginbotham MB, Cobb FR. The ventilatory threshold: quantitative analysis of reproducibility and relation to arterial lactate concentration in normal subjects and in patients with chronic congestive heart failure. Am J Cardiol 1988; 62: 100–7.
42. Randle PJ. Pyruvate-dehydrogenase complex: meticulous regulator of glucose disposal in animals. Trends Biochem Sci 1978; 3: 217–9.
43. Denton RM, Halestrop AF. Regulation of pyruvate metabolism in mammalian tissue. Essays Biochem 1979; 15: 37–9.

44. Bieber LL, Emaus R, Valkner K, Farrell S. Possible functions of short-chain and medium-chain carnitine acyltransferases. Fed Proc 1982; 41: 2858–62.
45. Alkonyi I, Kerner J, Sandor A. The possible role of carnitine and carnitine acetyl-transferase in the contractile frog skeletal muscle. FEBS Lett 1975; 52: 265–8.
46. Bremer J. Carnitine – metabolism and functions. Physiol Rev 1983; 63: 1420–80.
47. Bjorkman O. Fuel utilization during exercise. In: Benzi G, Packer L, Siliprandi N, editors. Biochemical aspects of physical exercise. Amsterdam: Elsevier, 1986: 245–60.
48. Di Lisa F, Bobyleva-Guarriero V, Jocelyn P, Toninello A, Siliprandi N. Stabilizing action of carnitine on energy linked processes in rat liver mitochondria. Biochem Biophys Res Commun 1985; 134: 968–73.
49. Fafournoux P, Remesy C, Demigne C. Propionate transport in rat liver cells. Biochim Biophys Acta 1985; 818: 73–80.
50. Brass BP, Fennessey PV, Miller LV. Inhibition of oxidative metabolism by propionic acid and its reversal by carnitine in isolated rat hepatocytes. Biochem J 1986; 236: 131–6.
51. Brass BP, Beyerinck RA. Interactions of propionate and carnitine metabolism in isolated rat hepatocytes. Metabolism 1987; 36: 781–7.
52. Sundqvist JJ, Peuhkurinen KJ, Hiltunen JK, Hassinen IE. Effect of acetate and octanoate on tricarboxylic acid cycle metabolite disposal during propionate oxidation in the perfused rat heart. Biochim Biophys Acta 1984; 801: 429–36.
53. Latipaa PM, Peuhkurinen KJ, Hiltunen JK, Hassinen IE. Regulation of pyruvate dehydrogenase during infusion of fatty acids of varying chain lengths in the perfused rat heart. J Mol Cell Cardiol 1985; 17: 1161–71.
54. Peuhkurinen KJ. Accumulation and disposal of tricarboxylic acid cycle intermediates during propionate oxidation in the isolated perfused rat heart. Biochim Biophys Acta 1982; 721: 124–34.
55. Bolukoglu H, Nellis SH, Liedtke AJ. Effects of propionate on mechanical and metabolic performance in aerobic rat hearts. Cardiovasc Drugs Ther 1991; 5: 37–44.
56. Mancini M, Rengo F, Lingetti M, Sorrentino GP, Nolfe G. Controlled study on the therapeutic efficacy of propionyl-L-carnitine in patients with congestive heart failure. Arzneimittelforsch 1992; 42: 1101–4.
57. Pucciarelli G, Mastursi M, Latte S et al. Effetti clinici ed emodinamici della propionil-L-carnitina nel trattamento dello scompenso cardiaco congestizio. Clin Ter 1992; 141: 379–84.
58. Caponnetto S. New pharmacologic approach to congestive heart failure. Proc Second Int Workshop on New Trends in Cardiovascular Therapy and Technology; 1991; Genoa.
59. Brunelli C, Canale C, Masperone MA, Terrachini V, Caponnetto S. Hemodynamic and metabolic effects of propionyl-L-carnitine in patients with chronic heart failure. Eur Heart J 1994. In press.

*Corresponding Author*: Professor Salvatore Caponnetto, Department of Internal Medicine, Division of Cardiology, University of Genoa, Viale Benedetto XV, 6, I-16123 Genoa, Italy

# 25. Carnitine metabolism in peripheral arterial disease

WILLIAM R. HIATT and ERIC P. BRASS

"Both L-carnitine and propionyl-L-carnitine appear to increase muscle carnitine content in peripheral arterial disease patients, and have favorable effects on muscle metabolism."

## Introduction

The atherosclerotic disease process may affect any artery, including the coronary, cerebral or peripheral vessels. In peripheral arterial disease (PAD), atherosclerotic occlusions in the major arteries feeding the lower extremities result in a restricted blood flow to skeletal muscle, particularly during exercise. During walking exercise, PAD patients develop muscle ischemia and the symptom of intermittent claudication that results in an objective impairment in walking ability. The claudication-limited peak oxygen consumption (measured with a graded treadmill exercise test) of PAD patients was typically 12 to 20 ml/kg/min [1, 2], or approximately 40–60% of the age predicted maximal oxygen consumption for normal subjects [3]. This profound exercise impairment limits the ability of these patients to perform activities essential for daily living [4]. For example, in a European study, up to 38% of patients under the age of 55 years seeking treatment for claudication considered themselves functionally disabled [5]. The yearly health care costs in the United States for symptomatic PAD averaged $ 3,100 per patient [6], but when interventional therapy was utilized, the costs increase by $11,000 for procedures such as surgery, angioplasty or amputation [7]. This epidemiology has greatly increased the interest in developing new, nonsurgical therapies for PAD [2, 8–10].

In patients with claudication, the hemodynamic severity of the underlying vascular disease (defined by peripheral blood flow or ankle pressure) was not well correlated with treadmill performance [11, 12]. For example, exercise training substantially improved exercise performance without an increase in calf blood flow [2], while an improvement in peripheral blood flow with

*J.W. de Jong and R. Ferrari (eds): The carnitine system*, 353–363.

bypass surgery did not normalize exercise performance [13]. Therefore in addition to the limited blood flow, other factors must contribute to the functional impairment in patients with PAD.

Several changes in skeletal muscle structure and metabolism have been observed in patients with PAD. For example, muscle denervation and a selective loss of type II fibers leads to muscle atrophy and weakness [14–17]. The loss of muscle strength contributes to the reduced exercise performance of the patient with PAD [16]. In addition to the structural changes, calf muscle ischemia may alter several important aspects of muscle metabolism. A number of studies have addressed potential changes in skeletal muscle oxidative and glycolytic enzyme activities, but no consistent alterations were observed [16, 18, 19]. In contrast, recent studies demonstrated a number of clinically important changes in muscle carnitine metabolism in PAD [20, 21], and carnitine supplementation improved exercise performance in patients with claudication [10, 22, 23]. Therefore, while the initial disease process is a reduced blood and oxygen delivery to skeletal muscle [24], metabolic sequelae contribute to the disease pathophysiology and clinical severity. These alterations in muscle metabolism have led to the development of several new therapies for claudication that target the metabolic, rather than hemodynamic abnormalities in PAD.

## Exercise metabolism in normal subjects

There is a marked increase in skeletal muscle metabolic rate with exercise. In normal individuals, low-intensity work loads can be defined as an exercise intensity where blood lactate concentration did not increase [25, 26]. Under these exercise conditions, fatty acids were the major substrate for energy metabolism [27, 28], and the respiratory exchange ratio (RER, an index of substrate utilization) was usually less than 0.90. With incremental increases in work load, normal subjects cross a transition into high intensity exercise, defined as the exercise intensity where there was a progressive increase in blood lactate concentration [26]. At high work loads, the activity of pyruvate dehydrogenase was greatly increased [29, 30], allowing glucose to contribute an increasing share of ATP production. These changes in carbohydrate metabolism were reflected by an RER value of 1.00 or greater.

Carnitine has several important functions in skeletal muscle intermediary metabolism during exercise. Carnitine is required for the transport of activated long-chain fatty acids into mitochondria for subsequent beta-oxidation to acetyl-CoA [31]. In addition, carnitine interacts with other metabolic pathways through the reversible transfer of acyl groups from acyl-CoA's to form the corresponding acylcarnitines. Acyl-CoA's are intermediates in the oxidation of fuel substrates, but the accumulation of unusual acyl-CoA's may impair cellular metabolism [32]. Under conditions in which short-chain acyl-CoA intermediates accumulate, short-chain acylcarnitines were formed, making unesterified CoA available for other metabolic reactions [33]. Thus, the formation of short-chain acylcarnitines may serve to modulate changes in the acyl-CoA pool under conditions of acute metabolic stress.

With prolonged durations (60 min) of low intensity exercise, muscle carnitine metabolism remains grossly unchanged, with most of the total carnitine pool present as unesterified carnitine [34]. In contrast, even brief periods of high intensity exercise were associated with the generation of large amounts of short-chain acylcarnitines, particularly acetylcarnitine, in skeletal muscle [34, 35]. With the generation of short-chain acylcarnitines, there was a reciprocal decrease in the unesterified carnitine content [30, 34]. These changes in muscle carnitine metabolism reflect similar changes in the acetyl-CoA/CoASH distribution in muscle [30, 36]. Also, there was a close correlation between the accumulation of acetylcarnitine and lactate in skeletal muscle during high-intensity exercise [37]. This correlation may reflect the dual actions of acetyl-CoA accumulation as a precursor of acetylcarnitine, and inhibitor of pyruvate dehydrogenase (resulting in lactate accumulation) [29]. These observations in normal subjects provide a foundation for the understanding muscle metabolism in PAD.

**Muscle metabolism and function in patients with peripheral arterial disease**

*Chronic changes in muscle metabolism and function.* Several abnormalities have been described in ischemic skeletal muscle from patients with PAD. Patients with claudication have a selective loss of type II glycolytic fibers relative to type I oxidative fibers in the gastrocnemius muscle of affected legs [16, 38]. This change in fiber type distribution was associated with muscle weakness and decreased exercise performance [16]. In addition, a chronic denervation occurs in ischemic skeletal muscle that was also correlated with muscle weakness and dysfunction [14, 16]. Thus, patients with PAD have a number of factors that may adversely affect muscle metabolism. These factors include inactivity, repeated episodes of muscle ischemia during ambulatory activity, muscle denervation, and a change in fiber type distribution from glycolytic to oxidative.

In patients with PAD, several changes in muscle enzyme activities have been reported. An increase in oxidative enzyme activity (a potentially adaptive response to decreased oxygen delivery) has been observed in some studies [19, 39]. However, other authors have observed decreased activities of oxidative enzymes, perhaps reflecting inactivity and muscle denervation [40, 41]. More importantly, there was no demonstrable correlation between oxidative enzyme activities and exercise performance in the patients [16, 40].

Patients with PAD have several alterations in muscle metabolism that are chronic; these changes are observed in the resting state. In skeletal muscle, patients with severe ischemic disease who required bypass surgery had a reduced total muscle carnitine content [21]. However, muscle total carnitine

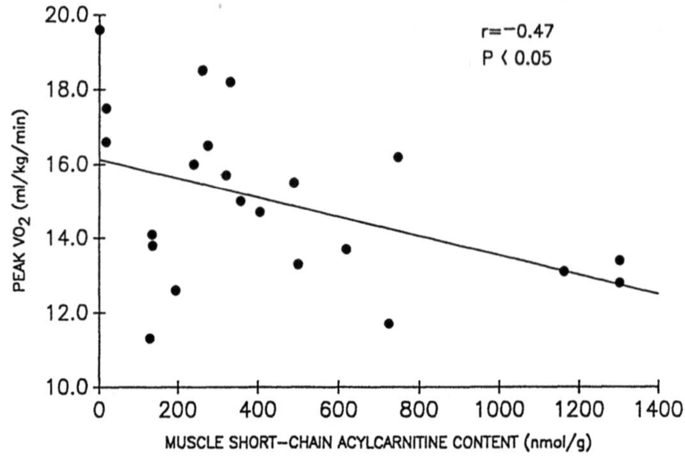


*Figure 1.* Correlation between resting gastrocnemius muscle short-chain acylcarnitine content and subsequent claudication-limited exercise performance. Maximal exercise performance was defined as the peak oxygen consumption ($VO_2$) during graded exercise testing. Each point represents an individual patient. Data were obtained from two previous studies [20, 42]. Muscle biopsies were performed in the affected leg of patients with unilateral arterial occlusive disease.

content was within the normal range in ambulatory patients with PAD who had claudication [20, 22]. Patients with claudication had an accumulation of lactate and short-chain acylcarnitines in ischemic skeletal muscle at rest [20]. Importantly, there was an inverse correlation between muscle short-chain acylcarnitine content in the diseased leg at rest and peak exercise performance. Thus, patients with PAD accumulate acylcarnitines in plasma [1] and ischemic muscle [20], and this accumulation was a marker of the functional disease severity (Figure 1). These observations suggest that there is an increase in the acyl-CoA/CoASH ratio at rest in legs affected by PAD. Further studies are needed to determine if the accumulating short-chain acylcarnitines are acetylcarnitine (as are generated during high-intensity exercise in normal subjects) or non-acetyl acylcarnitines.

The functional relationship between carnitine metabolism and exercise performance was further confirmed in a study of exercise training in patients with PAD [2]. Patients completing a 12-week training regimen significantly improved their treadmill walking capacity. The improvement in walking time was correlated to the decrease in the ratio of the resting plasma short-chain acylcarnitine/total acid-soluble carnitine ratio. These studies suggest that plasma and muscle acylcarnitine accumulation is associated with PAD, is a marker of functional disease severity, and can be modified by training.

*Acute changes in muscle metabolism with exercise.* With exercise, patients with PAD had an impaired increase in leg blood flow as compared to control subjects at comparable work loads [12, 42, 43]. Despite an increase in oxygen extraction across the exercising muscle bed, patients with PAD had a greater depletion in phosphocreatine than control subjects [44], suggesting that compensatory mechanisms were inadequate to maintain ATP production. As a result, the amount of muscular work was severely limited by muscle contractile dysfunction and the symptom of claudication pain [42, 43]. Also related to the reduced blood flow, lactate production was greatly increased from ischemic muscle at work loads not associated with lactate generation in normal subjects [12]. In contrast to the findings in normal subjects at similar work loads, patients with PAD had a lower leg respiratory quotient (predominately fatty acid oxidation) despite a higher lactate production [43]. Thus, the metabolic state of exercising muscle in patients with PAD is markedly different from normal subjects. At work loads that are low intensity for normal subjects, patients with PAD have an inadequate oxygen delivery, rapid depletion of high-energy phosphates, and large lactate generation. Despite this "high intensity" metabolic profile, fatty acids appear to remain a dominant oxidative substrate, as type I oxidative fibers are prevalent, and the respiratory quotient is lower than control values.

Additional metabolic abnormalities have been observed in patients with PAD. In a study of patients with unilateral arterial occlusions, the non-diseased leg served as a control for the changes observed in the leg with arterial disease [20]. In these patients, muscle biopsies were preformed in the gastrocnemius muscle at rest, and immediately after claudication-limited exercise. In the diseased legs, there was a large increase in muscle short-chain acylcarnitine content that was not related to a change in muscle lactate content. Rather, the increase in muscle short-chain acylcarnitine content in the ischemic muscle occurred at all work loads and correlated strongly with exercise duration. In contrast, changes in muscle carnitine metabolism in the non-diseased legs were that predicted from the findings in normal subjects [34]. With exercise, short-chain acylcarnitine content in the non-diseased legs changed in parallel with changes in muscle lactate content. Thus in the non-diseased leg, there was a metabolic-state dependent change in muscle carnitine metabolism with exercise. These findings demonstrate that acylcarnitines are generated and accumulate in under-perfused muscle during walking activities that are associated with claudication. Further, the exercise-induced changes in ischemic muscle carnitine metabolism were *not* associated with changes in muscle or systemic lactate metabolism. Thus, acylcarnitine accumulation may represent qualitatively distinct processes in the ischemic skeletal muscles of patients with PAD as compared with normal subjects.

**Carnitine supplementation**

*Animal studies.* In animals, administration of intravenous carnitine improved muscle contractile force and delayed fatigue in an in situ dog model [45]. However, since the animals had no increase in muscle carnitine content, the authors suggested that the benefit of carnitine on muscle fatigue was due to effects outside of the muscle cell. Subsequent in vitro studies in rats also demonstrated that carnitine delayed muscle fatigue [46]. This fatigue-limiting effect of carnitine was specific for type I, slow twitch fibers, with no benefit observed in type II glycolytic fibers. Also, the effect of carnitine on muscle fatigue was associated with a five- to six-fold increase in muscle total carnitine content that occurred with preincubation.

*Normal human studies.* In normal subjects, studies of the effects of carnitine administration on exercise performance have shown inconsistent results. In some studies, carnitine had no effect on exercise performance or maximal oxygen consumption [47–50], while in other studies there was a modest improvement in either the total work performed on a bicycle ergometer or maximal oxygen consumption [51–54]. The lack of effect of L-carnitine supplementation in normal subjects may be related to the fact that even large doses of intravenous L-carnitine given acutely did not increase total muscle carnitine content [50]. In addition, intravenous administration of L-carnitine at the start of high intensity exercise did not modify the accumulation of lactate or short-chain acylcarnitines in muscle. Therefore, the plasma and muscle carnitine compartments are highly segregated in normal humans. Any acute benefits of L-carnitine in healthy individuals may be modest, and not related to perturbation of muscle carnitine metabolism.

*Peripheral arterial disease.* In contrast to the findings in normal subjects, carnitine administration to patients with PAD has several important clinical effects. Oral administration of L-carnitine (2 g BID for 15 days) increased total muscle carnitine content [22]. In the same study, an acute intravenous dose of carnitine (3 g bolus followed by 2 mg/kg/min for 30 min) reduced popliteal vein lactate concentration and the lactate/pyruvate ratio during exercise as compared to placebo. The intravenous dose of L-carnitine had no effects on resting calf blood flow or ankle pressure, but did increase the blood flow response to reactive hyperemia [55]. Thus, the favorable metabolic effects of L-carnitine (reduced popliteal vein lactate concentration) may be due to either an improvement in ischemic muscle metabolism during exercise, or perhaps to increased blood flow.

In another study, administration of propionyl-L-carnitine (1.5 g bolus followed by 1 mg/kg/min for 30 min) to patients with severe PAD increased muscle total carnitine content and citrate synthase activity [21]. However, extrapolation of these observations must be done with caution because treatment effects were not assessed in the same patients before and after adminis-

*Table 1.* Clinical efficacy of L-carnitine and propionyl-L-carnitine.

| Authors | N | Drug | Design | Outcomes |
|---|---|---|---|---|
| Brevetti et al. 1988 [22] | 20 | *Oral*: L-carnitine 2 g BID for 3 weeks vs placebo for 3 weeks | Double-blind crossover | Oral L-carnitine improved ACD 75% compared to placebo |
| | 8 | *Intravenous*: L-carnitine 3 g bolus 2 mg/kg/min for 30 min | Double-blind crossover | IV L-carnitine improved ACD 67% compared to placebo |
| Brevetti et al. 1992 [10] | 14 | *Intravenous*: L-carnitine 500 mg bolus vs Propionyl-L-carnitine 600 mg bolus | Double-blind crossover | L-Carnitine improved ICD 17% and ACD 18% compared to placebo<br>PLC improved ICD 20% and ACD 28% compared to placebo PLC was more effective than L-carnitine on ACD |
| Coto et al. 1992 [23] | 300 | *Oral*: Propionyl-L-carnitine 2 g BID for 6 months vs placebo for 6 months | Double-blind parallel | PLC improved ICD 31% and ACD 30% compared to placebo |

Three placebo-controlled trials have evaluated the efficacy of L-carnitine and propionyl-L-carnitine (PLC) in patients with claudication. ICD = initial claudication distance, ACD = absolute claudication distance.

tration of propionyl-L-carnitine. However, both L-carnitine and propionyl-L-carnitine appear to increase muscle carnitine content in PAD patients, and have favorable effects on muscle metabolism. These results also suggest that ischemic muscle may be more responsive to the beneficial effects of carnitine supplementation than is normal skeletal muscle.

Three randomized controlled studies have evaluated the clinical efficacy of L-carnitine or propionyl-L-carnitine in patients with intermittent claudication (Table 1). The primary measures of outcome were changes in the onset of claudication pain (initial claudication distance or ICD) and the maximal walking distance (absolute claudication distance or ACD) on a constant-load treadmill protocol. In 20 patients, oral L-carnitine (2 g BID for three weeks) increased ACD by 75% as compared to placebo [22]. Interestingly in the same study, eight patients acutely treated with L-carnitine given as a 3 g intravenous bolus followed by 2 mg/kg/min infusion for 30 min increased their ACD by 67%.

In another study, a single intravenous dose of L-carnitine (500 mg) was compared to a similar dose of propionyl-L-carnitine (600 mg) given intravenously to PAD subjects [10]. The ICD was improved by a similar amount by both drugs. However, while the ACD improved 18% with L-carnitine, the improvement of 28% with propionyl-L-carnitine was greater than with L-carnitine.

The largest therapeutic trial of propionyl-L-carnitine to date was conducted in 300 patients with claudication. Oral propionyl-L-carnitine was given as 2 g BID for six months. Propionyl-L-carnitine increased the ICD by 31% and the ACD 30% as compared to placebo.

Taken together, these studies demonstrate that L-carnitine and propionyl-L-carnitine are effective in improving exercise performance in PAD patients with both chronic oral dosing as well as with acute intravenous administration. Thus, both L-carnitine and propionyl-L-carnitine are promising new agents for the treatment of claudication. Several studies are ongoing to evaluate the efficacy of propionyl-L-carnitine in patients with claudication.

## Conclusions

In normal humans, muscle energy demands are maintained by metabolic processes that differ according to the exercise intensity. With increasing work loads, there is a change in the profile of substrate utilization from predominantly fatty acid oxidation to an incremental reliance on glycolysis. Associated with these changes in substrate flux at high intensity work loads is an accumulation of lactate and acetyl-CoA in skeletal muscle. Carnitine serves as a buffer for the accumulating acetyl-CoA, forming acetylcarnitine and free coenzyme A. These changes in the muscle CoA and carnitine pools are highly compartmentalized, and are largely unaffected by exogenous carnitine administration.

In contrast to the findings in normal subjects, patients with PAD have a number of structural and metabolic abnormalities in their skeletal muscle. In particular, the accumulation of short-chain acylcarnitines in ischemic muscle has significance as a functional marker in that the greater the accumulation, the worse the exercise performance of the patient. Treatment of patients with L-carnitine has been shown to improve ischemic skeletal muscle metabolism during exercise. Finally, L-carnitine and propionyl-L-carnitine supplementation increase treadmill walking distance in patients with claudication. Future studies of propionyl-L-carnitine in large patient populations will better determine the magnitude of the improvement in exercise performance as well as the effects of the drug on other clinically relevant endpoints.

## Acknowledgement

Dr Hiatt is the recipient of an NIH Academic Award in Vascular Disease.

## References


1. Hiatt WR, Nawaz D, Brass EP. Carnitine metabolism during exercise in patients with peripheral vascular disease. J Appl Physiol 1987; 62: 2383–7.
2. Hiatt WR, Regensteiner JG, Hargarten ME, Wolfel EE, Brass EP. Benefit of exercise conditioning for patients with peripheral arterial disease. Circulation 1990; 81: 602–9.
3. Hossack KF, Bruce RA. Maximal cardiac function in sedentary normal men and women: Comparison of age-related changes. J Appl Physiol 1982; 53: 799–804.
4. Regensteiner JG, Steiner JF, Panzer RJ, Hiatt WR. Evaluation of walking impairment by questionnaire in patients with peripheral arterial disease. J Vasc Med Biol 1990; 2: 142–52.
5. Olsen PS, Gustafsen J, Rasmussen L, Lorentzen JE. Long-term results after arterial surgery for arteriosclerosis of the lower limbs in young adults. Eur J Vasc Surg 1988; 2: 15–8.
6. Stergachis A, Sheingold S, Luce BR, Psaty BM, Revicki DA. Medical care and cost outcomes after pentoxifylline treatment for peripheral arterial disease. Arch Intern Med 1992; 152: 1220–4.
7. Tunis SR, Bass EB, Steinberg EP. The use of angioplasty, bypass surgery, and amputation in the management of peripheral vascular disease. N Engl J Med 1991; 325: 556–62.
8. Lindgarde F, Jelnes R, Bjorkman H et al. Conservative drug treatment in patients with moderately severe chronic occlusive peripheral arterial disease. Scandinavian Study Group. Circulation 1989; 80: 1549–56.
9. Radack K, Wyderski RJ. Conservative management of intermittent claudication. Ann Intern Med 1990; 113: 135–46.
10. Brevetti G, Perna S, Sabba C et al. Superiority of L-propionyl carnitine vs L-carnitine in improving walking capacity in patients with peripheral vascular disease: An acute, intravenous, double-blind, cross-over study. Eur Heart J 1992; 13: 251–5.
11. Hiatt WR, Nawaz D, Regensteiner JG, Hossack KF. The evaluation of exercise performance in patients with peripheral vascular disease. J Cardiopulmonary Rehab 1988; 12: 525–32.
12. Pernow B, Zetterquist S. Metabolic evaluation of the leg blood flow in claudicating patients with arterial obstructions at different levels. Scand J Clin Lab Invest 1968; 21: 277–87.
13. Regensteiner JG, Hargarten ME, Rutherford RB, Hiatt WR. Functional benefits of peripheral vascular bypass surgery for patients with intermittent claudication. Angiology 1993; 44: 1–10.
14. England JD, Regensteiner JG, Ringel SP, Carry MR, Hiatt WR. Muscle denervation in peripheral arterial disease. Neurology 1992; 42: 994–9.
15. Hedberg B, Angquist KA, Henriksson-Larsen K, Sjostrom M. Fibre loss and distribution in skeletal muscle from patients with severe peripheral arterial insufficiency. Eur J Vasc Surg 1989; 3: 315–22.
16. Regensteiner JG, Wolfel EE, Brass EP et al. Chronic changes in skeletal muscle histology and function in peripheral arterial disease. Circulation 1993; 87: 413–21.
17. Gerdle B, Hedberg B, Angquist KA, Fugl-Meyer AR. Isokinetic strength and endurance in peripheral arterial insufficiency with intermittent claudication. Scand J Rehab Med 1986; 18: 9–15.
18. Bylund-Fellenius AC, Walker PM, Elander A, Schersten T. Peripheral vascular disease. Am Rev Respir Dis 1984; 129: S65–7.
19. Jansson E, Johansson J, Sylven C, Kaijser L. Calf muscle adaptation in intermittent claudication. Side-differences in muscle metabolic characteristics in patients with unilateral arterial disease. Clin Physiol 1988; 8: 17–29.
20. Hiatt WR, Wolfel EE, Regensteiner JG, Brass EP. Skeletal muscle carnitine metabolism in patients with unilateral peripheral arterial disease. J Appl Physiol 1992; 73: 346–53.
21. Brevetti G, Angelini C, Rosa M et al. Muscle carnitine deficiency in patients with severe peripheral vascular disease. Circulation 1991; 84: 1490–5.
22. Brevetti G, Chiariello M, Ferulano G et al. Increases in walking distance in patients with peripheral vascular disease treated with L-carnitine: A double-blind, cross-over study. Circulation 1988; 77: 767–73.

23. Coto V, D'Alessandro L, Grattarola G et al. Evaluation of the therapeutic efficacy and tolerability of levocarnitine propionyl in the treatment of chronic obstructive arteriopathies of the lower extremities: A multicentre controlled study vs. placebo. Drugs Exp Clin Res 1992; 18: 29–36.
24. Bylund-Fellenius AC, Walker PM, Elander A, Holm S, Holm J, Schersten T. Energy metabolism in relation to oxygen partial pressure in human skeletal muscle during exercise. Biochem J 1981; 200: 247–55.
25. Wasserman K. Determinants and detection of anaerobic threshold and consequences of exercise above it. Circulation 1987; 76: VI29–39.
26. Wasserman K, Beaver WL, Davis JA, Pu JZ, Heber D, Whipp BJ. Lactate, pyruvate and lactate-to-pyruvate ratio during exercise and recovery. J Appl Physiol 1985; 59: 935–40.
27. Jones NL, Heigenhauser GJ, Kuksis A, Matsos CG, Sutton JR, Toews CJ. Fat metabolism in heavy exercise. Clin Sci 1980; 59: 469–78.
28. Ravussin E, Bogardus C, Scheidegger K, LaGrange B, Horton ED, Horton ES. Effect of elevated FFA on carbohydrate and lipid oxidation during prolonged exercise in humans. J Appl Physiol 1986; 60: 893–900.
29. Ward GR, Sutton JR, Jones NL, Toews CJ. Activation by exercise of human skeletal muscle pyruvate dehydrogenase in vivo. Clin Sci 1982; 63: 87–92.
30. Constantin-Teodosiu D, Carlin JI, Cederblad G, Harris RC, Hultman E. Acetyl group accumulation and pyruvate dehydrogenase activity in human muscle during incremental exercise. Acta Physiol Scand 1991; 143: 367–72.
31. Bieber LL. Carnitine. Ann Rev Biochem 1988; 57: 261–83.
32. Brass EP, Fennessey PV, Miller LV. Inhibition of oxidative metabolism by propionic acid and its reversal by carnitine in isolated rat hepatocytes. Biochem J 1986; 236: 131–6.
33. Chalmers RA, Roe CR, Stacey TE, Hoppel CL. Urinary excretion of L-carnitine and acylcarnitines by patients with disorders of organic acid metabolism: Evidence for secondary insufficiency of l-carnitine. Pediatr Res 1984; 18: 1325–8.
34. Hiatt WR, Regensteiner JG, Wolfel EE, Ruff L, Brass EP. Carnitine and acylcarnitine metabolism during exercise in humans. Dependence on skeletal muscle metabolic state. J Clin Invest 1989; 84: 1167–73.
35. Harris RC, Foster CVL, Hultman E. Acetylcarnitine formation during intense muscular contraction in humans. J Appl Physiol 1987; 63: 440–2.
36. Spriet LL, Dyck DJ, Cederblad G, Hultman E. Effects of fat availability on acetyl-CoA and acetylcarnitine metabolism in rat skeletal muscle. Am J Physiol 1992; 263: C653–9.
37. Sahlin K. Muscle carnitine metabolism during incremental dynamic exercise in humans. Acta Physiol Scand 1990; 138: 259–62.
38. Makitie J. Peripheral neuromuscular system in peripheral arterial insufficiency. Scand J Rheumatology 1979; 30(Suppl): 157–62.
39. Lundgren F, Dahllof AG, Schersten T, Bylund-Fellenius AC. Muscle enzyme adaptation in patients with peripheral arterial insufficiency: Spontaneous adaptation, effect of different treatments and consequences on walking performance. Clin Sci 1989; 77: 485–93.
40. Clyne CAC, Mears H, Weller RO, O'Donnell TF. Calf muscle adaptation to peripheral vascular disease. Cardiovasc Res 1985; 19: 507–12.
41. Henriksson J, Nygaard E, Andersson J, Eklof B. Enzyme activities, fibre types and capillarization in calf muscles of patients with intermittent claudication. Scand J Clin Lab Invest 1980; 40: 361–9.
42. Sorlie D, Myhre K, Mjos OD. Exercise-and post-exercise metabolism of the lower leg in patients with peripheral arterial insufficiency. Scand J Clin Lab Invest 1978; 38: 635–42.
43. Lundgren F, Bennegard K, Elander A, Lundholm K, Schersten T, Bylund-Fellenius AC. Substrate exchange in human limb muscle during exercise at reduced blood flow. Am J Physiol 1988; 255: H1156–64.
44. Hands LJ, Bore PJ, Galloway G, Morris PJ, Radda GK. Muscle metabolism in patients with peripheral vascular disease investigated by 31P nuclear magnetic resonance spectroscopy. Clin Sci 1986; 71: 283–90.

45. Dubelaar ML, Lucas CM, Hülsmann WC. Acute effect of L-carnitine on skeletal muscle force tests in dogs. Am J Physiol 1991; 260: E189–93.
46. Brass EP, Scarrow AM, Ruff LJ, Masterson KA, Van Lunteren E. Carnitine delays rat skeletal muscle fatigue in vitro. J Appl Physiol 1993; 75: 1595–600.
47. Greig C, Finch KM, Jones DA, Cooper M, Sargeant AJ, Forte CA. The effect of oral supplementation with L-carnitine on maximum and submaximum exercise capacity. Eur J Appl Physiol 1987; 56: 457–60.
48. Soop M, Bjorkman O, Cederblad G, Hagenfeldt L, Wahren J. Influence of carnitine supplementation on muscle substrate and carnitine metabolism during exercise. J Appl Physiol 1988; 64: 2394–9.
49. Oyono-Enguelle S, Freund H, Ott C et al. Prolonged submaximal exercise and L-carnitine in humans. Eur J Appl Physiol 1988; 58: 53–61.
50. Brass EP, Hoppel CL, Hiatt WR. Effect of intravenous L-carnitine on carnitine homeostasis and fuel metabolism during exercise in humans. Clin Pharmacol Ther 1994; 55: 681–92.
51. Dal Negro R, Pomari G, Zoccatelli O, Turco P. Changes in physical performance of untrained volunteers: Effects of L-carnitine. Clin Trials J 1986; 23: 242–8.
52. Wyss V, Ganzit GP, Rienzi A. Effects of L-carnitine administration on VO2 max and the aerobic-anaerobic threshold in normoxia and acute hypoxia. Eur J Appl Physiol 1990; 60: 1–6.
53. Vecchiet L, Di Lisa F, Pieralisi G et al. Influence of L-carnitine administration on maximal physical exercise. Eur J Appl Physiol 1990; 61: 486–90.
54. Marconi C, Sassi G, Carpinelli A, Cerretelli P. Effects of L-carnitine loading on the aerobic and anaerobic performance of endurance athletes. Eur J Appl Physiol 1985; 54: 131–5.
55. Brevetti G, Attisano T, Perna S, Rossini A, Policicchio A, Corsi M. Effect of L-carnitine on the reactive hyperemia in patients affected by peripheral vascular disease: A double-blind, crossover study. Angiology 1989; 40: 857–62.

*Corresponding Author*: Professor William R. Hiatt, Section of Vascular Medicine, University of Colorado School of Medicine, 4200 E. Ninth Ave, Box B-180, Denver, CO 80262, USA

# 26. Effect of propionyl-L-carnitine on experimental models of peripheral arteriopathy in the rat

NERINA CORSICO and EDOARDO ARRIGONI-MARTELLI

"Propionyl-L-carnitine and pentoxifylline both provided protection against vascular damages caused by ergotamine tartrate administration, as evidenced by lower frequencies of epithelial tissue damage, at the site of injection, and a lower degree of ulceration and tissue necrosis. Propionyl-L-carnitine treatment (120 mg/kg per os) significantly restored the walking ability of the rats when the treatment started one week after streptozotocin injection, and lasted up to 9 weeks."

## Introduction

A common feature of several different vascular pathologies is an imbalance of energy metabolism in vascular and muscular tissues in the area affected, independent of the cause of the disease (vascular obstruction, diabetic status, sickle cell anemia, etc.). The energy production in these tissues is strongly dependent on the oxidation of fatty acids via the Krebs cycle. The crucial role of L-carnitine (L-C) in this process has been recognized for many years and explained by many authors [1–3]. L-C is a fundamental cofactor for fatty acid oxidation within the mitochondria; fatty acids, free or CoA-activated, cannot penetrate the mitochondrial membrane, while carnitine esters are transported through the membrane into the mitochondria by carnitine translocase. A beneficial effect of L-C in patients with peripheral vascular diseases (PVD) has already been shown in different clinical studies, as well as, secondary carnitine deficiency in the skeletal muscles of these subjects [4–6]. During the search for new derivatives with an improved activity in comparison to the original compound, an acyl-derivative of L-C, propionyl-L-carnitine (PLC), has been selected. Biochemical studies have indicated that PLC presents some advantages over L-C, mainly due to its better transport into the cells, related to the presence of the propionyl group [7–9]. PLC can be converted to propionyl-CoA by carnitine acetyltransferase (CAT) and further to methylmalonyl-CoA by mitochondrial propionyl-CoA carboxylase. Whereas methyl malonyl-CoA remains in the matrix for further conversion to succinyl-CoA and subsequently to succinate (thereby entering the Krebs

*J.W. de Jong and R. Ferrari (eds): The carnitine system*, 365–382.

cycle), PLC may move across the mitochondrial inner membrane by a reversible transport mechanism. As a consequence, PLC administration could result in an activation of the Krebs cycle through an "anaplerotic effect" with a subsequent increase of energy supply [10, 11]. This enhancement of energy production should be particularly useful in all those conditions in which a reduced oxygen availability is involved.

On the basis of these findings, PLC has been tested in vitro and in vivo in different models of peripheral vascular diseases, focusing on the effect of PLC on vascular and muscular tissues in terms of function, morphology and metabolism. An issue of major importance is the methodological approach. The most commonly used experimental models are ergotamine-induced tail necrosis and bilateral ligation of femoral arteries in the rat. Both methods have serious limitations and, apparently, are hardly predictive of the therapeutic activity of new compounds. Efforts have therefore been aimed at a new methodological approach, encompassing functional, morphological and biochemical alterations consequent to the reduction of blood flow in the hind limbs of rats. This review describes the findings obtained in different experimental models; they suggest that PLC, due to its particular profile of activity, is a good candidate for the treatment of peripheral vascular diseases.

## Ergotamine model

Ergotamine, an ergot alkaloid derivative, is a potent vasoconstrictor agent exerting its effect via adrenergic receptors. Subcutaneous injection of ergotamine tartrate into the root of the rat tail results in a long-lasting vasoconstriction, leading to decreased blood flow in the tail arteries and to thrombosis. Cyanotic areas appear, with skin dystrophy and necrosis at the site of injection, and necrosis of the terminal part of the tail [11, 12]. We administered PLC at doses of 100 and 300 mg/kg, and pentoxifylline (30 mg/kg), tested as reference drug and vehicle, by oral route before the injection of ergotamine tartrate. Treatment took place once a day for 18 days. Propionyl-L-carnitine and pentoxifylline both provided protection against vascular damages caused by ergotamine tartrate administration, as evidenced by lower frequencies of epithelial tissue damage at the site of injection, a lower degree of ulceration and tissue necrosis [11]. Since PLC does not interact with the adrenergic system (as shown by adrenergic receptor binding studies not reported here) a different mechanism has to be taken into account for explaining this PLC action. A direct protective effect on vasal endothelium cannot be ruled out, also considering in vitro studies on endothelial cells [13] that indicate that PLC can protect endothelial cell membranes from damage induced by peroxidative agents.

**Bilateral femoral artery occlusion model**

In this case a peripheral arterial insufficiency was induced by permanent ligature of both femoral arteries of the rat [11, 14]. Under barbiturate anesthesia femoral arteries were isolated just distal from their appearance from beneath the inguinal ligament and were ligated with a surgical thread. Sham-operated animals underwent anesthesia and the same surgical procedure without the ligation. As a consequence of the reduction of blood flow a severe impairment of the motor performance of the animals was observed by using the "treadmill test"; a motor-driven treadmill, constructed at the Sigma-Tau technical facilities, was used. Rats were forced to run the treadmill (fixed speed: 20 m/min and fixed inclination 15%). A shock was delivered to their paws every time the rats stopped walking on the treadmill; the number of shocks given to the animal in the test period (5 min) was considered an index of its walking capacity. Rats were tested before (basal), and four and eight days after surgery and the start of the treatment with PLC a dose of 100 mg/kg by gavage. The sham-treated animals had a stable baseline, with a number of shocks (n) not significantly different before ($n = 2$) and after the surgical intervention ($n = 3$). The administration of PLC (100 mg/kg os × 4 days) caused a significant improvement ($p < 0.05$) in the walking capacity of the animals (ligated control $n = 47 \pm 5$; ligated PLC $n = 21 \pm 4$). This effect is not associated with an analgesic effect, as in the Randall-Selitto test for analgesia no increase in pain threshold was observed after PLC treatment (data not reported). However, the model used presents some drawbacks, since following complete occlusion of the femoral arteries, collateral circulation develops very early. Therefore the experiment had to be ended four days after the ligature: a week later, control ligated rats were no longer significantly different from sham-operated rats. Anyway, these results can be considered indicative of the capacity of PLC to counteract a decrease in motor capability, induced by a drastic reduction of blood flow.

**Na-laurate model of peripheral arteriopathy**

A novel model of peripheral arteriopathy in the rat was established in our laboratory, since available experimental models were not considered completely satisfactory. Na-laurate is a detergent reported to induce vascular damage after its direct injection into rat arteries [12, 15, 16]. The endothelial tissue damage in turn causes platelet adhesion with attendant thrombus formation and blood flow reduction. Functional (walking ability), morphological (vascular and muscular tissue examination) and metabolic (muscle energy metabolism) alterations associated with the administration of the damaging agent were evaluated. Using this experimental procedure, the effect of short and long-term treatment with PLC was tested in different experiments. For comparison, pentoxifylline (a drug used clinically, for peri-

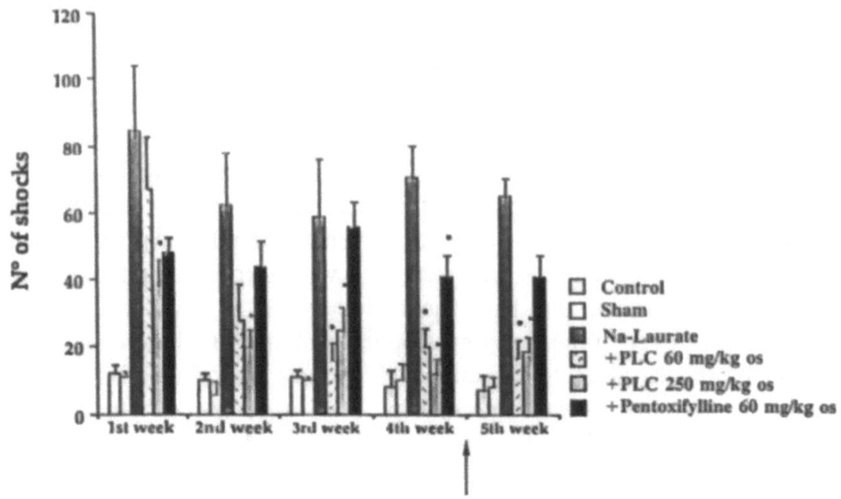


*Figure 1.* Long-term effect of propionyl-L-carnitine (PLC) on walking capacity ("treadmill test"). Groups of ten rats were treated daily by oral gavage and tested weekly, starting from 24 h after Na-laurate injection. The arrow indicates when the treatment was discontinued. The animals were tested again a week later. ●: $p < 0.05$ vs Na-laurate control (Dunnett test for unpaired data). Mean ± SE.

pheral arteriopathy therapy), L-carnitine, and the dextroisomer of propionyl-carnitine were used in some tests [17–19].

*Effect on walking ability*

Impairment of motor performance, caused by injection of 0.06 ml Na-laurate (5 mg/ml in 0.05% ethanol) into both femoral arteries, was evaluated by the treadmill test (see above). Before surgery, animals were selected for their ability to run the treadmill; unable rats were not admitted to the test. Propionyl-L-carnitine was administered by gavage per os at doses of 60 and 250 mg/kg/day, pentoxifylline was given at a dose of 60 mg/kg/day. Controls received water 5 ml/kg. Treatment lasted for 4 weeks and started the day after the surgery. The first walking test was performed one week after surgery and treatment, followed by tests at weekly intervals during the treatment period, and a week after its discontinuation. (The test was run once a week to avoid an exercise effect.) Results reported in Figure 1 show that the pattern of sham-operated animals was stable throughout the experiment and could be superimposed on that of control animals. There were no significant

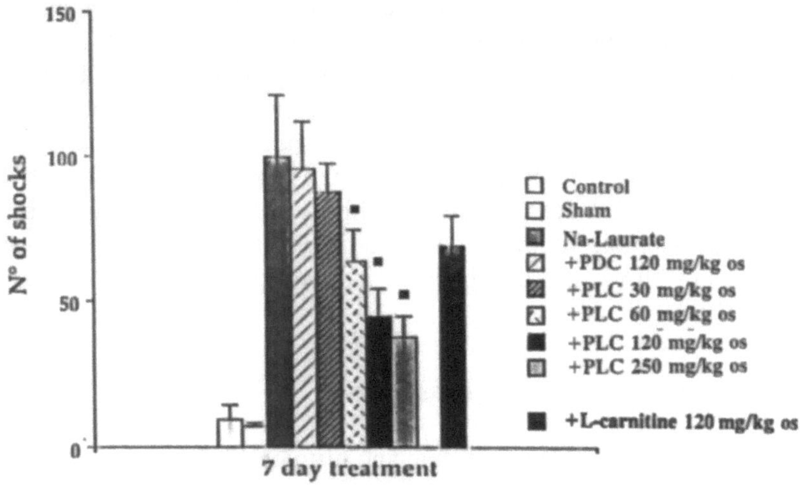


*Figure 2.* Effect on walking capacity ("treadmill test"): Dose-response relationship after 7 days of treatment. The duration of the test was 5 min, the treadmill speed was 20 m/min, and the inclination 15%. Ten rats/dose were used. ■: $p < 0.05$ vs Na-Laurate group (Dunnett *t* test for unpaired data). PDC and PLC, D- and L-form of propionylcarnitine, respectively. Mean ± SE.

differences between the two groups, indicating a good reliability of the experimental conditions. In contrast Na-laurate caused severe impairment of motor performance of the rats; the number of shocks taken during the 5-min test period was about 7–8× higher than that of sham-operated rats. The values were significantly different ($p < 0.05$) throughout the entire study period (5 weeks). Both PLC doses studied caused a significant ($p < 0.05$) improvement in the motor performance of the animals. The highest dosage was more effective (+38%) after one week of treatment; no clear-cut differences between the two dosages were recorded later on. When rats were tested one week after the discontinuation of the treatment, the improvement in walking capacity still persisted unmodified. The reference compound, pentoxifylline, tested only at the dose of 60 mg/kg, was effective but its protective effect was less than that of PLC administered at the same dosage.

The dose-dependency of PLC was assayed by using the same experimental conditions after a week of oral treatment. In the same paradigm the effect of the dextro-isomer propionyl-D-carnitine (PDC) was investigated. This compound is not recognized by CAT, the enzyme involved in the transport of PLC into the cells, and thus is metabolically inactive [3]. In addition, L-carnitine was tested for comparison. Figure 2 shows that in the Na-laurate injected rats the number of shocks delivered to the animals was significantly higher ($p < 0.05$) than in control or sham-operated rats, in agreement with the data obtained in the previous experiment. These findings indicate the good reproducibility and reliability of the model. The oral daily administration of PLC at 30, 60, 120 and 250 mg/kg caused a dose-dependent improvement in motor performance. The number of shocks received by the rats decreased by 19, 41, 64 and 71%, respectively, in comparison to Na-laurate controls. The differences were statistically significant ($p < 0.05$) except for the lowest dose. Treatment with L-carnitine at 120 mg/kg caused an effect comparable to that of PLC 60 mg/kg (the number of shocks was reduced by 40%). The metabolically inactive PDC was devoid of any beneficial effect.

*Effect on vascular and muscular damage*

Tissue damage at the macroscopic and histologic level was evaluated in the rats tested for their walking ability during long-term treatment.

*Macroscopic observation.* Visual observation for hind-leg damage was made at weekly intervals throughout the experiment from 7 days to 5 weeks after Na-laurate injection. A scoring system was adopted for the degree of severity ranging from 0 (normal) to 4 (mummification of half the paw). No detectable changes were observed in the hind limbs of sham-operated rats. In the Na-laurate group, the lesions progressed in time; in the first days after the damaging injection, swelling of the paws occurred. By the end of the first week, swelling disappeared and most of the rats walked leaning on the back of their hind paws (score 1). Afterwards there was a worsening of 60% in the rats in which mummification of the nails, toes and half of the paw occurred. PLC at both dosages studied (60 and 250 mg/kg for 4 weeks) induced a clear-cut reduction of the development of the lesions, both in terms of severity and incidence. These effects persisted one week after the treatment was discontinued. Comparable findings were obtained after treatment with pentoxifylline (60 mg/kg).

*Microscopic observation.* At the end of the experiment (5 weeks after Na-laurate injection, one week after discontinuation of oral treatment) femoral arteries, hind-limb muscles and hind paws were also sampled for histological evaluation. The samples were processed and stained either with hematoxilin or with Movat pentachromatic solution. An arbitrary score was given to each sample of artery, indicating the severity of the damage, ranging from 0 (normal) to 4 (thrombotic obstruction of the lumen). Damage to the muscles was graded from 0 (normal) to 4 (atrophy of bundles of muscular fibers substituted by fibroadipose tissue). The small size of the samples (five subjects/group) did not allow a statistical evaluation of the data; however, they can be considered indicative of the severity of the morphological changes observed:

*Femoral arteries*: Sham-control samples did not show differences of histological significance, indicating no influence of the surgical procedure. The injection of Na-laurate caused, on the contrary, thickening of the intima of different thickness. They consisted of juxtaposition of two or more smooth muscle cell layers, separated by small quantities of interstitial tissue. In some samples, the thickness of myointimal thickening exceeded that of the tunica media, causing a marked narrowing of the vasal lumen. No marked differences were detected among the Na-laurate control group and the groups treated in addition with PLC 60 mg/kg or pentoxyfilline; in the samples of the group treated with I PLC 250 mg/kg, the severity of the vascular damages was reduced, and the score was intermediate between the normal control and the Na-laurate control. Specimens obtained from the Na-laurate control and the PLC-treated are depicted in Figure 3.

*Hind-limb muscles*: No changes of muscular tissue, sampled at the thigh level, were detectable in any of the experimental groups, indicating no systemic diffusion of the damaging agent. The gastrocnemius muscle of Na-laurate injected rats was markedly injured; the damage consisted mainly of atrophied muscular fiber groups, with fibroadipose substitution. The extension of the lesions varied, and in some specimens the whole muscle was involved. Lesions of arteries with small to medium diameter, consisting of myointimal thickening or occluding organized thrombi, also occurred. These latter lesions were frequently associated with calcification of internal elastic lamina. Atrophic muscular lesions were topographically correlated with lesions of muscular vessels with small and medium size. These findings evidence that the damages have been developed slowly over time and are to be considered of ischemic origin. PLC oral treatment (60 and 250 mg/kg for 4 weeks) reduced the extent and severity of the muscular damage, as indicated by the lower scores attributed to specimens in these groups (control and sham-operated score: 0; Na-laurate score: 2.17; PLC 60 and 250 mg/kg score: 1.26 and 0.61, respectively). Pentoxifylline scored the same as the Na-laurate control group. Some specimens are displayed in Figure 4.

### *Effect on muscle energy metabolism*

Alterations of muscle cell energy metabolism were investigated ex vivo and in vivo by $^{31}$P-NMR spectroscopy.

*Ex vivo study*: The animals were subjected to the previously described procedure to induce peripheral vascular insufficiency by Na-laurate and were treated per os with PLC at the dose of 120 mg/kg/day. The treatment lasted 15 days and started 24 h after the surgical procedure, while another group was treated for 11 days starting four days after Na-laurate injection. PLC was also given at the same dose to a group of control rats. At the end of the treatment period, under anesthesia, the gastrocnemius muscles were

A

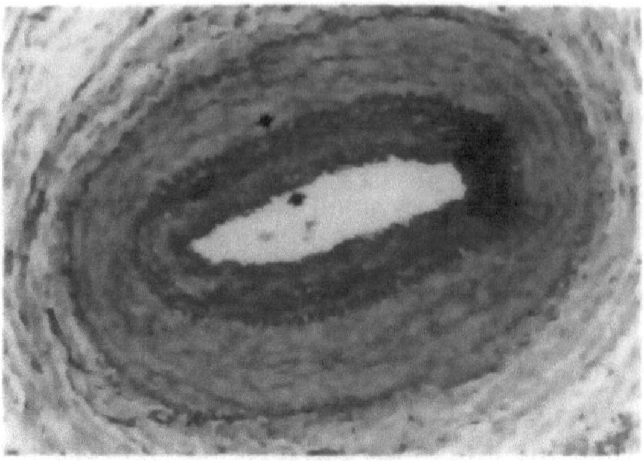

B

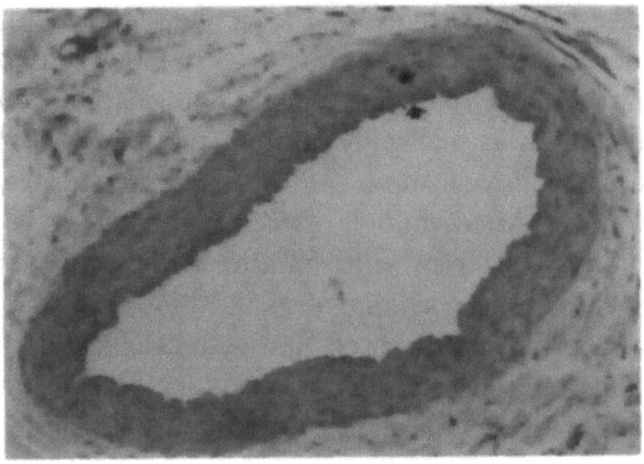

*Figure 3.* A: Femoral artery of a Na-laurate treated rat. The lumen of the artery is narrowed by a severe myointimal thickening (black arrows). The elastic lamellae are also visible. (Movat pentachromic staining. Magnification ×40.) B: Femoral artery of a rat treated with Na-laurate + PLC 250 mg/kg for 4 weeks. Note the presence of a mild muscular thickening of the tunica intima (black arrows). (Hematoxylin eosin staining [H&E]. Magnification ×40.)

removed and immediately frozen in liquid nitrogen. On the muscle extracts the following biochemical parameters were evaluated by means of $^{31}$P-NMR spectroscopy [20]: adenosine triphosphate (ATP), phosphocreatine (PCr), adenosine diphosphate (ADP), inorganic phosphate (Pi), the sum of adenosine monophosphate + inosine monophosphate (AMP + IMP), sugar phosphate (SP) and diphosphodiesters including nicotinamide coenzymes [NAD(P)/NAD(P)H] and others such as uridine diphospho-glucose and -galactose, reported as the sum of NAD(P) + DPDE. Concentrations of the metabolites examined were expressed as the percent of total phosphorus

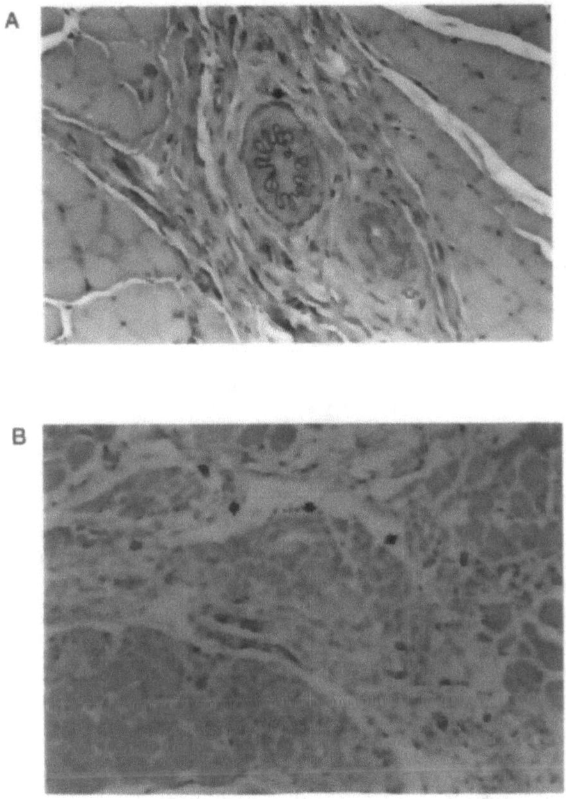

*Figure 4.* Changes in gastrocnemius muscle of a Na-laurate treated rat. (H&E. Magnification ×40.) A: Two small arteries are depicted, one of which is obliterated by an organized thrombus, the lumen of the other is narrowed by a myointimal thickening (white arrow). B: Large group atrophy of muscle fibers in the same muscle shown in A (black arrows).

calculated by measuring peak areas from NMR spectra. Single values were obtained as the ratio between the single peak area and the total area × 100. Muscular phosphorus content was expressed as μmol/mg protein and calculated by comparing the sum of peak areas of all the metabolites with the area of the trimer in the external tube. Significant and marked modifications in muscle metabolism were observed 15 days after Na-laurate injection. ATP and PCr levels were reduced by 42% and 25%, respectively, as compared to sham-operated rats, and changes were also observed in Pi (+64%), ADP (+138%) and AMP + IMP (+300%) levels. These alterations were counteracted by PLC treatment, started 24 h after Na-laurate injection. PCr and ATP levels in rat muscles treated with PLC were higher than those of Na-

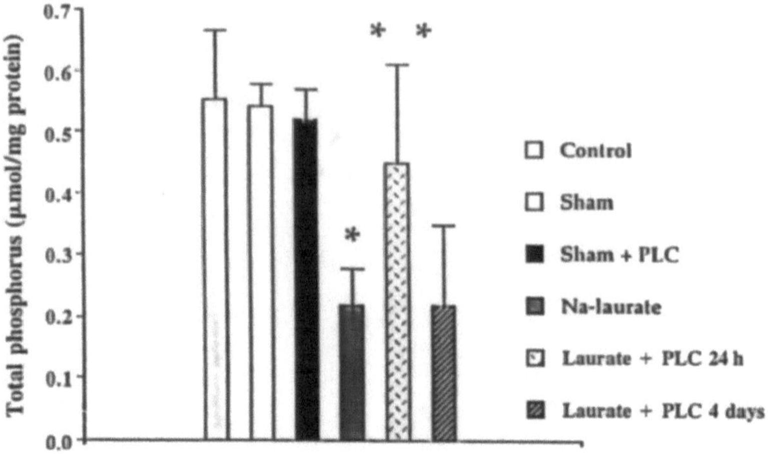


*Figure 5.* Total phosphorus content in gastrocnemius muscle aqueous extract. PLC was given at a dose of 120 mg/kg os for 15 days starting 24 h after Na-laurate (PLC 24 h) or for 11 days starting 4 days after Na-laurate (PLC 4 days). Sham animals were treated with the same dose for 15 days (sham + PLC). Mean ± SD (n = 5). * $p < 0.05$ vs sham-treated rats. ** $p < 0.05$ vs Na-laurate treated rats (Student's t-test).

laurate injected animals, and a parallel reduction in ADP, AMP + IMP, Pi and SP levels was also recorded. The same biochemical trend was also observed when PLC treatment started 4 days after the Na-laurate injection. However, in this group the levels of ADP and AMP + IMP were still significantly increased in comparison to the sham-operated group. No significant differences in the phosphorylated metabolite levels have been observed among control normal, sham-operated and PLC-treated sham-operated rats. These investigations provided additional important information. The treatment of normal rats with PLC did not result in any modification of the biochemical parameters measured, thus suggesting that this compound specifically restores deranged metabolic parameters.

The total phosphorus content provides a quantitative evaluation of the conditions of the muscular tissue; the marked decrease (−67%, $p < 0.05$) observed in the Na-laurate control indicates muscular-fiber necrosis in agreement with histological data. PLC treatment appears to have a protective effect on phosphorus content depletion when administered 24 h after the damaging agent, while it seemed to be ineffective when the delay between damage and pharmacological treatment was increased (Figure 5). This last finding suggests that muscular necrosis is too severe at this time and cannot be reversed.

*In vivo study*: In vivo NMR spectroscopy represents a powerful tool [21] to investigate the evolution of the metabolic alterations induced by Na-laurate, as it allows the sequential visualization of the muscle metabolic status under basal condition, and at different times after the injection of the damaging agent. After barbiturate anesthesia of the rat, $^{31}$P-NMR spectra were obtained on a Vivospec 4.7 T spectrometer using a 4-cm $^{31}$P-$^{1}$H double-resonance surface coil placed around the paws. Chemical shift values have been reported relative to the phosphocreatine signal set at 0 ppm. Evaluation of the spectral line area was performed by using a home-made simulation program which iteratively fits the spectrum with Lorentzian and/or Gaussian line shapes. All the animals were subjected to Na-laurate injection as described previously and divided into two groups. Starting on the day after damage induction, one group was treated with PLC per os (120 mg/kg/die, T group) for 23 days and the second was used as control (NT group). $^{31}$P-NMR spectra were obtained before the Na-laurate injection (control spectra), and 3, 10, 15 and 23 days later. Metabolite levels are reported in Table 1. A large increase of the Pi levels was evident in both the groups on the third day; it was followed in both cases by a decrease. The PLC-treated animals showed a significant difference in Pi versus the control group up to the 15th day. Conversely, the PCr levels were dramatically reduced in both groups on the third day, with a gradual recovery in the following days. A large increase of the PME levels was observed in both groups on the third day. Then the treated animals showed a gradual recovery, in contrast to the control group which recovered only partially ($p < 0.05$). The ATP levels changed little in both groups. Some spectra are depicted in Figure 6. The Pi/PCr ratio is close to 0 under normal physiological conditions; it was significantly increased at different times after Na-laurate injection. Figure 7 shows Pi/PCr ratios for both groups, which differ significantly from each other on the 3rd, 10th and 15th day, indicating a better recovery for the PLC group.

All the above reported findings show that the peripheral arteriopathy induced by Na-laurate injection into femoral arteries can be considered a reliable experimental model of peripheral vascular disease. Under our modified experimental conditions Na-laurate injection resulted in a long-lasting impairment of the walking capacity of rats, as evaluated by a treadmill test. Histological findings showed lesions of small, medium and large diameters, vessels associated with atrophic lesions of muscles, most likely consequent to the reduced blood flow. Nuclear magnetic spectroscopy revealed severe alterations of muscle energy metabolism. Phosphorylated metabolite levels were markedly reduced, probably due to an impairment in cell energy production. Muscular phosphorus content markedly decreased, suggesting fiber damage (atrophy and/or necrosis), in agreement with the histological findings. This pattern of lesion seems consistent with an ischemic pathogenesis more than with a direct detergent effect of Na-laurate on muscle plasma membrane. Repeated oral administration of propionyl-L-carnitine clearly

*Table 1.* Effect of propionyl-L-carnitine on rat skeletal muscle, damaged by Na-laurate treatment.

| Days | PME | | Pi | | PCr | | ATP | |
|---|---|---|---|---|---|---|---|---|
| Basal | 3.8 ± 1.4 | | 7.0 ± 1.4 | | 41.9 ± 1.8 | | 15.9 ± 1.3 | |
| | NT Group | T Group | NT Group | T Group | NT Group | T Group | NT Group | T Group |
| 3 | 12.1 ± 1.7 | 11.0 ± 16 | 31.5 ± 5.1 | 22.1 ± 0.5* | 12.1 ± 1.4 | 15.2 ± 2.9 | 15.4 ± 4.4 | 17.8 × .3 |
| 10 | 11.7 ± 3.1 | 9.6 ± 2.6 | 21.3 ± 4.2 | 13.1 ± 2.1* | 22.9 ± 2.8 | 27.4 ± 4.3 | 14.9 ± 4.4 | 18.5 ± 0.9 |
| 15 | 7.5 ± 3.2 | 7.3 ± 2.8 | 16.4 ± 3.4 | 11.0 ± 1.5* | 32.9 ± 2.6 | 36.6 ± 0.4* | 11.1 ± 2.9 | 13.7 ± 2.2 |
| 23 | 8.8 ± 3.4 | 3.7 ± 0.6* | 12.8 ± 3.6 | 11.8 ± 4.5 | 34.2 ± 4.1 | 35.4 ± 2.4 | 13.0 ± 2.2 | 14.1 ± 1.2 |

Values are expressed as percentage of the total phosphorus content (mean ± SD). Both groups of rats received the damaging treatment with Na-laurate after basal NMR measurements. Only group T was treated per os with propionyl-L-carnitine (120 mg/kg/die). PME = Phosphorous monoesters.
* $p < 0.05$ vs NT.

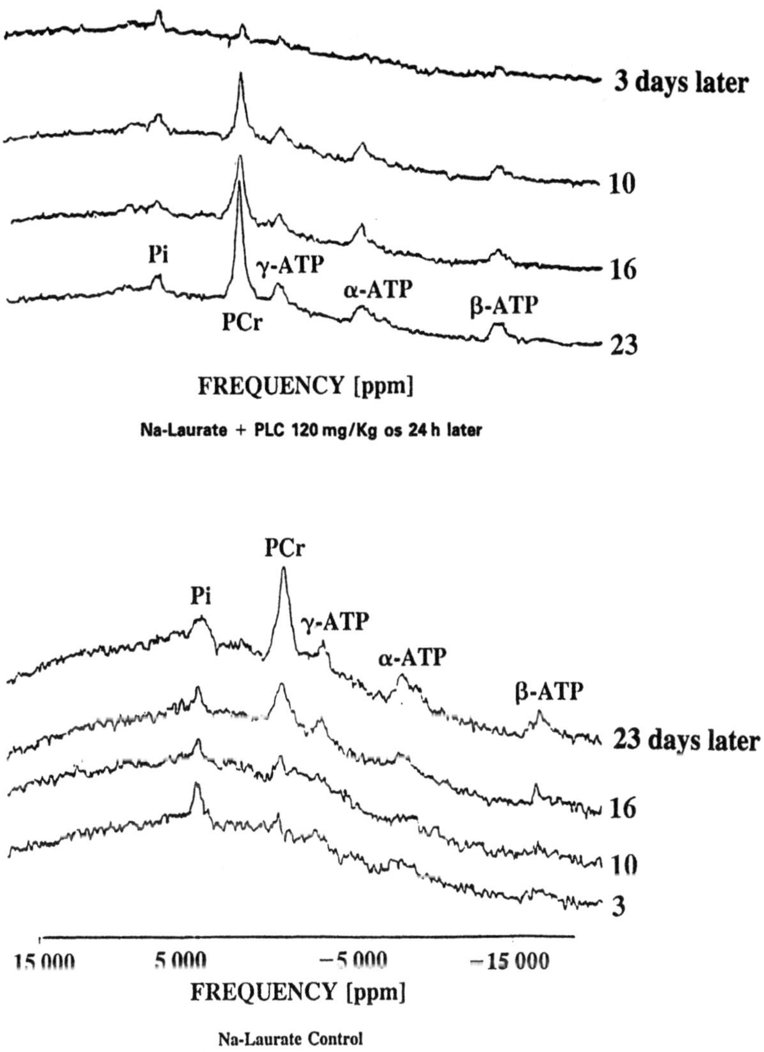


*Figure 6.* [31]P-NMR spectra obtained from both paw muscles at different times after Na-laurate injection in a control and in a PLC-treated rat.

had a beneficial effect on the functional, histological and metabolic parameters. The effect on function was dose-dependent. It persisted for at least a week after the discontinuation of the treatment. It is of interest that the dextro-isomer of propionylcarnitine was completely inactive. This inactivity of the dextro-isomer, which is not recognized by carnitine acetyltransferase, gives support to a metabolic action of PLC. Accordingly, NMR spectroscopy

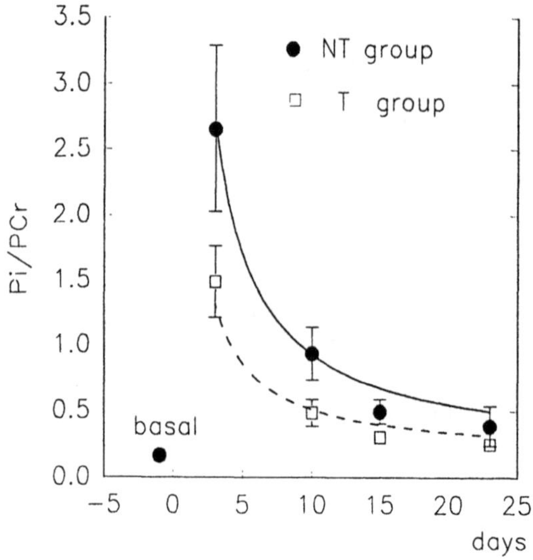


*Figure 7.* Pi/PCr ratios over time (mean ± SD). The two data sets can be fitted with hyperboles which have the basal value as the common asymptote. $r^2 = 0.873$ for NT (Na-laurate control) and $r^2 = 0.656$ for T (Na-laurate, PLC-treated).

showed that PLC has a positive effect on altered muscle energy metabolism, as shown by the recovery of high-energy phosphate levels such as ATP and PCr, even when the treatment was initiated four days after the Na-laurate injection.

**Diabetes-induced peripheral arteriopathy**

Alterations of the cardiovascular and peripheral nervous system observed in diabetes seem to be associated to vascular damages, mainly of the microcirculation; the diabetic rat can be therefore considered an experimental model of peripheral arteriopathy [22–24]. Diabetes was induced by injecting the rat tail vein with 45 mg/kg of streptozotocin (STZ). Control rats were injected with the vehicle only. One week after STZ injection, only those animals with a blood glucose ⩾500 mg/dl were used. Two weeks after STZ injection, rats were treated daily with either PLC (120 mg/kg) or vehicle by oral route. As an index of the occurrence of peripheral arteriopathy we measured motor performance of the rat by "treadmill test". The experimental conditions were the same as above reported. Before STZ injection rats were trained to walk the treadmill and unable animals were excluded from the study. The number

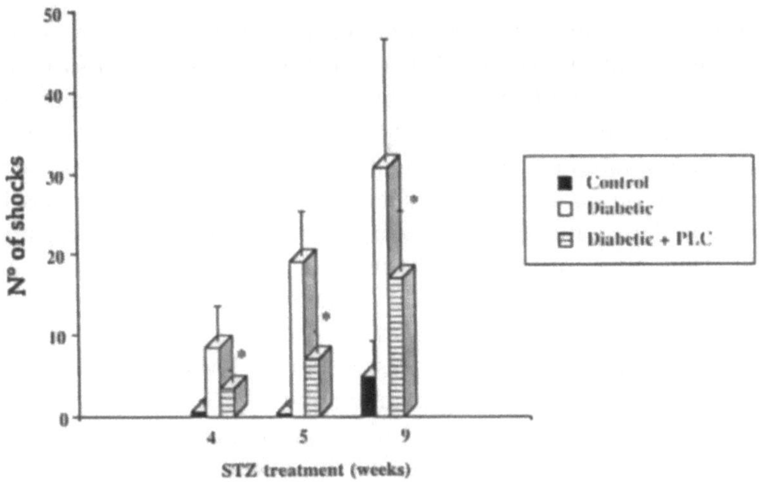


*Figure 8.* Walking ability ("treadmill test") in diabetic control and diabetic PLC-treated rats at different times after diabetes induction with streptozotocin (STZ; mean ± SE, n = 10). * $p < 0.05$ vs diabetic control (Dunnett test).

of shocks received by the animal in the test period was considered to be an index of its walking capacity. Rats were tested before STZ (basal) and at different times thereafter. Figure 8 indicates that in STZ rats with progressing diabetic status, motor performance on the "treadmill test" reduced, as indicated by the augmented number of shocks taken. Since our experimental procedures allowed testing of rats under basal conditions (before STZ injection) and at different times after diabetes induction, it was possible to follow both the development of the alterations associated with the diabetic status and the effect of the pharmacological treatment. PLC treatment (120 mg/kg per os) significantly restored the walking ability of the rats when the treatment started one week after STZ injection, and lasted up to 8 weeks. Our data show that in the experimental model of diabetes, induced by STZ, symptoms comparable to peripheral arteriopathy can be detected by using the "treadmill test" for motor performance. Since PLC has no hypoglycemic and aldose-reductase inhibitor effect (data not shown), it is not unlikely that the beneficial action of PLC is associated with a protective effect on the microcirculation. As in other models, a metabolic mechanism (preservation of energy cellular metabolism impaired by ischemic status) or a direct protective effect on endothelium seems possible.

## Conclusions

The earliest alterations observed in peripheral artery disease are of hemodynamic origin with reduction of blood supply and associated surrounding tissue hypoxia, followed, however, by several changes in skeletal muscle morphology and metabolism [25, 26]. Secondary carnitine deficiency, an accumulation of intermediates of oxidative metabolism (acyl-CoAs and acylcarnitines) and an impairment of energy metabolism in ischemic muscles is associated in these patients with muscle dysfunction and exercise impairment [27, 28]. Thus, while the primary disease process is hemodynamic, sequelae of metabolic events contribute to the disease pathophysiology. In the Na-laurate model of peripheral vascular insufficiency the evolution is comparable to the pattern clinically observed; vessel lesions develop in time, in association with atrophic muscle degeneration and necrosis, particularly of the surrounding area (findings consistent with an ischemic pathogenesis). In this model, PLC treatment reduced the severity of the vascular and muscular tissue damage. In the same model NMR spectroscopic studies ex vivo and in vivo showed that in the gastrocnemius muscle, PLC counteracted the significant reduction of phosphorylated metabolite levels caused by vascular insufficiency. This effect indicates that PLC can restore the deficit occurring in cell energy production. In rats made diabetic by STZ, PLC was able to reduce the impairment of motor performance (index of peripheral vascular insufficiency) caused by the diabetic status. These effects are independent from an hypoglycemic or aldose-reductase inhibition effect. It has been also observed [29] that in the STZ-diabetic rat a severe impairment of the phospholipid fatty acid turnover in the erythrocyte membrane occurs. The potential relevance of these membrane defects in the microvascular complication of diabetic disease is apparent. This metabolic derangement could be corrected by oral PLC administration [29]. All these findings indicate that PLC can exert a beneficial effect in different pathologies when an impaired cell energy production involves both the vascular and muscle tissues. Most likely, PLC activity is associated with its role in lipid metabolism, that is, stimulation of fatty acid oxidation and with its "anaplerotic action" (stimulation of citric acid cycle activity). Consequently, a direct protective effect on endothelium cannot be ruled out, since endothelial cells are strongly dependent on $\beta$-oxidation for their functioning; they lose carnitine under ischemic conditions [30]. In vitro, in endothelial cells PLC was able to counteract metabolic changes induced by hypoxia and subsequent reoxygenation [13]. A membrane stabilizing effect was also observed in human erythrocytes [31]. In vivo, oral administration of PLC to rats exerted a protective effect on tail lesions caused by a vasoconstrictor (ergotamine) in terms of reduced damage incidence and intensity. This effect is not related to an action on the adrenergic system, as it was shown that PLC does not interfere with this system, but most likely to an effect at the endothelial level. In conclusion, all these findings provide the rationale for testing PLC clinically in patients affected by peripheral vascular diseases of different etiopathologies.

## References


1. Bremer J. Carnitine – metabolism and function. Physiol Rev 1980; 63: 1420–80.
2. Bieber LL. Carnitine. Annu Rev Biochem 1988; 57: 261–83.
3. Fritz IB, Arrigoni Martelli E. Sites of action of carnitine and its derivatives on the cardiovascular system: interactions with membranes. Trends Pharmacol Sci 1993; 14: 355–60.
4. Hiatt WR, Narwaz D, Brass EP. Carnitine metabolism during exercise in patients with peripheral vascular diseases. J Appl Physiol 1987; 62: 2383–7.
5. Brevetti G, Angelini C, Rosa M et al. Muscle carnitine deficiency in patients with severe peripheral vascular disease. Circulation 1991; 84: 1490–5.
6. Brevetti G, Perna S. Metabolic and clinical effects of L-carnitine in peripheral vascular disease. In: Ferrari R, DiMauro S, Sherwood G, editors. L-Carnitine and its role in medicine: From function to therapy. London: Academic Press, 1992: 359–78.
7. Siliprandi N, Di Lisa F, Pivetta A, Miotto G, Siliprandi D. Transport and function of L-carnitine and L-propionylcarnitine: relevance to some cardiomyopathies and cardiac ischemia. Z Kardiol 1987; 76(Suppl 5): 34–40.
8. Paulson DJ, Traxler JS, Schmidt MJ, Noonan JJ, Shug AL. Protection of the ischaemic myocardium by L-propionylcarnitine: effects on the recovery of cardiac output after ischaemia and reperfusion, carnitine transport, and fatty acid oxidation. Cardiovasc Res 1986; 20: 536–41.
9. Siliprandi N, Di Lisa F, Menabò R. Propionyl-L-carnitine: biochemical significance and possible role in cardiac metabolism. Cardiovasc Drugs Ther 1991; 5(Suppl 1): 11–5.
10. Hülsmann WC. Biochemical profile of propionyl-L-carnitine. Cardiovasc Drugs Ther 1991; 5(Suppl 1): 7–9.
11. Corsico N, Casantini C, Arrigoni Martelli E. Study of the effect of propionyl-Lcarnitine on experimental models of peripheral arteriopathies. Pharm Res 1993; Suppl 27: 79–80.
12. Murai T, Muraoka K, Saga K et al. Effect of beraprost sodium on peripheral circulation insufficiency in rats and rabbits. Arzneimittelforsch 1989; 39: 856–9.
13. Bertelli A, Conte A, Ronca G, Segnini D, Yu G. Protective effect of propionylcarnitine against peroxidative damage to arterial endothelium membranes. Int J Tissue React 1991; 13: 41–3.
14. Mathien GM, Terjung RL. Influence of training following bilateral stenosis of the femoral artery in rats. Am J Physiol 1986; 250: H1050–9.
15. Ashida S, Ishihara M, Ogawa H, Abiko Y. Protective effect of ticlopidine on experimentally induced peripheral arterial occlusive disease in rats. Thromb Res 1980; 18: 55–67.
16. Hara H, Shimada H, Kitajima M, Tamao Y. Effect of (±)-2-(dimethylamino)-1-[[o-(m-methoxyphenethyl)phenoxy]methyl] ethyl hydrogen succinate on experimental models of peripheral obstructive disease. Arzneimittelforsch 1991; 41: 616–20.
17. Corsico N, Nardone A, Lucreziotti MR et al. Effect of propionyl-L-carnitine of peripheral arteriopathy: a functional, histologic, and NMR spectroscopic study. Cardiovasc Drugs Ther 1993; 7: 241–51.
18. Corsico N, Aureli T, Di Cocco ME, Nardone A, Arrigoni Martelli E, Conti F. Propionyl-L-carnitine restored impaired muscle energy metabolism in an experimental model of peripheral vascular insufficiency. A study by $^{31}$P-NMR spectroscopy. FASEB J 1992; 6: A1369 (Abstr).
19. Capuani G, Aureli T, Miccheli A et al. An investigation of experimental arteriopathy in

rat hind limb by "in vivo" $^{31}$P-NMR spectroscopy. Effect of propionyl-L-carnitine. Br J Pharmacol. In press.
20. Miccheli A, Aureli T, Delfini M et al. Study on influence of inactivation enzymes techniques and extraction procedures on cerebral phosphorylate metabolite level by $^{31}$P NMR spectroscopy. Cell Mol Biol 1988; 34: 591–603.
21. Challiss RAJ, Blackledge MJ, Radda GK. Spatially resolved changes in diabetic rat skeletal muscle metabolism in vivo studied by $^{31}$P-n.m.r. spectroscopy. Biochem J 1990; 268: 111–5.
22. Pugliese G, Tilton RG, Williamson JR. Glucose-induced metabolic imbalances in the pathogenesis of diabetic vascular disease. Diabetes 1990; 39: 312–23.
23. Williamson JR, Tilton RG, Kilo C. The polyol pathway and diabetic vascular complications. In: Conti F, Andreani D, Guerignian JL, Striker LGE, editors. Diabetic complications: epidemiology and pathogenic mechanism. New York: Raven Press, 1991: 45–58.
24. Corsico N, Cantagallo A, Nardone A, Morabito E, Serafini S, Arrigoni Martelli E. Vascular and cardiac alterations in diabetic rats: prevention by propionyl-L-carnitine. Proc Symp Vascular Complications of Diabetes Mellitus; Boston, 1993: 21 (Abstr).
25. Lassila R, Lepantalo M, Lindfors O. Peripheral arterial disease – natural outcome. Acta Med Scand 1986; 220: 295–301.
26. Regensteiner JG, Wolfel EE, Brass EP et al. Chronic changes in skeletal muscle histology and function in peripheral arterial disease. Circulation 1993; 87: 413–21.
27. Hiatt WR, Nawaz D, Brass EP. Carnitine metabolism during exercise in patients with peripheral vascular disease. J Appl Physiol 1987; 62: 2383–7.
28. Hiatt WR, Wolfel EE, Regensteiner JG, Brass EP. Skeletal muscle carnitine metabolism in patients with unilateral peripheral arterial disease. J Appl Phyiol 1992; 73: 346–53.
29. Arduini A, Molajoni F, Dottori S et al. Effect of oral propionyl-L-carnitine treatment on membrane phospholipid fatty acid turnover in diabetic rat erythrocytes. Diabetes. Accepted pending revision.
30. Hülsmann WC, Dubelaar ML. Carnitine requirement of vascular endothelial and smooth muscle cells in imminent ischemia. Mol Cell Biochem 1992; 116: 125–9.
31. Arduini A, Gorbunov N, Arrigoni Martelli E et al. Effect of L-carnitine and its acetate and propionate esters on the molecular dynamics of human erythrocyte membrane. Biochim Biophys Acta 1993; 1146: 229–35.

*Corresponding Author*: Dr Nerina Corsico, Pharmacology Department, Sigma-Tau S.p.A., Via Pontina Km. 30.400, I-00040 Pomezia (Rome), Italy

# 27. Effect of L-carnitine and propionyl-L-carnitine on cardiovascular diseases: a summary

JAN WILLEM DE JONG and ROBERTO FERRARI

## Deficiencies in carnitine biochemistry

Aberrations in carnitine biochemistry often lead to cardiovascular disorders. At the level of the heart, the abnormalities are probably due to the toxic effects of long-chain acylcarnitines [1]. Table 1 shows the effect of L-carnitine on patients with such deficiencies. Data from the literature indicate clearly the benefit of L-carnitine treatment for primary and secondary muscular and systemic carnitine deficiency linked to inborn errors of metabolism [2, 3]. It is one of the 30 drugs for life-threatening illnesses approved by the Food and Drug Administration in the 1980s [10].

## Acquired carnitine deficiency

Ischaemic and infarcted myocardium is characterized by a decrease in the tissue levels of carnitine; this has been shown in animals and humans where serum free carnitine levels and urinary total carnitine excretion are increased [11]. Plasma carnitine concentrations are elevated in patients with congestive heart failure as well as dilated and hypertrophic cardiomyopathies [12–15]. On the other hand, free carnitine and the ratio free to long-chain acylcarnitine in these hearts may be lowered [14–18]. Plasma carnitine and acylcarnitines are normal in patients with peripheral vascular disease [19], but free and total carnitine contents are lower in skeletal muscle of patients with severe peripheral arterial insufficiency [20]. Thus, it is difficult to estimate from plasma or serum carnitine levels the presence of an acquired carnitine deficiency in the tissues.

L-carnitine treatment appears to be a safe therapy. Studies convincingly support claims of successful treatment for patients with coronary heart disease and angina (see Chapter 16 and Table 2). No studies are presently available which compare the effects of L-carnitine with those of other well-known anti-anginal agents. It is advisable to use carnitine alongside regular medication.

There are interesting data which propose its use in the treatment of myo-

*J.W. de Jong and R. Ferrari (eds): The carnitine system*, 383–388.

*Table 1.* Effect of L-carnitine on patients with deficiency of carnitine biochemistry.

| Deficiency | Plasma [carnitine] | | Cardiovascular disorder | Response to carnitine treatment | Refs. |
|---|---|---|---|---|---|
| | Total | Esterified/total | | | |
| Carnitine transporter | Minute | Normal? | Dilated cardiomyopathy | Dramatic relief of symptoms, muscle carnitine not corrected | [2–4] |
| Carnitine palmitoyl-transferase I | High | Low | Heart failure | No response | [5] |
| Carnitine-acylcarnitine translocase | Low | High | Cardiomyopathy, auriculo ventricular block | No response | [6, 7] |
| Carnitine palmitoyl-transferase II | Low | High | Heart-beat disorders; cardiomyopathy (severe cases) | Recovery? | [8, 9] |

"Total" refers to the free *plus* acylcarnitine concentration, "Esterified/total" to the ratio [acylcarnitine]/[total carnitine].

*Table 2.* L-carnitine and cardiovascular diseases.

| Disorder | Effect of carnitine | Literature data [references] |
|---|---|---|
| Primary carnitine deficiency | Cardiomyopathy ↓; Myocardial function ↑ | Convincing [2, 21] |
| Coronary artery disease | Metabolism ↑; Exercise performance ↑; Arrhythmias ↓; Symptoms ↓ | Convincing [23–32] |
| Acute myocardial infarction | Cellular injury ↓; Arrhythmias ↓; Mortality ↓ ? | Promising [11, 33–35] |
| Peripheral vascular disease | Walking distance ↑ *; Intermittent claudication ↓ | Limited [36–38] |
| Congestive heart failure | Arrhythmias ↓; Physical performance ↑ | Preliminary [29, 31, 39, 40] |
| Anthracycline-induced cardiotoxicity | Cellular injury ↓ | Preliminary [21, 41] |

* Only studied after short-term administration.

cardial infarction. However, more extensive study will be needed to support this indication and such studies are presently in progress (see, e.g. Chapter 17 and ref. [22]). For other cardiovascular diseases, data are promising (Table 2) but more extensive trials will have to be carried out [21]. Carnitine could protect the heart against ischaemia induced during aortocoronary bypass grafting [42]. In chronic ischaemia, such as during ischaemic heart failure,

*Table 3.* Propionyl-L-carnitine and cardiovascular diseases.

| Disorder | Effect of carnitine | Literature data [references] |
|---|---|---|
| Coronary artery disease | Cardiac function ↑ ; Exercise tolerance ↑ | Limited [44–47] |
| Peripheral vascular disease | Walking distance ↑ ; Intermittent claudication ↓ | Reasonable [38, 48, 49] |
| Congestive heart failure | Exercise performance ↑ | Suggestive [50, 51] |

there is preliminary evidence that prolonged treatment with L-carnitine improves symptoms and cardiac function (Table 2).

For propionyl-L-carnitine, there are less published clinical data than for L-carnitine. The former compound could be more specific for skeletal and cardiac muscle than carnitine. In addition, it delivers propionate to the mitochondria. Propionyl-L-carnitine improves exercise performance (Chapters 23–25). Further proof of its efficacy in peripheral vascular disease (Chapter 25) and congestive heart failure (Chapter 24) will depend on ongoing studies (see, e.g. [43]). The literature provides reasonable evidence that propionyl-L-carnitine is effective in the treatment of coronary artery disease (Table 3).

The exact mechanism of action to explain the beneficial effect of L-carnitine and propionyl-L-carnitine in cardiovascular disease is lacking. Of interest are data showing [52, 53] that propionyl-L-carnitine but not L-carnitine replenishes citric acid cycle intermediates.

## Acknowledgement

We gratefully acknowledge the secretarial contribution of Carina D.M. Poleon-Weghorst.

## References

1. Corr PB, Creer MH, Yamada KA, Saffitz JE, Sobel BE. Prophylaxis of early ventricular fibrillation by inhibition of acylcarnitine accumulation. J Clin Invest 1989; 83: 927–36.
2. Treem WR, Stanley CA, Finegold DN, Hale DE, Coates PM. Primary carnitine deficiency due to a failure of carnitine transport in kidney, muscle, and fibroblasts. N Engl J Med 1988; 319: 1331–6.
3. Anonymous. Carnitine deficiency. Lancet 1990; 335: 631–3 (Editorial).
4. Marin-Garcia J, Goldenthal MJ. Cardiomyopathy and abnormal mitochondrial function. Cardiovasc Res 1994; 28: 456–63.
5. Vockley J. The changing face of disorders of fatty acid oxidation. Mayo Clin Proc 1994; 69: 249–57.
6. Stanley CA, Hale DE, Berry GT, Deleeuw S, Boxer J, Bonnefont J-P. Brief report: A deficiency of carnitine-acylcarnitine translocase in the inner mitochondrial membrane. N Engl J Med 1992; 327: 19–23.

7. Pande SV, Brivet M, Slama A, Demaugre F, Aufrant C, Saudubray J-M. Carnitine-acylcarnitine translocase deficiency with severe hypoglycemia and auriculo ventricular block. Translocase assay in permeabilized fibroblasts. J Clin Invest 1993; 91: 1247–52.
8. Demaugre F, Bonnefont J-P, Collonna M, Cepanec C, Leroux J-P, Saudubray J-M. Infantile form of carnitine palmitoyltransferase II deficiency with hepatomuscular symptoms and sudden death. Physiopathological approach to carnitine palmitoyltransferase II deficiencies. J Clin Invest 1991; 87: 859–64.
9. Taroni F, Verderio E, Fiorucci S et al. Molecular characterization of inherited carnitine palmitoyltransferase II deficiency. Proc Natl Acad Sci USA 1992; 89: 8429–33.
10. Anonymous. Drug development for life-threatening illnesses is taking under five years on average, FDA says; early NDAs before study completion are key. FDC Reports 1989; 51: 7–9.
11. Rizzon P, Biasco G, Di Biase M et al. High doses of L-carnitine in acute myocardial infarction: metabolic and antiarrhythmic effects. Eur Heart J 1989; 10: 502–8.
12. Tripp ME, Shug AL. Plasma carnitine concentrations in cardiomyopathy patients. Biochem Med 1984; 32: 199–206.
13. Conte A, Hess OM, Maire R et al. Klinische Bedeutung des Serumcarnitines für den Verlauf und die Prognose der dilatativen Kardiomyopathie. Z Kardiol 1987; 76: 15–24.
14. Regitz V, Shug AL, Fleck E. Defective myocardial carnitine metabolism in congestive heart failure secondary to dilated cardiomyopathy and to coronary, hypertensive and valvular heart diseases. Am J Cardiol 1990; 65: 755–60.
15. Pierpont ME, Judd D, Goldenberg IF, Ring WS, Olivari MT, Pierpont GL. Myocardial carnitine in end-stage congestive heart failure. Am J Cardiol 1989; 64: 56–60.
16. Regitz V, Bossaller C, Strasser R, Müller M, Shug AL, Fleck E. Metabolic alterations in end-stage and less severe heart failure – myocardial carnitine decrease. J Clin Chem Clin Biochem 1990; 28: 611–7.
17. Masumura Y, Kobayashi A, Yamazaki N. Myocardial free carnitine and fatty acylcarnitine levels in patients with chronic heart failure. Jpn Circ J 1990; 54: 1471–6.
18. Spagnoli LG, Corsi M, Villaschi S, Palmieri G, Maccari F. Myocardial carnitine deficiency in acute myocardial infarction. Lancet 1982; 1: 1419–20.
19. Hiatt WR, Nawaz D, Brass EP. Carnitine metabolism during exercise in patients with peripheral vascular disease. J Appl Physiol 1987; 62: 2383–7.
20. Brevetti G, Angelini C, Rosa M et al. Muscle carnitine deficiency in patients with severe peripheral vascular disease. Circulation 1991; 84: 1490–5.
21. Pepine CJ. The therapeutic potential of carnitine in cardiovascular disorders. Clin Ter 1991; 13: 2–21.
22. Iliceto S, D'Ambrosio G, Scrutinio D, Marangelli V, Boni L, Rizzon P. A digital network for long-distance echocardiographic image and data transmission in clinical trials: The CEDIM study experience. J Am Soc Echocardiogr 1993; 6: 583–92.
23. Kamikawa T, Suzuki Y, Kobayashi A et al. Effects of L-carnitine on exercise tolerance in patients with stable angina pectoris. Jpn Heart J 1984; 25: 587–97.
24. Ferrari R, Cucchini F, Visioli O. The metabolic effects of L-carnitine in angina pectoris. Int J Cardiol 1984; 5: 213–6.
25. Cherchi A, Lai C, Angelino F et al. Effects of L-carnitine on exercise tolerance in chronic stable angina: a multicenter, double-blind, randomized, placebo controlled crossover study. Int J Clin Pharmacol Ther Toxicol 1985; 23: 569–72.
26. Canale C, Terrachini V, Biagini A et al. Bicycle ergometer and echocardiographic study in healthy subjects and patients with angina pectoris after administration of L-carnitine: semiautomatic computerized analysis of M-mode tracing. Int J Clin Pharmacol Ther Toxicol 1988; 26: 221–4.
27. Sotobata I, Noda S, Hayashi H et al. Clinical evaluation of the effects of levocarnitine chloride (LC-80) on exercise tolerance in stable angina pectoris by the serial multistage treadmill exercise testing: a multicenter, double-blind study. Jpn J Clin Pharmacol Ther 1989; 20: 607–18.

28. Fujiwara M, Nakano T, Tamoto S et al. Effect of L-carnitine in patients with ischemic heart disease. J Cardiol 1991; 21: 493–504 (In Japanese).
29. Fernandez C, Proto C. La L-carnitina nel trattamento dell'ischemia miocardica cronica. Analisi dei risultati di tre studi multicentrici e rassegna bibliografica. Clin Ter 1992; 140: 353–77.
30. Cacciatore L, Cerio R, Ciarimboli M et al. The therapeutic effect of L-carnitine in patients with exercise-induced stable angina: a controlled study. Drugs Exp Clin Res 1991; 17: 225–35.
31. Kobayashi A, Masumura Y, Yamazaki N. L-Carnitine treatment for congestive heart failure – Experimental and clinical study. Jpn Circ J 1992; 56: 86–94.
32. Palazzuoli V, Modillo S, Faglia S, D'Aprile N, Camporeale A, Gennari C. Valutazione dell'attività antiaritmica della L-carnitina e del propafenone nella cardiopatia ischemica. Clin Ter 1993; 142: 155–9.
33. Di Biase M, Biasco G, Rizzon P. Ruolo della terapia metabolica nell'infarto miocardico. Cardiologia 1991; 36(Suppl 1): 389–92.
34. Davini P, Bigalli A, Lamanna F, Boem A. Controlled study on L-carnitine therapeutic efficacy in post-infarction. Drugs Exp Clin Res 1992; 18: 355–65.
35. Martina B, Zuber M, Weiss P, Burkart F, Ritz R. Antiarrhythmische Behandlung mit L-Carnitin beim akuten Myokardinfarkt. Schweiz Med Wochenschr 1992; 122: 1352–5.
36. Brevetti G, Chiariello M, Ferulano G et al. Increases in walking distance in patients with peripheral vascular disease treated with L-carnitine: a double-blind, cross-over study. Circulation 1988; 77: 767–73.
37. Brevetti G, Attisano T, Perna S, Rossini A, Policicchio A, Corsi M. Effect of L-carnitine on the reactive hyperemia in patients affected by peripheral vascular disease: a double-blind, crossover study. Angiology 1989; 40: 857–62.
38. Brevetti G, Perna S, Sabbà C et al. Superiority of L-propionylcarnitine vs L-carnitine in improving walking capacity in patients with peripheral vascular disease: an acute, intravenous, double-blind, cross-over study. Eur Heart J 1992; 13: 251–5.
39. Ghidini O, Azzurro M, Vita G, Sartori G. Evaluation of the therapeutic efficacy of L-carnitine in congestive heart failure. Int J Clin Pharmacol Ther Toxicol 1988; 26: 217–20.
40. Pucciarelli G, Mastursi M, Latte S et al. Effetti clinici ed emodinamici della propionil-L-carnitina nel trattamento dello scompenso cardiaco congestizio. Clin Ter 1992; 141: 379–84.
41. De Leonardis V, Neri B, Bacalli S, Cinelli P. Reduction of cardiac toxicity of anthracyclines by L-carnitine: preliminary overview of clinical data. Int J Clin Pharmacol Res 1985; 5: 137–42.
42. Demeyere R, Lormans P, Weidler B, Minten J, Van Aken H, Flameng W. Cardioprotective effects of carnitine in extensive aortocoronary bypass grafting: a double-blind, randomized, placebo-controlled clinical trial. Anesth Analg 1990; 71: 520–8.
43. Garg R, Yusuf S. Current and ongoing randomized trials in heart failure and left ventricular dysfunction. J Am Coll Cardiol 1993; 22(Suppl A): 194A–7A.
44. Chiddo A, Gaglione A, Musci S et al. Hemodynamic study of intravenous propionyl-L-carnitine in patients with ischemic heart disease and normal left ventricular function. Cardiovasc Drugs Ther 1991; 5(Suppl 1): 107–11.
45. Cherchi A, Lai C, Onnis E et al. Propionyl carnitine in stable effort angina. Cardiovasc Drugs Ther 1990; 4: 481–6.
46. La Gioia R, Scrutinio D, Mangini SG et al. Propionyl-L-carnitine: a new compound in the metabolic approach to the treatment of effort angina. Int J Cardiol 1992; 34: 167–72.
47. Bartels GL, Remme WJ, Pillay M et al. Acute improvement of cardiac function with intravenous L-propionylcarnitine in humans. J Cardiovasc Pharmacol 1992; 20: 157–64.
48. Coto V, D'Alessandro L, Grattarola G et al. Evaluation of the therapeutic efficacy and tolerability of levocarnitine propionyl in the treatment of chronic obstructive arteriopathies of the lower extremities: a multicentre controlled study vs. placebo. Drugs Exp Clin Res 1992; 18: 29–36.

49. Greco AV, Mingrone G, Bianchi M, Ghirlanda G. Effect of propionyl-L-carnitine in the treatment of diabetic angiopathy: controlled double blind trial versus placebo. Drugs Exp Clin Res 1992; 18: 69–80.
50. Mancini M, Rengo F, Lingetti M, Sorrentino GP, Nolfe G. Controlled study on the therapeutic efficacy of propionyl-L-carnitine in patients with congestive heart failure. Arzneimittelforsch 1992; 42: 1101–4.
51. Caponnetto S, Canale C, Masperone MA, Terracchini V, Valentini G, Brunelli C. Efficacy of L-propionylcarnitine treatment in patients with left ventricular dysfunction. Eur Heart J 1994; 15: 1267–73.
52. Di Lisa F, Menabò R, Siliprandi N. L-Propionyl-carnitine protection of mitochondria in ischemic rat hearts. Mol Cell Biochem 1989; 88: 169–73.
53. Tassani V, Cattapan F, Magnanimi L, Peschechera A. Anaplerotic effect of propionyl carnitine in rat heart mitochondria. Biochem Biophys Res Commun 1994; 199: 949–53.

*Corresponding Author*: Dr. Jan Willem de Jong, Cardiochemical Laboratory, Thoraxcenter Ee2371, Erasmus University Rotterdam, P.O. Box 1738, 3000 DR Rotterdam, The Netherlands

# Subject Index

# Developments in Cardiovascular Medicine

1. Ch.T. Lancée (ed.): *Echocardiology*. 1979 ISBN 90-247-2209-8
2. J. Baan, A.C. Arntzenius and E.L. Yellin (eds.): *Cardiac Dynamics*. 1980 ISBN 90-247-2212-8
3. H.J.Th. Thalen and C.C. Meere (eds.): *Fundamentals of Cardiac Pacing*. 1979 ISBN 90-247-2245-4
4. H.E. Kulbertus and H.J.J. Wellens (eds.): *Sudden Death*. 1980 ISBN 90-247-2290-X
5. L.S. Dreifus and A.N. Brest (eds.): *Clinical Applications of Cardiovascular Drugs*. 1980 ISBN 90-247-2295-0
6. M.P. Spencer and J.M. Reid: *Cerebrovascular Evaluation with Doppler Ultrasound*. With contributions by E.C. Brockenbrough, R.S. Reneman, G.I. Thomas and D.L. Davis. 1981 ISBN 90-247-2384-1
7. D.P. Zipes, J.C. Bailey and V. Elharrar (eds.): *The Slow Inward Current and Cardiac Arrhythmias*. 1980 ISBN 90-247-2380-9
8. H. Kesteloot and J.V. Joossens (eds.): *Epidemiology of Arterial Blood Pressure*. 1980 ISBN 90-247-2386-8
9. F.J.Th. Wackers (ed.): *Thallium-201 and Technetium-99m-Pyrophosphate. Myocardial Imaging in the Coronary Care Unit*. 1980 ISBN 90-247-2396-5
10. A. Maseri, C. Marchesi, S. Chierchia and M.G. Trivella (eds.): *Coronary Care Units*. Proceedings of a European Seminar (1978). 1981 ISBN 90-247-2456-2
11. J. Morganroth, E.N. Moore, L.S. Dreifus and E.L. Michelson (eds.): *The Evaluation of New Antiarrhythmic Drugs*. Proceedings of the First Symposium on New Drugs and Devices, held in Philadelphia, Pa., U.S.A. (1980). 1981 ISBN 90-247-2474-0
12. P. Alboni: *Intraventricular Conduction Disturbances*. 1981 ISBN 90-247-2483-X
13. H. Rijsterborgh (ed.): *Echocardiology*. 1981 ISBN 90-247-2491-0
14. G.S. Wagner (ed.): *Myocardial Infarction*. Measurement and Intervention. 1982 ISBN 90-247-2513-5
15. R.S. Meltzer and J. Roelandt (eds.): *Contrast Echocardiography*. 1982 ISBN 90-247-2531-3
16. A. Amery, R. Fagard, P. Lijnen and J. Staessen (eds.): *Hypertensive Cardiovascular Disease*. Pathophysiology and Treatment. 1982 IBSN 90-247-2534-8
17. L.N. Bouman and H.J. Jongsma (eds.): *Cardiac Rate and Rhythm*. Physiological, Morphological and Developmental Aspects. 1982 ISBN 90-247-2626-3
18. J. Morganroth and E.N. Moore (eds.): *The Evaluation of Beta Blocker and Calcium Antagonist Drugs*. Proceedings of the 2nd Symposium on New Drugs and Devices, held in Philadelphia, Pa., U.S.A. (1981). 1982 ISBN 90-247-2642-5
19. M.B. Rosenbaum and M.V. Elizari (eds.): *Frontiers of Cardiac Electrophysiology*. 1983 ISBN 90-247-2663-8
20. J. Roelandt and P.G. Hugenholtz (eds.): *Long-term Ambulatory Electrocardiography*. 1982 ISBN 90-247-2664-6
21. A.A.J. Adgey (ed.): *Acute Phase of Ischemic Heart Disease and Myocardial Infarction*. 1982 ISBN 90-247-2675-1
22. P. Hanrath, W. Bleifeld and J. Souquet (eds.): *Cardiovascular Diagnosis by Ultrasound*. Transesophageal, Computerized, Contrast, Doppler Echocardiography. 1982 ISBN 90-247-2692-1
23. J. Roelandt (ed.): *The Practice of M-Mode and Two-dimensional Echocardiography*. 1983 ISBN 90-247-2745-6
24. J. Meyer, P. Schweizer and R. Erbel (eds.): *Advances in Noninvasive Cardiology*. Ultrasound, Computed Tomography, Radioisotopes, Digital Angiography. 1983 ISBN 0-89838-576-8
25. J. Morganroth and E.N. Moore (eds.): *Sudden Cardiac Death and Congestive Heart Failure*. Diagnosis and Treatment. Proceedings of the 3rd Symposium on New Drugs and Devices, held in Philadelphia, Pa., U.S.A. (1982). 1983 ISBN 0-89838-580-6
26. H.M. Perry Jr. (ed.): *Lifelong Management of Hypertension*. 1983 ISBN 0-89838-582-2
27. E.A. Jaffe (ed.): *Biology of Endothelial Cells*. 1984 ISBN 0-89838-587-3

# Developments in Cardiovascular Medicine

28. B. Surawicz, C.P. Reddy and E.N. Prystowsky (eds.): *Tachycardias.* 1984
ISBN 0-89838-588-1
29. M.P. Spencer (ed.): *Cardiac Doppler Diagnosis.* Proceedings of a Symposium, held in Clearwater, Fla., U.S.A. (1983). 1983 ISBN 0-89838-591-1
30. H. Villarreal and M.P. Sambhi (eds.): *Topics in Pathophysiology of Hypertension.* 1984 ISBN 0-89838-595-4
31. F.H. Messerli (ed.): *Cardiovascular Disease in the Elderly.* 1984
Revised edition, 1988: see below under Volume 76
32. M.L. Simoons and J.H.C. Reiber (eds.): *Nuclear Imaging in Clinical Cardiology.* 1984 ISBN 0-89838-599-7
33. H.E.D.J. ter Keurs and J.J. Schipperheyn (eds.): *Cardiac Left Ventricular Hypertrophy.* 1983 ISBN 0-89838-612-8
34. N. Sperelakis (ed.): *Physiology and Pathology of the Heart.* 1984
Revised edition, 1988: see below under Volume 90
35. F.H. Messerli (ed.): *Kidney in Essential Hypertension.* Proceedings of a Course, held in New Orleans, La., U.S.A. (1983). 1984 ISBN 0-89838-616-0
36. M.P. Sambhi (ed.): *Fundamental Fault in Hypertension.* 1984 ISBN 0-89838-638-1
37. C. Marchesi (ed.): *Ambulatory Monitoring.* Cardiovascular System and Allied Applications. Proceedings of a Workshop, held in Pisa, Italy (1983). 1984
ISBN 0-89838-642-X
38. W. Kupper, R.N. MacAlpin and W. Bleifeld (eds.): *Coronary Tone in Ischemic Heart Disease.* 1984 ISBN 0-89838-646-2
39. N. Sperelakis and J.B. Caulfield (eds.): *Calcium Antagonists.* Mechanism of Action on Cardiac Muscle and Vascular Smooth Muscle. Proceedings of the 5th Annual Meeting of the American Section of the I.S.H.R., held in Hilton Head, S.C., U.S.A. (1983). 1984 ISBN 0-89838-655-1
40. Th. Godfraind, A.G. Herman and D. Wellens (eds.): *Calcium Entry Blockers in Cardiovascular and Cerebral Dysfunctions.* 1984 ISBN 0-89838-658-6
41. J. Morganroth and E.N. Moore (eds.): *Interventions in the Acute Phase of Myocardial Infarction.* Proceedings of the 4th Symposium on New Drugs and Devices, held in Philadelphia, Pa., U.S.A. (1983). 1984 ISBN 0-89838-659-4
42. F.L. Abel and W.H. Newman (eds.): *Functional Aspects of the Normal, Hypertrophied and Failing Heart.* Proceedings of the 5th Annual Meeting of the American Section of the I.S.H.R., held in Hilton Head, S.C., U.S.A. (1983). 1984
ISBN 0-89838-665-9
43. S. Sideman and R. Beyar (eds.): [3-D] *Simulation and Imaging of the Cardiac System.* State of the Heart. Proceedings of the International Henry Goldberg Workshop, held in Haifa, Israel (1984). 1985 ISBN 0-89838-687-X
44. E. van der Wall and K.I. Lie (eds.): *Recent Views on Hypertrophic Cardiomyopathy.* Proceedings of a Symposium, held in Groningen, The Netherlands (1984). 1985
ISBN 0-89838-694-2
45. R.E. Beamish, P.K. Singal and N.S. Dhalla (eds.), *Stress and Heart Disease.* Proceedings of a International Symposium, held in Winnipeg, Canada, 1984 (Vol. 1). 1985 ISBN 0-89838-709-4
46. R.E. Beamish, V. Panagia and N.S. Dhalla (eds.): *Pathogenesis of Stress-induced Heart Disease.* Proceedings of a International Symposium, held in Winnipeg, Canada, 1984 (Vol. 2). 1985 ISBN 0-89838-710-8
47. J. Morganroth and E.N. Moore (eds.): *Cardiac Arrhythmias.* New Therapeutic Drugs and Devices. Proceedings of the 5th Symposium on New Drugs and Devices, held in Philadelphia, Pa., U.S.A. (1984). 1985 ISBN 0-89838-716-7
48. P. Mathes (ed.): *Secondary Prevention in Coronary Artery Disease and Myocardial Infarction.* 1985 ISBN 0-89838-736-1
49. H.L. Stone and W.B. Weglicki (eds.): *Pathobiology of Cardiovascular Injury.* Proceedings of the 6th Annual Meeting of the American Section of the I.S.H.R., held in Oklahoma City, Okla., U.S.A. (1984). 1985 ISBN 0-89838-743-4

# Developments in Cardiovascular Medicine

50. J. Meyer, R. Erbel and H.J. Rupprecht (eds.): *Improvement of Myocardial Perfusion.* Thrombolysis, Angioplasty, Bypass Surgery. Proceedings of a Symposium, held in Mainz, F.R.G. (1984). 1985 ISBN 0-89838-748-5
51. J.H.C. Reiber, P.W. Serruys and C.J. Slager (eds.): *Quantitative Coronary and Left Ventricular Cineangiography.* Methodology and Clinical Applications. 1986 ISBN 0-89838-760-4
52. R.H. Fagard and I.E. Bekaert (eds.): *Sports Cardiology.* Exercise in Health and Cardiovascular Disease. Proceedings from an International Conference, held in Knokke, Belgium (1985). 1986 ISBN 0-89838-782-5
53. J.H.C. Reiber and P.W. Serruys (eds.): *State of the Art in Quantitative Cornary Arteriography.* 1986 ISBN 0-89838-804-X
54. J. Roelandt (ed.): *Color Doppler Flow Imaging and Other Advances in Doppler Echocardiography.* 1986 ISBN 0-89838-806-6
55. E.E. van der Wall (ed.): *Noninvasive Imaging of Cardiac Metabolism.* Single Photon Scintigraphy, Positron Emission Tomography and Nuclear Magnetic Resonance. 1987 ISBN 0-89838-812-0
56. J. Liebman, R. Plonsey and Y. Rudy (eds.): *Pediatric and Fundamental Electrocardiography.* 1987 ISBN 0-89838-815-5
57. H.H. Hilger, V. Hombach and W.J. Rashkind (eds.), *Invasive Cardiovascular Therapy.* Proceedings of an International Symposium, held in Cologne, F.R.G. (1985). 1987 ISBN 0-89838-818-X
58. P.W. Serruys and G.T. Meester (eds.): *Coronary Angioplasty.* A Controlled Model for Ischemia. 1986 ISBN 0-89838-819-8
59. J.E. Tooke and L.H. Smaje (eds.): *Clinical Investigation of the Microcirculation.* Proceedings of an International Meeting, held in London, U.K. (1985). 1987 ISBN 0-89838-833-3
60. R.Th. van Dam and A. van Oosterom (eds.): *Electrocardiographic Body Surface Mapping.* Proceedings of the 3rd International Symposium on B.S.M., held in Nijmegen, The Netherlands (1985). 1986 ISBN 0-89838-834-1
61. M.P. Spencer (ed.): *Ultrasonic Diagnosis of Cerebrovascular Disease.* Doppler Techniques and Pulse Echo Imaging. 1987 ISBN 0-89838-836-8
62. M.J. Legato (ed.): *The Stressed Heart.* 1987 ISBN 0-89838-849-X
63. M.E. Safar (ed.): *Arterial and Venous Systems in Essential Hypertension.* With Assistance of G.M. London, A.Ch. Simon and Y.A. Weiss. 1987 ISBN 0-89838-857-0
64. J. Roelandt (ed.): *Digital Techniques in Echocardiography.* 1987 ISBN 0-89838-861-9
65. N.S. Dhalla, P.K. Singal and R.E. Beamish (eds.): *Pathology of Heart Disease.* Proceedings of the 8th Annual Meeting of the American Section of the I.S.H.R., held in Winnipeg, Canada, 1986 (Vol. 1). 1987 ISBN 0-89838-864-3
66. N.S. Dhalla, G.N. Pierce and R.E. Beamish (eds.): *Heart Function and Metabolism.* Proceedings of the 8th Annual Meeting of the American Section of the I.S.H.R., held in Winnipeg, Canada, 1986 (Vol. 2). 1987 ISBN 0-89838-865-1
67. N.S. Dhalla, I.R. Innes and R.E. Beamish (eds.): *Myocardial Ischemia.* Proceedings of a Satellite Symposium of the 30th International Physiological Congress, held in Winnipeg, Canada (1986). 1987 ISBN 0-89838-866-X
68. R.E. Beamish, V. Panagia and N.S. Dhalla (eds.): *Pharmacological Aspects of Heart Disease.* Proceedings of an International Symposium, held in Winnipeg, Canada (1986). 1987 ISBN 0-89838-867-8
69. H.E.D.J. ter Keurs and J.V. Tyberg (eds.): *Mechanics of the Circulation.* Proceedings of a Satellite Symposium of the 30th International Physiological Congress, held in Banff, Alberta, Canada (1986). 1987 ISBN 0-89838-870-8
70. S. Sideman and R. Beyar (eds.): *Activation, Metabolism and Perfusion of the Heart.* Simulation and Experimental Models. Proceedings of the 3rd Henry Goldberg Workshop, held in Piscataway, N.J., U.S.A. (1986). 1987 ISBN 0-89838-871-6

71. E. Aliot and R. Lazzara (eds.): *Ventricular Tachycardias*. From Mechanism to Therapy. 1987 ISBN 0-89838-881-3
72. A. Schneeweiss and G. Schettler: *Cardiovascular Drug Therapoy in the Elderly*. 1988 ISBN 0-89838-883-X
73. J.V. Chapman and A. Sgalambro (eds.): *Basic Concepts in Doppler Echocardiography*. Methods of Clinical Applications based on a Multi-modality Doppler Approach. 1987 ISBN 0-89838-888-0
74. S. Chien, J. Dormandy, E. Ernst and A. Matrai (eds.): *Clinical Hemorheology*. Applications in Cardiovascular and Hematological Disease, Diabetes, Surgery and Gynecology. 1987 ISBN 0-89838-807-4
75. J. Morganroth and E.N. Moore (eds.): *Congestive Heart Failure*. Proceedings of the 7th Annual Symposium on New Drugs and Devices, held in Philadelphia, Pa., U.S.A. (1986). 1987 ISBN 0-89838-955-0
76. F.H. Messerli (ed.): *Cardiovascular Disease in the Elderly*. 2nd ed. 1988 ISBN 0-89838-962-3
77. P.H. Heintzen and J.H. Bürsch (eds.): *Progress in Digital Angiocardiography*. 1988 ISBN 0-89838-965-8
78. M.M. Scheinman (ed.): *Catheter Ablation of Cardiac Arrhythmias*. Basic Bioelectrical Effects and Clinical Indications. 1988 ISBN 0-89838-967-4
79. J.A.E. Spaan, A.V.G. Bruschke and A.C. Gittenberger-De Groot (eds.): *Coronary Circulation*. From Basic Mechanisms to Clinical Implications. 1987 ISBN 0-89838-978-X
80. C. Visser, G. Kan and R.S. Meltzer (eds.): *Echocardiography in Coronary Artery Disease*. 1988 ISBN 0-89838-979-8
81. A. Bayés de Luna, A. Betriu and G. Permanyer (eds.): *Therapeutics in Cardiology*. 1988 ISBN 0-89838-981-X
82. D.M. Mirvis (ed.): *Body Surface Electrocardiographic Mapping*. 1988 ISBN 0-89838-983-6
83. M.A. Konstam and J.M. Isner (eds.): *The Right Ventricle*. 1988 ISBN 0-89838-987-9
84. C.T. Kappagoda and P.V. Greenwood (eds.): *Long-term Management of Patients after Myocardial Infarction*. 1988 ISBN 0-89838-352-8
85. W.H. Gaasch and H.J. Levine (eds.): *Chronic Aortic Regurgitation*. 1988 ISBN 0-89838-364-1
86. P.K. Singal (ed.): *Oxygen Radicals in the Pathophysiology of Heart Disease*. 1988 ISBN 0-89838-375-7
87. J.H.C. Reiber and P.W. Serruys (eds.): *New Developments in Quantitative Coronary Arteriography*. 1988 ISBN 0-89838-377-3
88. J. Morganroth and E.N. Moore (eds.): *Silent Myocardial Ischemia*. Proceedings of the 8th Annual Symposium on New Drugs and Devices (1987). 1988 ISBN 0-89838-380-3
89. H.E.D.J. ter Keurs and M.I.M. Noble (eds.): *Starling's Law of the Heart Revisted*. 1988 ISBN 0-89838-382-X
90. N. Sperelakis (ed.): *Physiology and Pathophysiology of the Heart*. Rev. ed. 1988 3rd, revised edition, 1994: see below under Volume 151
91. J.W. de Jong (ed.): *Myocardial Energy Metabolism*. 1988 ISBN 0-89838-394-3
92. V. Hombach, H.H. Hilger and H.L. Kennedy (eds.): *Electrocardiography and Cardiac Drug Therapy*. Proceedings of an International Symposium, held in Cologne, F.R.G. (1987). 1988 ISBN 0-89838-395-1
93. H. Iwata, J.B. Lombardini and T. Segawa (eds.): *Taurine and the Heart*. 1988 ISBN 0-89838-396-X
94. M.R. Rosen and Y. Palti (eds.): *Lethal Arrhythmias Resulting from Myocardial Ischemia and Infarction*. Proceedings of the 2nd Rappaport Symposium, held in Haifa, Israel (1988). 1988 ISBN 0-89838-401-X
95. M. Iwase and I. Sotobata: *Clinical Echocardiography*. With a Foreword by M.P. Spencer. 1989 ISBN 0-7923-0004-1

# Developments in Cardiovascular Medicine

96. I. Cikes (ed.): *Echocardiography in Cardiac Interventions*. 1989
ISBN 0-7923-0088-2
97. E. Rapaport (ed.): *Early Interventions in Acute Myocardial Infarction*. 1989
ISBN 0-7923-0175-7
98. M.E. Safar and F. Fouad-Tarazi (eds.): *The Heart in Hypertension*. A Tribute to Robert C. Tarazi (1925-1986). 1989 ISBN 0-7923-0197-8
99. S. Meerbaum and R. Meltzer (eds.): *Myocardial Contrast Two-dimensional Echocardiography*. 1989 ISBN 0-7923-0205-2
100. J. Morganroth and E.N. Moore (eds.): *Risk/Benefit Analysis for the Use and Approval of Thrombolytic, Antiarrhythmic, and Hypolipidemic Agents*. Proceedings of the 9th Annual Symposium on New Drugs and Devices (1988). 1989 ISBN 0-7923-0294-X
101. P.W. Serruys, R. Simon and K.J. Beatt (eds.): *PTCA - An Investigational Tool and a Non-operative Treatment of Acute Ischemia*. 1990 ISBN 0-7923-0346-6
102. I.S. Anand, P.I. Wahi and N.S. Dhalla (eds.): *Pathophysiology and Pharmacology of Heart Disease*. 1989 ISBN 0-7923-0367-9
103. G.S. Abela (ed.): *Lasers in Cardiovascular Medicine and Surgery*. Fundamentals and Technique. 1990 ISBN 0-7923-0440-3
104. H.M. Piper (ed.): *Pathophysiology of Severe Ischemic Myocardial Injury*. 1990
ISBN 0-7923-0459-4
105. S.M. Teague (ed.): *Stress Doppler Echocardiography*. 1990 ISBN 0-7923-0499-3
106. P.R. Saxena, D.I. Wallis, W. Wouters and P. Bevan (eds.): *Cardiovascular Pharmacology of 5-Hydroxytryptamine*. Prospective Therapeutic Applications. 1990
ISBN 0-7923-0502-7
107. A.P. Shepherd and P.A. Öberg (eds.): *Laser-Doppler Blood Flowmetry*. 1990
ISBN 0-7923-0508-6
108. J. Soler-Soler, G. Permanyer-Miralda and J. Sagristà-Sauleda (eds.): *Pericardial Disease*. New Insights and Old Dilemmas. 1990 ISBN 0-7923-0510-8
109. J.P.M. Hamer: *Practical Echocardiography in the Adult*. With Doppler and Color-Doppler Flow Imaging. 1990 ISBN 0-7923-0670-8
110. A. Bayés de Luna, P. Brugada, J. Cosin Aguilar and F. Navarro Lopez (eds.): *Sudden Cardiac Death*. 1991 ISBN 0-7923-0716-X
111. E. Andries and R. Stroobandt (eds.): *Hemodynamics in Daily Practice*. 1991
ISBN 0-7923-0725-9
112. J. Morganroth and E.N. Moore (eds.): *Use and Approval of Antihypertensive Agents and Surrogate Endpoints for the Approval of Drugs affecting Antiarrhythmic Heart Failure and Hypolipidemia*. Proceedings of the 10th Annual Symposium on New Drugs and Devices (1989). 1990 ISBN 0-7923-0756-9
113. S. Iliceto, P. Rizzon and J.R.T.C. Roelandt (eds.): *Ultrasound in Coronary Artery Disease*. Present Role and Future Perspectives. 1990 ISBN 0-7923-0784-4
114. J.V. Chapman and G.R. Sutherland (eds.): *The Noninvasive Evaluation of Hemodynamics in Congenital Heart Disease*. Doppler Ultrasound Applications in the Adult and Pediatric Patient with Congenital Heart Disease. 1990
ISBN 0-7923-0836-0
115. G.T. Meester and F. Pinciroli (eds.): *Databases for Cardiology*. 1991
ISBN 0-7923-0886-7
116. B. Korecky and N.S. Dhalla (eds.): *Subcellular Basis of Contractile Failure*. 1990
ISBN 0-7923-0890-5
117. J.H.C. Reiber and P.W. Serruys (eds.): *Quantitative Coronary Arteriography*. 1991
ISBN 0-7923-0913-8
118. E. van der Wall and A. de Roos (eds.): *Magnetic Resonance Imaging in Coronary Artery Disease*. 1991 ISBN 0-7923-0940-5
119. V. Hombach, M. Kochs and A.J. Camm (eds.): *Interventional Techniques in Cardiovascular Medicine*. 1991 ISBN 0-7923-0956-1
120. R. Vos: *Drugs Looking for Diseases*. Innovative Drug Research and the Development of the Beta Blockers and the Calcium Antagonists. 1991 ISBN 0-7923-0968-5

# Developments in Cardiovascular Medicine

121. S. Sideman, R. Beyar and A.G. Kleber (eds.): *Cardiac Electrophysiology, Circulation, and Transport.* Proceedings of the 7th Henry Goldberg Workshop (Berne, Switzerland, 1990). 1991 ISBN 0-7923-1145-0
122. D.M. Bers: *Excitation-Contraction Coupling and Cardiac Contractile Force.* 1991 ISBN 0-7923-1186-8
123. A.-M. Salmasi and A.N. Nicolaides (eds.): *Occult Atherosclerotic Disease.* Diagnosis, Assessment and Management. 1991 ISBN 0-7923-1188-4
124. J.A.E. Spaan: *Coronary Blood Flow.* Mechanics, Distribution, and Control. 1991 ISBN 0-7923-1210-4
125. R.W. Stout (ed.): *Diabetes and Atherosclerosis.* 1991 ISBN 0-7923-1310-0
126. A.G. Herman (ed.): *Antithrombotics.* Pathophysiological Rationale for Pharmacological Interventions. 1991 ISBN 0-7923-1413-1
127. N.H.J. Pijls: *Maximal Myocardial Perfusion as a Measure of the Functional Significance of Coronary Arteriogram.* From a Pathoanatomic to a Pathophysiologic Interpretation of the Coronary Arteriogram. 1991 ISBN 0-7923-1430-1
128. J.H.C. Reiber and E.E. v.d. Wall (eds.): *Cardiovascular Nuclear Medicine and MRI.* Quantitation and Clinical Applications. 1992 ISBN 0-7923-1467-0
129. E. Andries, P. Brugada and R. Stroobrandt (eds.): *How to Face 'the Faces' of Cardiac Pacing.* 1992 ISBN 0-7923-1528-6
130. M. Nagano, S. Mochizuki and N.S. Dhalla (eds.): *Cardiovascular Disease in Diabetes.* 1992 ISBN 0-7923-1554-5
131. P.W. Serruys, B.H. Strauss and S.B. King III (eds.): *Restenosis after Intervention with New Mechanical Devices.* 1992 ISBN 0-7923-1555-3
132. P.J. Walter (ed.): *Quality of Life after Open Heart Surgery.* 1992 ISBN 0-7923-1580-4
133. E.E. van der Wall, H. Sochor, A. Righetti and M.G. Niemeyer (eds.): *What's new in Cardiac Imaging?* SPECT, PET and MRI. 1992 ISBN 0-7923-1615-0
134. P. Hanrath, R. Uebis and W. Krebs (eds.): *Cardiovascular Imaging by Ultrasound.* 1992 ISBN 0-7923-1755-6
135. F.H. Messerli (ed.): *Cardiovascular Disease in the Elderly.* 3rd ed. 1992 ISBN 0-7923-1859-5
136. J. Hess and G.R. Sutherland (eds.): *Congenital Heart Disease in Adolescents and Adults.* 1992 ISBN 0-7923-1862-5
137. J.H.C. Reiber and P.W. Serruys (eds.): *Advances in Quantitative Coronary Arteriography.* 1993 ISBN 0-7923-1863-3
138. A.-M. Salmasi and A.S. Iskandrian (eds.): *Cardiac Output and Regional Flow in Health and Disease.* 1993 ISBN 0-7923-1911-7
139. J.H. Kingma, N.M. van Hemel and K.I. Lie (eds.): *Atrial Fibrillation, a Treatable Disease?* 1992 ISBN 0-7923-2008-5
140. B. Ostadel and N.S. Dhalla (eds.): *Heart Function in Health and Disease.* Proceedings of the Cardiovascular Program (Prague, Czechoslovakia, 1991). 1992 ISBN 0-7923-2052-2
141. D. Noble and Y.E. Earm (eds.): *Ionic Channels and Effect of Taurine on the Heart.* Proceedings of an International Symposium (Seoul, Korea , 1992). 1993 ISBN 0-7923-2199-5
142. H.M. Piper and C.J. Preusse (eds.): *Ischemia-reperfusion in Cardiac Surgery.* 1993 ISBN 0-7923-2241-X
143. J. Roelandt, E.J. Gussenhoven and N. Bom (eds.): *Intravascular Ultrasound.* 1993 ISBN 0-7923-2301-7
144. M.E. Safar and M.F. O'Rourke (eds.): *The Arterial System in Hypertension.* 1993 ISBN 0-7923-2343-2
145. P.W. Serruys, D.P. Foley and P.J. de Feyter (eds.): *Quantitative Coronary Angiography in Clinical Practice.* With a Foreword by Spencer B. King III. 1994 ISBN 0-7923-2368-8

# Developments in Cardiovascular Medicine

146. J. Candell-Riera and D. Ortega-Alcalde (eds.): *Nuclear Cardiology in Everyday Practice.* 1994 (in prep.) ISBN 0-7923-2374-2
147. P. Cummins (ed.): *Growth Factors and the Cardiovascular System.* 1993 ISBN 0-7923-2401-3
148. K. Przyklenk, R.A. Kloner and D.M. Yellon (eds.): *Ischemic Preconditioning: The Concept of Endogenous Cardioprotection.* 1993 ISBN 0-7923-2410-2
149. T.H. Marwick: *Stress Echocardiography.* Its Role in the Diagnosis and Evaluation of Coronary Artery Disease. 1994 ISBN 0-7923-2579-6
150. W.H. van Gilst and K.I. Lie (eds.): *Neurohumoral Regulation of Coronary Flow.* Role of the Endothelium. 1993 ISBN 0-7923-2588-5
151. N. Sperelakis (ed.): *Physiology and Pathophysiology of the Heart.* 3rd rev. ed. 1994 ISBN 0-7923-2612-1
152. J.C. Kaski (ed.): *Angina Pectoris with Normal Coronary Arteries: Syndrome X.* 1994 ISBN 0-7923-2651-2
153. D.R. Gross: *Animal Models in Cardiovascular Research.* 2nd rev. ed. 1994 ISBN 0-7923-2712-8
154. A.S. Iskandrian and E.E. van der Wall (eds.): *Myocardial Viability.* Detection and Clinical Relevance. 1994 ISBN 0-7923-2813-2
155. J.H.C. Reiber and P.W. Serruys (eds.): *Progress in Quantitative Coronary Arteriography.* 1994 ISBN 0-7923-2814-0
156. U. Goldbourt, U. de Faire and K. Berg (eds.): *Genetic Factors in Coronary Heart Disease.* 1994 ISBN 0-7923-2752-7
157. G. Leonetti and C. Cuspidi (eds.): *Hypertension in the Elderly.* 1994 ISBN 0-7923-2852-3
158. D. Ardissino, S. Savonitto and L.H. Opie (eds.): *Drug Evaluation in Angina Pectoris.* 1994 ISBN 0-7923-2897-3
159. G. Bkaily (ed.): *Membrane Physiopathology.* 1994 ISBN 0-7923-3062-5
160. R.C. Becker (ed.): *The Modern Era of Coronary Thrombolysis.* 1994 ISBN 0-7923-3063-3
161. P.J. Walter (ed.): *Coronary Bypass Surgery in the Elderly.* Ethical, Economical and Quality of Life Aspects. With a foreword by N.K. Werger. 1995 ISBN 0-7923-3188-5
162. J.W. de Jong and R. Ferrari (eds.), *The Carnitine System.* A New Therapeutical Approach to Cardiovascular Diseases. 1995 ISBN 0-7923-3318-7
163. C.A. Neill and E.B. Clark: *The Developing Heart: A 'History' of Pediatric Cardiology.* 1995 ISBN 0-7923-3375-6
164. N. Sperelakis: *Electrogenesis of Biopotentials in the Cardiovascular System.* 1995 ISBN 0-7923-3398-5
165. M. Schwaiger (ed.): *Cardiac Positron Emission Tomography.* 1995 ISBN 0-7923-3417-5
166. E.E. van der Wall, P.K. Blanksma, M.G. Niemeyer and A.M.J. Paans (eds.): *Cardiac Positron Emission Tomography.* Viability, Perfusion, Receptors and Cardiomyopathy. 1995 ISBN 0-7923-3472-8

*Previous volumes are still available*

KLUWER ACADEMIC PUBLISHERS – DORDRECHT / BOSTON / LONDON

MIX
Papier aus verantwortungsvollen Quellen
Paper from responsible sources
FSC® C105338

If you have any concerns about our products,
you can contact us on
**ProductSafety@springernature.com**

In case Publisher is established outside the EU,
the EU authorized representative is:
**Springer Nature Customer Service Center GmbH**
**Europaplatz 3, 69115 Heidelberg, Germany**

Printed by Libri Plureos GmbH
in Hamburg, Germany